Policy Assessment for the Review of the Primary National Ambient Air Quality Standards for Oxides of Nitrogen

EPA-452/R-17-003
April 2017

Policy Assessment for the Review of the Primary National Ambient Air Quality Standards for Oxides of Nitrogen

U.S. Environmental Protection Agency
Office of Air Quality Planning and Standards
Health and Environmental Impacts Division
Research Triangle Park, NC

DISCLAIMER

This document has been prepared by staff in the U.S. Environmental Protection Agency's Office of Air Quality Planning and Standards (OAQPS). Conclusions do not necessarily reflect the views of the Agency. Mention of trade names or commercial products is not intended to constitute endorsement or recommendation for use. Questions or comments related to this document should be addressed to Ms. Breanna Alman, U.S. Environmental Protection Agency, Office of Air Quality Planning and Standards, C539-06, Research Triangle Park, North Carolina 27711 (email: alman.breanna@epa.gov) and Dr. Stephen Graham, U.S. Environmental Protection Agency, Office of Air Quality Planning and Standards, C539-06, Research Triangle Park, North Carolina 27711 (email: graham.stephen@epa.gov).

This page left blank intentionally

TABLE OF CONTENTS

APPENDICES

LIST OF FIGURES

LIST OF TABLES

LIST OF ACRONYMS AND ABBREVIATIONS

AADT	Annual Average Daily Traffic
Act	Clean Air Act
AQI	Air Quality Index
AQS	Air Quality System
AR	Airway responsiveness
ATS	American Thoracic Society
ATSDR	Agency for Toxic Substances and Disease Registry
BC	Black carbon
CAA	Clean Air Act
CASAC	Clean Air Scientific Advisory Committee
CBSA	Core-based statistical area
CDC	Centers for Disease Control
CFR	Code of Federal Regulations
CHS	Children's Health Study
CI	Confidence interval
CO	Carbon monoxide
COPD	Chronic obstructive pulmonary disease
DV	Design value
EC	Elemental carbon
ED	Emergency department
EGU	Electric power generating unit
eNO	Exhaled nitric oxide
EPA	Environmental Protection Agency
FEM	Federal Equivalent Method
FEV_1	Forced Expiratory Volume for 1 second
FR	Federal Register
FRM	Federal Reference Method
HA	Hospital Admission
HNO_3	Nitric acid
HNO_4	Peroxynitric acid
HONO	Nitrous acid
HPMS	Highway Performance Monitoring System
IDW	Inverse distance weighted
IgE	Immunoglobulin E

IRP	Integrated Review Plan
ISA	Integrated Science Assessment
LUR	Land use regression
m	meter
Max	maximum
MSA	Metropolitan Statistical Area
NAAQS	National ambient air quality standards
NAB	North American background
NCEA	National Center for Environmental Assessment
NEI	National Emissions Inventory
NO	Nitric oxide
NO_2	Nitrogen Dioxide
NO_3^-	Nitrate radicals
NO_X	Nitrogen Oxides
O_3	Ozone
OAR	Office of Air and Radiation
OAQPS	Office of Air Quality Planning and Standards
OC	Organic carbon
OMB	Office of Management and Budget
ORD	Office of Research and Development
PA	Policy Assessment
PD	Provocative dose
PAMS	Photochemical Assessment Monitoring Stations
PANs	Peroxyacetyl nitrates
ppb	Parts per billion
PM	Particulate matter
$PM_{2.5}$	Particles generally less than or equal to 2.5 μm in diameter
PM_{10}	Particles generally less than or equal to 10 micrometers (μm) in diameter
REA	Risk and Exposure Assessment
RTP	Research Triangle Park
SLAMS	State and Local Monitoring Stations
SO_2	Sulfur Dioxide
sRAW	Specific airways resistance
Th2	Type 2 T helper cell
TBD	To be determined

| U.S. | United States |
| VOCs | Volatile Organic Compounds |

EXECUTIVE SUMMARY

This Policy Assessment (PA) has been prepared by staff in the Environmental Protection Agency's (EPA) Office of Air Quality Planning and Standards (OAQPS) as part of the Agency's ongoing review of the primary (health-based) national ambient air quality standards (NAAQS) for oxides of nitrogen (referred to herein as the NO_2 NAAQS). It presents analyses and staff conclusions regarding the policy implications of the key scientific and technical information that informs this review. The PA is intended to "bridge the gap" between the relevant scientific evidence and technical information and the judgments required of the EPA Administrator in determining whether to retain or revise the current standards. Development of the PA is also intended to facilitate advice and recommendations on the standards to the Administrator from an independent scientific review committee, the Clean Air Scientific Advisory Committee (CASAC), as provided for in the Clean Air Act (CAA).

Staff's conclusions in this PA are informed by consideration of the scientific evidence summarized and assessed in the Integrated Science Assessment for Oxides of Nitrogen – Health Criteria (ISA) and updated analyses comparing ambient nitrogen dioxide (NO_2) concentrations to health-based benchmarks, included herein. Emphasis is given to considering the extent to which the evidence newly available since the last review alters conclusions drawn in the last review with regard to health effects related to NO_2 exposures, the exposure concentrations at which those effects occur, and populations that may be at increased risk for effects.

The overarching questions in this review, as in other NAAQS reviews, focus on the support provided by the available scientific and technical information for the adequacy of the current standards, and on the extent to which that scientific and technical information supports consideration of potential alternative standards. The analyses presented in this PA to address such questions lead to the staff conclusion that it is appropriate to consider retaining the current primary NO_2 standards, without revision, in this review. Accordingly, staff have not identified potential alternative standards for consideration. Advice and recommendations from CASAC, based on its review of the draft PA, and input from the public on the draft PA, have informed staff conclusions and the presentation of information in this final document.

History of the Primary NO_2 NAAQS

The NO_2 NAAQS was initially promulgated in 1971. At that time, the Administrator set a standard with an annual averaging time and a level of 53 ppb to protect against respiratory disease in children that had been reported in the available studies. In subsequent reviews of the NO_2 NAAQS, completed in 1985 and 1996, the annual standard was retained without revision.

The last review of the primary NO_2 NAAQS was completed in 2010. In that review, the EPA supplemented the existing annual NO_2 standard by establishing a new 1-hour standard. After considering an integrative synthesis of the body of evidence on human health effects related to NO_2 exposures and the available information from quantitative assessments of NO_2 exposures and health risks, the Administrator determined that the annual standard alone was not sufficient to protect the public health from the array of effects that could occur following short-term exposures to ambient NO_2. To increase protection against such exposures, the 1-hour NO_2 standard was set with a level of 100 ppb, based on the 3-year average of the 98th percentile of the annual distribution of daily maximum 1-hour concentrations. The EPA also retained the existing annual NO_2 standard with its level of 53 ppb.

In that review, the Administrator particularly noted the potential for adverse health effects to occur following exposures to elevated NO_2 concentrations that can occur around major roadways. Accordingly, the revisions to the primary NO_2 NAAQS in 2010 were accompanied by revisions to the ambient air monitoring and reporting requirements. States were required to locate monitors within 50 meters of heavily trafficked roadways in large urban areas and in other locations where maximum NO_2 concentrations were expected to occur. Near-road NO_2 monitors were initially required to become operational between January 1, 2014 and January 1, 2017. Currently, there are approximately 70 near-road monitors in operation in urban areas across the U.S., with approximately one to two years of data available from most of these monitors.

Scope and Approach in the Current Review

Consistent with the review completed in 2010, this review focuses on health effects associated with gaseous oxides of nitrogen and the protection afforded by the current primary NO_2 standards. The gaseous oxides of nitrogen include NO_2 and nitric oxide (NO), together referred to as NO_X, and their gaseous reaction products. Health effects and non-ecological welfare effects associated with particulate species (e.g., nitrates) are addressed in the review of the NAAQS for particulate matter (PM). Additionally, the EPA is separately reviewing the ecological welfare effects associated with oxides of nitrogen, oxides of sulfur, and PM, and the protection provided by the secondary NO_2, SO_2 and PM standards.

Staff's approach to reviewing the primary NO_2 NAAQS in the current review is focused on addressing a series of key policy-relevant questions. Consideration of these questions is intended to inform the Administrator's decisions as to whether, and if so, how, to revise the current primary NO_2 standards. In addressing these questions in this PA, we are mindful that the Administrator's ultimate judgments on the primary standards will most appropriately reflect an interpretation of the available scientific evidence and information that neither overstates nor understates the strengths and limitations of that evidence and information. This approach is consistent with the requirements of sections 108 and 109 of the CAA, as well as with how the EPA and the courts have historically interpreted the CAA.

Characterization of NO_X Emissions Sources and Trends in Ambient NO_2 Concentrations

As was the case in previous reviews, the major sources of NO_X emissions include highway vehicles, off-highway vehicles, and fuel combustion from utilities and other sources. Estimates indicate a 61% reduction in NO_X emissions across all source categories since 1980, and emissions are expected to decrease further as existing regulatory programs continue to be implemented. Reductions in NO_X emissions over past decades have occurred largely as the result of substantial decreases in emissions from mobile sources and from fuel combustion at utilities and other sources. Based on recent estimates, mobile sources remain the largest source of NO_X emissions in the U.S., with highway vehicles accounting for 34% of total NO_X emissions and off-highway vehicles and engines accounting for 23% of those emissions.

Consistent with reductions in NO_X emissions, ambient NO_2 concentrations have declined substantially since 1980 (i.e., by about 65% and 50% for annual and hourly concentrations, respectively). Based on recent data, all NO_2 monitors measure ambient concentrations that meet the existing NAAQS. Analyses of historical data indicate that monitoring sites meeting the current 1-hour NO_2 standard generally have corresponding annual average NO_2 concentrations below 35 ppb. Based on ongoing reductions in NO_X emissions, we anticipate that ambient NO_2 concentrations will continue to decline across most of the U.S.

Because mobile sources remain the largest contributors to NO_X emissions in the U.S., an important part of the current review is the evaluation of monitoring data from recently deployed near-road NO_2 monitors. Depending on local conditions, ambient NO_2 concentrations can be higher near roadways than at sites in the same area but farther removed from the road (and from other sources of NO_X emissions). Analyses included in this PA indicate that NO_2 concentrations are generally highest at sampling sites nearest to the road and decrease as distance from the road increases. This pattern of decreasing concentrations with increasing distance from the road has persisted over recent decades, though the absolute difference (in terms of ppb) between NO_2 concentrations close to roads and those farther from roads has declined over time.

Consistent with this analysis of historical air quality information, the limited amount of data available from recently deployed near-road monitors indicates that daily maximum 1-hour NO_2 concentrations are generally higher at near-road monitors than at the non-near-road monitors in the same area. This is the case in most of the CBSAs with near-road monitors, though these relationships vary across CBSAs and over the years with available data, particularly at the upper ends of the distributions of NO_2 concentrations (i.e., 98th, 99th percentiles). As more years of data from near-road monitors become available, we expect to gain an improved understanding of these relationships.

Health Effects Evidence and Review of the Primary NO_2 NAAQS

In this PA, we evaluate what the health effects evidence can tell us with regard to the adequacy of the public health protection provided by the current primary NO_2 NAAQS. In doing so, we consider the strength of the evidence for various effects and the extent to which that evidence indicates adverse effects attributable to NO_2 exposures at concentrations lower than previously identified or below the current standards.

As in the last review, the strongest evidence continues to come from studies examining respiratory effects following short-term NO_2 exposures (e.g., minutes up to one month). In particular, the ISA concludes that "[a] causal relationship exists between short-term NO_2 exposure and respiratory effects based on evidence for asthma exacerbation." (US EPA, 2016a, p. 1-17). The strongest support for this conclusion comes from controlled human exposure studies examining the potential for NO_2-induced increases in airway responsiveness (AR) (i.e., a hallmark of asthma) in individuals with asthma. Most of these studies were available in the last review. Together with an updated meta-analysis of their individual-level data, these studies indicate increases in AR in some people with asthma following resting exposures to NO_2 concentrations from 100 to 530 ppb. Important limitations in this evidence include the lack of an apparent dose-response relationship between NO_2 and AR and uncertainty in the adversity of the reported increases in AR. In addition, within the range of 100 to 530 ppb, the evidence for NO_2-induced increases in AR becomes less consistent across studies that examined the lower exposure concentrations, particularly 100 ppb.

Evidence supporting the ISA conclusion also comes from epidemiologic studies reporting associations between short-term NO_2 exposures and an array of respiratory outcomes related to asthma exacerbation. Such studies consistently report associations with several asthma-related outcomes, including asthma-related hospital admissions and emergency department visits in children and adults. The epidemiologic evidence that is newly available in the current review is consistent with evidence from the last review and does not fundamentally alter our understanding of respiratory effects related to short-term NO_2 exposures. While our fundamental understanding of such effects has not changed, recent epidemiologic studies do reduce some uncertainty from the last review regarding the extent to which effects may be independently related to short-term NO_2 exposures. This reduced uncertainty results from recent studies reporting health effect associations with short-term NO_2 exposures in co-pollutant models and from recent studies using improved exposure metrics.

In addition to the effects of short-term exposures, the ISA concludes that there is "likely to be a causal relationship" between long-term NO_2 exposures and respiratory effects, based on the evidence for asthma development in children. The strongest evidence supporting this conclusion comes from recent epidemiologic studies demonstrating associations between long-term NO_2 exposures and asthma incidence. Important uncertainties in these studies result from the methods used to assign NO_2 exposures, the high correlations between NO_2 and other traffic-related pollutants, and the lack of information regarding the extent to which reported effects are independently associated with NO_2 rather than the overall mixture of traffic-related pollutants. Additional support for the ISA conclusion comes from experimental studies supporting the biological plausibility of a potential mode of action by which NO_2 exposures could cause asthma development. These include studies that support a potential role for repeated short-term NO_2 exposures in the development of asthma.

While the overall evidence for NO_2-related respiratory effects supports a "causal" relationship with short-term NO_2 exposures and a "likely to be causal" relationship with long-term exposures, these studies do not provide evidence that calls into question the adequacy of the public health protection provided by current primary NO_2 NAAQS. In particular, compared to the last review when the 1-hour standard was set, evidence from controlled human exposure studies has not altered our understanding of the NO_2 exposure concentrations that cause increased AR. In addition, there remains uncertainty in this evidence due to the lack of an apparent dose-response relationship and uncertainty in the adversity of the response. These uncertainties are increasingly important for the lower NO_2 exposure concentrations evaluated (i.e., at and near 100 ppb), where the evidence across individual studies is less consistent. In addition, while epidemiologic studies report associations with asthma-related outcomes, these associations are generally in locations that would likely have violated one or both of the existing standards over at least part of the study periods. In the absence of studies reporting associations in locations meeting the current NO_2 standards, there is greater uncertainty regarding the extent to which serious asthma exacerbations (short-term exposures) or the development of asthma (long-term exposures) are caused by the NO_2 exposures that occur with air quality meeting those standards.

Comparisons of Ambient NO_2 Concentrations with Health-Based Benchmarks

Beyond our consideration of the scientific evidence, we also consider the extent to which quantitative analyses can inform conclusions on the adequacy of the public health protection provided by the current primary NO_2 standards. In particular, we have conducted updated analyses comparing NO_2 air quality with health-based benchmarks from 100 to 300 ppb to estimate the potential for exposures of public health concern that could be allowed by the current standards. Benchmarks are based on information from controlled human exposure studies indicating NO_2-induced increases in AR and on the meta-analysis of individual-level data from these studies.

Overall, these analyses indicate little potential for exposures to ambient NO_2 concentrations that would be of public health concern in locations meeting the current 1-hour standard. In particular, based on recent ambient measurements, all of which meet the current standards, analyses indicate almost no potential for 1-hour exposures to NO_2 concentrations at or above any of the benchmarks examined, even the lowest benchmark (i.e., 100 ppb). When air quality is adjusted upwards to simulate just meeting the current 1-hour NO_2 standard, there is also virtually no potential for exposures to the NO_2 concentrations that have been shown most consistently to increase AR in people with asthma (i.e., greater than 200 ppb), even under worst-case conditions across a variety of study areas with among the highest NO_X emissions in the U.S. Such NO_2 concentrations are not estimated to occur, even at monitoring sites adjacent to some of the most heavily trafficked roadways in the country. In addition, the current standard is estimated to limit exposures to NO_2 concentrations that have the potential to exacerbate asthma symptoms, but for which the evidence across studies is less consistent (i.e., 100 ppb). The results of these analyses, and the uncertainties inherent in their interpretation, suggest that there is little potential for exposures to ambient NO_2 concentrations that would be of public health concern in locations meeting the current 1-hour standard.

Staff Conclusions

Staff has reached the conclusion that the available scientific evidence, in combination with the available information from quantitative analyses, supports the adequacy of the public health protection provided by the current primary NO_2 standards. Staff further reaches the conclusion that it is appropriate to consider retaining the current standards, without revision, in this review. In light of this conclusion, we have not identified potential alternative standards for consideration in this PA. In its review of the draft PA, CASAC agreed with these conclusions, stating that it "concurs with the EPA that the current scientific literature does not support a revision to the primary NAAQS for nitrogen dioxide".

Staff additionally notes that the final decision on the adequacy of the current standards is a public health policy judgment to be made by the Administrator, drawing upon the scientific information as well as judgments about how to consider the range and magnitude of uncertainties that are inherent in this information. In this context, we recognize that the uncertainties and limitations associated with the estimated relationships between NO_2 exposures and adverse respiratory effects are amplified with consideration of increasingly lower NO_2 concentrations. In staff's view, there is appreciable uncertainty regarding the degree to which reductions in asthma exacerbations or asthma development would result from alternative NO$_2$ standards with levels lower than those of the current standards. Thus, the basis for any consideration of alternative lower standard levels would likely reflect different public health policy judgments as to the appropriate approach for weighing uncertainties in the evidence.

1   INTRODUCTION

1.1   PURPOSE

The U.S. Environmental Protection Agency (EPA) is conducting a review of the primary national ambient air quality standards (NAAQS) for Oxides of Nitrogen (referred to herein as the NO_2 NAAQS). An overview of the approach to reviewing the primary NO_2 NAAQS is presented in the *Integrated Review Plan for the Primary NAAQS for Nitrogen Dioxide* (IRP, U.S. EPA, 2014). The IRP discusses the key policy-relevant issues that frame the EPA's consideration of whether the current NO_2 NAAQS should be retained or revised and the planned approaches to be taken in developing key scientific, technical, and policy documents.

As part of the current review of the primary NO_2 NAAQS, staff in the EPA's Office of Air Quality Planning and Standards (OAQPS) has prepared this Policy Assessment (PA). The PA is intended to help bridge the gap between the relevant scientific and technical information and the judgments required of the EPA Administrator in determining whether, and if so how, it is appropriate to revise the NAAQS. This PA for NO_2 integrates and interprets information from the *Integrated Science Assessment for Oxides of Nitrogen – Health Criteria* (ISA, U.S. EPA, 2016a) and from available quantitative assessments (Chapter 4, below) to frame policy options for consideration by the Administrator. In doing so, we recognize that the selection of a specific approach to reaching final decisions on the primary NO_2 standards will reflect the judgments of the Administrator.

The development of the PA is also intended to facilitate advice to the Agency and recommendations to the Administrator from an independent scientific review committee, the Clean Air Scientific Advisory Committee (CASAC), as provided for in the Clean Air Act.[1] As discussed below in section 1.2.1, the CASAC is to advise not only on the Agency's assessment of the relevant scientific information, but also on the adequacy of the existing standards, and to make recommendations as to any revisions of the standards that may be appropriate. The EPA facilitates the CASAC's advice and recommendations, as well as public input, by requesting CASAC review and public comment on one or more drafts of the PA.[2] As such, the

---

[1] Beyond informing the EPA Administrator and facilitating the advice and recommendations of CASAC and the public, this PA is also intended to be a useful reference to all parties interested in the review of the primary NO_2 NAAQS. It is intended to serve as a single source of the most policy-relevant information that informs the Agency's review of the primary NO_2 NAAQS, and it is written to be understandable to a broad audience.

[2] The decision whether to prepare more than one draft of the PA is influenced by staff conclusions and associated CASAC advice and public input. Typically, a second draft PA is prepared in cases where the available information calls into question the adequacy of the current standard(s) and where staff analyses of potential alternative standards are developed. In such cases, a second draft PA includes staff conclusions regarding potential alternative standards and undergoes review by the CASAC and public comment prior to preparation of the final PA. When analyses of

considerations and conclusions in the final PA are informed by the advice and recommendations provided by CASAC, as well as by public input provided as part of the CASAC review process.

In this final PA, we take into account the available scientific and technical information as assessed in the ISA (U.S. EPA, 2016a). In so doing, we focus on information that is most relevant to evaluating the basic elements of NAAQS: indicator[3], averaging time, form[4], and level. These elements, which together serve to define each standard, must be considered collectively in evaluating the health protection afforded by the primary NO_2 standards. This final PA builds upon staff's preliminary conclusions, as presented in the draft PA (U.S. EPA, 2016b) and in the document titled *Review of the Primary National Ambient Air Quality Standards for Nitrogen Dioxide: Risk and Exposure Assessment Planning Document* (REA Planning Document, U.S. EPA, 2015). Staff's final conclusions in this PA have been informed by the advice received from the CASAC, based on its review of the draft PA and the REA Planning document (Diez Roux and Frey, 2015; Diez Roux and Sheppard, 2017), and by public input on those documents.

The remainder of this chapter summarizes the NAAQS legislative requirements and provides an overview of the history of the NO_2 NAAQS (Section 1.2), summarizes the scope of the review (Section 1.3), and provides an overview of the approach used to reaching decisions in the last review of the primary NO_2 standard and of our planned approach to reviewing the primary NO_2 standards in the current review (Section 1.4). Following Chapter 1, this PA presents an overview of the NO_2 monitoring network and of the available information on ambient NO_2 concentrations and trends (Chapter 2); staff's consideration of the available evidence for NO_2-attributable health effects and the NO_2 concentrations associated with those effects (Chapter 3); staff's consideration of quantitative analyses (Chapter 4); and staff's conclusions regarding the adequacy of the existing primary NO_2 standards (Chapter 5).

---

potential alternative standards are not undertaken, as is the case in this review of the primary NO_2 NAAQS (see Chapter 4 below), a second draft PA may not be warranted. For this PA, CASAC advised that, with its recommended revisions, another CASAC review of the draft PA would not be needed (Diez Roux and Sheppard, 2017). Staff similarly determined that a second draft PA is not warranted.

[3]The "indicator" of a standard defines the chemical species or mixture that is to be measured in determining whether an area attains the standard.

[4]The "form" of a standard defines the air quality statistic that is to be compared to the level of the standard in determining whether an area attains the standard.

1.2 BACKGROUND

1.2.1 Legislative Requirements

Two sections of the Clean Air Act (CAA) govern the establishment and revision of the NAAQS. Section 108 (42 U.S.C. 7408) directs the Administrator to identify and list certain "air pollutants" and then to issue air quality criteria for those pollutants that are listed. The Administrator is to list those air pollutants that, "in his judgment, cause or contribute to air pollution which may reasonably be anticipated to endanger public health or welfare;" "the presence of which in the ambient air results from numerous or diverse mobile or stationary sources;" and "for which...[the Administrator] plans to issue air quality criteria..." Air quality criteria are intended to "accurately reflect the latest scientific knowledge useful in indicating the kind and extent of all identifiable effects on public health or welfare which may be expected from the presence of [a] pollutant in the ambient air..." (42 U.S.C. 7408). Section 109 (42 U.S.C. 7409) directs the Administrator to propose and promulgate "primary" and "secondary" NAAQS for pollutants for which air quality criteria are issued. Section 109(b)(1) defines a primary standard as one "the attainment and maintenance of which in the judgment of the Administrator, based on such criteria and allowing an adequate margin of safety, are requisite to protect the public health."[5] A secondary standard, as defined in section 109(b)(2), must "specify a level of air quality the attainment and maintenance of which, in the judgment of the Administrator, based on such criteria, is requisite to protect the public welfare from any known or anticipated adverse effects associated with the presence of such air pollutant in the ambient air."[6] The secondary NO_2 standard is being reviewed separately.[7]

The requirement that primary standards provide an adequate margin of safety was intended to address uncertainties associated with inconclusive scientific and technical information available at the time of standard setting. It was also intended to provide a reasonable degree of protection against hazards that research has not yet identified. *See, e.g., State of Mississippi v. EPA*, 744 F. 3d 1334, 1353 (D.C. Cir. 2012); *Lead Industries Association v. EPA*,

---

[5]The legislative history of section 109 indicates that a primary standard is to be set at "the maximum permissible ambient air level . . . which will protect the health of any [sensitive] group of the population," and that for this purpose "reference should be made to a representative sample of persons comprising the sensitive group rather than to a single person in such a group" [S. Rep. No. 91-1196, 91st Cong., 2d Sess. 10 (1970)].

[6]Effects on welfare as defined in section 302(h) (42 U.S.C. 7602(h)) include, but are not limited to, "effects on soils, water, crops, vegetation, manmade materials, animals, wildlife, weather, visibility, and climate, damage to and deterioration of property, and hazards to transportation, as well as effects on economic values and on personal comfort and well-being."

[7] https://www.epa.gov/naaqs/nitrogen-dioxide-no2-and-sulfur-dioxide-so2-secondary-air-quality-standards

647 F.2d 1130, 1154 (D.C. Cir 1980), *cert. denied*, 449 U.S. 1042 (1980); *American Petroleum Institute v. Costle*, 665 F.2d 1176, 1186 (D.C. Cir. 1981), *cert. denied*, 455 U.S. 1034 (1982). Both types of uncertainties are components of the risk associated with pollution at levels below those at which human health effects can be said to occur with reasonable scientific certainty. Thus, in selecting primary standards that provide an adequate margin of safety, the Administrator is seeking not only to prevent pollution levels that have been demonstrated to be harmful but also to prevent lower pollutant levels that may pose an unacceptable risk of harm, even if the risk is not precisely identified as to nature or degree. The CAA does not require the Administrator to establish a primary NAAQS at a zero-risk level or at background concentration levels, *see Lead Industries Association v. EPA*, 647 F.2d at 1156 n. 51, but rather at a level that reduces risk sufficiently so as to protect public health with an adequate margin of safety.

In addressing the requirement for an adequate margin of safety, the EPA considers such factors as the nature and severity of the health effects involved, the size of the population(s) at risk, and the kind and degree of the uncertainties that must be addressed. The selection of any particular approach to providing an adequate margin of safety is a policy choice left specifically to the Administrator's judgment. *Lead Industries Association v. EPA*, 647 F.2d at 1161-62; *State of Mississippi*, 744 F. 3d at 1353.

In setting primary and secondary standards that are "requisite" to protect public health and welfare, respectively, as provided in section 109(b), the EPA's task is to establish standards that are neither more nor less stringent than necessary for these purposes. In so doing, the EPA may not consider the costs of implementing the standards. See generally, *Whitman v. America Trucking Associations*, 531 U.S. 457, 465-472, 475-76 (2001). Likewise, "[a]ttainability and technological feasibility are not relevant considerations in the promulgation of national ambient air quality standards." *American Petroleum Institute v. Costle*, 665 F. 2d at 1185.

Section 109(d)(1) requires that "not later than December 31, 1980, and at 5-year intervals thereafter, the Administrator shall complete a thorough review of the criteria published under section 108 and the national ambient air quality standards . . . and shall make such revisions in such criteria and standards and promulgate such new standards as may be appropriate . . . ." Section 109(d)(2) requires that an independent scientific review committee "shall complete a review of the criteria . . . and the national primary and secondary ambient air quality standards . . . and shall recommend to the Administrator any new . . . standards and revisions of existing criteria and standards as may be appropriate . . . ." This independent review function is now performed by the Clean Air Scientific Advisory Committee (CASAC) of EPA's Science Advisory Board.[8]

1.2.2   Previous NO₂ NAAQS Reviews

In 1971, the EPA added nitrogen oxides to the list of criteria pollutants under section 108(a)(1) of the CAA and issued the initial air quality criteria (36 FR 1515, January 30, 1971; U.S. EPA, 1971). Based on these air quality criteria, the EPA promulgated NAAQS for nitrogen oxides using NO_2 as the indicator (36 FR 8186, April 30, 1971). Both primary and secondary standards were set at 53 parts per billion (ppb),[9] annual average. Since then, the Agency has completed multiple reviews of the air quality criteria and primary NO_2 standards, as summarized in Table 1-1.

Table 1-1. Primary NO₂ NAAQS since 1971.

Final Rule/Decision	Indicator	Averaging Time	Level	Form
1971 36 FR 8186 April 30, 1971	NO_2	Annual	53 ppb[10]	Annual arithmetic average
1985 50 FR 25532 June 19, 1985	Primary NO₂ standards retained, without revision.			
1996 61 FR 52852 October 8, 1996	Primary NO₂ standards retained, without revision.			
2010 75 FR 6474 February 9, 2010	NO_2	1-hour	100 ppb	3-year average of the 98th percentile of the annual distribution of daily maximum 1-hour concentrations
	Primary annual NO₂ standard retained, without revision.			

---

[8] Lists of CASAC members and of members of the CASAC NO2 Review Panel are available at: http://yosemite.epa.gov/sab/sabpeople.nsf/WebExternalCommitteeRosters?OpenView&committee=CASAC&secondname=Clean%20Air%20Scientific%20Advisory%20Committee and http://yosemite.epa.gov/sab/sabpeople.nsf/WebCommitteesSubcommittees/CASAC%20Oxides%20of%20Nitrogen%20Primary%20NAAQS%20Review%20Panel%20(2013-2016), respectively.

[9] In 1971, primary and secondary NO_2 NAAQS were set at levels of 100 micrograms per cubic meter (µg/m3), which equals 0.053 parts per million (ppm) or 53 ppb.

[10] The official level of the annual NO_2 standard is 0.053 ppm, equal to 53 ppb, which is shown here for the purpose of clearer comparison to the 1-hour standard.

The EPA retained the primary NO_2 standard, without revision, in reviews completed in 1985 and 1996 (50 FR 25532, June 19, 1985; 61 FR 52852, October 8, 1996). In the latter of the two decisions, the EPA concluded that "the existing annual primary standard appears to be both adequate and necessary to protect human health against both long- and short-term NO_2 exposures" and that "retaining the existing annual standard is consistent with the scientific data assessed in the Criteria Document (U.S. EPA, 1993), the Staff Paper (U.S. EPA, 1995), and the advice and recommendations of [the] CASAC" (61 FR 52854, October 8, 1996).

The last review of the air quality criteria for oxides of nitrogen (health criteria) and the primary NO_2 standard was initiated in December 2005 (70 FR 73236, December 9, 2005).[11, 12] The EPA's plans for conducting that review were presented in the *Integrated Review Plan for the Primary National Ambient Air Quality Standard for Nitrogen Dioxide* (2007 IRP, U.S. EPA, 2007a), which included consideration of comments received during a CASAC consultation as well as public comment on a draft IRP. The scientific assessment for the review was described in the 2008 *Integrated Science Assessment for Oxides of Nitrogen – Health Criteria* (2008 ISA, U.S. EPA, 2008a), multiple drafts of which received review by the CASAC and the public. The EPA also conducted quantitative human risk and exposure assessments after consultation with the CASAC and after receiving public comment on an analysis plan (U.S. EPA, 2007b). These technical analyses were presented in the *Risk and Exposure Assessment to Support the Review of the NO_2 Primary National Ambient Air Quality Standard* (2008 REA, U.S. EPA, 2008b), multiple drafts of which received CASAC and public review.

In the course of reviewing the second draft REA in the last review, the CASAC expressed the view that the document would be incomplete without the addition of a policy assessment chapter presenting an integration of evidence-based considerations and risk and exposure assessment results. The CASAC stated that such a chapter would be "critical for considering options for the NAAQS for NO_2" (Samet, 2008a, p.4). In addition, within the period of the CASAC's review of the second draft REA, the EPA's Deputy Administrator indicated in a letter to the CASAC chair, addressing earlier CASAC comments on the NAAQS review process, that the risk and exposure assessment would include "a broader discussion of the science and how uncertainties may affect decisions on the standard" and "all analyses and approaches for considering the level of the standard under review, including risk assessment and weight of

---

[11] Documents related to the current review as well as reviews complete in 2010 and 1996 are available at: https://www.epa.gov/naaqs/nitrogen-dioxide-no2-primary-air-quality-standards

[12] The EPA conducted a separate review of the secondary NO_2 NAAQS jointly with a review of the secondary SO_2 NAAQS. The Agency retained those secondary standards, without revision, to address the direct effects on vegetation of exposure to gaseous oxides of nitrogen and sulfur (77 FR 20218, April 3, 2012).

evidence methodologies" (Peacock, 2008, p. 3). Accordingly, the final 2008 REA included a policy assessment chapter that considered the scientific evidence in the 2008 ISA and the exposure and risk results presented in other chapters of the 2008 REA as they related to the adequacy of the then current primary annual NO_2 standard and potential alternative standards for consideration (U.S EPA, 2008b, chapter 10).[13] The CASAC discussed the final version of the 2008 REA, with an emphasis on the policy assessment chapter, during a public teleconference on December 5, 2008 (73 FR 66895, November 12, 2008). Following that teleconference, the CASAC offered comments and advice on the primary NO_2 standard in a letter to the Administrator (Samet, 2008b)

In a notice published in the Federal Register on July 15, 2009, the EPA proposed to supplement the existing primary annual NO_2 standard by establishing a new short-term standard (74 FR 34404, July 15, 2009). After considering an integrative synthesis of the body of evidence on human health effects associated with the presence of NO_2 in the air and the exposure and risk information, the Administrator determined that the existing primary NO_2 NAAQS, based on an annual arithmetic average, was not sufficient to protect the public health from the array of effects that could occur following short-term exposures to ambient NO_2. In so doing, the Administrator particularly noted the potential for adverse health effects to occur following exposures to elevated NO_2 concentrations that can occur around major roads (75 FR 6482, February 9, 2012). In a notice published in the Federal Register on February 9, 2010, the EPA finalized a new short-term NO_2 standard with a level of 100 ppb, based on the 3-year average of the 98th percentile of the annual distribution of daily maximum 1-hour concentrations. The EPA also retained the existing primary annual NO_2 standard with a level of 53 ppb, annual average (75 FR 6474, February 9, 2010). The Agency's final decision included consideration of the CASAC's advice (Samet, 2009) and public comments on the proposed rule. The EPA's final rule was upheld against challenges in a decision issued by the U.S. Court of Appeals for the District of Columbia Circuit on July 17, 2012. *API v. EPA*, 684 F.3d 1342 (D.C. Cir. 2012).

Revisions to the NAAQS were accompanied by revisions to the data handling procedures, the ambient air monitoring and reporting requirements, and the Air Quality Index (AQI).[14] As described in Chapter 2, one aspect of the new monitoring network requirements

---

[13] Subsequent to the completion of the 2008 REA, the EPA Administrator Jackson called for additional key changes to the NAAQS review process including reinstating a policy assessment document that contains staff analysis of the scientific bases for alternative policy options for consideration by senior EPA management prior to rulemaking (Jackson, 2009).

[14] The current federal regulatory measurement methods for NO_2 are specified in 40 CFR part 50, Appendix F and 40 CFR part 53. Consideration of ambient air measurements with regard to judging attainment of the standards is specified in 40 CFR part 50, Appendix S. The NO_2 monitoring network requirements are specified in 40 CFR part 58, Appendix D, section 4.3. The EPA revised the AQI for NO_2 to be consistent with the revised primary NO_2 NAAQS as specified in 40 CFR part 58, Appendix G.

included requirements for states to locate monitors near heavily trafficked roadways in large urban areas and in other locations where maximum NO_2 concentrations can occur. Subsequent to the 2010 rulemaking, the Agency revised the deadlines by which the near-road monitors were to be operational in order to implement a phased deployment approach (78 FR 16184, March 14, 2013). The near-road NO_2 monitors were required to become operational between January 1, 2014 and January 1, 2017.

1.2.3   Current Review of the Primary NO_2 NAAQS

In February 2012, the EPA announced the initiation of the current periodic review of the air quality criteria for oxides of nitrogen and of the Primary NO_2 NAAQS and issued a call for information in the Federal Register (77 FR 7149, February 10, 2012). A wide range of external experts as well as EPA staff representing a variety of areas of expertise (e.g., epidemiology, human and animal toxicology, statistics, risk/exposure analysis, atmospheric science, and biology) participated in a workshop held by the EPA on February 29 to March 1, 2012 in Research Triangle Park, NC. The workshop provided an opportunity for a public discussion of the key policy-relevant issues around which the Agency would structure this primary NO_2 NAAQS review and the most meaningful new science that would be available to inform our understanding of these issues.

Based in part on the workshop discussions, the EPA developed a draft plan for the ISA and a draft IRP outlining the schedule, process, and key policy-relevant questions that would guide the evaluation of the air quality criteria for NO_2 and the review of the primary NO_2 NAAQS. The draft plan for the ISA was released in May of 2013 (78 FR 26026) and was the subject of a consultation with the CASAC on June 5, 2013 (78 FR 27234). Comments received from that consultation were considered in the preparation of first draft ISA, and preliminary drafts of key ISA chapters were reviewed by subject matter experts at a public workshop hosted by the EPA's National Center for Environmental Assessment (NCEA) in May 2013 (78 FR 27374). The first draft ISA was released in November 2013 (78 FR 70040). During this time, the draft IRP was also in preparation and was released in February 2014 (79 FR 7184). Both the draft IRP and first draft ISA were reviewed by the CASAC at a public meeting held in March 2014 (79 FR 8701), and the first draft ISA was further discussed at an additional teleconference held in May 2014 (79 FR 17538). The CASAC finalized its recommendations on the first draft

---

Certain topics related to implementation of the new standard were also discussed in the *Federal Register* notices for the proposed and final rules (74 FR 34404; 75 FR 6474).

ISA and the draft IRP in letters dated June 10, 2014, and the final IRP was released in June 2014 (79 FR 36801).

The EPA released the second draft ISA in January 2015 (80 FR 5110) and the REA Planning Document in May 2015 (80 FR 27304). These documents were review by the CASAC at a public meeting held in June 2015 (80 FR 22993). A follow-up teleconference with the CASAC was held in August 2013 (80 FR 43085) to finalize recommendations on the second draft ISA. The final ISA was released in January 2016 (81 FR 4910). The CASAC's recommendations on the draft REA Plan were provided to the EPA in a letter dated September 9, 2015.

After considering CASAC's advice and public comments, the EPA prepared a draft PA, which was released on September 23, 2016 (81 FR 65353). The draft PA was reviewed by the CASAC on November 9-10, 2016 (81 FR 68414), and a follow-up teleconference was held on January 24, 2017 (81 FR 95137). The CASAC's recommendations, based on its review of the draft PA, were provided in a letter to the EPA Administrator dated March 7, 2017 (Diez Roux and Sheppard, 2017). The EPA staff took into account these recommendations, as well as public comments provided on the draft PA, when developing this final PA.

In addition, in July 2016, several groups filed suit against the EPA for failure to complete its review of the primary NO_2 NAAQS within five years, as required by the CAA. *Center for Biological Diversity et al. v. McCarthy, (No. 4:16-cv-03796-VC, N.D. Cal., July 7, 2016).* A notice of a proposed consent decree to resolve this litigation was published in the Federal Register on January 17, 2017 (82 FR 4866). We anticipate that, as a result of this litigation, the court will establish deadlines for the EPA to take action in this review.

1.3 SCOPE OF THE CURRENT REVIEW

Consistent with the review completed in 2010, this review will focus on health effects associated with gaseous oxides of nitrogen and the protection afforded by the primary NO_2 standards. The gaseous oxides of nitrogen include NO_2 and nitric oxide (NO) as well as their gaseous reaction products. Total oxides of nitrogen include these gaseous species as well as particulate species (e.g., nitrates). Collectively, we refer to the total set of species as NO_Y (U.S. EPA, 2013b, Section 2.2, Figure 2-1). Health effects and non-ecological welfare effects associated with the particulate species are addressed in the review of the NAAQS for particulate matter (PM) (78 FR 30866, January 15, 2013; U.S. EPA, 2009).[15] The EPA is separately

---

[15] Additional information on the PM NAAQS is available at: https://www.epa.gov/naaqs/particulate-matter-pm-air-quality-standards.

reviewing the ecological welfare effects associated with oxides of nitrogen, oxides of sulfur, and PM, and the protection provided by the secondary NO_2, SO_2 and PM standards. (78 FR 53452, August 29, 2013).[16]

When referring to the group of gaseous oxidized nitrogen compounds as a whole, the ISA and other assessment documents developed in this review use the term "oxides of nitrogen." In the last review, the EPA used "NOx" as the abbreviation for oxides of nitrogen. However, based on the definition commonly used in the scientific literature, in this review, the abbreviation NO_X will refer specifically to the sum of NO_2 and NO concentrations, rather than all oxides of nitrogen (U.S. EPA, 2016).[17]

1.4   GENERAL APPROACH FOR REVIEW OF THE STANDARDS

As described in Section 1.1 above, this PA presents a transparent evaluation of the available scientific and technical information and staff's conclusions regarding the adequacy of the current primary NO_2 standards. Staff's considerations and conclusions in this document are based on the available body of scientific evidence assessed in the ISA (U.S. EPA, 2016a) and on the results of quantitative analyses comparing NO_2 air quality to NO_2 benchmarks based on the available health evidence (see Chapter 4 of this PA). Staff's considerations and conclusions have also been informed by the advice and recommendations received from CASAC during its review of the draft PA, and by public input received. Staff's considerations and conclusions in this final PA are intended to inform the Administrator's decision as to whether the existing primary NO_2 standards should be retained or revised.

Section 1.4.1 below summarizes the approach used by the Administrator in reaching conclusions in the last review of the primary NO_2 NAAQS. Building on this approach from the last review, section 1.4.2 summarizes staff's approach to informing the Administrator's decisions on the primary NO_2 NAAQS in the current review.

---

[16] Additional information on the ongoing and previous review of the secondary NO_2 and SO_2 NAAQS is available at: https://www.epa.gov/naaqs/nitrogen-dioxide-no2-and-sulfur-dioxide-so2-secondary-air-quality-standards.

[17] "…[T]he term "oxides of nitrogen" (NO_Y) refers to all forms of oxidized nitrogen (N) compounds, including nitric oxide (NO), nitrogen dioxide (NO_2), and all other oxidized N-containing compounds formed from NO and NO_2" (U.S. EPA, 2016, p. 2-1). "A large number of oxidized nitrogen species in the atmosphere are formed from the oxidation of NO and NO_2. These include nitrate radicals (NO_3), nitrous acid (HONO), nitric acid (HNO_3), dinitrogen pentoxide (N_2O_5), nitryl chloride ($ClNO_2$), peroxynitric acid (HNO_4), PAN and its homologues (PANs), other organic nitrates like alkyl nitrates [including isoprene nitrates(IN)], and pNO_3" (U.S. EPA, 2016, p. 2-2).

1.4.1 Approach Used in the Last Review

As noted above (Section 1.2.2), the last review of the primary NO$_2$ NAAQS was completed in 2010 (75 FR 6474, February 9, 2010). In that review, the EPA established a new 1-hour standard to provide increased public health protection, including for people with asthma and other at-risk populations,[18] against an array of adverse respiratory health effects that had been linked to short-term NO$_2$ exposures (75 FR 6498 to 6502; U.S. EPA, 2008a, Sections 3.1.7 and 5.3.2.1; Table 5.3-1). Specifically, the EPA established a short-term standard defined by the 3-year average of the 98th percentile of the annual distribution of daily maximum 1-hour NO$_2$ concentrations, with a level of 100 ppb. In addition to setting the new 1-hour standard, the EPA retained the existing annual standard with its level of 53 ppb (75 FR 6502, February 9, 2010). The Administrator concluded that, together, the two standards provide protection against adverse respiratory health effects associated with short-term exposures to NO$_2$ and effects potentially associated with long-term exposures. As discussed further in Chapter 2 below, in conjunction with the revised primary NO$_2$ NAAQS, the EPA also established a two-tiered monitoring network composed of (1) near-road monitors which would be placed near heavily trafficked roads in urban areas and (2) monitors located to characterize areas with the highest expected NO$_2$ concentrations at the neighborhood and larger spatial scales (also referred to as "area-wide" monitors) (75 FR 6505 to 6506, February 9, 2010).

Key aspects of the Administrator's approach to reaching these decisions are described below. Section 1.4.1.1 summarizes her approach to reaching the conclusion that it was appropriate to revise the primary NO$_2$ NAAQS. Section 1.4.1.2 summarizes her approach to considering the elements of a revised standard. Section 1.4.1.3 discusses the key uncertainties in the evidence and information identified in the last review.

1.4.1.1 Approach to Considering the Need for Revision

The 2010 decision to revise the existing primary NO$_2$ standard was based largely on the body of scientific evidence published through early 2008 and assessed in the 2008 ISA (U.S. EPA, 2008a); the quantitative exposure and risk analyses and the assessment of the policy-

---

[18] As used here and similarly throughout this document, the term *population* refers to persons having a quality or characteristic in common, such as a specific pre-existing illness or a specific age or lifestage. Lifestage refers to a distinguishable time frame in an individual's life characterized by unique and relatively stable behavioral and/or physiological characteristics that are associated with development and growth (i.e., children and older adults). Identifying at-risk populations includes consideration of intrinsic (e.g., genetic or developmental aspects) or acquired (e.g., disease or smoking status) factors that increase the risk of health effects due to exposure to oxides of nitrogen as well as extrinsic factors such as those related to socioeconomic status, reduced access to health care, or exposure. The ISA characterizes the strength of the evidence for various at-risk populations (U.S. EPA, 2016, Chapter 7).

relevant aspects of the evidence presented in the REA (U.S. EPA, 2008b);[19] the advice and recommendations of the CASAC (Samet, 2008); and public comments on the proposal.

As an initial consideration in reaching this decision, the Administrator noted that the evidence relating short-term (minutes to hours) NO_2 exposures to respiratory morbidity was judged in the ISA to be "sufficient to infer a likely causal relationship" (75 FR 6489, February 9, 2010; U.S. EPA, 2008a, Sections 3.1.7 and 5.3.2.1).[20] The scientific evidence included controlled human exposure studies providing evidence of increases in airway responsiveness in people with asthma following short-term exposures to NO_2 concentrations as low as 100 ppb[21] and epidemiologic studies reporting associations between short-term NO_2 exposures and respiratory effects in locations that would have met the annual standard.

The quantitative analyses presented in the 2008 REA included exposure and risk estimates for air-quality adjusted to just meet the annual standard. The Administrator took note of the REA conclusion that risks estimated for air quality adjusted upward to simulate just meeting the current standard could reasonably be concluded to be important from a public health perspective, while additionally recognizing the uncertainties associated with adjusting air quality in such analyses (75 FR 6489, February 9, 2010). For air quality adjusted to just meet the existing annual standard, the REA findings given particular attention by the Administrator included the following: "a large percentage (8 to 9%) of respiratory-related emergency department visits in Atlanta could be associated with short-term NO_2 exposures; most asthmatics in Atlanta could be exposed on multiple days per year to NO_2 concentrations at or above 300 ppb; and most locations evaluated could experience on-/near-road NO_2 concentrations above 100 ppb on more than half of the days in a given year" (75 FR 6489, February 9, 2010; U.S. EPA, 2008b, Section 10.3.2).

In reaching the conclusion on adequacy of the annual standard alone, the Administrator also considered advice received from the CASAC. In its advice, the CASAC agreed that the primary concern in the review was to protect against health effects that have been associated with short-term NO_2 exposures. The CASAC also agreed that the annual standard alone was not

---

[19] As discussed in the IRP for NO_2 (U.S. EPA, 2014, section 1.3), due to changes in the NAAQS process, the last review of the NO_2 NAAQS did not include a separate Policy Assessment document. Rather, the REA for that review included a policy assessment chapter.

[20] In contrast, the evidence relating long-term (weeks to years) NO_2 exposures to adverse health effects was judged to be either "suggestive but not sufficient to infer a causal relationship" (respiratory morbidity) or "inadequate to infer the presence or absence of a causal relationship" (mortality, cancer, cardiovascular effects, reproductive/developmental effects) (75 FR 6478, February 9, 2010). The causal framework used in the ISA for the current review is discussed below in Chapter 3.

[21] Transient increases in airway responsiveness have the potential to increase asthma symptoms and worsen asthma control (74 FR 34415, July 15, 2009; U.S. EPA, 2008a, sections 5.3.2.1 and 5.4).

sufficient to protect public health against the types of exposures that could lead to these health effects. As noted in its letter to the EPA Administrator, "[The] CASAC concurs with EPA's judgment that the current NAAQS does not protect the public's health and that it should be revised" (Samet, 2008, p. 2).

Based on the considerations summarized above, the Administrator concluded that the annual NO_2 NAAQS alone was not requisite to protect public health with an adequate margin of safety and that the standard should be revised in order to provide increased public health protection against respiratory effects associated with short-term exposures, particularly for at-risk populations and lifestages such as asthmatics, children, and older adults (75 FR 6490, February 9, 2010). Upon consideration of approaches to revising the standard, the Administrator concluded that it was appropriate to set a new short-term standard, in addition to the existing annual standard with its level of 53 ppb, as described below.

1.4.1.2 *Approach to Considering the Elements of a Revised Standard*

In considering appropriate revisions in the last review, each of the four basic elements of the NAAQS (indicator, averaging time, level, and form) was evaluated. The sections below summarize the approaches used by the Administrator, and her final decisions, on each of those elements.

Indicator

In the review completed in 2010, as well as in previous reviews, the EPA focused on NO_2 as the most appropriate indicator for oxides of nitrogen because the available scientific information regarding health effects was largely indexed by NO_2. Controlled human exposure studies and animal toxicological studies provided specific evidence for health effects following exposures to NO_2. In addition, epidemiologic studies typically reported effects associated with NO_2 concentrations[22] (75 FR 6490, February, 9, 2010; U.S. EPA 2008b, Section 2.2.3). Based on the information available in the last review, and consistent with the views of the CASAC (Samet, 2008, p.2; Samet, 2009, p.2), the EPA concluded it was appropriate to continue to use NO_2 as the indicator for a standard that was intended to address effects associated with exposure to NO_2, alone or in combination with other gaseous oxides of nitrogen. In so doing, the EPA recognized that measures leading to reductions in population exposures to NO_2 will also reduce exposures to other oxides of nitrogen (75 FR 6490, February, 9, 2010).

Averaging time

---

[22] The degree to which monitored NO_2 reflected actual NO_2 concentrations, as opposed to NO_2 plus other gaseous oxides of nitrogen, was recognized as an uncertainty (75 FR 6490, February, 9, 2010; U.S. EPA 2008b, section 2.2.3).

In considering the most appropriate averaging time(s) for the primary NO_2 NAAQS, the Administrator noted the available scientific evidence as assessed in the ISA, the air quality analyses presented in the REA, the conclusions of the policy assessment chapter of the REA, and recommendations from the CASAC.[23] Her key considerations are summarized below.

When considering averaging time, the Administrator first noted that the evidence relating short-term (minutes to hours) NO_2 exposures to respiratory morbidity was judged in the ISA to be "sufficient to infer a likely causal relationship" (U.S. EPA, 2008a, section 5.3.2.1) while the evidence relating long-term (weeks to years) NO_2 exposures to adverse health effects was judged to be either "suggestive but not sufficient to infer a causal relationship" (respiratory morbidity) or "inadequate to infer the presence or absence of a causal relationship" (mortality, cancer, cardiovascular effects, reproductive/developmental effects) (U.S. EPA, 2008a, Sections 5.3.2.4-5.3.2.6). The Administrator concluded that these judgments most directly supported an averaging time that focused protection on effects associated with short-term exposures to NO_2.

In considering the level of support available for specific short-term averaging times, the Administrator noted that the policy assessment chapter of the REA considered evidence from both experimental and epidemiologic studies. Controlled human exposure studies and animal toxicological studies provided evidence that NO_2 exposures from less than 1 hour up to 3 hours can result in respiratory effects such as increased airway responsiveness and inflammation (U.S. EPA, 2008a, Section 5.3.2.7). She specifically noted the ISA conclusion that exposures of asthmatic adults to 100 ppb NO_2 for 1-hour (or 200 to 300 ppb for 30 minutes) can result in small but significant increases in nonspecific airway responsiveness (U.S. EPA, 2008a, Section 5.3.2.1). In addition, the epidemiologic evidence provided support for short-term averaging times ranging from approximately 1 hour up to 24 hours (U.S. EPA, 2008a, Section 5.3.2.7). Based on this, the Administrator concluded that a primary concern with regard to averaging time is the degree of protection provided against effects associated with 1-hour NO_2 concentrations. Based on REA analyses of ratios between 1-hour and 24-hour NO_2 concentrations (U.S. EPA, 2008b, Section 10.4.2), she further concluded that a standard based on 1-hour daily maximum NO_2 concentrations could also be effective at protecting against effects associated with 24-hour NO_2 exposures.

Based on the above, the Administrator judged that it was appropriate to set a new NO_2 standard with a 1-hour averaging time. She concluded that such a standard would be expected to effectively limit short-term (e.g., 1- to 24-hours) exposures that have been linked to adverse respiratory effects. She also retained the existing annual standard to continue to provide protection against effects potentially associated with long-term exposures to oxides of nitrogen

---

[23] She also considered public comments received on the proposal (75 FR 6490, February, 9, 2010)

(75 FR 6502, February, 9, 2010). These decisions were consistent with CASAC advice to establish a short-term primary standard for oxides of nitrogen based on using 1-hour maximum NO_2 concentrations and to retain the current annual standard (Samet, 2008, p. 2; Samet, 2009, p. 2).

Level

With consideration of the available health effects evidence, exposure and risk analyses, and air quality information, the Administrator set the level of the new 1-hour NO_2 standard at 100 ppb. This standard was focused on limiting the *maximum* 1-hour NO_2 concentrations in ambient air (75 FR 6474, February, 9, 2010).[24] In establishing this new standard, the Administrator emphasized the importance of protecting against exposures to peak concentrations of NO_2, such as those that can occur around major roadways. Available evidence and information suggested that roadways account for the majority of exposures to peak NO_2 concentrations and, therefore, are important contributors to NO_2-associated public health risks (U.S. EPA, 2008b, Figures 8-17 and 8-18).

In setting the level of the new 1-hour standard at 100 ppb, the Administrator noted that there is no bright line clearly directing the choice of level. Rather, the choice of what is appropriate is a public health policy judgment entrusted to the Administrator. This judgment must include consideration of the strengths and limitations of the evidence and the appropriate inferences to be drawn from the evidence and the exposure and risk assessments.

The Administrator judged that the existing evidence from controlled human exposure studies supported the conclusion that the NO_2-induced increase in airway responsiveness at or above 100 ppb presented a risk of adverse effects for some asthmatics, especially those with more serious (i.e., more than mild) asthma. The Administrator noted that the risks associated with increased airway responsiveness could not be fully characterized based on available controlled human exposure studies, and thus she was not able to determine whether the increased airway responsiveness experienced by asthmatics in these studies was an adverse health effect. However, the Administrator concluded that asthmatics, particularly those suffering from more severe asthma, warrant protection from the risk of adverse effects associated with the NO_2-induced increase in airway responsiveness. Therefore, the Administrator concluded that the controlled human exposure evidence supported setting a standard level no higher than 100 ppb to reflect a cautious approach to the uncertainty regarding the adversity of the effect. However, those uncertainties led her to also conclude that this evidence did not support setting a standard level lower than 100 ppb (75 FR 6500-6501, February, 9, 2010).

---

[24] In conjunction with this new standard, the Administrator established a 2-tiered monitoring network that included monitors sited to measure the maximum NO_2 concentrations near major roadways, as well as monitors sited to measure maximum area-wide NO_2 concentrations.

The Administrator also considered the more serious health effects reported in NO_2 epidemiologic studies. She noted that a new standard focused on protecting against maximum 1-hour NO_2 concentrations in ambient air anywhere in an area, with a level of 100 ppb and an appropriate form (as discussed below), would be expected to limit area-wide[25] NO_2 concentrations to below those in locations where epidemiologic studies had reported associations with respiratory-related hospital admissions or emergency department visits. The Administrator also concluded that such a 1-hour standard would be consistent with the REA conclusions based on the NO_2 exposure and risk information (75 FR 6501, February, 9, 2010).

Given the above considerations and the comments received on the proposal, and considering the entire body of evidence and information before her, as well as the related uncertainties, the Administrator judged it appropriate to set a 1-hour standard reflecting the maximum allowable NO_2 concentrations that can occur anywhere in an area, with a level of 100 ppb. Specifically, she concluded that such a standard, with an appropriate form as discussed below, would provide a significant increase in public health protection compared to that provided by the annual standard alone and would be expected to protect against the respiratory effects that have been linked with NO_2 exposures in both controlled human exposure and epidemiologic studies. This includes limiting exposures at and above 100 ppb for the vast majority of people, including those in at-risk groups, and maintaining area-wide NO_2 concentrations below those in locations where key U.S. epidemiologic studies had reported that ambient NO_2 was associated with clearly adverse respiratory health effects, as indicated by increased hospital admissions and emergency department visits. The Administrator also noted that a standard level of 100 ppb was consistent with the consensus recommendation of the CASAC. (75 FR 6501, February, 9, 2010).

In setting the standard level at 100 ppb rather than at a lower level, the Administrator also acknowledged the uncertainties associated with the scientific evidence. She noted that a 1-hour standard with a level lower than 100 ppb would only result in significant further public health protection if, in fact, there is a continuum of serious, adverse health risks caused by exposure to NO_2 concentrations below 100 ppb and/or associated with area-wide NO_2 concentrations well below those in locations where key U.S. epidemiologic studies had reported associations with respiratory-related emergency department visits and hospital admissions. Based on the available evidence, the Administrator did not believe that such assumptions were warranted. Taking into account the uncertainties that remained in interpreting the evidence from available controlled human exposure and epidemiologic studies, the Administrator observed that the likelihood of obtaining benefits to public health with a standard set below 100 ppb decreased, while the

---

[25]As discussed below in Chapter 2, area-wide concentrations refer to those measured by monitors that have been sited to characterize ambient concentrations at the neighborhood and larger spatial scales.

likelihood of requiring reductions in ambient concentrations that go beyond those that are needed to protect public health increased. (75 FR 6501-02, February, 9, 2010).

Form

The "form" of a standard defines the air quality statistic that is to be compared to the level of the standard in determining whether an area attains the standard. The Administrator recognized that for short-term standards, concentration-based forms that reflect consideration of a statistical characterization of an entire distribution of air quality data, with a focus on a single statistical metric such as the 98th or 99th percentile, can better reflect pollutant-associated health risks than forms based on expected exceedances. This is the case because concentration-based forms give proportionally greater weight to days when pollutant concentrations are well above the level of the standard than to days when the concentrations are just above the level of the standard.[26] In addition, she concluded that when averaged over three years, these concentration-based forms provide an appropriate balance between limiting peak pollutant concentrations and providing a stable regulatory target, facilitating the development of stable implementation programs (75 FR 6492, February, 9, 2010).

In the last review, the EPA considered two specific concentration-based forms (i.e., the 98th and 99th percentile concentrations), averaged over 3 years, for the new 1-hour NO_2 standard. The focus on the upper percentiles of the distribution was based, in part, on evidence of health effects associated with short-term NO_2 exposures from experimental studies which provided information on specific exposure concentrations that were linked to respiratory effects. In a letter to the Administrator following issuance of the Agency's proposed rule, the CASAC recommended a form based on the 3-year average of the 98th percentile of the distribution of 1-hour daily maximum NO_2 concentrations (Samet, 2009, p. 2). In making this recommendation, the CASAC noted the potential for instability in the higher percentile concentrations and the absence of data from the near-road monitoring network.

Given the limited available information on the variability in peak NO_2 concentrations near important sources of NO_2 such as near major roadways, and given the recommendation from the CASAC regarding the potential for instability in the 99th percentile concentrations, the Administrator judged it appropriate to set the form based on the 3-year average of the 98th percentile of the annual distribution of daily maximum 1-hour NO_2 concentrations. In addition, consistent with the CASAC's advice (Samet, 2008, p. 2; Samet, 2009, p.2), the EPA retained the form of the annual standard (75 FR 6502, February, 9, 2010).

1.4.1.3 Areas of Uncertainty in Last Review

While the available scientific information informing the last review was stronger and more consistent than in previous reviews and provided a strong basis for decision making in that

---

[26] Compared to an exceedance-based form, a concentration-based form reflects the magnitude of the exceedance of a standard level not just the fact that such an exceedance occurred.

review, the Agency recognized that areas of uncertainty remained. These were generally related to the following: (1) understanding the role of NO_2 in the complex ambient mixture which includes a range of co-occurring pollutants (e.g., $PM_{2.5}$, CO and other traffic-related pollutants; ozone (O_3), SO_2,) (e.g., 75 FR 6485 February 9, 2010); (2) understanding the extent to which monitored ambient NO_2 concentrations used in epidemiologic studies reflect exposures in study populations and the range of ambient concentrations over which we continue to have confidence in the health effects observed in the epidemiologic studies (e.g., 75 FR 6501, February 9, 2010); (3) understanding the magnitude and potential adversity of NO_2-induced respiratory effects reported in controlled human exposure studies (e.g., 75 FR 6500, February 9, 2010); and (4) understanding the NO_2 concentration gradients around important sources, such as major roads, and relating those gradients to broader ambient monitoring concentrations (e.g., 75 FR 6479, February 9, 2010).

1.4.2   General Approach for the Current Review

Staff's approach to reviewing the primary NO_2 standards in the current review builds off the approach taken in the last review and reflects the updated scientific and technical information now available, as assessed in the 2016 ISA. Our considerations and conclusions related to the primary NO_2 standards in the current review are framed by a series of key policy-relevant questions, expanding upon those presented in the IRP at the outset of this review (U.S. EPA, 2014). Our consideration of these questions is intended to inform the Administrator's decisions as to whether, and if so how, to revise the current NO_2 standards.

In reaching conclusions on options for the Administrator's consideration, we note that the final decision to retain or revise the current primary NO_2 standard is a public health policy judgment to be made by the Administrator. This final decision by the Administrator will draw upon the available scientific evidence for NO_2-attributable health effects and on information from available quantitative analyses, including judgments about the appropriate weight to assign the range of uncertainties inherent in the evidence and analyses. Our general approach in the current review to informing these decisions recognizes that the available health effects evidence reflects a continuum from relatively higher NO_2 concentrations, at which scientists generally agree that health effects are likely to occur, through lower concentrations, at which the likelihood and magnitude of a response become increasingly uncertain. In developing conclusions in this PA, we are mindful that the Administrator's ultimate judgments on the primary standard will most appropriately reflect an interpretation of the available scientific evidence and information that neither overstates nor understates the strengths and limitations of that evidence and information. This approach is consistent with the requirements of sections 108 and 109 of the CAA, as well as with how the EPA and the courts have historically interpreted the CAA.

Figure 1-1 below provides an overview of our approach in this review. We believe that the general approach outlined in Figure 1-1 provides a comprehensive basis to help inform the judgments required of the Administrator in reaching decisions about the current and, if appropriate, potential alternative primary NO_2 standards.

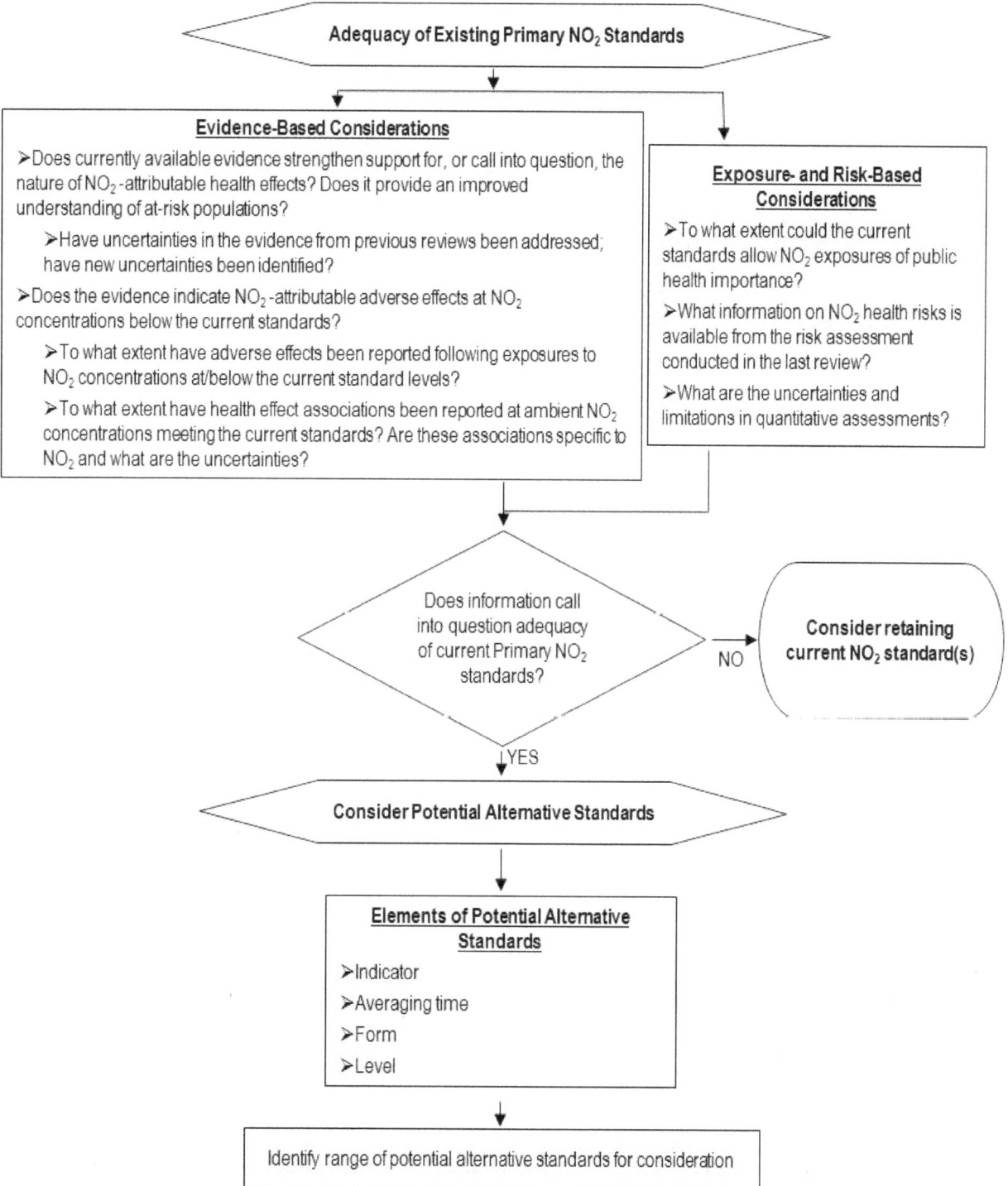

Figure 1-1. Overview of the Approach to Reviewing the Primary NO₂ NAAQS.

1.5 REFERENCES

Diez Roux A and Frey H.C. (2015). Letter from Drs. Ana Diez Roux, Chair and H. Christopher Frey, Immediate Past Chair, Clean Air Scientific Advisory Committee to EPA Administrator Gina McCarthy. CASAC Review of the EPA's Review of the Primary National Ambient Air Quality Standards for Nitrogen Dioxide: Risk and Exposure Assessment Planning Document. EPA-CASAC-15-002. September 9, 2015. Available at: https://yosemite.epa.gov/sab/sabproduct.nsf/264cb1227d55e02c85257402007446a4/A7922887D5BDD8D485257EBB0071A3AD/$File/EPA-CASAC-15-002+unsigned.pdf

Diez Roux A and Sheppard, E (2017). Letter form Dr. Elizabeth A. (Lianne) Sheppard, Chair, Clean Air Scientific Advisory Committee to EPA Administrator E. Scott Pruitt. CASAC Review of the EPA's Policy Assessment for the Review of the Primary National Ambient Air Quality Standards for Nitrogen Dioxide (External Review Draft- September 2016). EPA-CASAC-17-001. March 7th, 2017. Available at: https://yosemite.epa.gov/sab/sabproduct.nsf/LookupWebProjectsCurrentCASAC/7C2807D0D9BB4CC8852580DD004EBC32/$File/EPA-CASAC-17-001.pdf

Peacock M (2008). Letter from Marcus Peacock, Deputy Administrator, U.S. EPA to Dr. Rogene Henderson, Chair, Clean Air Scientific Advisory Committee. September 8, 200. Available at: http://yosemite.epa.gov/sab/sabproduct.nsf/WebCASAC/CASAC_09-08-08/$File/CASAC%20Letter%20to%20Dr%20Rogene%20Henderson.pdf.

Samet J (2008a). Letter from Dr. Jonathan M. Samet, Chair, Clean Air Scientific Advisory Committee to EPA Administrator Stephen Johnson. Clean Air Scientific Advisory Committee's (CASAC) Peer Review of Draft Chapter 8 of EPA's Risk and Exposure Assessment to Support the Review of the NO_2 Primary National Ambient Air Quality Standard. EPA-CASAC-09-001. October 28, 2008. Available at: http://yosemite.epa.gov/sab/sabproduct.nsf/264cb1227d55e02c85257402007446a4/87D38275673D66B8852574F00069D45E/$File/EPA-CASAC-09-001-unsigned.pdf.

Samet J (2008b). Letter from Dr. Jonathan M. Samet, Chair, Clean Air Scientific Advisory Committee to EPA Administrator Stephen Johnson. Clean Air Scientific Advisory Committee's (CASAC) Review comments on EPA's Risk and Exposure Assessment to Support the Review of the NO_2 Primary National Ambient Air Quality Standard. EPA-CASAC-09-003. December 16, 2008. Available at: http://yosemite.epa.gov/sab/sabproduct.nsf/264cb1227d55e02c85257402007446a4/9C4A540D86BFB67A852575210074A7AE/$File/EPA-CASAC-09-003-unsigned.pdf.

Samet J (2009). Letter from Dr. Jonathan M. Samet, Chair, Clean Air Scientific Advisory Committee to EPA Administrator Lisa P. Jackson. Comments and Recommendations Concerning EPA's Proposed Rule for the Revision of the National Ambient Air Quality Standards (NAAQS) for Nitrogen Dioxide. EPA-CASAC-09-014. September 9, 2009. Available at: http://yosemite.epa.gov/sab/sabproduct.nsf/264cb1227d55e02c85257402007446a4/0067573718EDA17F8525762C0074059E/$File/EPA-CASAC-09-014-unsigned.pdf.

U.S. EPA (1971). Air Quality Criteria for Nitrogen Oxides. U.S. Environmental Protection Agency. Air Pollution Control Office, Washington, D.C. January 1971. Air Pollution Control Office Publication No. AP-84.

U.S. EPA (1993). Air Quality Criteria for Oxides of Nitrogen. Office of Health and Environmental Assessment, Environmental Criteria and Assessment Office. Research Triangle Park, NC. EPA–600/8–91–049aF–cF, August 1993. Available at: http://cfpub.epa.gov/ncea/cfm/recordisplay.cfm?deid=40179.

U.S. EPA (1995). Review of the National Ambient Air Quality Standards for Nitrogen Oxides: Assessment of Scientific and Technical Information, OAQPS Staff Paper. US EPA, Office of Air Quality Planning and Standards, Research Triangle Park, NC. EPA–452/R–95–005, September 1995. Available at: http://www.epa.gov/ttn/naaqs/standards/nox/data/noxsp1995.pdf.

U.S. EPA (2007a). Integrated Review Plan for the Primary National Ambient Air Quality Standard for Nitrogen Dioxide. U.S. EPA. National Center for Environmental Assessment and Office of Air Quality Planning and Standards. Research Triangle Park, NC. August 2007. Available at: http://www.epa.gov/ttn/naaqs/standards/nox/data/20070823_nox_review_plan_final.pdf.

U.S. EPA (2007b). Nitrogen Dioxide Health Assessment Plan: Scope and Methods for Exposure and Risk Assessment. U.S. EPA, Office of Air Quality Planning and Standards, Research Triangle Park, NC. Draft September 2007. Available at:http://www.epa.gov/ttn/naaqs/standards/nox/data/20070927_risk_exposure_scope.pdf.

U.S. EPA (2008a). Integrated Science Assessment for Oxides of Nitrogen – Health Criteria. U.S. EPA, National Center for Environmental Assessment and Office, Research Triangle Park, NC. EPA/600/R-08/071. July 2008. Available at: http://cfpub.epa.gov/ncea/cfm/recordisplay.cfm?deid=194645.

U.S. EPA (2008b). Risk and Exposure Assessment to Support the Review of the NO2 Primary National Ambient Air Quality Standard. U.S. EPA, Office of Air Quality Planning and Standards. Research Triangle Park, NC. EPA 452/R-08-008a/b. November 2008. Available at: http://www.epa.gov/ttn/naaqs/standards/nox/s_nox_cr_rea.html.

U.S. EPA. (2014). Integrated Review Plan for the Primary National Ambient Air Quality Standards for Nitrogen Dioxide. U.S. EPA, National Center for Environmental Assessment and Office of Air Quality Planning and Standards, Research Triangle Park, NC. EPA-452/R-14-003. June 2014. Available at: http://www.epa.gov/ttn/naaqs/standards/nox/data/201406finalirpprimaryno2.pdf.

U.S. EPA. (2015). Review of the Primary National Ambient Air Quality Standards for Nitrogen Dioxide: Risk and Exposure Assessment Planning Document. U.S. EPA, Office of Air Quality Planning and Standards, Research Triangle Park, NC. EPA-452/D-15-001. May 13, 2015. Available at: https://www3.epa.gov/ttn/naaqs/standards/nox/data/20150504reaplanning.pdf

U.S. EPA. (2016a). Integrated Science Assessment for Oxides of Nitrogen – Health Criteria (2016 Final Report). U.S. EPA, National Center for Environmental Assessment, Research Triangle Park, NC. EPA/600/R-15/068. January 2016. Available at: https://cfpub.epa.gov/ncea/isa/recordisplay.cfm?deid=310879.

U.S. EPA. (2016b). Policy Assessment for the Review of the Primary National Ambient Air Quality Standards for Nitrogen Dioxide – External Review Draft. U.S. EPA, Office of Air Quality Planning and Standards, Research Triangle Park, NC. EPA-452/P-16-001. September 2016. Available at: https://www3.epa.gov/ttn/naaqs/standards/nox/data/20160927-no2-pa-external-review-draft.pdf

2   NO$_2$ AIR QUALITY

This chapter presents information on NO$_2$ atmospheric chemistry, monitoring, and ambient concentrations,[27] with a focus on information that is most relevant for our review of the primary NO$_2$ standards. It is intended as a prologue for detailed discussions on the evidence for health effects and exposures to NO$_2$ that follow in the subsequent chapters, and as a source of information to help interpret those effects in the context of air quality. We generally focus on NO$_2$ in this chapter, as this is the indicator for oxides of nitrogen and most relevant to the evaluation of health evidence. The ISA presents a more thorough characterization of air quality for the broader category of oxides of nitrogen (U.S. EPA, 2016, Chapter 2).

In this chapter, Section 2.1 provides an overview of the atmospheric chemistry of NO$_2$ formation and the NO$_X$ emissions that contribute to ambient NO$_2$. Section 2.2 discusses NO$_2$ ambient monitoring methods and provides an overview of the U.S. ambient monitoring network for NO$_2$. Section 2.3 summarizes information on recent ambient concentrations of NO$_2$, including information from the near-road monitoring network, and on long-term temporal trends in NO$_2$ air quality.

2.1   NO$_2$ ATMOSPHERIC CHEMISTRY AND NO$_X$ EMISSIONS

2.1.1   Atmospheric Chemistry

The overall chemistry of reactive, oxidized nitrogen compounds in the atmosphere is summarized in Figure 2-1. Ambient concentrations of NO$_2$ are influenced by both direct NO$_2$ emissions and by emissions of nitric oxide (NO), with the subsequent conversion of NO to NO$_2$ primarily though reaction with ozone (O$_3$). The initial reaction between NO and O$_3$ to form NO$_2$ occurs fairly quickly during the daytime, with reaction times on the order of minutes. However, NO$_2$ can also be photolyzed to reform NO, creating new O$_3$ in the process (U.S. EPA, 2016, Section 2.2). A large number of oxidized nitrogen species in the atmosphere are formed from the oxidation of NO and NO$_2$. These include nitrate radicals (NO$_3$), nitrous acid (HONO), nitric acid (HNO$_3$), dinitrogen pentoxide (N$_2$O$_5$), nitryl chloride (ClNO$_2$), peroxynitric acid (HNO$_4$), peroxyacetyl nitrate and its homologues (PANs), other organic nitrates, such as alkyl nitrates (including isoprene nitrates), and pNO$_3$. The sum of these reactive oxidation products

---

[27] This chapter focuses on monitored ambient NO$_2$ concentrations. In addition, we have developed an approach to adjust ambient NO$_2$ concentrations in order to estimate the concentrations that could occur if urban areas were to "just meet" the current NO$_2$ NAAQS. These estimated concentrations are based on statistical adjustments of existing air quality concentrations and are discussed in detail in chapter 4 and appendix B of this PA.

(collectively referred to as NO_Z) and NO plus NO_2 (i.e., referred to as NO_X) comprise NO_Y (Figure 2-1).

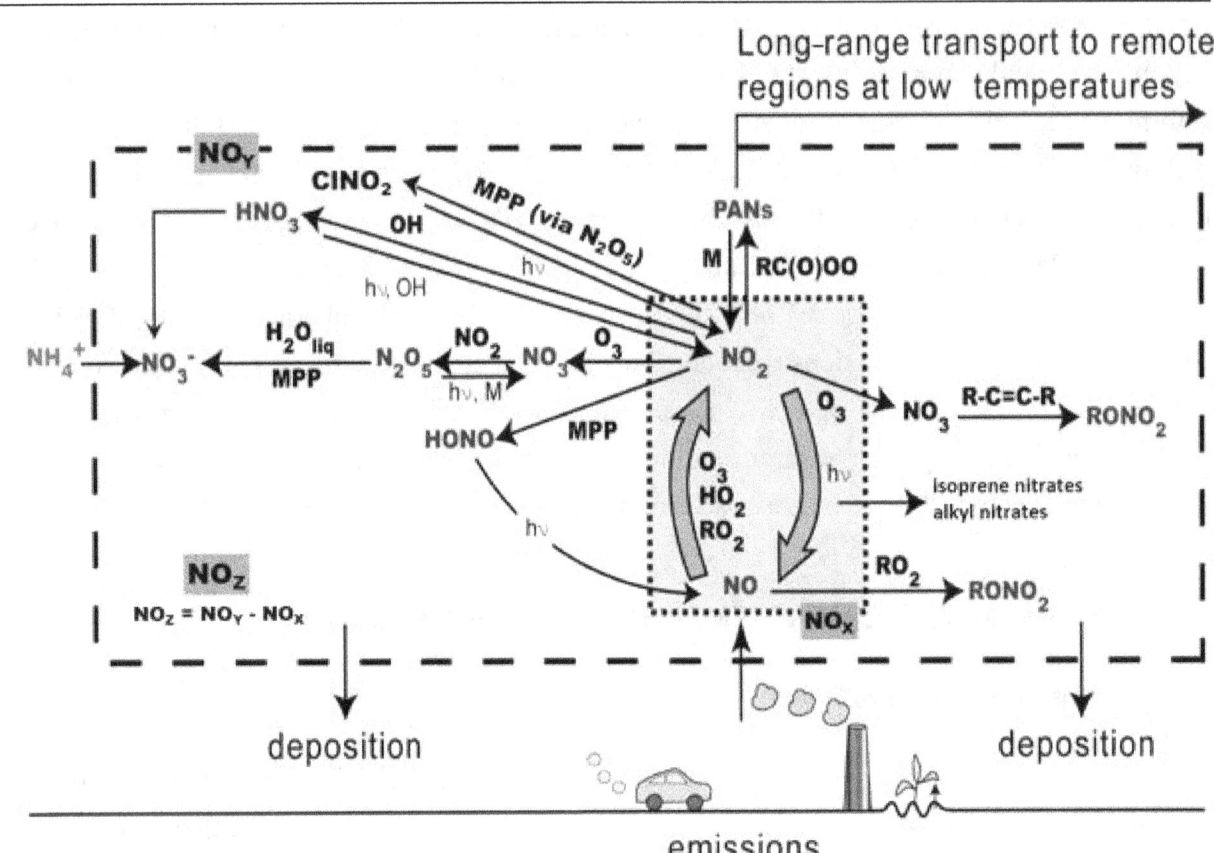

Note: The inner shaded box contains NO_X (= NO + NO_2). The outer box contains other species (NO_Z) formed from reactions of NO_X. All species shown in the outer and inner boxes are collectively referred to as NO_Y by the atmospheric sciences community.

hv = solar photon, M = species transferring/removing enough energy to cause a molecule to decompose/stabilize, MPP = multiphase processes, R = organic radical.

Source: National Center for Environmental Assessment.

Figure 2-1   Schematic diagram of the cycle of reactive, oxidized nitrogen species in the atmosphere.

Due to the close relationship between NO and NO_2, and their ready interconversion, these species are often grouped together and referred to as NO_X. The majority of NO_X emissions are in the form of NO. For example, 90% or more of tail-pipe NO_X emissions are in the form of NO, with only about 2 to 10% emitted as NO_2 (Itano et al., 2014; Kota et al., 2013; Jimenez et al., 2000; Richmond-Bryant et al., 2016). As noted above, NO_X emissions require time and sufficient O_3 concentrations for the conversion of NO to NO_2. Higher temperatures and concentrations of reactants result in shorter conversion times (e.g., less than one minute under some conditions), while dispersion and depletion of reactants results in longer conversion times. The time required to transport emissions away from a roadway can vary from less than one minute (e.g., under open conditions) to about one hour (e.g., for certain urban street canyons) (Düring et al., 2011; Richmond-Bryant and Reff, 2012). These factors can affect the locations where the highest NO_2 concentrations occur. In particular, while ambient NO_2 concentrations are often elevated near important sources of NO_X emissions, such as major roadways, the highest measured ambient concentrations in a given urban area may not always occur immediately adjacent to those sources.[28]

The near-road environment provides a clear example of the interplay between NO_X emissions, meteorology, and the atmospheric chemistry that impacts ambient NO_2 concentrations. Vehicular emissions tend to peak during the morning and afternoon commutes, while peak O_3 concentrations generally occur in the late morning to early evenings. In addition, atmospheric mixing tends to be the strongest during the daytime, rapidly diluting roadway emissions. Given the relative timing of O_3 availability and peak atmospheric mixing conditions, the highest near-road NO_2 concentrations often occur during the early morning hours (i.e., before atmospheric mixing can rapidly dilute emissions) (Kimbrough et al., 2016; Richmond-Bryant et al., 2016).[29]

Oxidized nitrogen compounds are ultimately lost from the atmosphere by wet and dry deposition to the Earth's surface. Soluble species are taken up by aqueous aerosols and cloud droplets and are removed by wet deposition by rainout (i.e., incorporation into cloud droplets that eventually coagulate into falling rain drops). Both soluble and insoluble species are removed by washout (i.e., impaction with falling rain drops, another component of wet deposition), and by dry deposition (i.e., impaction with the surface and gas exchange with plants). NO and NO_2 are not very soluble, and therefore wet deposition is not a major removal process for them. However, a major NO_X reservoir species, HNO_3, is extremely soluble, and its deposition (both wet and dry) represents a major sink for NO_Y.

---

[28] Ambient NO_2 concentrations around stationary sources of NO_X emissions are similarly impacted by the availability of O_3 and by meteorological conditions, although surface-level NO_2 concentrations can be less impacted in cases where stationary source NO_X emissions are emitted from locations elevated substantially above ground level.

[29] The conversion of NO_X into the species that make up NO_Z typically takes place on a much longer time scale than do interconversions between NO and NO_2 (e.g., Ren et al., 2013). NO_X emitted during morning rush hour by vehicles can be converted almost completely to products by late afternoon during warm, sunny conditions.

2.1.2 Emissions

The National Emissions Inventory (NEI)[30] is a national compilation of emissions sources collected from state, local, and tribal air agencies, as well as emission estimates developed by the EPA from data on specific source sectors. According to the NEI, anthropogenic sources account for a large majority of NO_X emissions in the U.S., with highway vehicles, off-highway vehicles, and fuel combustion identified the largest contributors. More specifically, highway vehicles include all on-road vehicles, including light duty as well as heavy duty vehicles, both gasoline- and diesel-powered. Off-highway vehicles and engines include aircraft, commercial marine vessels, locomotives, and non-road equipment. Fuel combustion-utilities includes electric power generating units (EGUs), which derive their power generation from all types of fuels. EGU emissions are dominated by coal combustion, which accounts for 86% of all NO_X emissions from utilities in the 2014 NEI. The fuel combustion-other category includes commercial/institutional, industrial, and residential combustion of biomass, coal, natural gas, oil, and other fuels. Other anthropogenic sources include field burning, prescribed fires, and various industrial processes (e.g., cement manufacturing, oil and gas production). On a national scale, agricultural field burning and prescribed fires are the greatest contributors to the Other Anthropogenic sources category. Biogenics and Wildfires include emissions estimates for plants and soil (i.e., biogenics) and for wildfires.

Nationwide estimates indicate a 61% decrease in total NO_X emissions from 1980 to 2016 (Figure 2-2) as a result of multiple regulatory programs. These include an assortment of key rules implemented over multiple decades, some of which are highlighted in Table 2-1.

---

[30] The NEI may be found at: https://www.epa.gov/air-emissions-inventories

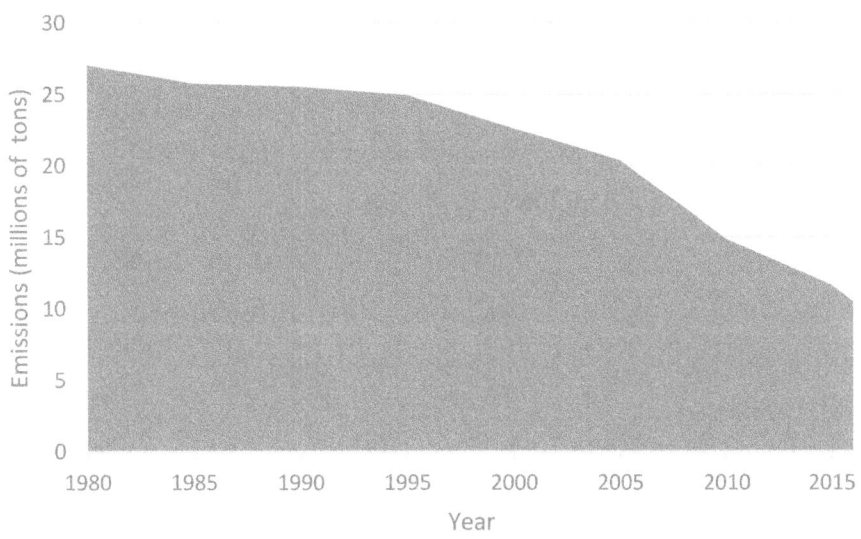

Figure 2-2. U.S. national average NOx emissions from 1980 to 2016. [31]

Table 2-1. Select Programs and Rules that have contributed to NOx reductions over time.

Key Programs Leading to Reductions in NOx Emissions	Year Program or Rule was Created
Clean Air Act – included NOx reduction mandates	1970
Energy Policy Conservation Act – Established Corporate Average Fuel Economy (CAFÉ) standards	1975
1990 Clean Air Act amendments – Reasonably Available Control Technology for stationary sources; Lowered emissions standards for mobile sources; Acid Rain program	1990
NOx SIP Call	1997
Ozone Transport Commission – NOx Budget Program	1999
Tier 2 Light Duty emissions rule	1999
Clean Air Nonroad Diesel rule	2004
Clean Air Interstate Rule	2009
Cross State Air Pollution Rule	2011
Tier 3 Light duty vehicle emission and fuel standards	2014

---

[31] Emissions trends information is based on a rolling 5-year average to smooth out variance caused by methodology differences across certain transition years. This figure reflects an update to the information included in Figure 2-2 of the ISA (U.S. EPA, 2016). Underlying data can be found at: http://www.epa.gov/air-emissions-inventories/air-pollutant-emissions-trends-data

The overall decrease in NO$_X$ emissions has been driven primarily by decreases from the four largest emissions sectors. Specifically, compared to the 1980 NEI, estimates for 2016 indicate a 69% reduction in NO$_X$ emissions from highway vehicles, a 28% reduction from off-highway vehicles and engines, an 85% reduction from fuel combustion-utilities, and a 60% reduction from fuel combustion-other (see Figure 2-3, below).

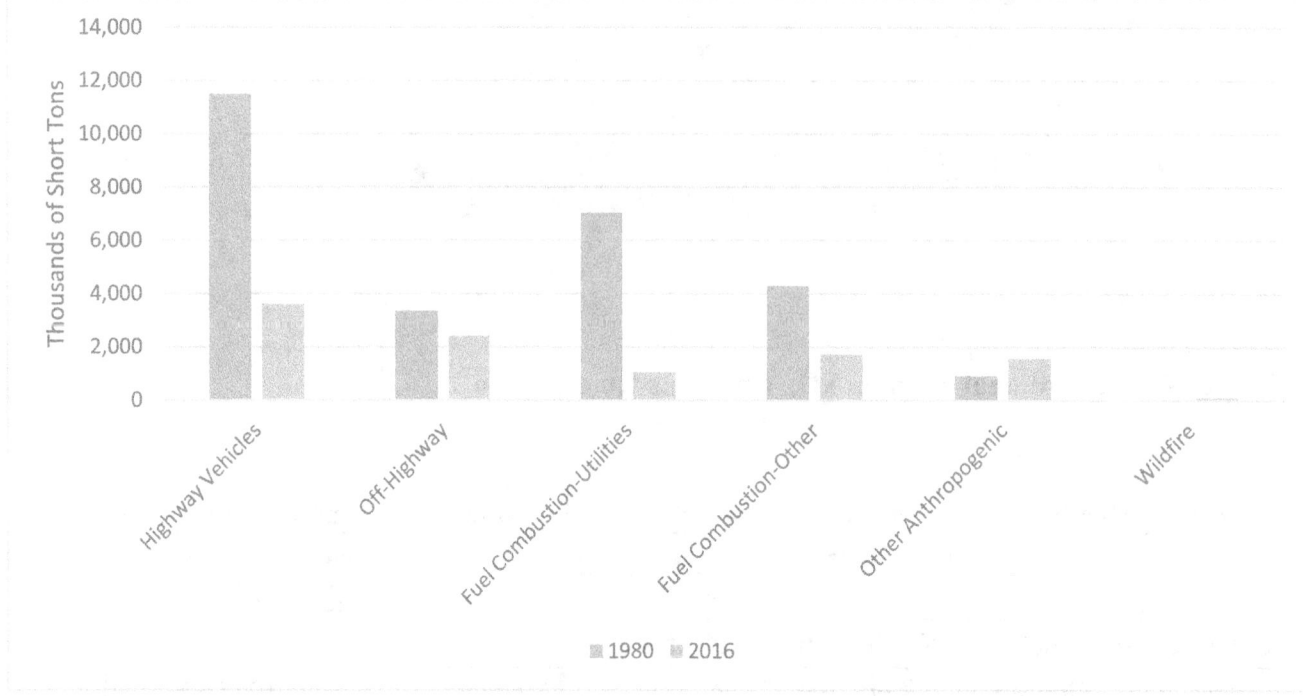

Figure 2-3. Major source sectors of NO$_X$ emissions in the U.S. from the 1980 and 2014 National Emissions Inventories.[32]

Despite substantial reductions, mobile source-related emissions still dominate the NO$_X$ emissions inventory. Highway vehicles are the largest source NO$_X$ emissions in the U.S., contributing 34% of the total NO$_X$ emissions. Off-highway vehicles and engines account for 23% of emissions, EGUs for 10%, fuel combustion-other for 16%, other anthropogenic sources for 15%, and wildfires for 1%.[33]

In contrast to the reductions estimated across the largest categories of NO$_X$ emitters, estimated NO$_X$ emissions for the other anthropogenic category increased by 72% (Figure 2-3, above), with the greatest increases observed for oil and gas production, agricultural field burning, prescribed fires, and mining. While the fraction of total NO$_X$ emissions that comes from oil and gas production is only about 8% nationwide, regional and local contributions from this industry

---

[32] 2016 emissions estimates are based on projections from the 2014 NEI (see https://www.epa.gov/air-emissions-inventories/air-pollutant-emissions-trends-data).

[33] See http://www.epa.gov/air-emissions-inventories/air-pollutant-emissions-trends-data

can be much higher. For example, estimates in New Mexico, Oklahoma, and Texas indicate that up to about 18 to 20% of state NO_X emissions come from oil and gas operations.

2.2   AMBIENT NO₂ MONITORING

2.2.1   NO₂ Methods

Ambient NO_2 concentrations are measured by monitoring networks operated by state, local, and tribal air agencies, which are typically funded in part by the EPA. The main network of monitors providing ambient data for use in implementation activities related to the NAAQS is the State and Local Air Monitoring Stations (SLAMS) network. This network relies on a chemiluminescent Federal Reference Method (FRM) and on Federal Equivalent Methods (FEM) that use either chemiluminescence or direct measurement methods of NO_2. Chemiluminescent-based FRMs only detect NO in the sample stream. Therefore, a two-step process is employed to measure NO_2, based on the subtraction of NO from oxidized nitrogen.[34] Data produced by chemiluminescent analyzers include NO, NO_2, and NO_X measurements, which are all routinely logged by state and local air monitoring agencies. Hourly average values are typically reported to the EPA's Air Quality System (AQS).

As discussed in the ISA (U.S. EPA, 2016, p. 2-24) the traditional chemiluminescence FRM is subject to potential measurement biases resulting from interference by NO_Z species. These potential biases are measurement uncertainties that can impact exposure analyses. However, within metropolitan areas, where a majority of the NO_2 monitoring network is located and is influenced by strong NO_X sources, the potential for NO_Z related bias is relatively small. There have been recent advances in methods that provide measurements of NO_2 with less potential for interference. These newer methods include photolytic-chemiluminescent methods that rely on photodissociation of NO_2 using specific wavelengths of light, and direct measurements of NO_2, including cavity attenuated phase shift [CAPS] spectrometry and cavity ring-down spectroscopy. It should be noted that the direct measurement methods do not provide

---

[34] First, the analyzer determines the amount of NO in the sample air. Second, the analyzer re-routes air flow so that the sample air stream passes over a heated molybdenum oxide catalytic converter reducing a large majority (if not all) of the oxidized nitrogen species present in the sample stream to NO, before again measuring the amount of NO in the sample. The analyzer then subtracts the measured, actual ambient NO, determined in the first step, from the amount measured in the second step, allowing for the determination of NO, NO_2, and NO_X (where $NO_X = NO + NO_2$). The catalytic converter can convert nitric acid (HNO_3) and peroxyacetyl nitrate to NO, which would subsequently be counted as NO_2. Photolytic-chemiluminescence FEM carries out the reduction of NO_2 to NO in a photolytic converter with a known converter efficiency rate, which is specific to NO_2 and, thus, is not subject to the same positive bias potential as the chemiluminescent FRM.

NO or NO$_X$ data (U.S. EPA, 2016, Section 2.4). These newer methods are expected to gradually replace the older FRMs as the older monitors age.

2.2.2   Ambient Monitoring Network

Ambient NO$_2$ monitors in the SLAMS network began operating in the late 1970s and have been used to make measurements supporting NAAQS compliance, the Photochemical Assessment Monitoring Station (PAMS) program, and other objectives at the national, state, and local levels. As of January 2016, approximately 484 NO$_2$ monitors were in operation across the nation and reporting data to AQS. The network has grown and contracted in size since its initiation in the 1970's, in response to changing objectives, priorities of federal, state, and local air monitoring agencies, and resources. Currently, the network is growing due to the addition of near-road monitors (discussed below) and as part of the revisions to the PAMS requirements (80 FR 65291, December 28, 2015).

In consideration of the location and measurements taken, each monitor is assigned a spatial scale associated with the size of the area that it represents. The monitor spatial scales are defined in 40 CFR 58 appendix D as:

1. *Microscale*: area dimensions ranging from several meters up to about 100 meters.
2. *Middle scale*: areas up to several city blocks in size with dimensions ranging from about 100 meters to 0.5 kilometer.
3. *Neighborhood scale*: extended city area with relatively uniform land use and dimensions in the 0.5 to 4.0 kilometers range.
4. *Urban scale*: area of city-like dimensions, on the order of 4 to 50 kilometers. Within a city, the geographic placement of sources may result in there being no single site that can be said to represent air quality on an urban scale.
5. *Regional scale*: rural area of reasonably homogeneous geography without large sources, and extends from tens to hundreds of kilometers.
6. *National and global scales*: concentrations characterizing the nation and the globe as a whole.

At the time of the last review of the primary NO$_2$ NAAQS, the majority of NO$_2$ monitors were sited to represent the neighborhood scale. We used the term "area-wide" to refer to monitors sited at neighborhood, urban, and regional scales, as well as those monitors sited at either micro-

or middle-scale that are representative of many such locations in the same core-based statistical area (CBSA)[35] (75 FR 6474, February 9, 2010).

In the 2010 review of the primary NO_2 NAAQS, consideration of population exposures was focused on major roadways. Due to the lack of monitors specifically sited near major roadways, new near-road monitoring requirements were promulgated (75 FR 6474, February 9, 2010). At that time, one near-road monitor was required in any CBSA with a population of 500,000 or more. An additional near-road monitor was required in CBSAs with populations of at least 2,500,000 and in CBSAs with populations of at least 500,000 with roadway segments carrying traffic volumes of at least 250,000 vehicles per day.

The near-road network has been implemented in phases. The first phase included CBSAs with populations greater than 1,000,000 and was required to be operational as of January 2014. The second phase included CBSAs with populations greater than 2,500,000 and CBSAs having at least 500,000 people that also had one or more road segments with an AADT of at least 250,000, and was required to be in operation starting in January 2015.[36] As of January 2017, the EPA estimates that 70 near-road monitors are in operation and reporting data to AQS. These monitors are sited as close as two meters from the target road, with several monitors within 10 m. The EPA maintains a database containing the meta-data for near-road monitors,[37] including the relevant CBSA, population estimates, the AQS ID of the near-road site, latitude and longitude, site installation date, the name of the target road of individual near-road monitors, traffic volume data of the target road, the distance to the target road, the height of the monitor probe, and information on other pollutant monitoring occurring at the site.

2.3 NO_2 MONITORING DATA TRENDS AND AIR QUALITY RELATIONSHIPS

This section presents information on ambient NO_2 concentrations. Section 2.3.1 presents data on national trends in ambient NO_2 concentrations, section 2.3.2 presents data on the NO_2 concentrations measured by recently deployed near-road monitors, section 2.3.3 presents data on the relationships between 1-hour and annual NO_2 concentrations, and section 2.3.4 discusses background NO_2 concentrations.

---

[35] A CBSA is a geographic area defined by the Office of Management and Budget (OMB) that consists of one or more counties anchored by an urban core with a population $\geq 10,000$. CBSAs have replaced metropolitan statistical areas (MSAs) that were previously used by OMB.

[36] The EPA, through public notice and comment rulemaking, removed requirements for Phase 3 near-road monitors, effective December 30, 2016 (81 FR 96381, December 30, 2016).

[37] The database is found at: http://www3.epa.gov/ttn/amtic/nearroad.html

2.3.1 National Trends in Ambient NO₂ Concentrations

The metric used to determine whether areas meet or exceed the NAAQS is called a design value (DV).[38] In the case of NO_2, there are 2 types of DVs: the annual DV and the hourly DV. The annual DV for a particular year is the average of all hourly values within that calendar year. The hourly DV is the three-year average of the 98th percentiles of the annual distributions of daily maximum 1-hour NO_2 concentrations. DVs are considered to be valid if the monitoring data used to calculate them meet completeness criteria described in 40 CFR 50.11 and appendix S to Part 50.[39]

The long-term trends in DVs across the U.S. are displayed in Figures 2-4 and 2-5. The distributions of valid[40] DVs across the country as a function of time are shown in Figure 2-4. Figure 2-4 shows that DVs across the country have been, on average, declining since 1980.

---

[38] A design value is a statistic that describes the air quality status of a given area relative to the NAAQS. Design values are typically used to classify nonattainment areas, assess progress towards meeting the NAAQS, and develop control strategies. See http://epa.gov/airtrends/values.html for guidance on how these values are defined.

[39] See 40 CFR 50.11 (available at http://www.ecfr.gov/cgi-bin/text-idx?SID=86b930e674d72c8e0e14bb65c51a0047&mc=true&node=se40.2.50_111&rgn=div8) and appendix S to Part 50 for more information on the calculation of DVs.

[40] 40 CFR part 50 appendix S states that a year is considered complete when all 4 quarters have at least 75 percent of the sampling days, with a sampling day requiring coverage of 75 percent of the hours in the day. The 1-hour DV requires 3 years of complete data.

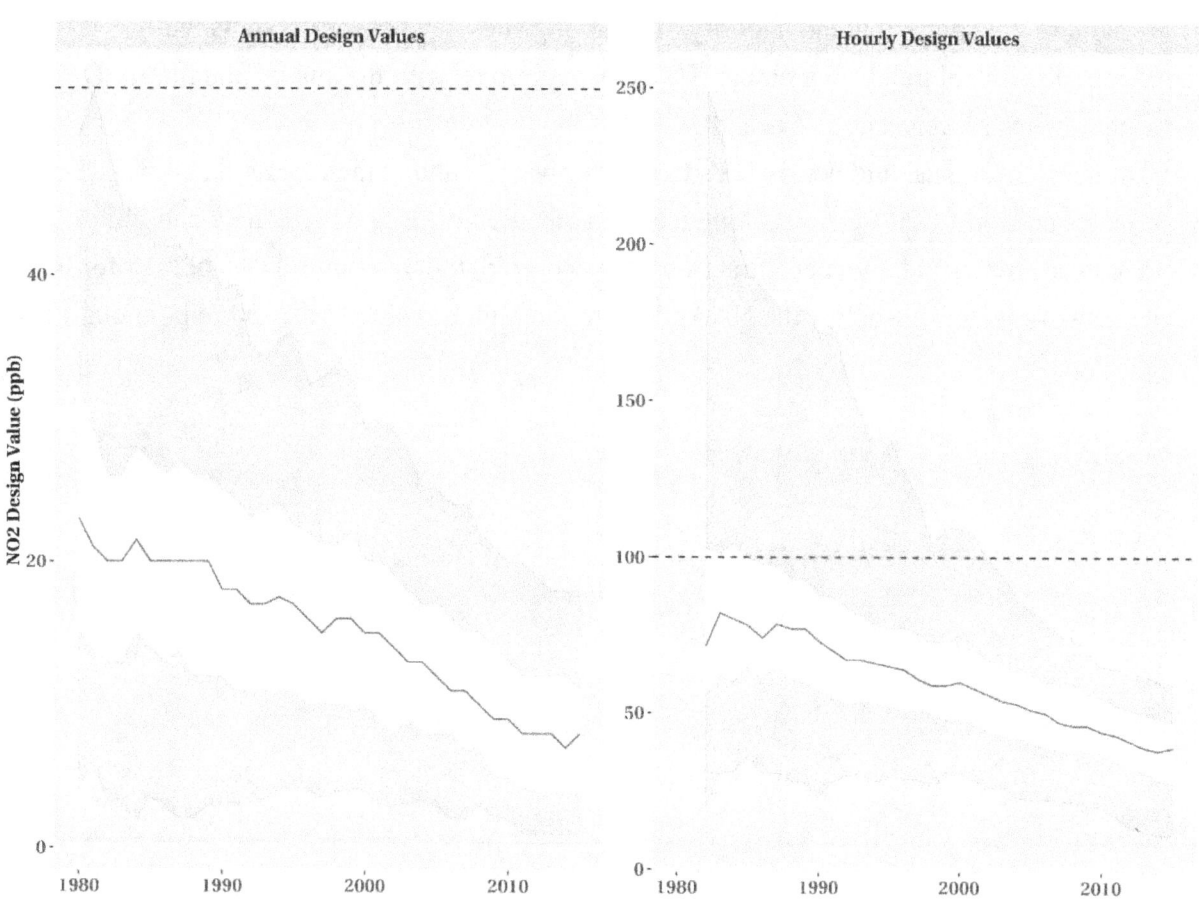

Figure 2-4. Distributions of NO₂ design values across the U.S. from 1980- 2015. The middle lines represent the median, the middle white band extends from the 25th to the 75th percentile, and the outer colored band extends from the 5th to the 95th percentile.

Figure 2-5 shows maps of the NO₂ monitoring network, with the direction of the symbol and color of each point indicating the long term (1980-2015) trend direction.[41] Since 1980, there has been data collected at 2099 sites in the U.S., cumulatively, with individual sites having a wide range in duration of operations across multiple decades. However, only sampling sites with data sufficient to produce at least 5 valid DVs, whenever they might have occurred, were considered in this analysis. After this screen, 647 and 433 monitors were used to determine trends of annual and hourly DVs, respectively.

---

[41] These directions were determined using the sign of spearman correlation coefficient between DV and year. Only DVs determined to be valid by the completeness criteria in CFR 40 Appendix S were included in the calculation. Trend directions were determined to be insignificant if the associated p-values were greater than 0.05 (95% confidence level).

Figure 2-5 shows that the majority of sampling sites have observed statistically significant downward trends in ambient NO_2 concentrations, with the annual and hourly DVs showing downward trends at 61.5% and 74.8% of monitoring sites, respectively.[42] At 3.9% and 1.8% of sites, the annual and hourly DVs trended upward,[43] and at the remaining 34.6% and 23.3% sites no significant trend was found. Even considering the fact there are a handful of sites where upward trends in NO_2 concentrations have occurred, the maximum DVs in 2015 for the whole network were well-below the NAAQS, with the highest values being 30 ppb (annual) and 72 ppb (hourly).

---

[42] Since this analysis required 5 valid DVs, and near-road monitors have not been in operation long enough to calculate 5 DVs, these trends do not reflect the near-road monitoring network.

[43] It is not clear what specific sources may be responsible for these upward trends in ambient NO_2 concentrations. As discussed above (section 2.1.2), since 1980 increases in NO_X emissions have been observed for several types of sources, including oil and gas production, agricultural field burning, prescribed fires and mining. Though relatively small contributors nationally, emissions from these sources can be substantial in some areas (e.g., see Table 2-2 in U.S. EPA, 2016).

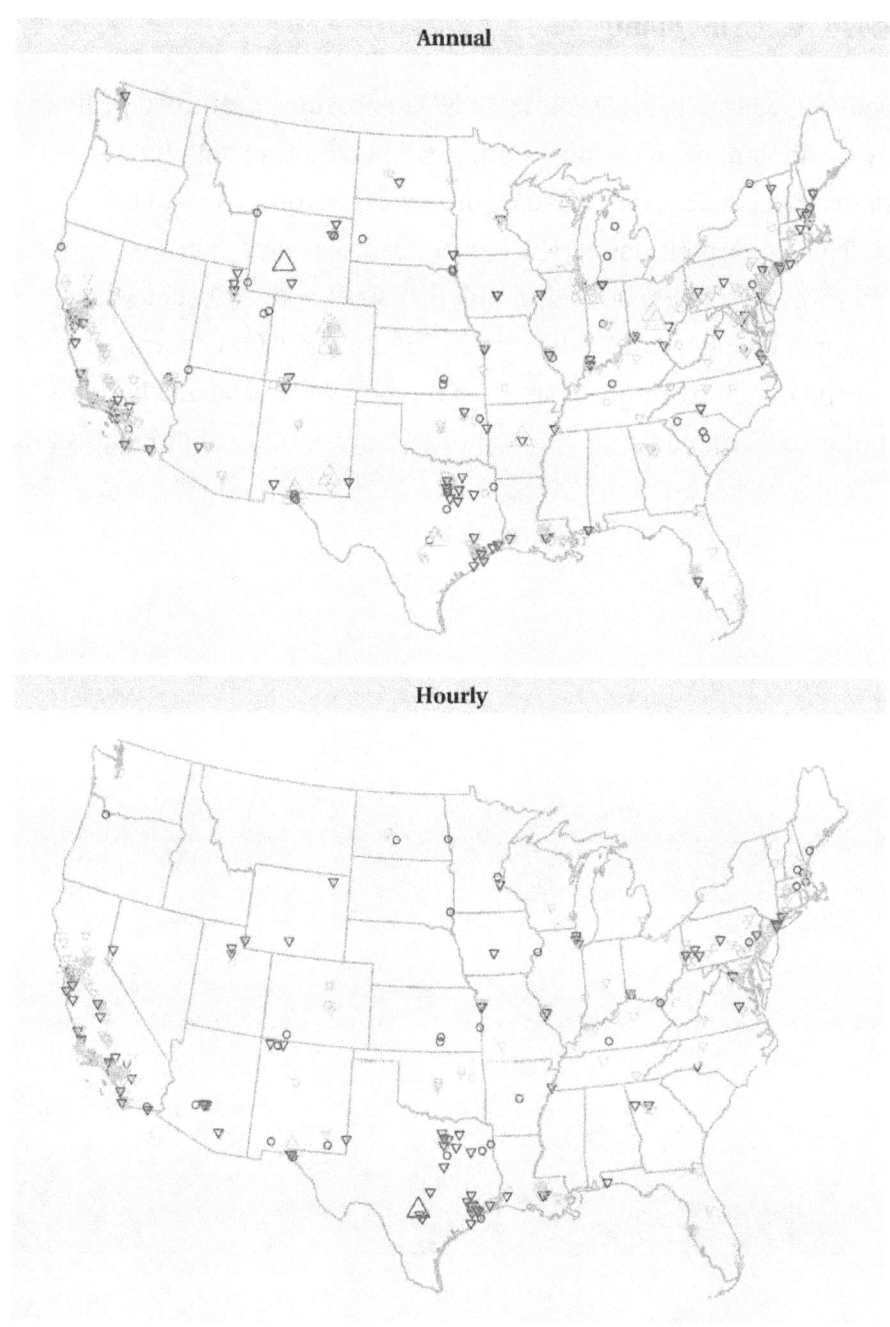

Annual

Hourly

- First Valid DV from 2000 or later   First Valid DV from before 2000

Direction of DV Trend: ▽ Decreasing   △ Increasing   ○ Insignificant

Figure 2-5. Trend directions of NO₂ design values for 1980-2015 at U.S. sampling sites.

2.3.2   Near-Road NO$_2$ Air Quality

As discussed above, the largest single source of NO$_X$ emissions is on-road vehicles, and emissions are primarily in the form of NO, with NO$_2$ formation requiring both time and sufficient O$_3$ concentrations. Depending on local meteorological conditions and O$_3$ concentrations, ambient NO$_2$ concentrations can be higher near roadways than at sites in the same area but farther removed from the road (and from other sources of NO$_X$ emissions). To better understand the historical relationships between distributions of NO$_2$ concentrations at monitors near roadways and monitors further away from roads,[44] the annual and hourly DVs from 1980 to 2015 are plotted by decade, as a function of distance from road in Figures 2-6.[45] This analysis focused on monitors located inside the boundary of CBSAs. In all graphs, the color is mapped to the number of sites included in each boxplot.

---

[44] As defined by the 2012 HPMS shapefile used to determine road locations, located at http://www.fhwa.dot.gov/policyinformation/hpms/shapefiles.cfm. This file contains main roads that are part of the National Highway System. See Appendix A for more details.

[45] NO$_2$ monitors meeting the near-road siting requirements set forth in the 2010 NO$_2$ NAAQS were not available in most CBSAs prior to 2014. In particular, monitors were not sited within 50 m of the most heavily trafficked roads in an area. Thus, the historical relationships reflected in Figure 2-6 do not reflect the relationships that existed between NO$_2$ concentrations and distance from the most heavily trafficked roads. However, for the 2010 to 2015 bin, data from recently deployed near-road monitors are included, to the extent available, even when completeness criteria for calculating DVs are not met.

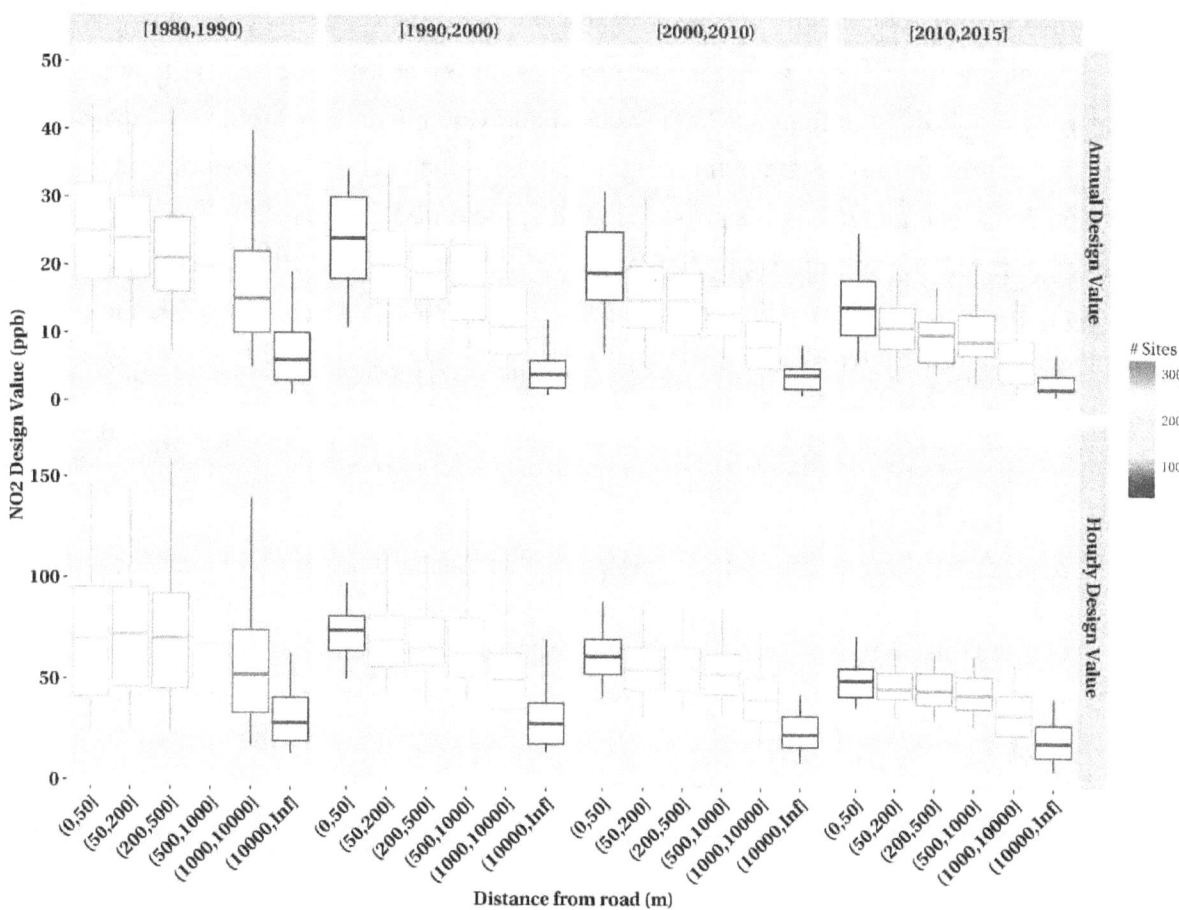

Figure 2-6. Distributions by decade of NO₂ design values for six different bins of distances from major roads in CBSAs. The middle lines represent the median, box edges represent the 25th and 75th percentiles, and whisker ends represent the 5th and 95th percentiles.

Figure 2-6 indicates that NO₂ DVs are generally highest at sampling sites nearest to the road (less than 50 meters) and decrease as distance from the road increases. This relationship is more pronounced for annual DVs than for hourly DVs. The general pattern of decreasing DVs with increasing distance from the road has persisted over time, though the absolute difference (in terms of ppb) between NO₂ concentrations close to roads and those farther from roads has generally decreased over time (i.e., compare 1980-1990 DVs with more recent DVs). This decrease is likely due to the concurrent decrease in mobile source NO_X emissions that occurred over the same time period discussed above (Figure 2-3).

Figures 2-7, 2-8, 2-9, and 2-10 further explore the relationships between NO₂ concentrations measured by newly deployed near-road monitors and those measured by non-near-road monitors (generally area-wide[46]) in the same CBSA. For the years 2013, 2014, and 2015, we identified CBSAs with complete NO_2 data from at least one near-road and one non-near-road monitor.[47] For the year 2016 we used all available data in AQS, as 4th quarter data were not yet due into AQS at the time of the analysis. Each near-road monitor was paired with the non-near-road monitor in the same CBSA that measured the highest 98th percentile NO_2 concentrations.[48] Distributions of daily maximum 1-hour NO_2 concentrations from these monitor pairs are presented for 2013 (Figure 2-7), 2014 (Figure 2-8), 2015 (Figure 2-9), and the partial data available for 2016 (Figure2-10).[49]

---

[46] Non-near-road monitors can generally be considered area-wide, but in some cases, non-near-road monitors can be located close to other sources of NO_X emissions (e.g., ports, railyards).

[47] As indicated, 40 CFR part 50 appendix S states that a year is considered complete when all 4 quarters have at least 75 percent of the sampling days, with a sampling day requiring coverage of 75 percent of the hours in the day.

[48] 98th percentiles from non-near-road monitors were based on the same years that the near-road monitor was in operation.

[49] The data from 2016 is based on what was in AQS in January 2017 and the data have not been certified.

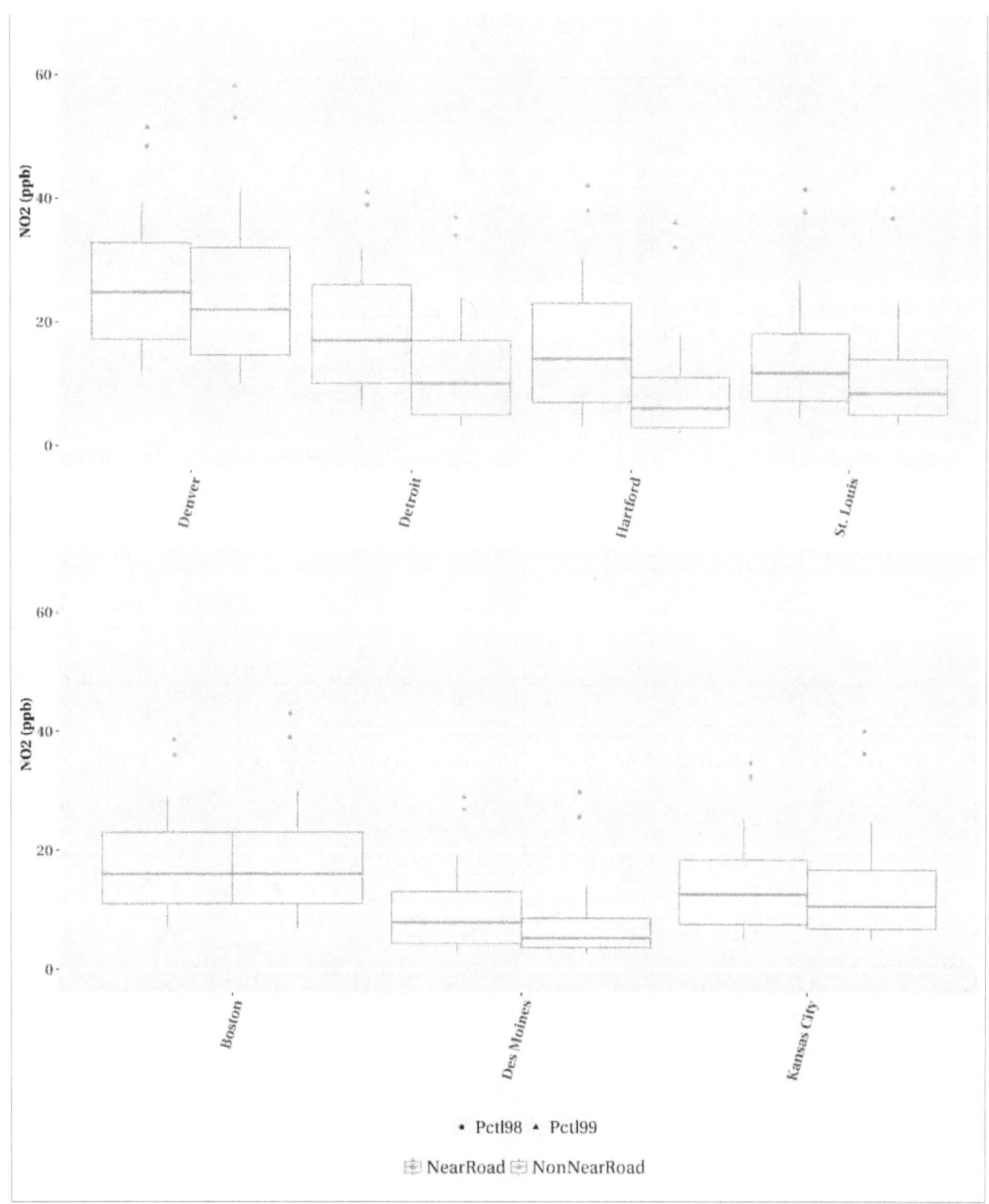

Figure 2-7. Distributions of the near-road and non-near-road maximum 1-hr daily NO₂ concentrations from 2013. The middle lines represent the median, box edges represent the 25th and 75th percentiles, and whisker ends represent the 5th and 95th percentiles. Circles represent 98th percentiles and triangles represent 99th percentiles.

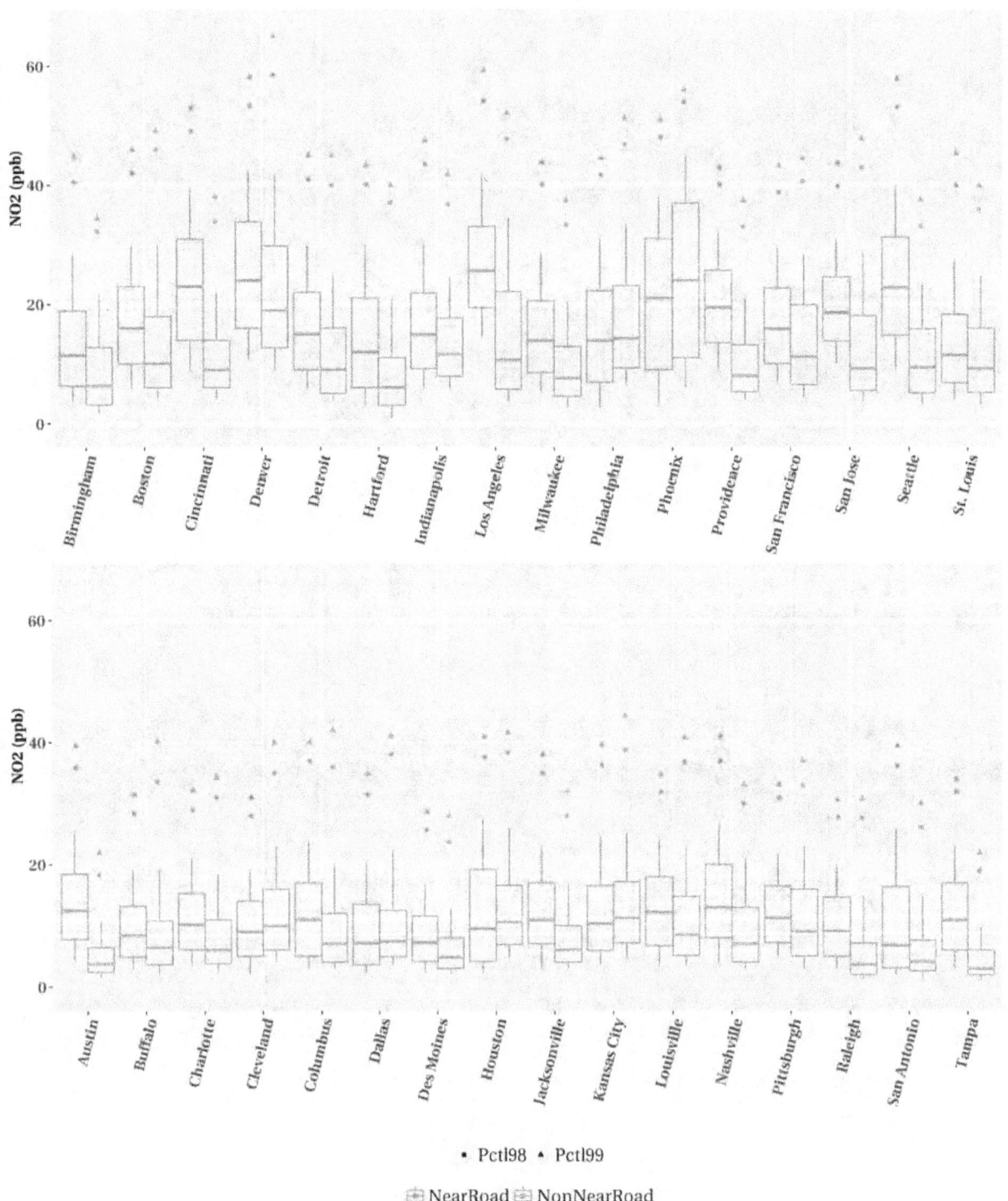

Figure 2-8. Distributions of the near-road and non-near-road maximum 1-hr daily NO₂ concentrations from 2014. The middle lines represent the median, box edges represent the 25th and 75th percentiles, and whisker ends represent the 5th and 95th percentiles. Circles represent 98th percentiles and triangles represent 99th percentiles.

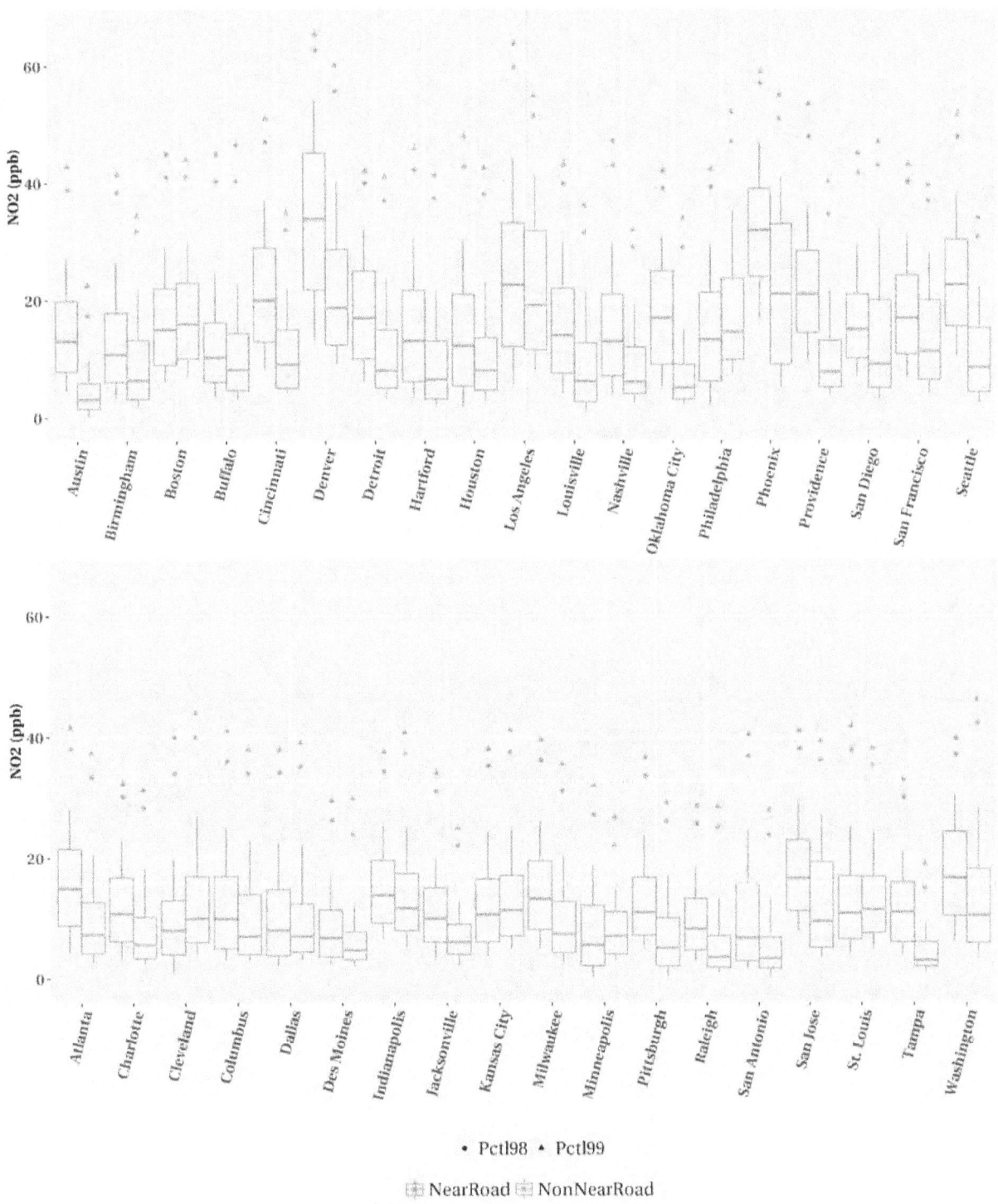

Figure 2-9. Distributions of the near-road and non-near-road maximum 1-hr daily NO₂ concentrations from 2015. The middle lines represent the median, box edges represent the 25th and 75th percentiles, and whisker ends represent the 5th and 95th percentiles. Circles represent 98th percentiles and triangles represent 99th percentiles.

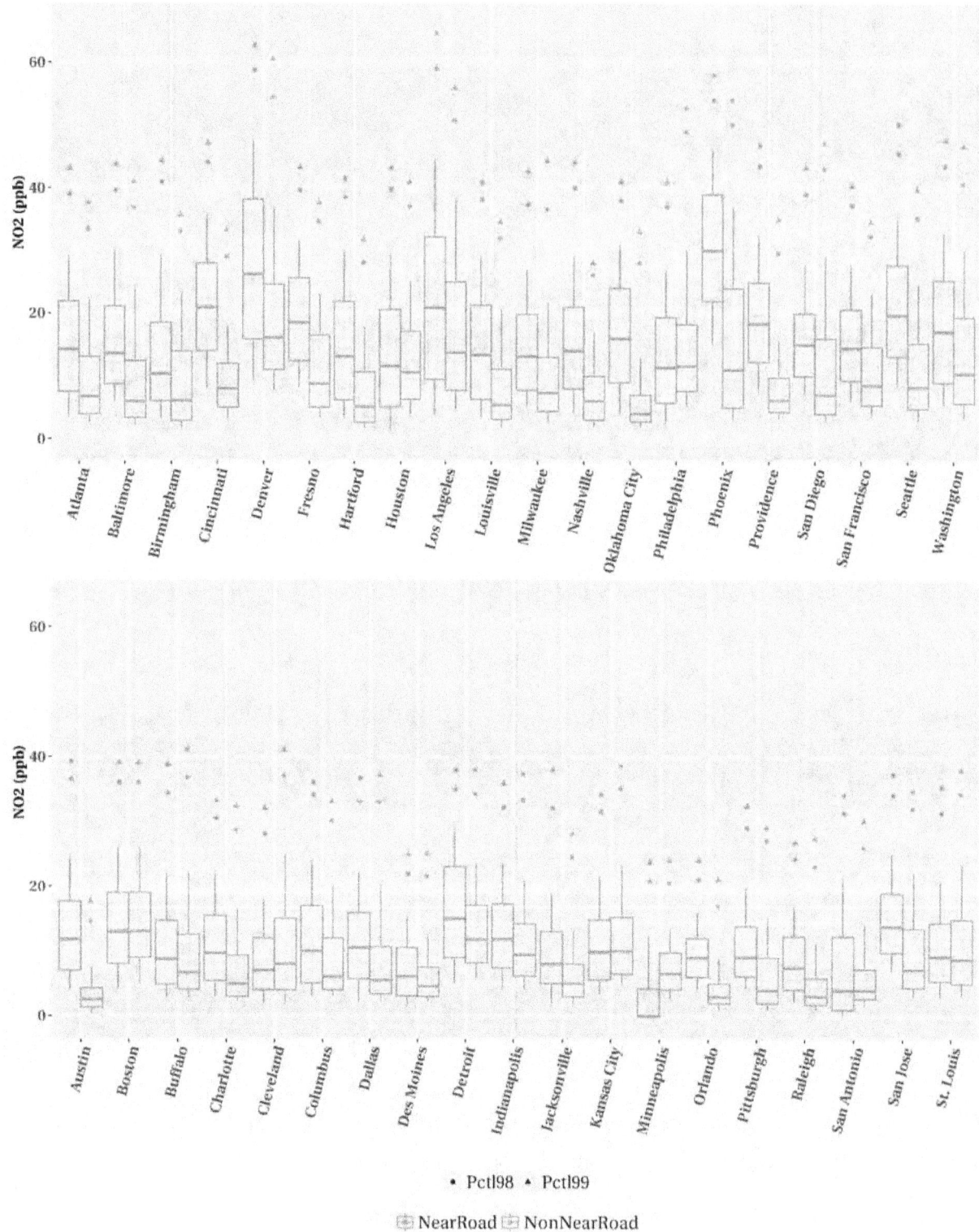

Figure 2-10. Distributions of the available near-road and non-near-road maximum 1-hr daily NO₂ concentration data from 2016. The middle lines represent the median, box edges represent the 25th and 75th percentiles, and whisker ends represent the 5th and 95th percentiles. Circles represent 98th percentiles and triangles represent 99th percentiles.

For the 4 years of available data, Figures 2-7 to 2-10 indicate that daily maximum 1-hour NO_2 concentrations are generally higher at near-road monitors than at non-near-road monitors in the same CBSA. The 98th percentiles of 1-hour daily maximum concentrations (the statistic most relevant to the current standard) were highest at near-road monitors (i.e., higher than all non-near-road monitors in the same CBSA) in 58-77% of the CBSAs evaluated, depending on the year. Near road monitors reported higher annual average NO_2 concentrations in virtually all instances (Appendix A, Figure A-1).

2.3.3 Relationships between Hourly and Annual NO_2 Concentrations

As discussed above, control programs have resulted in substantial reductions in NO_X emissions since the 1980s. These reductions in NO_X emissions have decreased both short-term peak NO_2 concentrations and annual average concentrations. Figure 2-10 illustrates the relationship between 1-hour and annual DVs at individual monitors across the U.S., with data segregated by decade

When considering the change from the 1980-1990 bin to the 2010-2015 bin, the median annual DV has decreased by about 65% (i.e., from ~23 ppb to ~8 ppb) and the median 1-hour DV has decreased by about 50% (i.e., from ~74 ppb to ~37 ppb) (Figure 2-10). At various times in the past, a number of sites would have violated the 1-hour standard without violating the annual standard; however, no sites would have violated the annual standard without also violating the 1-hour standard. Furthermore, these data indicate that 1-hour DVs at or below 100 ppb generally correspond to annual DVs below 35 ppb. CASAC noted this relationship, stating that "attainment of the 1-hour standard corresponds with annual design value averages of 30 ppb NO_2" (Diez Roux and Sheppard, 2017). Thus, meeting the 1-hour standard with its level of 100 ppb would be expected to maintain annual average NO_2 concentrations well-below the 53 ppb level of the annual standard.

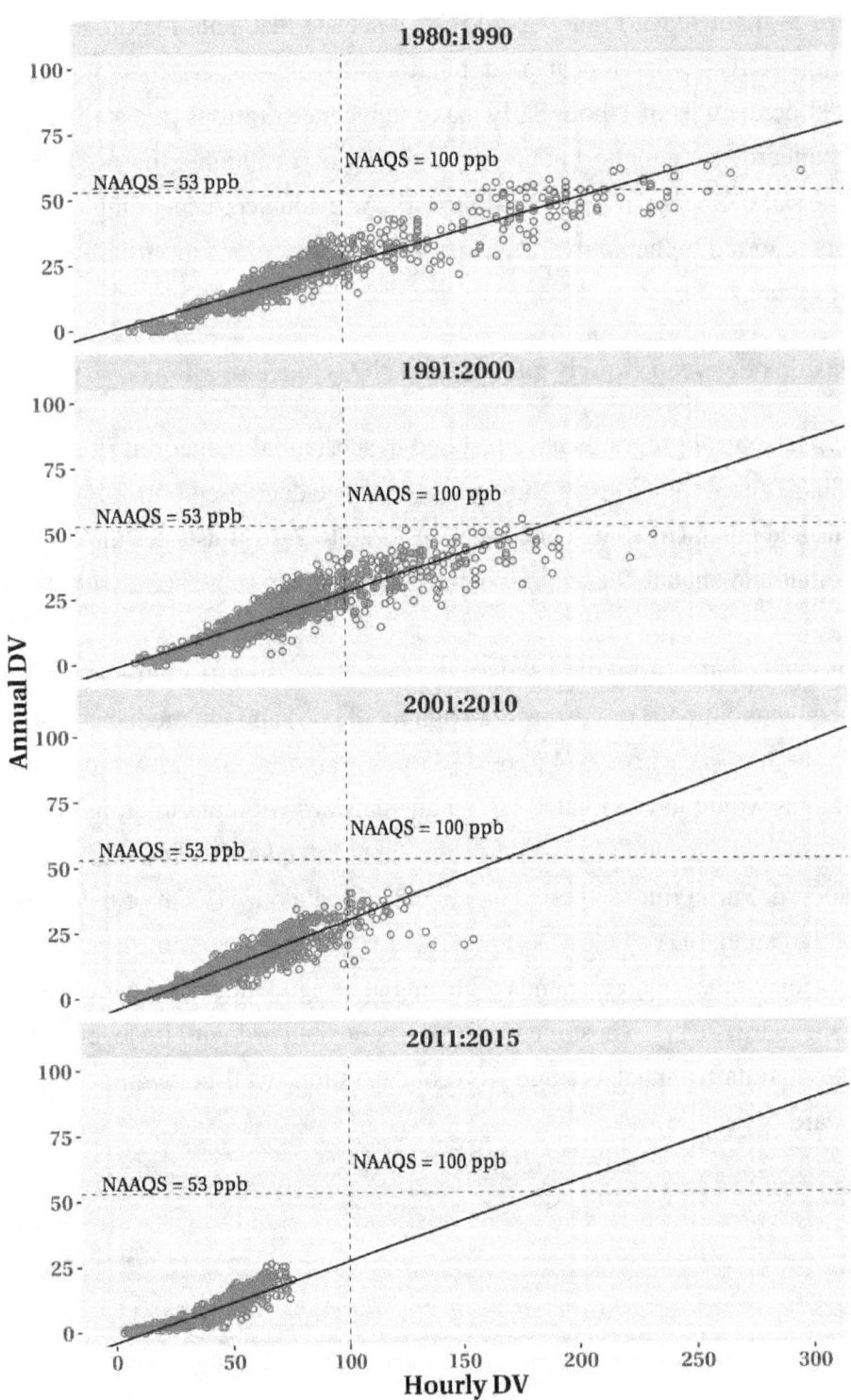

Figure 2-11. Relationships between annual and hour DVs from 1980 to 2015. Hourly and annual DVs are plotted for various decades. Linear regression lines are shown. Near-road monitors are not included in this analysis due to the limited amount of data available.

2.3.4 Background NO₂ Concentrations

In the context of the review of a NAAQS, the EPA generally defines background concentrations in a way that distinguishes among concentrations that result from precursor emissions that are relatively less controllable from those that are relatively more controllable through U.S. policies or through international agreements. One approach to considering background concentrations is to estimate the pollutant concentrations that would exist in the absence of anthropogenic emissions from the U.S., Canada, and Mexico. Such background concentrations are referred to as North American Background (NAB). NAB includes contributions resulting from emissions by natural sources (e.g., soils, wildfires, and lightning around the world and by anthropogenic sources outside of the U.S., Canada, and Mexico.[50]

NO₂ background concentrations are much lower than the NO₂ concentrations currently measured in the ambient air (and much lower than current standard levels). In particular, as discussed in the 2008 ISA, NAB is less than 300 ppt over most of the continental U.S. and less than 100 ppt in the eastern U.S. (U.S. EPA, 2008, Figure 2.4-18). The distribution of background concentrations in the 2008 ISA was shown to reflect the distribution of soil NO emissions and lightning, with some local increases due to biomass burning, mainly in the western U.S. In the northeastern U.S., where present-day NO₂ concentrations are highest, NAB contributes <1% to the total NO₂ concentration (U.S. EPA, 2016, Section 2.5.6).

---

[50] Other approaches to defining background include U.S. background (USB), which includes contributions from emissions from natural sources and from anthropogenic sources outside the U.S., and natural background, which includes only contributions from emissions from natural sources.

2.4   REFERENCES

Düring, I; Bächlin, W; Ketzel, M; Baum, A; Friedrich, U; Wurzler, S. (2011). A new simplified NO/NO2 conversion model under consideration of direct NO2-emissions. Meteor Z 20: 67-73. http://dx.doi.org/10.1127/0941-2948/2011/0491

Hameed, S; Pinto, JP; Stewart, RW. (1979). Sensitivity of the predicted CO-OH-CH4 perturbation to tropospheric NOx concentrations. J Geophys Res Oceans 84: 763-768. http://dx.doi.org/10.1029/JC084iC02p00763

Itano Y et al. (2014). Estimation of Primary NO2/NOx Emission Ratio from Road Vehicles Using Ambient Monitoring Data. *Studies in Atm Sci*, 1-7

Jimenez J.L. et al. (2000). Remote sensing of NO and NO2 emissions from heavy-duty diesel trucks using tunable diode lasers. *Environ Sci Technol*, 2380-2387.

Kimbrough S et al. (2016). Near-road NO to NO2 Conversion Rate Study using Field Measurements and Dispersion Modeling Techniques in Las Vegas, NV. *Submitted*.

Kota S. H. et al. (2013). Simulating near-road reactive dispersion of gaseous air pollutants using a three-dimensional eulerian model. *Sci Total Environ*, Simulating near-road reactive dispersion of gaseous air pollutants using a three-dimensional eulerian model.

Ren, X; van Duin, D; Cazorla, M; Chen, S; Mao, J; Zhang, L, i; Brune, WH; Flynn, JH; Grossberg, N; Lefer, BL; Rappenglueck, B; Wong, K; Tsai, C; Stutz, J; Dibb, JE; Jobson, BT; Luke, WT; Kelley, P. (2013). Atmospheric oxidation chemistry and ozone production: Results from SHARP 2009 in Houston, Texas. J Geophys Res Atmos 118: 5770-5780. http://dx.doi.org/10.1002/jgrd.50342

Richmond-Bryant, J; Reff, A. (2012). Air pollution retention within a complex of urban street canyons: A two-city comparison. Atmos Environ 49: 24-32. http://dx.doi.org/10.1016/j.atmosenv.2011.12.036

Richmond-Bryant J et al. (2016). Estimation of on-road NO_2 concentrations, NO_2/NO_X ratios, and related roadway gradients from near-road monitoring data. *Submitted to Air Quality, Atm and Health* .

U.S. EPA. (2008). Integrated Science Assessment for Oxides of Nitrogen – Health Criteria. U.S. EPA, National Center for Environmental Assessment and Office, Research Triangle Park, NC. EPA/600/R-08/071. July 2008. Available at: http://cfpub.epa.gov/ncea/cfm/recordisplay.cfm?deid=194645.

U.S. EPA. (2016). Integrated Science Assessment for Oxides of Nitrogen – Health Criteria (2016 Final Report). U.S. EPA, National Center for Environmental Assessment, Research Triangle Park, NC. EPA/600/R-15/068. January 2016. Available at: https://cfpub.epa.gov/ncea/isa/recordisplay.cfm?deid=310879.

Valin, LC; Russell, AR; Cohen, RC. (2013). Variations of OH radical in an urban plume inferred from NO 2 column measurements. Geophys Res Lett 40: 1856-1860. http://dx.doi.org/10.1002/grl.50267

3   CONSIDERATION OF EVIDENCE FOR NO$_2$-RELATED HEALTH EFFECTS

In this chapter, the scientific evidence on health effects attributable to short or long-term NO$_2$ exposure is discussed, with a focus on the most policy relevant information. Staff has drawn from the EPA's synthesis and assessment of the scientific evidence presented in the *Integrated Science Assessment for Oxides of Nitrogen – Health Criteria* (ISA) (U.S. EPA, 2016). In this chapter, section 3.1 summarizes the weight of evidence approach used in evaluating and integrating scientific evidence in the ISA. Section 3.2 discusses the evidence for health effects attributable to short-term NO$_2$ exposures, and section 3.3 discusses the evidence for health effects attributable to long-term NO$_2$ exposures. Section 3.4 discusses the potential public health implications of NO$_2$-attributable effects.

3.1   WEIGHT OF EVIDENCE IN THE ISA

In the current review of the primary NO$_2$ NAAQS, the Agency has used two frameworks: one for characterizing the strength of the available scientific evidence for health effects attributable to NO$_2$ exposures and the other a recently developed framework to classify evidence for factors that may increase risk in some populations or lifestages[51] (U.S. EPA, 2015, Preamble, Section 6). These frameworks provide the basis for robust, consistent, and transparent evaluation of the scientific evidence, including uncertainties in the evidence, and for drawing conclusions on air pollution-related health effects and at-risk populations.

With regard to characterization of the health effects, the ISA uses a five-level hierarchy to classify the overall weight of evidence into one of the following categories: causal relationship; likely to be a causal relationship; suggestive of, but not sufficient to infer, a causal relationship; inadequate to infer a causal relationship; and not likely to be a causal relationship (U.S. EPA, 2015, Preamble Table II). In using the weight-of-evidence approach to inform judgments about the likelihood that various health effects are caused by exposure to NO$_2$, the ISA notes that confidence in the relationship increases when the evidence base is large and consistently supports a relationship with a particular health endpoint. In addition, biological plausibility, strength, and coherence in the evidence are important aspects considered in making judgments regarding causality of relationships. Conclusions about biological plausibility, consistency, and coherence of NO$_2$-related health effects are drawn from the integration of multiple lines of evidence including epidemiologic, controlled human exposure, and animal

---

[51] As defined in Chapter 1, the term "population" refers to people having a quality or characteristic in common, including a specific pre-existing illness or a specific age or lifestage.

toxicological studies as discussed in the ISA (U.S. EPA, 2015, Preamble, Section 5.c.) and further described below. In this PA, we consider the full body of health evidence, placing the greatest emphasis on the effects for which the evidence has been judged in the ISA to demonstrate a "causal" or a "likely to be a causal" relationship with NO_2 exposures.

Controlled human exposure studies can provide direct evidence of relationships between pollutant exposures and human health effects (U.S. EPA, 2015, Preamble Section 4.c). Because data on health effects in these studies are collected under closely monitored conditions, this type of evidence can provide information on exposure concentrations, durations, and ventilation rates under which effects can occur, as well as information on exposure-response relationships. Further, as discussed in the ISA, controlled human exposure studies can provide clear and compelling evidence for an array of human health effects that are directly attributable to acute exposures to NO_2 *per se* (i.e., as opposed to other oxides of nitrogen species, for which NO_2 is an indicator, or other co-occurring pollutants) (U.S. EPA, 2015, Preamble Section 4.c). In addition, exposure concentrations used in some controlled human exposure studies are near those found in the ambient air and results are not subject to uncertainties related to inter-species variation.

Toxicological studies in animals provide another line of experimental evidence that can inform understanding of effects related to NO_2 exposures, particularly the biological action of a pollutant under controlled and monitored exposure circumstances. Compared to controlled human exposure studies, animal toxicological studies can examine more severe outcomes, invasive endpoints (i.e., pathology), and effects of long-term exposures. However, results from animal studies are subject to uncertainty due to inter-species variation.[52] Also, animal studies are often conducted with NO_2 concentrations well above those in ambient air. Although some of these high concentrations are considered to be ambient-relevant because of dosimetric considerations (U.S. EPA, 2016, Section 1.2), results from animal studies are subject to uncertainties regarding the likelihood that such effects could occur with ambient exposures in humans. Nonetheless, evidence from animal studies can provide support for effects observed in human studies. Together, evidence from human and animal studies can provide information on and confidence regarding key events in the proposed mode(s) of action, which informs biological plausibility for health effects observed in epidemiologic studies.

---

[52] "The differences between humans and other species have to be considered, including metabolism, hormonal regulation, breathing pattern, and differences in lung structure and anatomy. Given these differences, uncertainties are associated with quantitative extrapolations of observed pollutant-induced pathophysiological alterations between laboratory animals and humans, as those alterations are under the control of widely varying biochemical, endocrine, and neuronal factors." (U.S. EPA, 2016, pp. liii).

Epidemiologic studies provide information on associations between variability in short-term and long-term average ambient NO_2 concentrations and various health outcomes, including those related to asthma exacerbation and incidence (i.e., airway responsiveness, lung function decrements, respiratory symptoms, pulmonary inflammation, hospital admissions, emergency department visits, and asthma incidence) (U.S. EPA, 2016, Chapters 5 and 6). Epidemiologic studies can inform our understanding of the effects in the study population of real-world exposures to the range of NO_2 concentrations in ambient air, and can provide evidence of associations between exposures to ambient NO_2 and serious acute and chronic health effects that cannot be assessed in controlled human exposure studies. Moreover, epidemiologic studies often include populations or lifestages that may have increased risk for pollutant-related health effects (e.g., individuals with pre-existing disease, children, and older adults). In evaluating epidemiologic studies, it is important to consider the degree of uncertainty introduced by potential confounding variables (e.g., other pollutants, temperature) and other factors (e.g., study design, exposure assessment, statistical methods) affecting the level of confidence that the observed health effects are independently related to ambient exposure to NO_2.

The ISA also includes an evaluation and synthesis of evidence across scientific disciplines to inform whether specific populations or lifestages may be at increased risk of a health effect related to NO_2 exposures. The ISA characterizes the evidence for a number of "factors", including both intrinsic (i.e., biologic, such as pre-existing disease or lifestage) and extrinsic (i.e., non-biologic, such as diet or socioeconomic status) factors. The categories considered in classifying the evidence for these potential at-risk factors are "adequate evidence," "suggestive evidence," "inadequate evidence," and "evidence of no effect." These categories are discussed in more detail in the ISA (U.S. EPA, 201, Section 5.c, Table II). In this PA, we focus our consideration of potential at-risk populations and lifestages on those factors for which the ISA judges there is "adequate" evidence (U.S. EPA, 2016, Table 7-27). The primary NAAQS are set requisite to protect the public health with an adequate margin of safety, including the health of populations at increased risk for pollutant-related health effects, and thus, identifying at-risk populations and lifestages is a critical part of this review. At-risk populations and potential public health implications are discussed in more detail in Section 3.4.

3.2 EFFECTS OF SHORT-TERM NO₂ EXPOSURES

This section discusses the nature of the health effects that have been shown to occur following short-term NO_2 exposures (Section 3.2.1) and the NO_2 concentrations at which those effects have been demonstrated to occur (Section 3.2.2).

3.2.1 Nature of Effects

Across previous reviews of the primary NO_2 NAAQS (U.S. EPA, 1993; U.S. EPA, 2008), evidence has consistently demonstrated respiratory effects attributable to short-term NO_2 exposures. In the last review, the 2008 ISA concluded that evidence was "sufficient to infer a likely causal relationship between short-term NO_2 exposure and adverse effects on the respiratory system" based on the large body of epidemiologic evidence demonstrating positive associations with respiratory symptoms and hospitalization or ED visits as well as supporting evidence from controlled human exposure and animal studies (U.S. EPA, 2008, p. 5-6). Evidence for cardiovascular effects and mortality attributable to short-term NO_2 exposures was weaker and was judged "inadequate to infer the presence or absence of a causal relationship" and "suggestive of, but not sufficient to infer, a causal relationship," respectively. The 2008 ISA noted an overarching uncertainty in determining the extent to which NO_2 is independently associated with effects or if NO_2 is a marker for the effects of another traffic-related pollutant or mix of pollutants (U.S. EPA, 2008, Section 5.3.2.2 to 5.3.2.6).

For the current review, there is newly available evidence for both respiratory effects and other health effects critically evaluated in the ISA as part of the full body of evidence informing the nature of the relationship between health effects and short-term exposures to NO_2 (U.S. EPA, 2016). In characterizing the available evidence and the causal determinations presented in the ISA, this section poses the following policy-relevant questions:

- **To what extent does the currently available scientific evidence strengthen, or otherwise alter, our conclusions from the last review regarding health effects attributable to short-term NO_2 exposure? Have previously identified uncertainties been reduced? What important uncertainties remain and have new uncertainties been identified?**

As discussed above, causal determinations for health effects related to short-term NO_2 exposures are presented in the ISA, which classifies short-term exposures as those that are one month or less (U.S. EPA, 2016). Table 3-1, below, lists the causal determinations from the ISA for the current review as well as those from the previous review for respiratory and cardiovascular health effects, and mortality.[53] It is noteworthy that the causal determinations for respiratory and cardiovascular health effects have been strengthened in the current review due, in part, to more explicit consideration of the evidence integrated for specific outcomes (e.g., asthma exacerbation for respiratory) rather than broad outcome categories (e.g., all respiratory effects).

---

[53] Short-term exposure studies on reproductive and birth health effects and cancer are considered below in the context of long-term exposures.

The evidence informing these determinations, including uncertainties in that evidence, is summarized below.

Table 3-1. ISA causal determinations for health effects related to short-term nitrogen dioxide (NO₂) exposures

Health effect	Review completed 2010	Current Review
Respiratory	Sufficient to infer a likely causal relationship	Causal relationship
Cardiovascular	Inadequate to infer the presence or absence of a causal relationship	Suggestive of, but not sufficient to infer, a causal relationship
Total Mortality	Suggestive of, but not sufficient to infer, a causal relationship	Suggestive of, but not sufficient to infer, a causal relationship

Respiratory

The ISA concludes that evidence for respiratory effects related to short-term NO₂ exposures indicates that there is a causal relationship, primarily based on evidence for asthma exacerbation. This conclusion is strengthened from the last review "because epidemiologic, controlled human exposure, and animal toxicological evidence together can be linked in a coherent and biologically plausible pathway to explain how NO₂ exposure can trigger an asthma exacerbation" (U.S. EPA, 2016, p. 1-17). The 2008 ISA described much of the same evidence and determined it was "sufficient to infer a likely causal relationship," citing uncertainty as to whether the epidemiologic results for NO₂ primarily reflected the effects of other traffic-related pollutants. The 2008 ISA did not explicitly evaluate the extent to which various lines of evidence supported effects on asthma attacks. In contrast, in the current review the ISA states that "the determination of a causal relationship is not based on new evidence as much as it is on the integrated findings for asthma attacks with due weight given to experimental studies" (U.S. EPA, 2016, p. lxxxiii).[54] When taken together, the epidemiologic evidence for asthma attacks and controlled human exposure study findings for increased airway responsiveness (AR)[55] and allergic inflammation demonstrate that short-term NO₂ exposure has an independent relationship

---

[54] As noted above, experimental studies such as controlled human exposure studies provide support for effects of exposures to NO₂ itself, and generally do not reflect the complex atmospheres to which people are exposed.

[55] The ISA states that airway responsiveness is "inherent responsiveness of the airways to challenge by bronchoconstricting agents" (U.S. EPA, 2016, p. 5-9). More specifically, airway responsiveness refers to increased sensitivity of the airways to an inhaled bronchoconstricting agent. This is most often quantified as the dose of challenge agent that results in a 20% reduction in FEV_1, but some studies report the change in FEV_1 for a specified dose of challenge agent. The change in specific airways resistance (sRaw) is also used to quantify AR.

with respiratory effects, specifically with asthma exacerbation, and is not just an indicator for other traffic-related pollutants.

The evaluation of controlled human exposure studies in the ISA focuses on results from a recently published meta-analysis of NO_2-induced increases in AR by Brown (2015). AR has been the key respiratory outcome from controlled human exposures in the previous and current reviews of the primary NO_2 NAAQS, and the ISA specifically notes that "airway hyperresponsiveness can lead to poorer control of symptoms and is a hallmark of asthma" (U.S. EPA, 2016, p. 1-18). Brown (2015) examined the relationship between AR and NO_2 exposures in subjects with asthma across the large body of controlled human exposure studies,[56] most of which were available in the last review (Tables 3-2 and 3-3). More specifically, the meta-analysis identified the fraction of individuals having an increase in AR following NO_2 exposure, compared to the fraction having a decrease, across studies. The meta-analysis also stratified results to consider the influence of factors that may affect results including exercise/rest and non-specific/specific challenge agents.[57]

The results from the meta-analysis demonstrate that the majority of study volunteers with asthma experienced increased AR following resting exposure to NO_2 concentrations ranging from 100 to 530 ppb, relative to filtered air. While results from individual studies do not always demonstrate NO_2-induced increases in AR, particularly for exposure concentrations between 100 and 200 ppb (U.S. EPA, 2016, Table 5-1; Section 3.2.2.1, below), and important uncertainties remain due to the lack of an apparent a dose-response relationship (see below), the meta-analysis indicates that when data are pooled, a statistically significant majority of study volunteers experienced increases in nonspecific airway responsiveness. Significant majorities experienced such increases following (1) 20 to 60-minute exposures to 400-530 ppb NO_2, (2) 30-minute exposures to 250 to 300 ppb NO_2, and (3) 60-minute exposures to 100 to 200 ppb NO_2. When comparing results across the three exposure categories, the fractions of individuals with increased AR (out of those who experienced either an increase or a decrease) were 73%, 78%,

---

[56] While these controlled human exposure studies were conducted in people with asthma, a group at increased risk for NO_2-related effects, they generally did not evaluate individuals with severe asthma (Brown, 2015). These studies also did not evaluate people in other potentially at-risk groups.

[57] "Bronchial challenge agents can be classified as nonspecific (e.g., histamine; SO_2; cold air) or specific (i.e., an allergen). Nonspecific agents can be differentiated between "direct" stimuli (e.g., histamine, carbachol, and methacholine) which act on airway smooth muscle receptors and "indirect" stimuli (e.g., exercise, cold air) which act on smooth muscle through intermediate pathways, especially via inflammatory mediators. Specific allergen challenges (e.g., house dust mite, cat allergen) also act "indirectly" via inflammatory mediators to initiate smooth muscle contraction and bronchoconstriction." (U.S. EPA, 2016, p. 5-8)

and 67%, respectively, and all were statistically significant with p-values < 0.05 (U.S. EPA, 2016, Table 5-3).

In considering the potential for these increases in AR to be adverse, we note the ISA's characterization of their clinical relevance. The ISA uses an approach to characterizing clinical relevance that is based on guidelines from the American Thoracic Society (ATS) and the European Respiratory Society (ERS) for the assessment of therapeutic agents (Reddel et al., 2009). Specifically, based on individual-level responses reported in a subset of studies, the ISA considered a halving of the provocative dose (PD) to indicate responses that may be clinically relevant.[58, 59] With regard to this approach, the ISA notes that "in a joint statement of the [ATS] and [ERS], one doubling dose change in PD is recognized as a potential indicator, although not a validated estimate, of clinically relevant changes in airway responsiveness (Reddel et al., 2009)" (U.S. EPA, 2016, p. 5-12). While there is uncertainty in using this approach to characterize whether a particular response in an individual is "adverse," it can provide insight into the potential for adversity, particularly when applied to a population of exposed individuals.[60]

Based on a subset of studies, Brown (2015) shows that NO_2 exposures from 100 to 530 ppb resulted in a halving of the dose of a challenge agent required to increase airway responsiveness (i.e., a halving of the PD) for about a quarter of study volunteers. While these results support the potential for clinically relevant increases in AR in some individuals with asthma following NO_2 exposures within the range of 100 to 530 ppb, uncertainty remains given that this analysis is limited to a small subset of the studies included in the broader meta-analysis and given the lack of an apparent dose-response relationship.[61] In addition, compared to conclusions based on the entire range of NO_2 exposure concentrations evaluated (i.e., 100 to 530 ppb), there is greater uncertainty in reaching conclusions about the potential for clinically relevant effects at any particular NO_2 exposure concentration within this range.

---

[58] PD is the dose of challenge agent required to elicit a particular magnitude of change in FEV_1 or other measure of lung function.

[59] The ISA's characterization of a clinically relevant response is based on evidence from controlled human exposure studies evaluating the efficacy of inhaled corticosteroids that are used to prevent bronchoconstriction and airway responsiveness as described by Reddell et al. (2009). Generally, a change of at least one doubling dose is considered to be an indication of clinical relevance. Based on this, a halving of the PD is taken in the ISA to represent an increase in AR that indicates a clinically relevant response.

[60] Based on recommendations from the ATS, if a population is exposed, NO_2-induced increases in AR may be considered adverse at the population level. This is because the increases in AR could increase the proportion of the population with clinically important changes that can contribute to the exacerbation of asthma (U.S. EPA, 2016, section 1.6.5). Chapter 4 (below) presents the results of analyses evaluating the degree to which populations in U.S. urban areas could experience NO_2 exposures shown to increase AR.

[61] Section 3.2.2.1 below includes additional discussion of these uncertainties.

Controlled human exposure studies also evaluated a range of other respiratory effects, including lung function decrements, respiratory symptoms, and pulmonary inflammation. The evidence does not consistently demonstrate these effects following exposures to NO_2 concentrations at or near those found in the ambient air in the U.S. However, a subset of studies using exposures to 260 ppb for 15-30 min or 400 ppb for up to 6 hours provide evidence that study volunteers with asthma and allergy can experience increased inflammatory responses following allergen challenge. Evidence for pulmonary inflammation was more mixed across studies that did not use an allergen challenge following NO_2 exposures ranging from 300-1,000 ppb (U.S. EPA, 2016, Section 5.2.2.5).

In addition to this evidence for NO_2-induced increases in AR and allergic inflammation in controlled human exposure studies, the ISA also describes consistent evidence from epidemiologic studies for positive associations between short-term NO_2 exposures and an array of respiratory outcomes related to asthma. Thus, coherence and biological plausibility is demonstrated in the evidence integrated between controlled human exposure studies and the various asthma-related outcomes examined in epidemiologic studies. The ISA indicates that epidemiologic studies consistently demonstrate NO_2-health effect associations with asthma hospital admissions and ED visits among subjects of all ages and children, and with asthma symptoms in children (U.S. EPA, 2016, Sections 5.2.2.4 and 5.2.2.3). The robustness of the evidence is demonstrated by associations found in studies conducted in diverse locations in the U.S., Canada, and Asia, including several multicity studies. The evidence for asthma exacerbation is substantiated by several recent studies with strong exposure assessment characterized by measuring NO_2 concentrations in subjects' location(s). Epidemiologic studies also demonstrated associations between short-term NO_2 exposures and respiratory symptoms, lung function decrements, and pulmonary inflammation, particularly for measures of personal total and ambient NO_2 exposures and NO_2 measured outside schools. This is important because there is considerable spatial variability in NO_2 concentrations, and measurements in subjects' locations may better represent this variability in ambient NO_2 exposures, compared to measurements at central site monitors (U.S. EPA, 2016, Sections 2.5.3 and 3.4.4). Epidemiologic studies generally did not find NO_2-associated changes in inflammatory cell counts in populations with asthma; however, they did consistently indicate ambient or personal NO_2-associated increases in exhaled nitric oxide (eNO, a marker of airway inflammation), which is coherent with experimental findings for allergic inflammation (U.S. EPA, 2016, Section 5.2.2.6).

In assessing the evidence from epidemiologic studies, the ISA not only considers the consistency of effects across studies, but also evaluates other study attributes that affect study quality, including potential confounding and exposure assignment. Regarding potential confounding, the ISA notes that NO_2 associations with asthma-related effects persist with adjustment for temperature; humidity; season; long-term time trends; and PM_{10}, SO_2, or O_3. Recent studies also add findings for NO_2 associations that generally persist with adjustment for key copollutants, including $PM_{2.5}$ and traffic-related copollutants such as elemental carbon (EC) or black carbon (BC), ultra-fine particles (UFPs), or carbon monoxide (CO) (examined in few studies). Confounding by organic carbon (OC), PM metal species, or volatile organic compounds (VOCs) is poorly studied, but NO_2 associations with asthma exacerbation tend to persist in the few available copollutant models. We recognize, however, that copollutant models have inherent limitations and cannot conclusively rule out confounding (U.S. EPA, 2015, Preamble, Section 4.b). Recent epidemiologic results also suggest asthma exacerbation in relation to indices that combine NO_2 with EC, $PM_{2.5}$, O_3, and/or SO_2 concentrations, but neither epidemiologic nor experimental studies strongly indicate synergistic effects between NO_2 and copollutants (U.S. EPA, 2016, Section 5.2.9).

The ISA also notes that results based on personal exposures or pollutants measured at people's locations provide support for NO_2 associations that are independent of $PM_{2.5}$, EC/BC, OC, or UFPs. Compared to ambient NO_2 concentrations measured at central-site monitors, personal NO_2 exposure concentrations and indoor NO_2 concentrations exhibit lower correlations with many traffic-related copollutants (e.g., $r = -0.37$ to 0.31). Thus, these health effect associations with personal and indoor NO_2 may be less prone to confounding by these traffic-related copollutants (U.S. EPA, 2016, Section 1.4.3).

Overall, in consideration of this evidence in answering the question posed above, we note that for respiratory effects, the strongest evidence supporting the conclusion of the causal relationship determined in the ISA comes from controlled human exposure studies demonstrating NO_2-induced increases in AR in individuals with asthma, with supporting evidence for a range of respiratory effects from epidemiologic studies. The conclusion of a causal relationship in the ISA is based on this evidence, and its explicit integration within the context of effects related to asthma exacerbation. Most of the controlled human exposure studies assessed in the ISA were available in the last review, particularly studies of non-specific AR, and thus, do not themselves provide substantively new information. However, by pooling data from a subset of studies, the newly available meta-analysis by Brown (2015) has partially addressed an uncertainty from the last review by demonstrating the potential for clinically relevant increases in AR following exposures to NO_2 concentrations in the range of 100 to 530 ppb. Similarly, the epidemiologic evidence that is newly available in the current review is consistent with evidence from the last review and does not alter our understanding of respiratory effects related to ambient NO_2 exposures. New epidemiologic evidence does, however, reduce some uncertainty from the last review regarding the extent to which effects may be independently related to NO_2 as there is more evidence from studies using measures that better capture personal exposure as well as a more robust evidence base examining copollutant confounding. Some uncertainty remains in the epidemiologic evidence regarding confounding by the most relevant copollutants (i.e., those from traffic).

Cardiovascular

The evidence for cardiovascular health effects and short-term NO_2 exposures in the 2016 ISA was judged "suggestive of, but not sufficient to infer, a causal relationship" (U.S. EPA, 2016, Section 5.3.11), which is stronger than the conclusion in the last review that the evidence was "inadequate to infer the presence or absence of a causal relationship." The more recent causal determination was primarily supported by consistent epidemiologic evidence from multiple new studies indicating associations for triggering of a myocardial infarction. However, further evaluation and integration of evidence points to uncertainty related to exposure measurement error and potential confounding by traffic-related pollutants. There is consistent evidence demonstrating NO_2-associated hospital admissions and ED visits for ischemic heart disease, myocardial infarction, and angina as well as all cardiovascular diseases, which is coherent with evidence from other studies indicating NO_2-associated repolarization abnormalities and cardiovascular mortality. There are experimental studies that provide some evidence for effects on key events in the proposed mode of action (e.g., systemic inflammation), but these studies do not provide evidence that is sufficiently coherent with the epidemiologic studies to help rule out chance, confounding, and other biases. In particular, the ISA concludes that "[t]here continues to be a lack of experimental evidence that is coherent with the epidemiologic studies to strengthen the inference of causality for NO_2-related cardiovascular effects, including [myocardial infarction]" (U.S. EPA, 2016, p. 5-335). Beyond evidence for myocardial infarction, there were studies examining other cardiovascular health effects, but results across these outcomes are inconsistent. Thus, while the evidence is stronger in the current review than in the last review, important uncertainties remain regarding the independent effects of NO_2.

Mortality

The ISA concludes that the evidence for short-term NO_2 exposures and total mortality is "suggestive of, but not sufficient to infer, a causal relationship" (U.S. EPA, 2016, Section 5.4.8), which is the same conclusion reached in the last review (U.S. EPA, 2008). Several recent multicity studies add to the evidence base for the current review and demonstrate associations that are robust in copollutant models with PM_{10}, O_3, or SO_2. However, confounding by traffic-related copollutants, which is of greatest concern, is not examined in the available copollutant models for NO_2-associated mortality. Overall, the recent evidence assessed in the ISA builds upon and supports conclusions in the last review, but key limitations across the evidence include a lack of biological plausibility as experimental studies and epidemiologic studies on cardiovascular morbidity, a major cause of mortality, do not clearly provide a mechanism by which NO_2-related effects could lead to mortality. In addition, important uncertainties remain regarding the independent effect of NO_2.

3.2.2 Consideration of NO_2 Concentrations: Health Effects of Short-Term NO_2 Exposures

In evaluating the NO_2 exposure concentrations associated with health effects within the context of the adequacy of the current standard, we consider the following specific question:

- **To what extent does the evidence indicate adverse respiratory effects attributable to short-term exposures to NO_2 concentrations lower than previously identified or below the existing standards?**

In addressing this question, we further consider the extent to which NO_2-induced adverse effects have been reported over the ranges of NO_2 exposure concentrations evaluated in controlled human exposure studies and the extent to which NO_2-associated effects have been reported for distributions of ambient NO_2 concentrations in epidemiologic study locations meeting existing standards. Each of these is discussed below.

3.2.2.1 *NO_2 Concentrations in Controlled Human Exposure Studies*

As discussed in detail in the ISA (U.S. EPA, 2016) and summarized above in section 3.2.1, controlled human exposure studies, most of which were available and considered in the last review, have evaluated various respiratory effects following short-term NO_2 exposures. These include AR, inflammation and oxidative stress, respiratory symptoms, and lung function decrements. Generally, when considering respiratory effects from controlled human exposure studies in healthy adults without asthma, evidence does not indicate respiratory symptoms or lung function decrements following NO_2 exposures below 4,000 ppb and limited evidence indicates airway inflammation following exposures below 1,500 ppb (U.S. EPA, 2016, Section 5.2.7).[62] There is a substantial body of evidence demonstrating increased AR in healthy adults with exposures in the range of 1,500-3,000 ppb.

Evidence for respiratory effects following exposures to NO_2 concentrations at or near those found in the ambient air is strongest for AR in individuals with asthma (U.S. EPA, 2016, Section 5.2.2 p. 5-7). In contrast, controlled human exposure studies evaluated in the ISA do not

---

[62] Exposure durations were from one to three hours in studies evaluating AR and respiratory symptoms, and up to five hours in studies evaluating lung function decrements.

provide consistent evidence for respiratory symptoms, lung function decrements, or pulmonary inflammation in adults with asthma following exposures to NO_2 concentrations at or near those in ambient air (i.e., <1,000 ppb; U.S. EPA, 2016, Section 5.2.2). There is some indication of allergic inflammation in adults with allergy and asthma following exposures to 260-1,000 ppb. However, evidence across studies is inconsistent, making it difficult to interpret the likelihood that these effects could potentially occur following NO_2 exposures at or below the level of the current standard.

Thus, in considering evidence from controlled human exposure studies to address the above question, we focus on the body of evidence for NO_2-induced increases AR in adults with asthma. In evaluating the NO_2 exposure concentrations at which increased AR is observed, we consider both the group mean results reported in individual studies and the results evaluated across studies in a recent meta-analysis (Brown, 2015; U.S. EPA, 2016, Section 5.2.2.1). Group mean responses in individual studies, and the variability in those responses, can provide insight into the extent to which observed changes in AR are due to NO_2 exposures, rather than to chance alone, and have the advantage of being based on the same exposure conditions. The meta-analysis by Brown (2015) can aid in identifying trends in individual-level responses across studies and can have the advantage of increased power to detect effects, even in the absence of statistically significant effects in individual studies.

Tables 3-2 and 3-3 (adapted from the ISA; U.S. EPA, 2016, Tables 5-1 and 5-2) provide details for the studies examining AR in individuals with asthma at rest and with exercise, respectively. These tables note various study details including the exposure concentration, duration of exposure, type of challenge (nonspecific or specific[63]), number of study subjects, number of subjects having an increase or decrease in AR following NO_2 exposure, average provocative dose (PD; dose of challenge agent required to elicit a particular magnitude of change in FEV_1 or other measure of lung function) across subjects, and the statistical significance of the change in AR following NO_2 exposures.

---

[63] As previously described, bronchial challenge agents can be classified as nonspecific (e.g., histamine; sulfur dioxide, SO_2; cold air) or specific (i.e., an allergen). Nonspecific agents can be differentiated between "direct" stimuli (e.g., histamine, carbachol, and methacholine) and "indirect" stimuli (e.g., exercise, cold air) (U.S. EPA, 2016)

Table 3-2.   Resting exposures to nitrogen dioxide and airway responsiveness in individuals with asthma.[a]

Reference	NO$_2$ ppb	Exp. (min)	Challenge Type	N	Change in AR[b] +	Change in AR[b] −	Average PD ± SE[c] Air	Average PD ± SE[c] NO$_2$	p-value[d]
Ahmed et al., 1983a	100	60	Non-specific, CARB	20	13	7	6.0 ± 2.4	2.7 ± 0.8	NA
Orehek et al., 1976	100	60	Non-specific, CARB	20	14	3	0.56 ± 0.08	0.36 ± 0.05	<0.01[e]
Hazucha et al., 1983	100	60	Non-specific, METH	15	6	7	1.9 ± 0.4	2.0 ± 1.0	n.s.
Ahmed et al., 1983b	100	60	Specific, RAG	20	10	8	9.0 ± 5.7	11.7 ± 7.6	n.s.
Tunnicliffe et al., 1994	100	60	Specific, HDM	8	3	5	−14.62 ΔFEV$_1$	−14.41 ΔFEV$_1$	n.s.
Bylin et al., 1988	140	30	Non-specific, HIST	20	14	6	0.39 ± 0.07	0.28 ± 0.05	0.052[f]
Orehek et al., 1976	200	60	Non-specific, CARB	4	3	0	0.60 ± 0.10	0.32 ± 0.02	n.s.
Jörres et al., 1990	250	30	Non-specific, SO$_2$	14	11	2	46.5 ± 5.1	37.7 ± 3.5	<0.01
Barck et al., 2002	260	30	Specific, BIR, TIM	13	5	7	−5 ± 2 ΔFEV$_1$	−4 ± 2 ΔFEV$_1$	n.s.
Strand et al., 1997	260	30	Specific, BIR, TIM	18	9	9	860 ± 450	970 ± 450	n.s.
Strand et al, 1998	260	30	Specific, BIR	16	11	4	−0.1 ± 0.8 ΔFEV$_1$	−2.5 ± 1.0 ΔFEV$_1$	0.03
Bylin et al., 1988	270	30	Non-specific, HIST	20	14	6	0.39 ± 0.07	0.24 ± 0.04	<0.01
Tunnicliffe et al., 1994	400	60	Specific, HDM	8	8	0	−14.62 ΔFEV$_1$	−18.64 ΔFEV$_1$	0.009
Bylin et al, 1985	480	20	Non-specific, HIST	8	5	0	>30	>20	0.04
Mohsenin et al., 1987	500	60	Non-specific, METH	10	7	2	9.2 ± 4.7	4.6 ± 2.6	0.042
Bylin et al., 1988	530	30	Non-specific, HIST	20	12	7	0.39 ± 0.07	0.34 ± 0.08	n.s.

AR = airway responsiveness; BIR = birch; CARB = carbachol; Exp. = exposure; HDM = house dust mite allergen; HIST = histamine; METH = methacholine; NA = not available; NO$_2$ = nitrogen dioxide; n.s. = less than marginal statistical significance (i.e., $p > 0.10$); RAG = ragweed; SO$_2$ = sulfur dioxide; TIM = timothy

[a] Adapted from Table 5-1 in Integrated Science Assessment for Oxides of Nitrogen (Health) – Final (U.S. EPA, 2016, Section 5.2.2.1)

[b] Change in AR: number of individuals showing increased (+) or decreased (−) airway responsiveness after NO$_2$ exposure compared to air.

[c] PD ± SE: arithmetic or geometric mean provocative dose (PD) ± standard error (SE). See individual papers for PD calculation and dosage units. ΔFEV$_1$ indicates the change in FEV$_1$ response at a constant challenge dose.

[d] Statistical significance of increase in AR to bronchial challenge following NO$_2$ exposure compared to filtered air as reported in the original study unless otherwise specified. Statistical tests varied between studies, e.g., sign test, t-test, and analysis of variance.

[e] Statistical significance for all individuals with asthma from analysis by Dawson et al. (1979). Orehek et a. (1976) only tested for differences in sub-sets of individuals classified as "responders" and "non-responders."

[f] This p-value from p. 609 of Bylin et al. (1988) corrects the "n.s." indicated in the 2016 ISA and Brown (2015)

Table 3-3. **Exercising exposures to nitrogen dioxide and airway responsiveness in individuals with asthma.[a]**

Reference	NO_2 ppb	Exp. (min)	Challenge Type	N	Change in AR[b] +	Change in AR[b] −	Average PD ± SE[c] Air	Average PD ± SE[c] NO_2	p-value[d]
Roger et al., 1990	150	80	Non-specific, METH	19	10[d]	7[d]	3.3 ± 0.7	3.1 ± 0.7	n.s.
Kleinman et al., 1983	200	120	Non-specific, METH	31	20	7	8.6 ± 2.9	3.0 ± 1.1	<0.05
Jenkins et al., 1999	200	360	Specific, HDM	11	6	5	2.94	2.77	n.s.
Jörres et al., 1991	250	30	Non-specific, METH	11	6	5	0.41 ± 1.6	0.41 ± 1.6	n.s.
Strand et al., 1996	260	30	Non-specific, HIST	19	13	5	296 ± 76	229 ± 56	0.08
Avol et al., 1988	300	120	Non-specific, COLD	37	11[d]	16[d]	−8.4 ± 1.8 ΔFEV_1	−10.7 ± 2.0 ΔFEV_1	n.s.
Avol et al., 1989	300	180	Non-specific, COLD	34	12[d]	21[d]	−5 ± 2 ΔFEV_1	−4 ± 2 ΔFEV_1	n.s.
Bauer et al., 1986	300	30	Non-specific, COLD	15	9	3	0.83 ± 0.12	0.54 ± 0.10	<0.05
Morrow et al., 1989	300	240	Non-specific, CARB	20	7[e]	2[e]	3.31 ± 8.64[e] ΔFEV_1	−6.98 ± 3.35[e] ΔFEV_1	n.s.
Roger et al., 1990	300	80	Non-specific, METH	19	8[d]	9[d]	3.3 ± 0.7	3.3 ± 0.8	n.s.
Rubinstein et al., 1990	300	30	Non-specific, SO_2	9	4	5	1.25 ± 0.23	1.31 ± 0.25	n.s.
Riedl et al., 2012	350	120	Non-specific, METH	15	6	7	7.5 ± 2.6	7.0 ± 3.8	n.s.
Riedl et al., 2012	350	120	Specific, CA	15	4	11	−6.9 ± 1.7 ΔFEV_1	−0.5 ± 1.7 ΔFEV_1	<0.05[f]
Jenkins et al., 1999	400	180	Specific, HDM	10	7	3	3.0	2.78	0.018
Witten et al., 2005	400	180	Specific, HDM	15	8	7	550 ± 240	160 ± 60	n.s.
Avol et al., 1988	600	120	Non-specific, COLD	37	13[e]	16[e]	−8.4 ± 1.8 ΔFEV_1	−10.4 ± 2.2 ΔFEV_1	n.s.
Roger et al., 1990	600	80	Non-specific, METH	19	11[d]	8[d]	3.3 ± 0.7	3.7 ± 1.1	n.s.

AR = airway responsiveness; BIR = birch; CARB = carbachol; Exp. = exposure; HDM = house dust mite allergen; HIST = histamine; METH = methacholine; NA = not available; NO_2 = nitrogen dioxide; n.s. = less than marginal statistical significance, $p > 0.10$; RAG = ragweed; SO_2 = sulfur dioxide; TIM = timothy

[a] Adapted from Table 5-2 in Integrated Science Assessment for Oxides of Nitrogen (Health) – Final (U.S. EPA, 2016, Section 5.2.2.1)

[b] Change in AR: number of individuals showing increased (+) or decreased (−) airway responsiveness after NO_2 exposure compared to air.

[c] PD ± SE: arithmetic or geometric mean provocative dose (PD) ± standard error (SE). See individual papers for PD calculation and dosage units. ΔFEV_1 indicates the change in FEV_1 response at a constant challenge dose.

[d] Statistical significance of increase in AR to bronchial challenge following NO_2 exposure compared to filtered air as reported in the original study unless otherwise specified. Statistical tests varied between studies, e.g., sign test, t-test, analysis of variance.

[e] Statistical significance for all individuals with asthma from analysis by Dawson et al. (1979). Orehek et a. (1976) only tested for differences in sub-sets of individuals classified as "responders" and "non-responders."

Consideration of group mean results from individual studies

In first considering studies conducted at rest, we note that the lowest NO_2 concentration to which individuals with asthma have been exposed is 100 ppb, with an exposure duration of 60 minutes in all studies. Of the five studies conducted at 100 ppb, a statistically significant increase in AR following exposure to NO_2 was only observed in the study by Orehek et al. (1976) (N = 20). Of the four studies that did not report statistically significant increases in AR following exposures to 100 ppb NO_2, three reported weak trends towards decreased AR (n = 20, Ahmed et al., 1983b; n = 15, Hazucha et al., 1983; n = 8, Tunnicliffe et al., 1994), and one reported a trend towards increased AR (n = 20, Ahmed et al., 1983a). Resting exposures to 140 ppb NO_2 resulted in increases in AR that reached marginal statistical significance (n = 20; Bylin et al., 1988). In addition, the one study conducted at 200 ppb demonstrated a trend towards increased AR, but this study was small and results were not statistically significant (n = 4; Orehek et al., 1976). Thus, individual controlled human exposure studies have generally not reported statistically significant increases in AR following resting exposures to NO_2 concentrations from 100 to 200 ppb. Group mean responses in these studies suggest a trend towards increased AR following exposures to 140 and 200 ppb NO_2, while trends in the direction of group mean responses were inconsistent following exposures to 100 ppb NO_2.

In next considering studies in individuals with asthma conducted with exercise, we note that three studies evaluated NO_2 exposure concentrations between 150 and 200 ppb (n = 19, Roger et al., 1990; n = 31, Kleinman et al., 1983; n = 11, Jenkins et al., 1999). Of these studies, only Kleinman et al. (1983) reported a statistically significant increase in AR following NO_2 exposure (i.e., at 200 ppb). Roger et al. (1990) and Jenkins et al. (1999) did not report statistically significant increases, but showed weak trends for increases in AR following exposures to 150 ppb and 200 ppb NO_2, respectively. Thus, as with studies of resting exposures, studies that evaluated exposures to 150 to 200 ppb NO_2 with exercise report trends toward increased AR, though results are generally not statistically significant.

Several studies evaluated exposures of individuals with asthma to NO_2 concentrations above 200 ppb. Of the five studies that evaluated 30-minute resting exposures to NO_2 concentrations from 250 to 270 ppb, NO_2-induced increases in AR were statistically significant in three (n = 14, Jörres et al., 1990; n = 18, Strand et al., 1988; n = 20, Bylin et al., 1988). Statistically significant increases in airway responsiveness are also more consistently reported across studies that evaluated resting exposures to 400-530 ppb NO_2, with three of four studies reporting a statistically significant increase in airway responsiveness following such exposures.

However, studies conducted with exercise do not indicate consistent increases in AR following exposures to NO_2 concentrations from 300 to 600 ppb (Table 3-3).[64]

Consideration of results from the meta-analysis

As discussed above in Section 3.2.1, the ISA assessment of the evidence for AR in individuals with asthma also focuses on a recently published meta-analysis (Brown, 2015) investigating individual-level data from the studies included in Tables 3-2 and 3-3. While individual controlled human exposure studies can lack statistical power to identify effects, the meta-analysis of individual-level data combined from multiple studies (Brown, 2015) has greater statistical power due to increased sample size. The meta-analysis considered individual-level responses, specifically whether individual study subjects experienced an increase or decrease in AR following NO_2 exposure compared to air exposure, combining information from the studies presented in Tables 3-2 and 3-3. Evidence was evaluated together across all studies and also stratified for exposures conducted with exercise and at rest, and for measures of specific and non-specific AR. The ISA notes that these methodological differences may have important implications with regard to results (U.S. EPA, 2016; Brown, 2015; Goodman et al., 2009), contributing to the ISA's emphasis on studies of resting exposures and non-specific challenge agents. Overall, the meta-analysis presents the fraction of individuals having an increase in AR following exposure to various NO_2 concentrations (i.e., 100 ppb, 100 ppb to < 200 ppb, 200 ppb up to and including 300 ppb, and above 300 ppb) (U.S. EPA, 2016, Section 5.2.2.1).[65] The number of participants in each study and the number having an increase or decrease in AR is indicated in Tables 3-2 and 3-3.

We first consider the meta-analysis results across all exposure conditions (i.e., resting, exercising, non-specific challenge, and specific challenge). For 100 ppb NO_2 exposures, Brown (2015) reported that, of the study participants who experienced either an increase or decrease in AR following NO_2 exposures, 61% experienced an increase ($p = 0.08$). For 100 to < 200 ppb NO_2 exposures, 62% of study subjects experienced an increase in AR following NO_2 exposures ($p = 0.014$). For 200 to 300 ppb NO_2 exposures, 58% of study subjects experienced an increase in AR following NO_2 exposures ($p = 0.008$). For exposures above 300 ppb NO_2, 57% of study

---

[64] There are eight additional studies with exercising exposures to 300-350 ppb NO_2 as presented in Table 3-3, with exposure durations ranging from 30-240 minutes. Results across these studies are inconsistent, with only two of eight reporting significant results. Only one of four studies with exercising exposures of 400 or 600 ppb reported statistically significant increases in airway responsiveness.

[65] Brown et al. (2015) compared the number of study participants who experienced an increase in AR following NO_2 exposures to the number who experienced a decrease in AR. Study participants who experienced no change in AR were not included in comparisons. *P*-value refers to the significance level of a two-tailed sign test.

subjects experienced an increase in AR following NO_2 exposures, though this fraction was not statistically different than the fraction experiencing a decrease.

We also consider the results of Brown (2015) for various subsets of the available studies, based on the exposure conditions evaluated (i.e., resting, exercising) and the type of challenge agent used (specific, non-specific). For exposures conducted at rest, across all exposure concentrations (i.e., 100-530 ppb NO_2, n = 139; Table 3-2), Brown (2015) reported that a statistically significant fraction of study participants (71%, $p < 0.001$) experienced an increase in AR following NO_2 exposures, compared to the fraction that experienced a decrease in AR. The meta-analysis also presented results for various concentrations or ranges of concentrations. Following resting exposure to 100 ppb NO_2, 66% of study participants experienced increased non-specific AR. For exposures to concentrations of 100 ppb to < 200 ppb, 200 ppb up to and including 300 ppb, and above 300 ppb, increased non-specific AR was reported in 67%, 78%, and 73% of study participants, respectively.[66] For non-specific challenge agents, the differences between the fractions of individuals who experienced increased AR following resting NO_2 exposures and the fraction who experienced decreased AR reached statistical significance for all of the ranges of exposures concentrations evaluated ($p < 0.05$).

In contrast to the results from studies conducted at rest, the fraction of individuals having an increase in AR following NO_2 exposures with exercise was not consistently greater than 50%, and none of the results were statistically significant (Brown, 2015). Across all NO_2 exposures with exercise, measures of non-specific AR were available for 241 individuals, 54% of whom experienced an increase in AR following NO_2 exposures relative to air controls. There were no studies in this group conducted at 100 ppb, and for exercising exposures to 150-200 ppb, 250-300 ppb, and 350-600 ppb, the fraction of individuals with increased AR was 59%, 55%, and 49%, respectively.

In addition to examining results from studies of non-specific AR, the meta-analysis also considered results from studies that evaluated changes in specific AR (i.e., AR following an allergen challenge; n=130; Table 3-3) following NO_2 exposures. The results do not indicate statistically significant fractions of individuals having an increase in specific AR following exposure to NO_2 at concentrations below 400 ppb, even when considering resting and exercising exposures separately (Brown, 2015). Of the three studies that evaluated specific AR at concentrations of 400 ppb, one was conducted at rest (Tunnicliffe et al., 1994). This study

---

[66] For the exposure category of "above 300 ppb", exposures included 400, 480, 500, and 530 ppb. No studies used concentrations between 300 and 400 ppb.

reported that all individuals experienced increased AR following 400 ppb NO_2 exposures (Brown, 2015, Table 4). In contrast, for exposures during exercise, most study subjects did not experience NO_2-induced increases in specific AR.[67] Overall, results across studies are less consistent for increases in specific AR following NO_2 exposures.

Uncertainties in evidence for airway responsiveness

When considering the evidence for NO_2-induced increases in AR in individuals with asthma, there are important uncertainties that should be considered. Both the meta-analysis by Brown (2015) and an additional meta-analysis and meta-regression by Goodman et al. (2009) conclude that there is no indication of a dose-response relationship for exposures between 100 and 500 ppb NO_2 and increased AR in individuals with asthma. A dose-response relationship generally increases confidence that observed effects are due to pollutant exposures rather than to chance; however, the lack of a dose-response relationship does not necessarily indicate that there is no relationship between the exposure and effect, particularly in these analyses based on between-subject comparisons (i.e., as opposed to comparisons within the same subject exposed to multiple concentrations). As discussed in the ISA, there are a number of methodological differences across studies that could contribute to between-subject differences and that could obscure a dose-response relationship between NO_2 and AR. These include subject activity level (rest vs. exercise) during NO_2 exposure, asthma medication usage, choice of airway challenge agent (e.g., direct and indirect non-specific stimuli), method of administering the bronchoconstricting agents, and physiological endpoint used to assess AR. Such methodological differences across studies likely contribute to the variability and uncertainty in results across studies and complicate interpretation of the overall body of evidence for NO_2-induced AR. Thus, while the lack of an apparent dose-response relationship adds uncertainty to our interpretation of controlled human exposure studies of AR, it does not necessarily indicate the lack of an NO_2 effect.

An additional uncertainty in interpreting these studies within the context of the adequacy of the protection provided by the NO_2 NAAQS is the potential adversity of the reported NO_2-induced increases in AR. As discussed above (section 3.2.1), the meta-analysis by Brown (2015) used an approach that is consistent with guidelines from the ATS and the ERS for the assessment of therapeutic agents (Reddel et al., 2009) to assess the potential for clinical relevance of these responses. Specifically, based on individual-level responses reported in a subset of studies,

---

[67] Forty-eight percent experienced increased AR and 52% experienced decreased AR, based on individual-level data for study participants exposed to 350 ppb (Riedl et al., 2012) or 400 ppb (Jenkins et al., 1999; Witten et al., 2005) NO_2.

Brown (2015) considered a halving of the provocative dose to indicate responses that may be clinically relevant.[68] With regard to this approach, the ISA notes that "one doubling dose change in PD is recognized as a potential indicator, although not a validated estimate, of clinically relevant changes in airway responsiveness (Reddel et al., 2009)" (U.S. EPA, 2016, p. 5-12). While there is uncertainty in using this approach to characterize whether a particular response in an individual is "adverse," it can provide insight into the potential for adversity, particularly when applied to a population of exposed individuals.[69]

Only five studies provided data for each individual's provocative dose. These five studies provided individual-level data for a total of 72 study participants (116 AR measurements) and eight NO_2 exposure concentrations, for resting exposures and non-specific bronchial challenge agents. Across exposures to 100, 140, 200, 250, 270, 480, 500, and 530 ppb NO_2, 24% of study participants experienced a halving of the provocative dose (indicating increased AR) while 8% showed a doubling of the provocative dose (indicating decreased AR). The relative distributions of the provocative doses at different concentrations were similar, with no dose-response relationship indicated (Brown, 2015). While these results support the potential for clinically relevant increases in AR in some individuals with asthma following NO_2 exposures within the range of 100 to 530 ppb, uncertainty remains given that this analysis is limited to a small subset of studies and given the lack of an apparent dose-response relationship. In addition, compared to conclusions based on the entire range of NO_2 exposure concentrations evaluated (i.e., 100 to 530 ppb), there is greater uncertainty in reaching conclusions about the potential for clinically relevant effects at any particular NO_2 exposure concentration within this range.

Conclusion

As in the last review, a meta-analysis of individual-level data supports the potential for increased AR in individuals with asthma following 30 minute to 1 hour exposures to NO_2 concentrations from 100 to 530 ppb, particularly for resting exposures and measures of non-specific AR (N = 33 to 70 for various ranges of NO_2 exposure concentrations). Individual studies most consistently report statistically significant NO_2-induced increases in AR following exposures to NO_2 concentrations at or above 250 ppb. Individual studies (N = 4 to 20) generally

---

[68] More specifically, clinical relevance in the ISA is based on evidence from controlled human exposure studies evaluating efficacy of inhaled corticosteroids that are used to prevent bronchoconstriction and airway responsiveness as described by Reddell et al. (2009). Generally, a change of at least one doubling dose is considered to be an indication of clinical relevance (this represents a decline in AR as the dose to induce AR is doubled). Based on this, a halving of the provocative dose is taken in the ISA to represent an increase in AR that is an indication of clinical relevance.

[69] As noted above, the degree to which populations in U.S. urban areas have the potential for such NO_2 exposures is evaluated in Chapter 4.

do not report statistically significant increases in AR following exposures to NO_2 concentrations at or below 200 ppb, though the evidence suggests a trend toward increased AR following NO_2 exposures from 140 to 200 ppb. In contrast, individual studies do not indicate a consistent trend towards increased AR following 1-hour exposures to 100 ppb NO_2. Important limitations in this evidence include the lack of a dose-response relationship between NO_2 and AR and uncertainty in the adversity of the reported increases in AR. These limitations become increasingly important at the lower NO_2 exposure concentrations (i.e., at or near 100 ppb), where the evidence for NO_2-induced increases in AR is not consistent across studies.

3.2.2.2 *Concentrations in Locations of Epidemiologic Studies*

We next consider distributions of ambient NO_2 concentrations in locations where epidemiologic studies have examined NO_2 associations with asthma-related hospital admissions or emergency department (ED) visits. These outcomes are clearly adverse and study results comprise a key line of epidemiologic evidence in the determination of a causal relationship in the ISA (U.S. EPA, 2016, Section 5.2.9). As in other NAAQS reviews (U.S. EPA, 2014; U.S. EPA, 2011), when considering epidemiologic studies within the context of the adequacy of the current standard, we emphasize those studies conducted in the U.S. and Canada.[70] For short-term exposures to NO_2, we emphasize studies reporting associations with effects judged in the ISA to be robust to confounding by other factors, including co-occurring air pollutants. In addition, we consider the statistical significance and precision of study results, and the inclusion of at-risk populations for which the NO_2-health effect associations may be larger. These considerations help inform the range of ambient NO_2 concentrations over which we have the most confidence in NO_2-asssociated health effects and the range of concentrations over which our confidence in such effects is appreciably lower. In our consideration of these issues, we specifically focus on the following question:

- **To what extent have U.S. and Canadian epidemiologic studies reported associations between asthma-related hospital admissions or emergency department visits and short-term NO_2 concentrations in study areas that would have met the current 1-hour NO_2 standard during the study period?**

Addressing this question can provide important insights into the extent to which NO_2-health effect associations are present for distributions of ambient NO_2 concentrations that would be allowed by the current standards. The presence of such associations would support the potential for the current standards to allow the NO_2-associated effects indicated by epidemiologic

---

[70] Such studies are likely to reflect air quality and exposure patterns that are generally applicable to the U.S. In addition, air quality data corresponding to study locations and study time periods is often readily available for studies conducted in the U.S. and Canada. Nonetheless, we recognize the importance of all studies, including other international studies, in the ISA's assessment of the weight of the evidence that informs causal determinations.

studies. To the degree studies have not reported associations in locations meeting the current NO_2 standards, there is greater uncertainty regarding the potential for the reported effects to occur following NO_2 exposures associated with air quality meeting those standards.

In addressing the question above and considering the available evidence, we place the greatest emphasis on studies reporting positive, and relatively precise, health effect associations. In evaluating whether such associations are likely to reflect NO_2 concentrations meeting the existing 1-hour standard, we consider the 1-hour ambient NO_2 concentrations measured at monitors in study locations during study periods. We also consider what additional information is available to inform our understanding of the ambient NO_2 concentrations that could have been present in the study locations during the study periods (e.g., around major roads). When considered together, this information can provide important insights into the extent to which NO_2 health effect associations have been reported for NO_2 air quality concentrations that likely would have met the current 1-hour NO_2 standard.

We have identified U.S. and Canadian studies of respiratory-related hospital admissions and emergency department (ED) visits, with a focus on studies of asthma-related effects (studies identified from ISA Table 5-10).[71] For each NO_2 monitor in the locations evaluated by these studies and the ranges of years encompassed by studies, we have identified the 3-year averages of the 98th percentiles of the annual distributions of daily maximum 1-hour NO_2 concentrations.[72] These concentrations approximate the design values (DVs) that are used when determining whether an area meets the primary NO_2 NAAQS.[73] Thus, these "DVs" can provide perspective on whether study areas would likely have met or exceeded the primary 1-hour NO_2 NAAQS

---

[71] Studies of asthma-related hospital admissions and ED visits were identified in the ISA as comprising an important line of epidemiologic evidence to support the conclusion that there is a "causal" relationship between short-term NO_2 exposures and respiratory effects (U.S. EPA, 2016). Strong support was also provided by epidemiologic studies for respiratory symptoms, but the majority of studies on respiratory symptoms were only conducted over part of a year, complicating the evaluation of a DV based on data from 3 years of monitoring data relative to the respective health effect estimates. For more information on these studies and the DVs in the study locations, see Appendix A.

[72] All study locations had maximum annual design values below 53 ppb (Appendix A).

[73] As described in Chapter 2, a design value is a statistic that describes the air quality status of a given area relative to the NAAQS and that is typically used to classify nonattainment areas, assess progress towards meeting the NAAQS, and develop control strategies. For the 1-hour NO_2 standard, the DV is calculated at individual monitors and based on 3 consecutive years of data collected from that site. In the case of the 1-hour NO_2 standard, the design value for a monitor is based on the 3-year average of the 98th percentile of the annual distribution of daily maximum 1-hour NO_2 concentrations. For more information on these studies and the calculation of study area "DVs," see Appendix A.

during the study periods.[74] Based on this approach, study locations could have met the current 1-hour standard over the entire study period if all of the hourly DVs were at or below 100 ppb.

A key limitation in these analyses of NO_2 DVs is that currently required near-road NO_2 monitors were not in place during study periods. The studies evaluated (see Figure 3-1 below) were based on air quality from 1980-2006, with most studies spanning the 1990s to early 2000s. As discussed above in Chapter 2, there were no specific near-road monitoring network requirements during these years, and most areas did not have monitors sited to measure NO_2 concentrations near the most heavily-trafficked roadways. In addition, mobile source NO_x emissions were considerably higher during the time periods of available epidemiologic studies than in more recent years, suggesting that the NO_2 concentration gradients around major roads were likely more pronounced than indicated by data from recently deployed near-road monitors (Figure 2-6).[75] This information suggests that if the current near-road monitoring network had been in operation during study periods, NO_2 DVs measured at near-road monitors would likely have been higher than the DVs reflected in Figure 3-1 below. This uncertainty particularly limits the degree to which we can draw strong conclusions based on study areas with DVs that are at or just below 100 ppb.

With this key limitation in mind, we consider what the available epidemiologic evidence can tell us with regard to the adequacy of the public health protection provided by the current 1-hour standard against short-term NO_2 exposures. Figure 3-1, below, highlights the epidemiologic studies examining associations between asthma hospitalizations or ED visits and short-term exposures to ambient NO_2 that were conducted in the U.S. and Canada. These studies were identified and evaluated in the ISA and include both the few recently published studies and the studies that were available in the previous review. Figure 3-1 depicts the range of associations

---

[74] The DVs indicated in Figure 3-1 are different from the NO_2 concentrations reported in the studies themselves, which are often averaged across monitors in the study areas and can reflect averaging periods other than 1-hour. Thus, the concentrations reported in the studies are not appropriate for direct comparison to the level of the 1-hour standard. We are, however, providing them for additional context. The NO_2 concentrations reported in studies are as follows: Steib et al., 2009: 24-h average: 9.3-22.7ppb; 75th percentile: 12.3-27.6ppb; Linn et al., 2000, 24-h average 3.4ppb; Peel et al., 2005, 1-h max: 45.9ppb; Ito et al., 2007, 24-h average: 31.1ppb; Villeneuve et al., 2007, 24-h average: 17.5ppb (Summer) and 28.5ppb (Winter); Burnett et al., 2009, 24-h average: 25.2ppb; Strickland et al., 2010, 1-h max: 23.3ppb; ATSDR, 2006, 24-h average: 36ppb (Bronx) and 31ppn (Manhattan); Jaffe et al., 2003, 24-h average: 50ppb (Cincinnati) and 48ppb (Cleveland); Li et al., 2011: 24-h average: 15.7ppb, 75th percentile: 21.2ppb, Max: 55.2ppb

[75] Recent data indicate that, for most near-road monitors, measured 1-hour NO_2 concentrations are higher than those measured at all of the non-near-road monitors in the same CBSA (Section 2.3.2).

across U.S. and Canadian studies and also indicates maximum and mean hourly DVs for the study locations and years.[76]

---

[76] Similar analyses of study area air quality were presented in the 2008 REA. However, because the 1-hour standard was set in the 2010 final decision, the methods for calculating 1-hour NO_2 DVs had not been established at the time of the development of the 2008 REA. Therefore, the study area NO_2 concentrations identified in the 2008 REA did not correspond to 1-hour DVs for the current 1-hour NO_2 standard. As a result, even when the same study is evaluated, study area NO_2 concentrations are not identical in the 2008 REA and in Figure 3-1 of this PA.

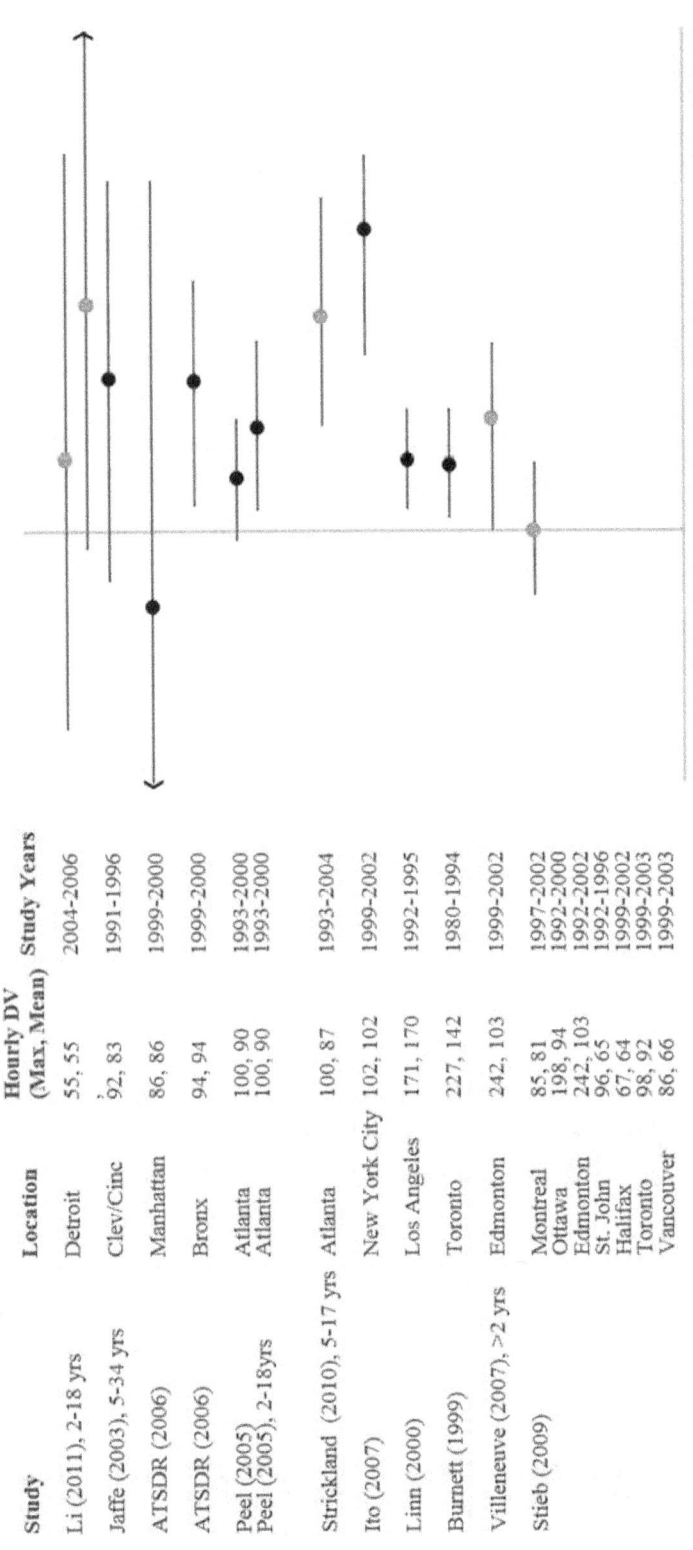

Study	Location	Hourly DV (Max, Mean)	Study Years
Li (2011), 2-18 yrs	Detroit	55, 55	2004-2006
Jaffe (2003), 5-34 yrs	Clev/Cinc	92, 83	1991-1996
ATSDR (2006)	Manhattan	86, 86	1999-2000
ATSDR (2006)	Bronx	94, 94	1999-2000
Peel (2005)	Atlanta	100, 90	1993-2000
Peel (2005), 2-18yrs	Atlanta	100, 90	1993-2000
Strickland (2010), 5-17 yrs	Atlanta	100, 87	1993-2004
Ito (2007)	New York City	102, 102	1999-2002
Linn (2000)	Los Angeles	171, 170	1992-1995
Burnett (1999)	Toronto	227, 142	1980-1994
Villeneuve (2007), >2 yrs	Edmonton	242, 103	1999-2002
Stieb (2009)	Montreal	85, 81	1997-2002
	Ottawa	198, 94	1992-2000
	Edmonton	242, 103	1992-2002
	St. John	96, 65	1992-1996
	Halifax	67, 64	1999-2002
	Toronto	98, 92	1999-2003
	Vancouver	86, 66	1999-2003

Figure 3-1.   U.S. and Canadian epidemiologic studies of short-term NO$_2$ exposures and asthma hospital admissions and emergency department visits. Study locations and years are reported with hourly DVs for studies asthma hospital admissions and ED visits with effect estimates and 95% confidence intervals, standardized as described in the ISA (U.S. EPA, 2016). Effect estimates in blue represent studies that are new in the current review. Clev = Cleveland; Cinc = Cincinnati. If ages are not specified after the study, then hospital admissions and ED visits for all ages were included. Li et al. reports an effect estimate from a time-series analysis and case-crossover analysis. Because hourly DVs are based on 3 years of data, DVs for the first 2 years of a study period were not considered. The ATSDR study did not include 3 years, thus DVs reported for these locations include data from a year preceding the study (1998).

Figure 3-1 includes both multi-city and single city studies. In considering the information in Figure 3-1, we note that multi-city studies tend to have greater power to detect associations. The one multi-city study that has become available since the last review (Stieb et al., 2009) reported a null association with asthma ED visits, based on study locations with maximum DVs ranging from 67-242 ppb (six of seven study cities had maximum DVs at or above 85 ppb). Of the single city studies in Figure 3-1, those reporting positive and relatively precise (i.e., relatively narrow 95% CIs) associations were conducted in locations with maximum, and often mean, DVs at or above 100 ppb (i.e., Linn et al., 2000; Peel et al., 2005; Ito et al., 2007; Villeneuve et al., 2007; Burnett et al., 1999; Strickland et al., 2010). For the other single city studies in Figure 3-1, two reported more mixed results in locations with maximum DVs around 90 ppb (Jaffe et al., 2003; ATSDR, 2006).[77] Associations in these studies were generally not statistically significant, were less precise (i.e., wider 95% CIs), and included a negative association (Manhattan, NY). One single city study was conducted in a location with 1-hour DVs well-below 100 ppb (Li et al, 2011), though the reported associations were not statistically significant and were relatively imprecise. Thus, of the U.S. and Canadian studies that can most clearly inform our consideration of the adequacy of the current NO_2 standards, the lone multicity study did not report a positive health effect association and the single-city studies reporting positive, and relatively precise, associations were generally conducted in locations with maximum 1-hour DVs at or above 100 ppb. The evidence for associations in locations with maximum DVs below 100 ppb is more mixed, and reported associations are generally less precise.

An uncertainty in this body of evidence is the potential for copollutant confounding. When pollutants are highly correlated, it can be difficult to determine the independent effects of single pollutants from other pollutants in the mixture. Copollutant (two-pollutant) models can be used in epidemiologic studies in an effort to disentangle such independent effects, though there can be limitations in these models due to differential exposure measurement error and high correlations with traffic-related copollutants. For NO_2, the copollutants that are most relevant to consider are those from traffic sources such as CO, EC/BC, UFP, and VOCs such as benzene as well as $PM_{2.5}$ and PM_{10} (ISA, Section 3.5). Of the studies examining asthma-related hospital admissions and ED visits in the U.S. and Canada in Figure 3-1, three examined effect estimates from copollutant models (Ito et al., 2007; Villeneuve et al., 2007; Strickland et al., 2010)). Ito et al. (2007) found that in copollutant models with $PM_{2.5}$, SO_2, CO, or O_3, NO_2 consistently had the

---

[77] The study by the U.S. Agency for Toxic Substances and Disease Registry (ATSDR) was not published in a peer-review journal. Rather, it was a report prepared by New York State Department of Health's Center for Environmental Health, the New York State Department of Environmental Conservation and Columbia University in the course of performing work contracted for and sponsored by the New York State Energy Research and Development Authority and the ATSDR.

strongest effect estimates that were robust to the inclusion of other pollutants. Villeneuve et al. (2007) utilized a model including NO_2 and CO (r = 0.74) for ED visits in the warm season and reported that associations for NO_2 were robust to CO. Strickland et al. (2010) found that the relationship between ambient NO_2 and asthma ED visits in Atlanta, GA was robust in models including O_3, but copollutant models were not analyzed for other pollutants and the correlations between NO_2 and other pollutants were not reported. Taken together, these studies provide some evidence for independent effects of NO_2 for asthma ED visits, but some important traffic-related copollutants (e.g. EC/BC, VOCs) have not been examined in this body of evidence and the limitations of copollutant models in demonstrating an independent association, particularly for NO_2, are well recognized (U.S. EPA, 2016).

Conclusions

Considering this evidence together, we note the following observations. First, the only recent multicity study evaluated, which had maximum DVs ranging from 67 to 242 ppb, did not report a positive association between NO_2 and ED visits (Stieb et al., 2009). In addition, of the single-city studies in Figure 3-1 reporting positive and relatively precise associations between NO_2 and asthma hospital admissions and ED visits, most locations likely had NO_2 concentrations above the current 1-hour NO_2 standard over at least part of the study period. Although maximum DVs for the studies conducted in Atlanta were 100 ppb, it is likely that those DVs would have been higher than 100 ppb had currently required near-road monitors been in place. For the study locations with maximum DVs below 100 ppb, mixed results are reported with associations that are generally statistically non-significant and imprecise, indicating that associations between NO_2 concentrations and asthma-related ED visits are more uncertain in locations that could have met the current standards. Given that near-road monitors were not in operation during study periods, it is not clear that even these DVs below 100 ppb indicate study areas that would have met the current 1-hour standard. Thus, when considering our analyses of study area NO_2 concentrations in light of uncertainties related to roadway NO_2 concentrations and copollutant confounding, we reach the conclusion that available U.S. and Canadian epidemiologic studies do not provide support for NO_2-associated hospital admissions or ED visits in locations with NO_2 concentrations that would have clearly met the current 1-hour NO_2 standard.

3.3 EFFECTS OF LONG-TERM NO₂ EXPOSURES

3.3.1 Nature of Effects

In the last review of the primary NO_2 NAAQS, evidence for health effects related to long-term ambient NO_2 exposure was judged "suggestive of, but not sufficient to infer" or "inadequate to infer the presence or absence of" a causal relationship across health effect categories. These included respiratory, cardiovascular, and reproductive and developmental effects as well as cancer and total mortality. In the current review, new epidemiologic evidence, in conjunction with explicit integration of evidence across related outcomes, has resulted in strengthening of some of the causal determinations. Though the evidence of health effects associated with long-term exposure to NO_2 is more robust than in previous reviews, there are still a number of uncertainties limiting our understanding of the role of long-term NO_2 exposures in causing health effects. We focus our discussion of evidence available in the current review for health effects related to long-term NO_2 exposures, including strengths and limitations, on the following overarching questions:

- **To what extent does the currently available scientific evidence alter or strengthen our conclusions from the last review regarding health effects attributable to long-term NO_2 exposures? Have previously identified uncertainties been reduced? What important uncertainties remain and have new uncertainties been identified?**

Table 3-4 provides an overview of the causal determinations for the previous and current reviews for long-term NO_2 exposures and various health effect categories including respiratory, cardiovascular, and reproductive and developmental effects, as well as mortality and cancer. In particular, the causal determination between long-term NO_2 exposures and respiratory effects was strengthened to "likely to be a causal relationship." The evidence on which these causal judgments are based is summarized below.

Table 3-4. Causal determinations for long-term nitrogen dioxide (NO_2) exposure and health effects evaluated in the ISA for Oxides of Nitrogen in the previous and current review

Health effect	Review completed 2010	Current Review
Respiratory	Suggestive of, but not sufficient to infer, a causal relationship	Likely to be a causal relationship
Cardiovascular and Diabetes[78]	Inadequate to infer the presence or absence of a causal relationship	Suggestive of, but not sufficient to infer, a causal relationship
Total Mortality	Inadequate to infer the presence or absence of a causal relationship	Suggestive of, but not sufficient to infer, a causal relationship
Reproductive and Developmental–Fertility, Reproduction, Pregnancy		Inadequate to infer a causal relationship
Reproductive and Developmental – Birth Outcomes	Inadequate to infer the presence or absence of a causal relationship[a]	Suggestive of, but not sufficient to infer, a causal relationship
Reproductive and Developmental–Postnatal Development		Inadequate to infer a causal relationship

[78] The addition of diabetes to the ISA causal determination is new to this review.

| Cancer | Inadequate to infer the presence or absence of a causal relationship | Suggestive of, but not sufficient to infer, a causal relationship |

Respiratory

The 2016 ISA concluded that there is "likely to be a causal relationship" between long-term NO_2 exposure and respiratory effects, based primarily on evidence integrated across disciplines for a relationship with asthma development in children.[79] Evidence for other outcomes integrated across epidemiologic and experimental studies, including decrements in lung function and partially irreversible decrements in lung development, respiratory disease severity, chronic bronchitis/asthma incidence in adults, COPD hospital admissions, and respiratory infections, is less consistent and has larger uncertainty in whether there is an independent effect of long-term NO_2 exposure (U.S. EPA, 2016, Section 6.2.9).

The conclusion of a "likely to be a causal relationship" in the current review represents a change from 2008 ISA conclusion that the evidence was "suggestive of, but not sufficient to infer, a causal relationship" (U.S. EPA, 2008, Section 5.3.2.4). The epidemiologic evidence base has expanded since the last review. This expanded evidence includes several recently published longitudinal studies that indicate positive associations between asthma incidence in children and long-term NO_2 exposures, with improved exposure assessment in some studies based on NO_2 modeled estimates for children's homes or NO_2 measured near children's homes or schools. Associations were observed across various periods of exposure, including first year of life, year prior to asthma diagnosis, and cumulative exposure. In addition, the ISA notes several other strengths of the evidence base including the general timing of asthma diagnosis and relative confidence that the NO_2 exposure preceded asthma development in longitudinal studies, more reliable estimates of asthma incidence based on physician-diagnosis in children older than 5 years of age from parental report or clinical assessment, as well as residential NO_2 concentrations estimated from land use regression (LUR) models with good NO_2 prediction in some studies.

While the causal determination has been strengthened in this review, important uncertainties remain. For example, the ISA notes that as in the last review, a "key uncertainty that remains when examining the epidemiologic evidence alone is the inability to determine whether NO_2 exposure has an independent effect from that of other pollutants in the ambient

---

[79] Elsewhere referred to as "asthma incidence." Asthma development and asthma incidence refer to the onset of the disease rather than the exacerbation of existing disease.

mixture" (U.S. EPA, 2016, Section 6.2.2.1, p. 6-21). While a few studies have included copollutant models for respiratory effects other than asthma development, the ISA states that "[e]pidemiologic studies of asthma development in children have not clearly characterized potential confounding by $PM_{2.5}$ or traffic-related pollutants [e.g., CO, BC/EC, volatile organic compounds (VOCs)]" (U.S. EPA, 2016, p. 6-64). The ISA further notes that "[i]n the longitudinal studies, correlations with $PM_{2.5}$ and BC were often high (e.g., $r = 0.7–0.96$), and no studies of asthma incidence evaluated models to address copollutant confounding, making it difficult to evaluate the independent effect of NO_2" (U.S. EPA, 2016, p. 6-64). High correlations between NO_2 and other traffic-related pollutants were based on modeling, and studies of asthma incidence that used monitored NO_2 concentrations as an exposure surrogate did not report such correlations (US EPA, 2016, Section 6). This uncertainty is important to consider when interpreting the epidemiologic evidence regarding the extent to which NO_2 is independently related to asthma development.

For additional context, the ISA also evaluated copollutant confounding in long-term studies beyond asthma incidence to examine whether studies of other respiratory effects could provide information on the potential for confounding by traffic-related copollutants. Several studies examined correlations between NO_2 and traffic-related copollutants and found them to be relatively high in many cases, ranging from 0.54-0.95 for $PM_{2.5}$, 0.54-0.93 for BC/EC, 0.2-0.95 for PM_{10}, and 0.64-0.86 for OC (U.S. EPA, 2016, Tables 6-1 and 6-3). While these correlations are often based on model estimates, some are based on monitored pollutant concentrations (i.e., McConnell et al. (2003) reported correlations of 0.54 with $PM_{2.5}$ and EC) (U.S. EPA, 2016, Table 6-3). Additionally, three studies (McConnell et al., 2003; MacIntyre et al., 2014; Gehring et al., 2013) evaluated co-pollutant models with NO_2 and $PM_{2.5}$, and some findings suggest that associations for NO_2 with bronchitic symptoms, lung function, and respiratory infection are not robust because effect estimates decreased in magnitude and became imprecise when a copollutant was added in the model. Overall, examination of evidence from studies of other respiratory effects indicates moderate to high correlations between long-term NO_2 concentrations and traffic-related copollutants, with very limited evaluation of the potential for confounding. Thus, when considering the collective evidence, it is difficult to disentangle the independent effect of NO_2 from other traffic-related pollutants or mixtures in epidemiologic studies (U.S. EPA, 2016, Sections 3.4.4 and 6.2.9.5).

While this uncertainty continues to apply to the epidemiologic evidence for asthma incidence in children, the ISA describes that the uncertainty is partly reduced by the coherence of findings from experimental studies and epidemiologic studies. Experimental studies demonstrate effects on key events in the mode of action proposed for the development of asthma and provide biological plausibility for the epidemiologic evidence. For example, one study demonstrated that airway hyperresponsiveness was induced in guinea pigs after long-term exposure to NO_2 [1,000–4,000 ppb; (Kobayashi and Miura, 1995)]. Other experimental studies examining oxidative stress report mixed results, but some evidence from short-term studies supports a relationship between NO_2 exposure and increased pulmonary inflammation in healthy humans. The ISA also points to supporting evidence from studies demonstrating that short-term exposure repeated over several days (260-1,000 ppb) and long-term NO_2 exposure (2,000-4,000 ppb) can induce Th2 skewing/allergic sensitization in healthy humans and animal models by showing increased Th2 cytokines, airway eosinophils, and IgE-mediated responses (U.S. EPA, 2016, Sections 4.3.5 and 6.2.2.3). Epidemiologic studies also provide some supporting evidence for these key events in the mode of action. Some evidence from epidemiologic studies also demonstrates associations between short-term ambient NO_2 concentrations and increases in pulmonary inflammation in healthy children and adults (U.S. EPA, 2016, Section 5.2.2.5). Overall, evidence from experimental and epidemiologic studies provide support for a role of NO_2 in asthma development by describing a potential role for repeated exposures to lead to recurrent inflammation and allergic responses.

In addressing the questions posed at the beginning of this section, we note that there is new evidence available that strengthens conclusions from the last review regarding respiratory health effects attributable to long-term ambient NO_2-exposure. The majority of new evidence is from epidemiologic studies of asthma incidence in children with improved exposure assessment (i.e., measured or modeled at or near children's homes or schools), which builds upon previous evidence for associations of long-term NO_2 and asthma incidence and also partly reduces uncertainties related to measurement error. Explicit integration of evidence for individual outcome categories (e.g. asthma incidence, respiratory infection) provides improved characterization of biological plausibility and mode of action, including some new evidence from studies of short-term exposure supporting an effect on asthma development. Although this partly reduces the uncertainty regarding independent effects of NO_2, because of the high correlation with other traffic-related copollutants and the general lack of copollutant model results in epidemiologic studies, the potential for confounding remains a concern when interpreting epidemiologic studies of NO_2 and asthma development. In particular, it remains unclear the degree to which NO_2 itself may be causing the development of asthma versus serving primarily as a surrogate for the broader traffic-pollutant mix.

Cardiovascular and Diabetes

In the previous review, the 2008 ISA stated that the evidence for cardiovascular effects attributable to long-term ambient NO_2 exposure was "inadequate to infer the presence or absence of a causal relationship." The epidemiologic and experimental evidence was limited, with uncertainties related to traffic-related copollutant confounding (U.S. EPA, 2008). For the current review, the body of epidemiologic evidence available is substantially larger than that in the last review and includes evidence for diabetes. The conclusion on causality is stronger in the current review with regard to the relationship between long-term exposure to NO_2 and cardiovascular effects and diabetes as the ISA judged the evidence to be "suggestive, but not sufficient to infer" a causal relationship (U.S. EPA, 2016, Section 6.3). The strongest evidence comes from recent epidemiologic studies reporting positive associations of NO_2 with heart disease and diabetes with improved exposure assessment (i.e., residential estimates from models that well predict NO_2 concentrations in the study areas), but the evidence across experimental studies remains limited and inconsistent and does not provide sufficient biological plausibility for effects observed in epidemiologic studies. Specifically, the ISA concludes that "[e]pidemiologic studies have not adequately accounted for confounding by $PM_{2.5}$, noise, or traffic-related copollutants, and there is limited coherence and biological plausibility for NO_2-related development of heart disease" (U.S. EPA, 2016, p. 6-98) or "for NO_2-related development of diabetes" (U.S. EPA, 2016, p. 6-99). Thus, substantial uncertainty exists regarding the independent effect of NO_2 and the total evidence is "suggestive of, but not sufficient to infer, a causal relationship" between long-term NO_2 exposure and cardiovascular effects and diabetes (U.S. EPA, 2016, Section 6.3.9).

Reproductive and Developmental Effects

In the previous review, a limited number of epidemiologic and toxicological studies had assessed the relationship between long-term NO_2 exposure and reproductive and developmental effects. The 2008 ISA concluded that there was not consistent evidence for an association between NO_2 and birth outcomes and that evidence was "inadequate to infer the presence or absence of a causal relationship" with reproductive and developmental effects overall (U.S. EPA, 2008). In the ISA for the current review, a number of recent studies added to the evidence base, and reproductive effects were considered as three separate categories: birth outcomes; fertility, reproduction, and pregnancy; and postnatal development (U.S. EPA, 2016, Section 6.4). Overall, the evidence is "suggestive of, but not sufficient to infer, a causal relationship" between long-term exposure to NO_2 and birth outcomes and is "inadequate to infer the presence or absence of a causal relationship" between exposure to NO_2 and fertility, reproduction and pregnancy as well as postnatal development. Evidence for effects on fertility, reproduction, and pregnancy and for effect on postnatal development is inconsistent across both epidemiologic and toxicological studies. Additionally, there are few toxicological studies available. The ISA concludes the change in the causal determination for birth outcomes reflects the large number of studies that generally observed associations with fetal growth restriction and the improved outcome assessment (e.g., measurements throughout pregnancy via ultrasound) and exposure assessment (e.g., well-validated LUR models) employed by many of these studies (U.S. EPA, 2016, Section 6.4.5). For birth outcomes, there is uncertainty in whether the epidemiologic findings reflect an independent effect of NO_2 exposure.

Total Mortality

In the 2008 ISA, a limited number of epidemiologic studies assessed the relationship between long-term exposure to NO_2 and mortality in adults. The 2008 ISA concluded that the scarce amount of evidence was "inadequate to infer the presence or absence of a causal relationship" (U.S. EPA, 2008c). The ISA for the current review concludes that evidence is "suggestive of, but not sufficient to infer, a causal relationship" between long-term exposure to NO_2 and mortality among adults (U.S. EPA, 2016, Section 6.5.3). This causal determination is based on evidence from recent studies demonstrating generally positive associations between long-term exposure to NO_2 and total mortality from extended analyses of existing cohorts as well as original results from new cohorts. In addition, there is evidence for associations between long-term NO_2 exposures and mortality due to respiratory and cardiovascular causes. However, there were several studies that did not observe an association between long-term exposure to NO_2 and mortality.

Some recent studies examined the potential for copollutant confounding by $PM_{2.5}$, BC, or measures of traffic proximity or density in copollutant models with results from these models generally showing attenuation of the NO_2 effect with the adjustment for $PM_{2.5}$ or BC. It remains difficult to disentangle the independent effect of NO_2 from the potential effect of the traffic-related pollution mixture or other components of that mixture. Further, as described above, there is large uncertainty whether long-term NO_2 exposure has an independent effect on the cardiovascular and respiratory morbidity outcomes that are major underlying causes of mortality. Thus, it is not clear by what biological pathways NO_2 exposure could lead to mortality. In conclusion, the generally positive epidemiologic evidence with uncertainty regarding an independent NO_2 effect is "suggestive of, but not sufficient to infer, a causal relationship" between long-term exposure to NO_2 and total mortality (U.S. EPA, 2016, 6.5.3).

Cancer

The evidence evaluated in the 2008 ISA was judged "inadequate to infer the presence or absence of a causal relationship" (U.S. EPA, 2008c) based on a few epidemiologic studies indicating associations between long-term NO_2 exposure and lung cancer incidence but lack of toxicological evidence demonstrating that NO_2 induces tumors. In the current review, the integration of recent and older studies on long-term NO_2 exposure and cancer yielded an evidence base judged "suggestive of, but not sufficient to infer, a causal relationship" (U.S. EPA, 2016, Section 6.6.9). This conclusion is based primarily on recent epidemiologic evidence, some of which shows NO_2-associated lung cancer incidence and mortality but does not address confounding by traffic-related copollutants, and is also based on previous toxicological evidence that implicates NO_2 in tumor promotion (U.S. EPA, 2016, Section 6.6.9).

3.3.2 Consideration of NO₂ Concentrations: Health Effects of Long-Term NO₂ Exposures

In evaluating the adequacy of the current NO_2 standards to protect against long-term NO_2 exposures, we consider the following question:

- **To what extent does the evidence support the occurrence of NO₂-attributable asthma development in children at NO₂ concentrations below the existing standards?**

To address this question, we consider (1) the extent to which epidemiologic studies indicate associations between long-term NO_2 exposures and asthma development for distributions of ambient NO_2 concentrations that would likely have met the existing standards and (2) the extent to which effects related to asthma development have been reported following the range of NO_2 exposure concentrations examined in experimental studies. Each of these is discussed below.

3.3.2.1 Ambient NO₂ Concentrations in Locations of Epidemiologic Studies

As discussed above for short-term exposures (Section 3.2.2.2), when considering epidemiologic studies within the context of the adequacy of the current NO_2 standards, we emphasize studies conducted in the U.S. and Canada.[80] We consider the extent to which these studies report positive and relatively precise associations with long-term NO_2 exposures, and the extent to which important uncertainties could impact our emphasis on particular studies. For the studies with potential to inform our conclusions on adequacy, we also evaluate available air quality information in study locations, focusing on DVs over the course of study periods.

In first considering the availability of studies that could inform our conclusions on adequacy, we focus the following specific questions:

---

[80] As indicated in Section 3.2.2.2, studies from the U.S. and Canada are likely to reflect air quality and exposure patterns that are generally applicable to the U.S. In addition, air quality data corresponding to study locations and study time periods is often readily available for studies conducted in the U.S. and Canada. Nonetheless, we recognize the importance of all studies, including other international studies, in the ISA's assessment of the weight of the evidence that informs causal determinations.

- **To what extent do U.S. and Canadian epidemiologic studies report positive, and relatively precise, associations between long-term NO₂ exposures and asthma development? What are the important uncertainties in these studies?**

The epidemiologic studies available in the current review that evaluate associations between long-term NO_2 exposures and asthma incidence are summarized in Table 6-1 of the ISA (U.S. EPA, 2016, pp. 6-7). There are six longitudinal epidemiologic studies conducted in the U.S. and Canada that vary in terms of the populations examined and methods used. Of the six studies, the ISA identifies three as key studies supporting the causal determination (Carlsten et al., 2011; Clougherty et al., 2007; Jerrett et al., 2008). The other three studies, not identified as key studies in the ISA causality determination, had a greater degree of uncertainty inherent in their characterizations of NO_2 exposures (Clark et al., 2010; McConnell et al., 2010, Nishimura et al., 2013). In evaluating the adequacy of the current NO_2 standards, we place the greatest emphasis on the three U.S. and Canadian studies identified in the ISA as providing key supporting evidence for the causal determination. However, we also consider what the additional three U.S. and Canadian studies can tell us about the adequacy of the current standards, while noting the increased uncertainty in these studies.

Effect estimates in U.S. and Canadian studies are generally positive and, in some cases, statistically significant and relatively precise (U.S. EPA, 2016, Table 6-1; Figure 3-2, below). However, there are important uncertainties in this body of evidence for asthma incidence, limiting the extent to which these studies can inform our consideration of the adequacy of the current NO_2 standards to protect against long-term NO_2 exposures. For example, there is uncertainty in the degree to which reported associations are specific to NO_2, rather than reflecting associations with another traffic-related copollutant or the broader mix of pollutants. Overall, the potential for copollutant confounding has not been well studied in this body of evidence. As described above (Section 3.3.1), the ISA concludes that "[e]pidemiologic studies of asthma development in children have not clearly characterized potential confounding by $PM_{2.5}$ or traffic-related pollutants [e.g., CO, BC/EC, volatile organic compounds (VOCs)]" (U.S. EPA, 2016, p. 6-64). The ISA further notes that "[i]n the longitudinal studies, correlations with $PM_{2.5}$ and BC were often high (e.g., $r = 0.7-0.96$), and no studies of asthma incidence evaluated copollutant models to address copollutant confounding, making it difficult to evaluate the independent effect of NO_2" (U.S. EPA, 2016, p. 6-64). Of the U.S. and Canadian studies, Carlsten et al. (2011) reported correlations between NO_2 and traffic-related pollutants (0.7 for $PM_{2.5}$, 0.5 for BC based on land use regression). Other U.S. and Canadian studies did not report quantitative results, but generally reported "moderate" to "high" correlations between NO_2 and other pollutants (U.S. EPA, 2016, Table 6-1). Given the relatively high correlations for NO_2 with co-occurring pollutants, study authors often interpreted associations with NO_2 as reflecting associations with traffic-related pollution more broadly (e.g., Jerrett et al., 2008; McConnell et al., 2010).[81]

Another important uncertainty is the potential for exposure measurement error in these epidemiologic studies. The ISA states that "a key issue in evaluating the strength of inference about NO_2-related asthma development from epidemiologic studies is the extent to which the NO_2 exposure assessment method used in a study captured the variability in exposure among study subjects" (U.S. EPA, 2016, pp. 6-16). We note that the ISA conclusion of a "likely to be a causal relationship" is based on the total body of evidence, with the strongest basis for inferring associations of NO_2 with asthma incidence coming from studies that "estimated residential NO_2 from LUR models that were demonstrated to predict well the variability in NO_2 in study locations or examined NO_2 measured at locations 1-2 km of subjects' school or home" (U.S. EPA, 2016, pp. 6-21). The studies that meet this criterion were mostly conducted outside of the U.S. or Canada, with the exception of Carlsten et al. (2011), which used a LUR model with good predictive capacity. The other U.S. and Canadian studies employed LUR models with unknown validation or central-site measurements that have well-recognized limitations in reflecting variability in ambient NO_2 concentrations in a community and may not well represent variability in NO_2 exposure among subjects. Thus, the extent to which these U.S. and Canadian studies provide reliable estimates of asthma incidence for particular NO_2 concentrations is unclear.

Overall, in revisiting the question posed above, we note that U.S. and Canadian epidemiologic studies report positive, and in some cases relatively precise, associations between long-term NO_2 exposure and asthma incidence in children. While it is appropriate to consider what these studies can tell us with regard to the adequacy of the existing NO_2 standards (see below), the emphasis that we place on these considerations will reflect important uncertainties related to the potential for confounding by traffic-related copollutants and for exposure measurement error.

While keeping in mind these uncertainties, we next consider the ambient NO_2 concentrations present at monitoring sites in locations and time periods of key U.S. and Canadian epidemiologic studies. We specifically consider the following question:

---

[81] For example, McConnell et al. (2010) reported that "modeled exposures reflect the mixture of multiple pollutants from nearby traffic, and the high correlation of pollutants in the mixture precludes identifying the effect of any specific pollutant in the mixture" (p. 1023 in published article).

- **To what extent do U.S. and Canadian epidemiologic studies report associations with long-term NO₂ in locations likely to have met the current NO₂ standards?**

As discussed above (Section 3.2.2), addressing this question can provide important insights into the extent to which NO₂-health effect associations are present for distributions of ambient NO₂ concentrations that would be allowed by the current standards. The presence of such associations would support the potential for the current standards to allow the NO₂-associated asthma development indicated by epidemiologic studies. To the degree studies have not reported associations in locations meeting the current NO₂ standards, there is greater uncertainty regarding the potential for the development of asthma to result from the NO₂ exposures associated with air quality meeting those standards.

To evaluate this issue, we compare NO₂ DVs in study areas to the levels of the current NO₂ standards. In additional to comparing annual DVs to the level of the annual standard, support for consideration of 1-hour DVs comes from the ISA's integrated mode of action information describing the biological plausibility for development of asthma (Section 3.1, above). In particular, studies demonstrate the potential for repeated short-term NO₂ exposures to induce pulmonary inflammation and development of allergic responses. The ISA states that "findings for short-term NO₂ exposure support an effect on asthma development by describing a potential role for repeated exposures to lead to recurrent inflammation and allergic responses," which are "identified as key early events in the proposed mode of action for asthma development" (U.S. EPA, 2016, p. 6-66 and p. 6-64). More specifically, the ISA states the following (U.S. EPA, 2016, p. 4-64):

> The initiating events in the development of respiratory effects due to long-term NO₂ exposure are recurrent and/or chronic respiratory tract inflammation and oxidative stress. These are the driving factors for potential downstream key events, allergic sensitization, airway inflammation, and airway remodeling, that may lead to the endpoint [airway hyperresponsiveness]. The resulting outcome may be new asthma onset, which presents as an asthma exacerbation that leads to physician-diagnosed asthma.

Thus, when considering the protection provided by the current standards against NO₂-associated asthma development, we consider the combined protection afforded by the 1-hour and annual standards.[82]

---

[82] It is also the case that broad changes in NO₂ concentrations will affect both hourly and annual metrics. This is discussed in more detail in section 2.3.3 above, and in CASAC's letter to the Administrator (Diez Roux and Sheppard, 2017). Thus, as in the recent review of the O₃ NAAQS (80 FR 65292, October 26, 2015), it is appropriate

To inform our consideration of whether a study area's air quality could have met the current NO_2 standards during study periods, we have calculated DVs based on the NO_2 concentrations measured at existing monitors during the years over which the epidemiologic studies of long-term NO_2 exposures were conducted.[83] The DVs for the epidemiologic studies of asthma incidence conducted in the U.S. and Canada are presented below in Figure 3-2. Mean DVs represent the average DVs across study periods and maximum DVs represent the year (annual standard) or 3-year period (1-hour standard) with the highest DV during the study period. Study locations could have met the current standards for the respective study periods if all of the annual averages were at or below 53 ppb and all of the 3-year averages of the 98th percentiles of daily maximum 1-hour NO_2 concentrations were at or below 100 ppb.

In interpreting these comparisons of DVs with the NO_2 standards, we also consider uncertainty in the extent to which identified DVs represent the higher NO_2 concentrations likely to have been present near major roads during study periods (see Section 3.2.2). In particular, as discussed above for short-term exposures, study area DVs are based on NO_2 concentrations from the generally area-wide NO_2 monitors that were present during study periods. Calculated DVs could have been higher if the near-road monitors that are now required in major U.S. urban areas had been in place. On this issue, we note that the published scientific literature supports the occurrence of higher NO_2 concentrations near roadways and that recent air quality information from the new near-road NO_2 monitoring network generally indicates higher NO_2 concentrations at near-road monitoring sites than at non-near road monitors in the same CBSA (Section 2.3.2). In addition, mobile source NO_X emissions were substantially higher during the majority of study periods (1986-2006) than they are today (Section 2.1.2), and NO_2 concentration gradients around roadways were generally more pronounced during study periods than indicated by recent air quality information. Thus, even in cases where DVs during study periods are at or somewhat below the levels of current standards, it is not clear that study areas would have met the standards if the currently required near-road monitors had been in place.

---

to consider the extent to which a short-term standard could provide protection against longer-term pollutant exposures.

[83] As discussed above for short-term exposures, the "DVs" reported here are meant to approximate the values that are used when determining whether an area meets the primary NO_2 NAAQS (see Appendix A).

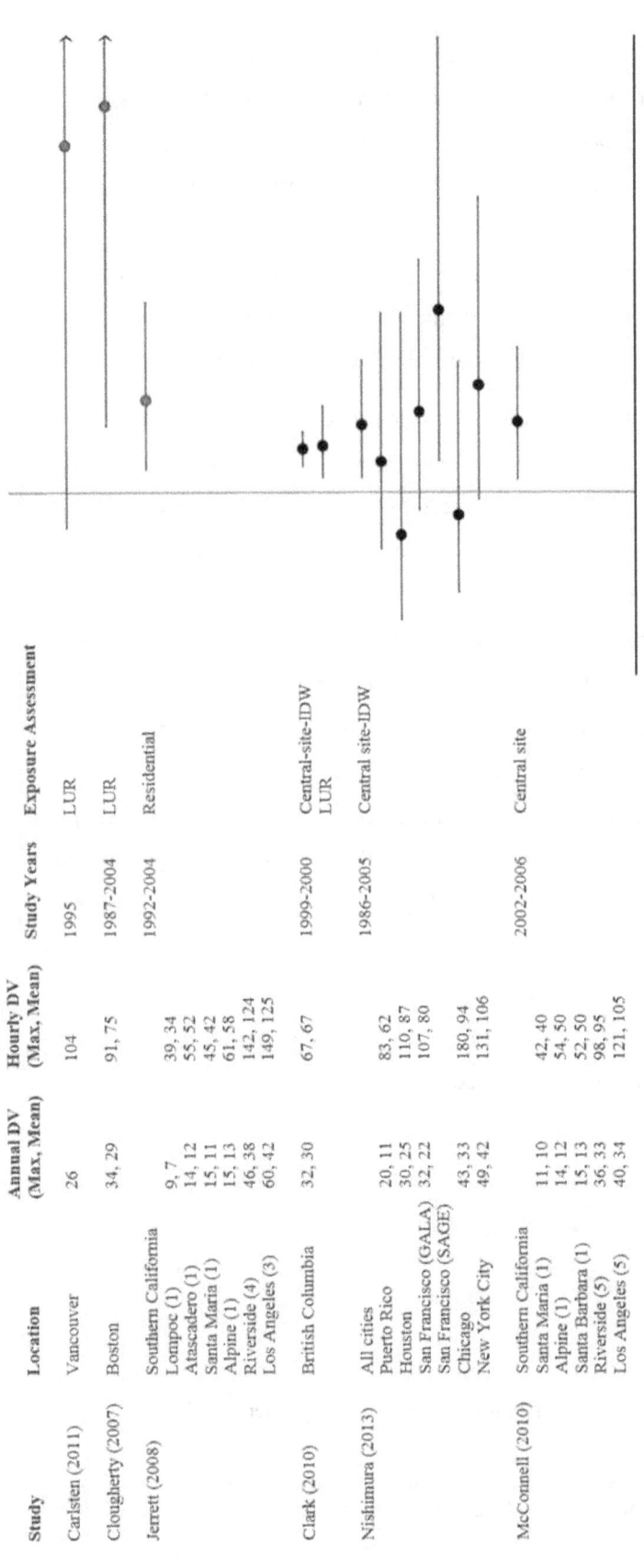

Study	Location	Annual DV (Max, Mean)	Hourly DV (Max, Mean)	Study Years	Exposure Assessment
Carlsten (2011)	Vancouver	26	104	1995	LUR
Clougherty (2007)	Boston	34, 29	91, 75	1987–2004	LUR
Jerrett (2008)	Southern California			1992–2004	Residential
	Lompoc (1)	9, 7	39, 34		
	Atascadero (1)	14, 12	55, 52		
	Santa Maria (1)	15, 11	45, 42		
	Alpine (1)	15, 13	61, 58		
	Riverside (4)	46, 38	142, 124		
	Los Angeles (3)	60, 42	149, 125		
Clark (2010)	British Columbia	32, 30	67, 67	1999–2000	Central-site-IDW LUR
Nishimura (2013)	All cities			1986–2005	Central site-IDW
	Puerto Rico	20, 11	83, 62		
	Houston	30, 25	110, 87		
	San Francisco (GALA)	32, 22	107, 80		
	San Francisco (SAGE)				
	Chicago	43, 33	180, 94		
	New York City	49, 42	131, 106		
McConnell (2010)	Southern California			2002–2006	Central site
	Santa Maria (1)	11, 10	42, 40		
	Alpine (1)	14, 12	54, 50		
	Santa Barbara (1)	15, 13	52, 50		
	Riverside (5)	36, 33	98, 95		
	Los Angeles (5)	40, 34	121, 105		

Risk or Odds Ratio

Figure 3-2. U.S. and Canadian epidemiologic studies of long-term NO₂ exposures and asthma incidence. Study locations and years are reported with annual and hourly DVs for studies of asthma incidence with risk/odds ratios and 95% confidence intervals standardized to 10 ppb increment in NO₂, as presented in the ISA (U.S. EPA, 2016). Effect estimates in red are for studies identified as key evidence in the ISA (U.S. EPA, 2016). IDW = inverse distance weighted; LUR = land use regression; GALA and SAGE are different study cohorts. For Carlsten et al. (2011), 1995 was the birth year, for which exposure was estimated. The cohort was followed to age 7. For Jerrett et al. (2008) and McConnell et al. (2010), parentheses indicate the number of communities represented by the DVs. For the Riverside and Los Angeles CBSAs, all monitors within the CBSAs were considered given the likelihood that they represent study communities within those CBSAs due to their close proximity, and DVs for the highest monitor are reported. Community-specific monitors were selected for the other communities as CBSA-wide monitors had wide spatial distribution and were not likely to represent the respective study locations (Lompoc, Atascadero, Santa Maria, Alpine, and Santa Barbara). For more details on DV calculation and study communities, see Appendix A.

In considering the studies in Figure 3-2, we first note the information from the key studies as identified in the ISA (Jerrett et al., 2008; Carlsten et al., 2011, Clougherty et al., 2007). Jerrett et al. (2008) reported positive and relatively precise associations with asthma incidence, based on analyses across several communities in Southern California.[84] Of the 11 study communities evaluated by Jerrett et al. (2008), most (i.e., seven) had maximum annual DVs that were near (i.e., 46 ppb for the four communities represented by the Riverside DVs) or above (i.e., 60 ppb for the three communities represented by the Los Angeles DVs) 53 ppb.[85] These seven communities also had 1-hour DVs (max and mean) that were well-above 100 ppb. The other key studies (i.e., Carlsten et al., 2011; Clougherty et al., 2007), conducted in single cities, reported positive but statistically imprecise associations. The annual DVs in locations of these studies during study years were below 53 ppb, but maximum 1-hour DVs were near (Clougherty)[86] or above (Carlsten) 100 ppb.[87]

We also consider the information from the other U.S. and Canadian studies available that, due to additional uncertainties, were not identified as key studies in the ISA (Clark et al., 2010; McConnell et al., 2010; Nishimura et al., 2013). The multi-city study by Nishimura et al. (2013) reports a positive and relatively precise association with asthma incidence, based on five U.S. cities and Puerto Rico (see "combined" estimate in Figure 3-2). Annual DVs in all study cities were below 53 ppb, while maximum 1-hour DVs were above 100 ppb in four of the five study cities (mean 1-hour DVs were also near or above 100 ppb in most study cities). Nishimura et al. (2013) also reported mixed results in city-specific effects estimates. McConnell et al. (2010) also conducted a multi-community study in Southern California and reported a positive and relatively precise association between asthma incidence and long-term NO_2 exposures based on central-site measurements. This study encompasses some of the same communities as Jerrett et al. (2008),

---

[84] The multi-community studies by Jerrett et al. (2008) and McConnell et al. (2010) did not include community-specific analyses.

[85] For the studies by Jerrett et al. (2008) and McConnell et al. (2010), the majority of communities were located within the Los Angeles and Riverside CBSAs. Because of this, and because community-specific NO_2 monitoring data were often not available in these areas (Appendix A), DVs for the Los Angeles and Riverside CBSAs were used to represent multiple study communities.

[86] As noted above, even in cases where DVs during study periods are at or somewhat below the levels of current standards, it is not clear that study areas would have met the standards if the currently required near-road monitors had been in place.

[87] As discussed above, DVs are different from the NO_2 concentrations reported in the studies themselves, which are often averaged across study areas and, in some cases, are based on methods other than ambient monitoring. The annual mean (SD) NO_2 concentrations reported in these studies, which are not appropriate for direct comparison to the current NO_2 NAAQS, are as follows: Jerrett et al., 2008: 9.6 (2.5)- 51.3 (4.4) ppb; Carlsten et al., 2011: 17.3 (13.1) ppb; Clougherty: 27.5 (4.3) ppb.

and while the annual DVs for these study years are more mixed, the 1-hour DVs representing 10 of 13 communities are near or above 100 ppb. Finally, Clark et al. (2010) reported a relatively precise and statistically significant association in a study conducted over a two-year period in British Columbia, with annual and hourly DVs of 32 ppb and 67 ppb, respectively. However, this result, as noted previously, was based on central-site NO_2 measurements that have well-recognized limitations in reflecting variability in ambient NO_2 concentrations in a community and variability in NO_2 exposure among subjects.[88]

Conclusions

Based on the information discussed above, we reach the conclusion that the available evidence from epidemiologic studies does not provide support for NO_2-associated asthma development in locations that would have clearly met the existing annual and 1-hour NO_2 standards. This conclusion stems from our consideration of the available evidence from U.S. and Canadian studies for NO_2-associated asthma incidence, our consideration of the ambient NO_2 concentrations present in study locations during study periods, and the uncertainties and limitations inherent in the evidence and in our analysis of study area DVs.

With regard to uncertainties in the evidence, we particularly note the potential for confounding by co-occurring pollutants, as described above, given the following: (1) the relatively high correlations observed between long-term concentrations of NO_2 and long-term concentrations of other roadway-associated pollutants; (2) the general lack of information from copollutant models on the potential for NO_2 associations that are independent of another traffic-related pollutant or mix of pollutants. This uncertainty limits what these studies can tell us with regard to the adequacy of the public health protection provided by the current NO_2 standards.

Even if we were to dismiss this fundamental uncertainty in the epidemiologic evidence, our analysis of study area DVs does not provide support for the occurrence of NO_2-associated asthma incidence in locations with ambient NO_2 concentrations clearly meeting the current NAAQS. In particular, for most of the study locations evaluated in the lone key U.S. multi-community study (Jerrett et al., 2008), 1-hour DVs were above 100 ppb and annual DVs were near or above 53 ppb. In addition, the two key single-city studies evaluated reported positive, but relatively imprecise, associations in locations with 1-hour DVs near (Clougherty et al., 2007 in Boston) or above (Carlston et al., 2011 in Vancouver) 100 ppb. Had currently required near-road monitors been in operation during study periods, DVs in U.S. study locations would likely have been higher. Other U.S. and Canadian studies evaluated were subject to greater uncertainties in

---

[88] Annual mean (SD) NO_2 concentrations reported in these studies are as follows: Clark et al., 2010: 16.3 (12.3) ppb; McConnell et al., 2010: 20.4 (8.7-23.6) ppb; Nishimura et al., 2013: 9.9 (2.9) – 32.1 (5.7) ppb.

the characterization of NO_2 exposures. Given these additional uncertainties, the degree to which these studies can inform our consideration of the adequacy of the current NO_2 NAAQS is limited.

3.3.2.2 NO_2 Concentrations in Long-Term Experimental Studies

In addition to the evidence from epidemiologic studies, we also consider evidence from experimental studies in animals and humans.[89] In assessing the evidence for respiratory morbidity related to long-term NO_2 exposures, we consider the following specific question regarding exposure concentrations in experimental studies:

- **To what extent do experimental studies demonstrate effects plausibly related to the development of asthma following exposures to NO_2 lower than previously observed or at concentrations below the levels of the existing standards?**

Experimental studies examining asthma-related effects attributable to long-term NO_2 exposures are largely limited to animals exposed to NO_2 concentrations well-above those found in the ambient air (i.e. $\geq 1,000$ ppb). As discussed above, the ISA indicates evidence from these animal studies supports the causal determination by characterizing "a potential mode of action linking NO_2 exposure with asthma development" (U.S. EPA, 2016, p. 1-20). In particular, there is limited evidence for airway responsiveness in guinea pigs with exposures to 1,000-4,000 ppb for 6-12 weeks. There is inconsistent evidence for pulmonary inflammation across all studies, though effects were reported following NO_2 exposures of 500-2,000 ppb for 12 weeks. Despite providing support for the "likely to be a causal" relationship, evidence from these experimental studies, by themselves, does not provide insight into the occurrence of adverse health effects following exposures below the levels of the existing NO_2 standards.[90]

Overall Conclusions

Taking all of the evidence and information together, including important uncertainties, we revisit the question posed at the beginning of this section:

- **To what extent does the evidence support the occurrence of NO_2-attributable asthma development in children at NO_2 concentrations below the existing standards?**

Based on the considerations discussed above, we reach the conclusion that the available evidence does not provide support for asthma development attributable to long-term exposures to NO_2 concentrations that would meet the existing annual and 1-hour NO_2 standards. This

---

[89] While there are not controlled human exposure studies for long-term exposures, we consider the extent to which evidence from short-term studies can provide support for effects observed in long-term studies.

[90] In addition, the ISA draws from short-term experimental evidence to support the biological plausibility of asthma development. Consideration of the NO_2 exposure concentrations evaluated in these studies is discussed in Section 3.3.2.

conclusion recognizes the NO_2 air quality relationships, which indicate that meeting the 1-hour NO_2 standard would be expected to limit annual NO_2 concentrations to well-below the level of the current annual standard (Section 2.3.3, above). This conclusion also recognizes the uncertainties in interpreting the epidemiologic evidence within the context of evaluating the existing standards due to the lack of near-road monitors during study periods and due to the potential for confounding by co-occurring pollutants. Thus, we conclude that epidemiologic studies of long-term NO_2 exposures and asthma development do not provide a clear basis for concluding that ambient NO_2 concentrations allowed by the current standards are independently (i.e., independent of co-occurring roadway pollutants) associated with the development of asthma. In addition, while experimental studies provide support for NO_2-attributable effects that are plausibly related to asthma development, the relatively high NO_2 exposure concentrations used in these studies do not provide insight into whether such effects would occur at NO_2 exposure concentrations that would be allowed by the current standards.

3.4   POTENTIAL PUBLIC HEALTH IMPLICATIONS

Evaluation of the public health protection provided against ambient NO_2 exposures requires consideration of populations and lifestages that may be at greater risk of experiencing NO_2-attributable health effects. In the last review, the 2008 ISA for Oxides of Nitrogen noted that a considerable fraction of the U.S. population lives, works, or attends school near major roadways, where ambient NO_2 concentrations are often elevated (U.S. EPA, 2008a, Section 4.3). Of this population, the 2008 ISA concluded that "those with physiological susceptibility will have even greater risks of health effects related to NO_2" (U.S. EPA, 2008, p. 4-12). With regard to susceptibility, the 2008 ISA concluded that "[p]ersons with preexisting respiratory disease, children, and older adults may be more susceptible to the effects of NO_2 exposure" (U.S. EPA, 2008, p. 4-12).

In the current review, the ISA again notes because of the large populations attending school, living, working, and commuting on or near roads, where ambient NO_2 concentrations can be higher than in many other locations (Section 2.5.3),[91] there is widespread potential for elevated ambient NO_2 exposures. For example, Rowangould et al. (2013) found that over 19% of the U.S. population lives within 100 m of roads with an annual average daily traffic (AADT) of 25,000 vehicles, and 1.3% lives near roads with AADT greater than 200,000. The proportion is much larger in certain parts of the country, mostly coinciding with urban areas. Among California residents, 40% live within 100 m of roads with AADT of 25,000 (Rowangould, 2013).

---

[91] The ISA specifically notes that a zone of elevated NO_2 concentrations typically extends 200 to 500 m from roads with heavy traffic (U.S. EPA, 2016, Section 2.5.3).

In addition, 7% of U.S. schools serving a total of 3,152,000 school children are located within 100 m of a major roadway, and 15% of U.S. schools serving a total of 6,357,000 school children are located within 250 m of a major roadway (Kingsley et al., 2014). Thus, as in the last review, the available information indicates that large proportions of the U.S. population potentially have elevated NO_2 exposures as a result of living, working, attending school, or commuting on or near roadways.

The impacts of exposures to elevated NO_2 concentrations, such as those that can occur around roadways, are of particular concern for populations at increased risk of experiencing adverse effects. In the current review, our consideration of potential at-risk populations draws from the 2016 ISA's assessment of the evidence (U.S. EPA, 2016, Chapter 7). The ISA uses a systematic approach to evaluate factors that may increase risks in a particular population or during a particular lifestage, noting that increased risk could be due to "intrinsic or extrinsic factors, differences in internal dose, or differences in exposure" (U.S. EPA, 2016, p. 7-1).

The ISA evaluates the evidence for a number of potential at-risk factors, including pre-existing diseases like asthma (U.S. EPA, 2016, Section 7.3), genetic factors (U.S. EPA, 2016, Section 7.4), sociodemographic factors (U.S. EPA, 2016, Section 7.5), and behavioral and other factors (U.S. EPA, 2016, Section 7.6). The ISA then uses a systematic approach for classifying the evidence for each potential at-risk factor (U.S. EPA, 2015, Preamble, Section 6.a, Table III). The categories considered are "adequate evidence," "suggestive evidence," "inadequate evidence," and "evidence of no effect" (U.S. EPA, 2016, Table 7-1). Consistent with other recent NAAQS reviews (e.g., 80 FR 65292, October 26, 2015), we focus our consideration of potential at-risk populations on those factors for which the ISA determines there is "adequate" evidence (U.S. EPA, 2016, Table 7-27). In the case of NO_2, this includes people with asthma, children and older adults (U.S. EPA, 2016, Table 7-27), based primarily on evidence for asthma exacerbation or asthma development as evidence for other health effects is more uncertain.

Our consideration of the evidence supporting these at-risk populations specifically focuses on the following question:

- **To what extent does the currently available scientific evidence expand our understanding of populations and/or lifestages that may be at greater risk for NO_2-related health effects?**

In addressing this question, we consider the evidence for effects in people with asthma (Section 3.4.1), children (Section 3.4.2), and older adults (Section 3.4.3) (U.S. EPA, 2016, Chapter 7, Table 7-27). Section 3.4.4 presents our overall conclusions regarding the populations at increased risk of NO_2-related effects.

3.4.1 People with Asthma

Approximately 8.0% of adults and 9.3% of children (age <18 years) in the U.S. currently have asthma (Blackwell et al., 2014; Bloom et al., 2013), and it is the leading chronic illness affecting children (U.S. EPA, 2016, Section 7.3.1). Individuals with pre-existing diseases like asthma may be at greater risk for some air pollution-related health effects if they are in a compromised biological state. The 2008 ISA for Oxides of Nitrogen (U.S. EPA, 2008) concluded that those with pre-existing pulmonary conditions, especially asthma, were likely to be at greater risk for NO_2-related respiratory effects.

As in the last review, controlled human exposure studies demonstrating NO_2-induced increases in AR provide key evidence that people with asthma are more sensitive than people without asthma to the effects of short-term NO_2 exposures. In particular, a meta-analysis conducted by Folinsbee et al. (1992) demonstrates that NO_2 exposures from 100 to 300 ppb increased AR in the majority of adults with asthma, while AR in adults without asthma was increased only for NO_2 exposure concentrations greater than 1,000 ppb (U.S. EPA, 2016, Section 7.3.1). Brown (2015) showed that following resting exposures to NO_2 concentrations in the range of 100 to 530 ppb, about a quarter of individuals with asthma experience clinically relevant increases in AR to non-specific bronchial challenge. Results of epidemiologic studies are less clear regarding potential differences between populations with and without asthma (U.S. EPA, 2016, Section 7.3.1). Additionally, studies of activity patterns do not clearly indicate difference in time spent outdoors to suggest differences in NO_2 exposure. However, the meta-analysis of information from controlled human exposure studies clearly demonstrates increased sensitivity of adults with asthma compared to healthy adults.[92] Thus, consistent with observations made in the 2008 ISA (U.S. EPA, 2008a), in the current review the ISA determines that the "evidence is adequate to conclude that people with asthma are at increased risk for NO_2-related health effects" (U.S. EPA, 2016, p. 7-7).

3.4.2 Children

According to the 2010 census, 24% of the U.S. population is less than 18 years of age, with 6.5% less than age 6 years (Howden and Meyer, 2011). The National Human Activity Pattern Survey shows that children spend more time than adults outdoors (Klepeis et al., 1996), and a longitudinal study in California showed a larger proportion of children reported spending time engaged in moderate or vigorous outdoor physical activity (Wu et al., 2011b). In addition, children have a higher propensity than adults for oronasal breathing (U.S. EPA, 2016, Section 4.2.2.3) and the human respiratory system is not fully developed until 18–20 years of age (U.S.

---

[92] Though, as discussed above (Section 3.2), there is uncertainty in the extent to which increases in AR following exposures to NO_2 concentrations near those found in the ambient air (i.e., around 100 ppb) would be adverse.

EPA, 2016, Section 7.5.1). All of these factors could contribute to children being at higher risk than adults for effects attributable to ambient NO_2 exposures (U.S. EPA, 2016, Section 7.5.1.1).

Epidemiologic evidence across diverse locations (U.S., Canada, Europe, Asia, Australia) consistently demonstrates adverse effects of both short- and long-term NO_2 exposures in children. In particular, short-term increases in ambient NO_2 concentrations are consistently associated with larger increases in asthma-related hospital admissions, ED visits or outpatient visits in children than in adults (U.S. EPA, 2016, Section 7.5.1.1, Table 7-13). In general, these results indicate NO_2-associated impacts that are 1.8 to 3.4-fold larger in children (Son et al., 2013; Ko et al., 2007; Atkinson et al., 1999; Anderson et al., 1998). In addition, asthma development in children has been reported to be associated with long-term NO_2 exposures, based on exposure periods spanning infancy to adolescence (U.S. EPA, 2016, Section 6.2.2.1). Given the consistent epidemiologic evidence for associations between ambient NO_2 and asthma-related outcomes, including the larger associations with short-term exposures observed in children, the ISA concludes the evidence "is adequate to conclude that children are at increased risk for NO_2-related health effects" (U.S. EPA, 2016, p. 7-32).

3.4.3 Older adults

According to the 2012 National Population Projections issued by the U.S. Census Bureau, 13% of the U.S. population was age 65 years or older in 2010, and by 2030, this fraction is estimated to grow to 20% (Ortman et al., 2014). The 2008 ISA (U.S. EPA, 2008) indicated that older adults may be at increased risk for NO_2-related respiratory effects and mortality, and recent epidemiologic findings add to this body of evidence (US EPA, 2016 Table 7-15). While it is not clear that older adults experience greater NO_2 exposures or doses, epidemiologic evidence generally indicates greater risk of NO_2-related health effects in older adults compared with younger adults. For example, comparisons of older and younger adults with respect to NO_2-related asthma exacerbation generally show larger (one to threefold) effects in adults ages 65 years or older than among individuals ages 15−64 years or 15−65 years (Ko et al., 2007; Villeneuve et al., 2007; Migliaretti et al., 2005; Anderson et al., 1998). Results for all respiratory hospital admissions combined also tend to show larger associations with NO_2 among older adults ages 65 years or older (Arbex et al., 2009; Wong et al., 2009; Hinwood et al., 2006; Atkinson et al., 1999). The ISA determined that, overall, the consistent epidemiologic evidence for asthma-related hospital admissions and ED visits "is adequate to conclude that older adults are at increased risk for NO_2-related health effects" (U.S. EPA, 2016, p. 7-37).

3.4.4 Conclusions

Consistent with the last review, the ISA determined that the available evidence is adequate to conclude that people with asthma, children, and older adults are at increased risk for

NO$_2$-related health effects. The large proportions of the U.S. population that encompass each of these groups and lifestages (i.e., 8% adults and 9.3% children with asthma, 24% children, 13% older adults) underscores the potential for important public health impacts attributable to NO$_2$ exposures. These impacts are of particular concern for members of these populations and lifestages who live, work, attend school or otherwise spend a large amount of time in locations of elevated ambient NO$_2$, including near heavily trafficked roadways.

3.5 REFERENCES

Ahmed, T; Dougherty, R; Sackner, MA. (1983a). Effect of 0.1 ppm NO2 on pulmonary functions and non-specific bronchial reactivity of normals and asthmatics [final report]. (CR-83/11/BI). Warren, MI: General Motors Research Laboratories.

Ahmed, T; Dougherty, R; Sackner, MA. (1983b). Effect of NO2 exposure on specific bronchial reactivity in subjects with allergic bronchial asthma [final report]. (CR-83/07/BI). Warren, MI: General Motors Research Laboratories.

Anderson, HR; Ponce de Leon, A; Bland, JM; Bower, JS; Emberlin, J; Strachen, DP. (1998). Air pollution, pollens, and daily admissions for asthma in London 1987-92. Thorax 53: 842-848. http://dx.doi.org/10.1136/thx.53.10.842

Arbex, MA; de Souza Conceição, GM; Cendon, SP; Arbex, FF; Lopes, AC; Moysés, EP; Santiago, SL; Saldiva, PHN; Pereira, LAA; Braga, ALF. (2009). Urban air pollution and chronic obstructive pulmonary disease-related emergency department visits. J Epidemiol Community Health 63: 777-783. http://dx.doi.org/10.1136/jech.2008.078360

Atkinson, RW; Anderson, HR; Strachan, DP; Bland, JM; Bremner, SA; Ponce de Leon, A. (1999a). Short-term associations between outdoor air pollution and visits to accident and emergency departments in London for respiratory complaints. Eur Respir J 13: 257-265.

ATSDR (Agency for Toxic Substances and Disease Registry). (2006). A study of ambient air contaminants and asthma in New York City: Part A and B. Atlanta, GA: U.S. Department of Health and Human Services. http://permanent.access.gpo.gov/lps88357/ASTHMA_BRONX_FINAL_REPORT.pdf

Avol, EL; Linn, WS; Peng, RC; Valencia, G; Little, D; Hackney, JD. (1988). Laboratory study of asthmatic volunteers exposed to nitrogen dioxide and to ambient air pollution. Am Ind Hyg Assoc J 49: 143-149. http://dx.doi.org/10.1080/15298668891379530

Avol, EL; Linn, WS; Peng, RC; Whynot, JD; Shamoo, DA; Little, DE; Smith, MN; Hackney, JD. (1989). Experimental exposures of young asthmatic volunteers to 0.3 ppm nitrogen dioxide and to ambient air pollution. Toxicol Ind Health 5: 1025-1034.

Barck, C; Sandstrom, T; Lundahl, J; Hallden, G; Svartengren, M; Strand, V; Rak, S; Bylin, G. (2002). Ambient level of NO2 augments the inflammatory response to inhaled allergen in asthmatics. Respir Med 96: 907-917. http://dx.doi.org/10.1053/rmed.2002.1374

Bauer, MA; Utell, MJ; Morrow, PE; Speers, DM; Gibb, FR. (1986). Inhalation of 0.30 ppm nitrogen dioxide potentiates exercise-induced bronchospasm in asthmatics. Am Rev Respir Dis 134: 1203-1208.

Blackwell, DL; Lucas, JW; Clarke, TC. (2014). Summary health statistics for U.S. adults: National health interview survey, 2012. In Vital and health statistics. Hyattsville, MD: National Center for Health Statistics, U.S Department of Health and Human Services. http://www.cdc.gov/nchs/data/series/sr_10/sr10_260.pdf

Bloom, B; Jones, LI; Freeman, G. (2013). Summary health statistics for U.S. children: National health interview survey, 2012. In Vital and health statistics. Hyattsville, MD: National Center for Health Statistics, U.S Department of Health and Human Services. http://www.cdc.gov/nchs/data/series/sr_10/sr10_258.pdf

Brown, JS. (2015). Nitrogen dioxide exposure and airway responsiveness in individuals with asthma. Inhal Toxicol 27: 1-14. http://dx.doi.org/10.3109/08958378.2014.979960

Burnett, RT; Smith-Doiron, M; Stieb, D; Cakmak, S; Brook, JR. (1999). Effects of particulate and gaseous air pollution on cardiorespiratory hospitalizations. Arch Environ Health 54: 130-139. http://dx.doi.org/10.1080/00039899909602248

Bylin, G; Hedenstierna, G; Lindvall, T; Sundin, B. (1988). Ambient nitrogen dioxide concentrations increase bronchial responsiveness in subjects with mild asthma. Eur Respir J 1: 606-612.

Carlsten, C; Dybuncio, A; Becker, A; Chan-Yeung, M; Brauer, M. (2011). Traffic-related air pollution and incident asthma in a high-risk birth cohort. Occup Environ Med 68: 291-295. http://dx.doi.org/10.1136/oem.2010.055152

Clark, NA; Demers, PA; Karr, CJ; Koehoorn, M; Lencar, C; Tamburic, L; Brauer, M. (2010). Effect of early life exposure to air pollution on development of childhood asthma. Environ Health Perspect 118: 284-290. http://dx.doi.org/10.1289/ehp.0900916

Clougherty, JE; Levy, JI; Kubzansky, LD; Ryan, PB; Suglia, SF; Canner, MJ; Wright, RJ. (2007). Synergistic effects of traffic-related air pollution and exposure to violence on urban asthma etiology. Environ Health Perspect 115: 1140-1146. http://dx.doi.org/10.1289/ehp.9863

Dawson, SV; Schenker, MB. (1979). Health effects of inhalation of ambient concentrations of nitrogen dioxide [Editorial]. Am Rev Respir Dis 120: 281-292.

Folinsbee, LJ. (1992). Does nitrogen dioxide exposure increase airways responsiveness? Toxicol Ind Health 8: 273-283.

Gehring, U; Gruzieva, O; Agius, RM; Beelen, R; Custovic, A; Cyrys, J; Eeftens, M; Flexeder, C; Fuertes, E; Heinrich, J; Hoffmann, B; de Jongste, JC; Kerkhof, M; Klümper, C; Korek, M; Mölter, A; Schultz, ES; Simpson, A; Sugiri, D; Svartengren, M; von Berg, A; Wijga, AH; Pershagen, G; Brunekreef, B. (2013). Air pollution exposure and lung function in children: the ESCAPE project. Environ Health Perspect 121: 1357-1364. http://dx.doi.org/10.1289/ehp.1306770

Goodman, JE; Chandalia, JK; Thakali, S; Seeley, M. (2009). Meta-analysis of nitrogen dioxide exposure and airway hyper-responsiveness in asthmatics. Crit Rev Toxicol 39: 719-742. http://dx.doi.org/10.3109/10408440903283641

Hazucha, MJ; Ginsberg, JF; McDonnell, WF; Haak, ED, Jr; Pimmel, RL; Salaam, SA; House, DE; Bromberg, PA. (1983). Effects of 0.1 ppm nitrogen dioxide on airways of normal and asthmatic subjects. J Appl Physiol Respir Environ Exerc Physiol 54: 730-739.

Hinwood, AL; De Klerk, N; Rodriguez, C; Jacoby, P; Runnion, T; Rye, P; Landau, L; Murray, F; Feldwick, M; Spickett, J. (2006). The relationship between changes in daily air pollution and hospitalizations in Perth, Australia 1992-1998: A case-crossover study. Int J Environ Health Res 16: 27-46. http://dx.doi.org/10.1080/09603120500397680

Howden, LM; Meyer, JA. (2011). Age and sex composition: 2010. (2010 Census Briefs, C2010BR-03). Washington, DC: U.S. Department of Commerce, Economics and Statistics Administration, U.S. Census Bureau. http://www.census.gov/prod/cen2010/briefs/c2010br-03.pdf

Ito, K; Mathes, R; Ross, Z; Nádas, A; Thurston, G; Matte, T. (2011). Fine particulate matter constituents associated with cardiovascular hospitalizations and mortality in New York City. Environ Health Perspect 119: 467-473. http://dx.doi.org/10.1289/ehp.1002667

Jaffe, DH; Singer, ME; Rimm, AA. (2003). Air pollution and emergency department visits for asthma among Ohio Medicaid recipients, 1991-1996. Environ Res 91: 21-28. http://dx.doi.org/10.1016/S0013-9351(02)00004-X

Jenkins, HS; Devalia, JL; Mister, RL; Bevan, AM; Rusznak, C; Davies, RJ. (1999). The effect of exposure to ozone and nitrogen dioxide on the airway response of atopic asthmatics to inhaled allergen: Dose- and time-dependent effects. Am J Respir Crit Care Med 160: 33-39. http://dx.doi.org/10.1164/ajrccm.160.1.9808119

Jerrett, M; Shankardass, K; Berhane, K; Gauderman, WJ; Künzli, N; Avol, E; Gilliland, F; Lurmann, F; Molitor, JN; Molitor, JT; Thomas, DC; Peters, J; McConnell, R. (2008). Traffic-related air pollution and asthma onset in children: A prospective cohort study with individual exposure measurement. Environ Health Perspect 116: 1433-1438. http://dx.doi.org/10.1289/ehp.10968

Jörres, R; Magnussen, H. (1990). Airways response of asthmatics after a 30 min exposure, at resting ventilation, to 0.25 ppm NO2 or 0.5 ppm SO2. Eur Respir J 3: 132-137.

Jörres, R; Magnussen, H. (1991). Effect of 0.25 ppm nitrogen dioxide on the airway response to methacholine in asymptomatic asthmatic patients. Lung 169: 77-85. http://dx.doi.org/10.1007/BF02714145

Kingsley, SL; Eliot, MN; Carlson, L; Finn, J; Macintosh, DL; Suh, HH; Wellenius, GA. (2014). Proximity of US schools to major roadways: a nationwide assessment. J Expo Sci Environ Epidemiol 24: 253-259. http://dx.doi.org/10.1038/jes.2014.5

Kleinman, MT; Bailey, RM; Linn, WS; Anderson, KR; Whynot, JD; Shamoo, DA; Hackney, JD. (1983). Effects of 0.2 ppm nitrogen dioxide on pulmonary function and response to bronchoprovocation in asthmatics. J Toxicol Environ Health 12: 815-826. http://dx.doi.org/10.1080/15287398309530472

Klepeis, NE; Tsang, AM; Behar, JV. (1996). Analysis of the national human activity pattern survey (NHAPS) respondents from a standpoint of exposure assessment [EPA Report]. (EPA/600/R-96/074). Washington, DC: U.S. Environmental Protection Agency. http://exposurescience.org/pub/reports/NHAPS_Report1.pdf#....Local SettingsTemporary Internet FilesContent.Outlook3JQ221FPB_Approaches_Population_Tables.docx

Ko, FWS; Tam, W; Wong, TW; Lai, CKW; Wong, GWK; Leung, TF; Ng, SSS; Hui, DSC. (2007). Effects of air pollution on asthma hospitalization rates in different age groups in Hong Kong. Clin Exp Allergy 37: 1312-1319. http://dx.doi.org/10.1111/j.1365-2222.2007.02791.x

Li, S; Batterman, S; Wasilevich, E; Wahl, R; Wirth, J; Su, FC; Mukherjee, B. (2011). Association of daily asthma emergency department visits and hospital admissions with ambient air pollutants among the pediatric Medicaid population in Detroit: Time-series and time-stratified case-crossover analyses with threshold effects. Environ Res 111: 1137-1147. http://dx.doi.org/10.1016/j.envres.2011.06.002

Linn, WS; Szlachcic, Y; Gong, H, Jr; Kinney, PL; Berhane, KT. (2000). Air pollution and daily hospital admissions in metropolitan Los Angeles. Environ Health Perspect 108: 427-434.

MacIntyre, EA; Gehring, U; Mölter, A; Fuertes, E; Klümper, C; Krämer, U; Quass, U; Hoffmann, B; Gascon, M; Brunekreef, B; Koppelman, GH; Beelen, R; Hoek, G; Birk, M; de Jongste, JC; Smit, HA; Cyrys, J; Gruzieva, O; Korek, M; Bergström, A; Agius, RM; de Vocht, F; Simpson, A; Porta, D; Forastiere, F; Badaloni, C; Cesaroni, G; Esplugues, A; Fernández-Somoano, A; Lerxundi, A; Sunyer, J; Cirach, M; Nieuwenhuijsen, MJ; Pershagen, G; Heinrich, J. (2014). Air pollution and respiratory infections during early childhood: an analysis of 10 European birth cohorts within the ESCAPE Project. Environ Health Perspect 122: 107-113. http://dx.doi.org/10.1289/ehp.1306755

McConnell, R; Islam, T; Shankardass, K; Jerrett, M; Lurmann, F; Gilliland, F; Gauderman, J; Avol, E; Künzli, N; Yao, L; Peters, J; Berhane, K. (2010). Childhood incident asthma and traffic-related air pollution at home and school. Environ Health Perspect 118: 1021-1026. http://dx.doi.org/10.1289/ehp.0901232

Migliaretti, G; Cadum, E; Migliore, E; Cavallo, F. (2005). Traffic air pollution and hospital admission for asthma: a case-control approach in a Turin (Italy) population. Int Arch Occup Environ Health 78: 164-169. http://dx.doi.org/10.1007/s00420-004-0569-3

Mohsenin, V. (1987). Airway responses to nitrogen dioxide in asthmatic subjects. J Toxicol Environ Health 22: 371-380. http://dx.doi.org/10.1080/15287398709531080

Nishimura, KK; Galanter, JM; Roth, LA; Oh, SS; Thakur, N; Nguyen, EA; Thyne, S; Farber, HJ; Serebrisky, D; Kumar, R; Brigino-Buenaventura, E; Davis, A; LeNoir, MA; Meade, K; Rodriguez-Cintron, W; Avila, PC; Borrell, LN; Bibbins-Domingo, K; Rodriguez-Santana, JR; Sen, S; Lurmann, F; Balmes, JR; Burchard, EG. (2013). Early-life air pollution and asthma risk in minority children: The GALA II and SAGE II studies. Am J Respir Crit Care Med 188: 309-318. http://dx.doi.org/10.1164/rccm.201302-0264OC

Orehek, J; Massari, JP; Gayrard, P; Grimaud, C; Charpin, J. (1976). Effect of short-term, low-level nitrogen dioxide exposure on bronchial sensitivity of asthmatic patients. J Clin Invest 57: 301-307. http://dx.doi.org/10.1172/JCI108281

Ortman, JM; Velkoff, VA; Hogan, H. (2014). An aging nation: The older population in the United States (pp. 1-28). (P25-1140). United States Census Bureau. http://www.census.gov/library/publications/2014/demo/p25-1140.html

Peel, JL; Tolbert, PE; Klein, M; Metzger, KB; Flanders, WD; Todd, K; Mulholland, JA; Ryan, PB; Frumkin, H. (2005). Ambient air pollution and respiratory emergency department visits. Epidemiology 16: 164-174. http://dx.doi.org/10.1097/01.ede.0000152905.42113.db

Reddel, HK; Taylor, DR; Bateman, ED; Boulet, LP; Boushey, HA; Busse, WW; Casale, TB; Chanez, P; Enright, PL; Gibson, PG; de Jongste, JC; Kerstjens, HA; Lazarus, SC; Levy, ML; O'Byrne, PM; Partridge, MR; Pavord, ID; Sears, MR; Sterk, PJ; Stoloff, SW; Sullivan, SD; Szefler, SJ; Thomas, MD; Wenzel, SE.

(2009). An official American Thoracic Society/European Respiratory Society statement: Asthma control and exacerbations: Standardizing endpoints for clinical asthma trials and clinical practice. Am J Respir Crit Care Med 180: 59-99. http://dx.doi.org/10.1164/rccm.200801-060ST

Riedl, MA; Diaz-Sanchez, D; Linn, WS; Gong, H, Jr; Clark, KW; Effros, RM; Miller, JW; Cocker, DR; Berhane, KT. (2012). Allergic inflammation in the human lower respiratory tract affected by exposure to diesel exhaust [HEI] (pp. 5-43; discussion 45-64). (ISSN 1041-5505 Research Report 165). Boston, MA: Health Effects Institute. http://pubs.healtheffects.org/view.php?id=373

Roger, LJ; Horstman, DH; McDonnell, W; Kehrl, H; Ives, PJ; Seal, E; Chapman, R; Massaro, E. (1990). Pulmonary function, airway responsiveness, and respiratory symptoms in asthmatics following exercise in NO2. Toxicol Ind Health 6: 155-171. http://dx.doi.org/10.1177/074823379000600110

Rowangould, GM. (2013). A census of the US near-roadway population: Public health and environmental justice considerations. Transport Res Transport Environ 25: 59-67. http://dx.doi.org/10.1016/j.trd.2013.08.003

Rubinstein, I; Bigby, BG; Reiss, TF; Boushey, HA, Jr. (1990). Short-term exposure to 0.3 ppm nitrogen dioxide does not potentiate airway responsiveness to sulfur dioxide in asthmatic subjects. Am Rev Respir Dis 141: 381-385. http://dx.doi.org/10.1164/ajrccm/141.2.381

Son, JY; Lee, JT; Park, YH; Bell, ML. (2013). Short-term effects of air pollution on hospital admissions in Korea. Epidemiology 24: 545-554. http://dx.doi.org/10.1097/EDE.0b013e3182953244

Stieb, DM; Szyszkowicz, M; Rowe, BH; Leech, JA. (2009). Air pollution and emergency department visits for cardiac and respiratory conditions: A multi-city time-series analysis. Environ Health 8. http://dx.doi.org/10.1186/1476-069X-8-25

Strand, V; Salomonsson, P; Lundahl, J; Bylin, G. (1996). Immediate and delayed effects of nitrogen dioxide exposure at an ambient level on bronchial responsiveness to histamine in subjects with asthma. Eur Respir J 9: 733-740. http://dx.doi.org/10.1183/09031936.96.09040733

Strand, V; Rak, S; Svartengren, M; Bylin, G. (1997). Nitrogen dioxide exposure enhances asthmatic reaction to inhaled allergen in subjects with asthma. Am J Respir Crit Care Med 155: 881-887. http://dx.doi.org/10.1164/ajrccm.155.3.9117021

Strand, V; Svartengren, M; Rak, S; Barck, C; Bylin, G. (1998). Repeated exposure to an ambient level of NO2 enhances asthmatic response to a nonsymptomatic allergen dose. Eur Respir J 12: 6-12. http://dx.doi.org/10.1183/09031936.98.12010006

Strickland, MJ; Darrow, LA; Klein, M; Flanders, WD; Sarnat, JA; Waller, LA; Sarnat, SE; Mulholland, JA; Tolbert, PE. (2010). Short-term associations between ambient air pollutants and pediatric asthma emergency department visits. Am J Respir Crit Care Med 182: 307-316. http://dx.doi.org/10.1164/rccm.200908-1201OC

Tunnicliffe, WS; Burge, PS; Ayres, JG. (1994). Effect of domestic concentrations of nitrogen dioxide on airway responses to inhaled allergen in asthmatic patients. Lancet 344: 1733-1736. http://dx.doi.org/10.1016/s0140-6736(94)92886-x

U.S. EPA (1993). Air Quality Criteria for Oxides of Nitrogen. Office of Health and Environmental Assessment, Environmental Criteria and Assessment Office. Research Triangle Park, NC. EPA–600/8–91–049aF–cF, August 1993. Available at: http://cfpub.epa.gov/ncea/cfm/recordisplay.cfm?deid=40179.

U.S. EPA (2008). Integrated Science Assessment for Oxides of Nitrogen – Health Criteria. U.S. EPA, National Center for Environmental Assessment, Research Triangle Park, NC. EPA/600/R-08/071. July 2008. Available at: http://cfpub.epa.gov/ncea/cfm/recordisplay.cfm?deid=194645.

U.S. EPA (2011). Policy Assessment for the Review of the Particulate Matter National Ambient Air Quality Standards. Office of Air Quality Planning and Standards, U.S. Environmental Protection Agency, Research Triangle Park, NC. EPA 452/R-11-003. April 2011. Available at: https://www3.epa.gov/ttn/naaqs/standards/pm/s_pm_2007_pa.html

U.S. EPA (2014). Policy Assessment for the Review of the Ozone National Ambient Air Quality Standards. Office of Air Quality Planning and Standards, U.S. Environmental Protection Agency, Research Triangle Park,

NC. EPA 452/R-14-006. August 2014. Available at:
https://www3.epa.gov/ttn/naaqs/standards/ozone/s_o3_2008_pa.html

U.S. EPA (2015). Preamble to the Integrated Science Assessments. U.S. EPA, Washington, DC, EPA/600/R-15/067. November 2015. Available at: https://cfpub.epa.gov/ncea/isa/recordisplay.cfm?deid=310244

U.S. EPA (2016). Integrated Science Assessment for Oxides of Nitrogen – Health Criteria (2016 Final Report). U.S. EPA, National Center for Environmental Assessment, Research Triangle Park, NC. EPA/600/R-15/068. January 2016. Available at: https://cfpub.epa.gov/ncea/isa/recordisplay.cfm?deid=310879.

Villeneuve, PJ; Chen, L; Rowe, BH; Coates, F. (2007). Outdoor air pollution and emergency department visits for asthma among children and adults: A case-crossover study in northern Alberta, Canada. Environ Health 6: 40. http://dx.doi.org/10.1186/1476-069X-6-40

Witten, A; Solomon, C; Abbritti, E; Arjomandi, M; Zhai, W; Kleinman, M; Balmes, J. (2005). Effects of nitrogen dioxide on allergic airway responses in subjects with asthma. J Occup Environ Med 47: 1250-1259. http://dx.doi.org/10.1097/01.jom.0000177081.62204.8d

Wong, CM; Yang, L; Thach, TQ; Chau, PY; Chan, KP; Thomas, GN; Lam, TH; Wong, TW; Hedley, AJ; Peiris, JS. (2009). Modification by influenza on health effects of air pollution in Hong Kong. Environ Health Perspect 117: 248-253. http://dx.doi.org/10.1289/ehp.11605

4 CONSIDERATION OF NO₂ AIR QUALITY-, EXPOSURE- AND RISK-BASED INFORMATION

Beyond our consideration of the scientific evidence, discussed above in Chapter 3, we also consider the extent to which quantitative analyses of NO₂ air quality, exposures or health risks could inform conclusions on the adequacy of the public health protection provided by the current primary NO₂ standards. Such quantitative analyses, if supported, could inform judgments about the public health impacts of NO₂-related health effects and could help to place the evidence for specific effects into a broader public health context. To this end, in the REA Planning Document (U.S. EPA, 2015) and in this PA, we have evaluated the potential support for conducting new or updated analyses of NO₂ air quality concentrations, exposures and health risks. In doing so, we have carefully considered the assessments developed as part of the last review of the primary NO₂ NAAQS (U.S. EPA, 2008a) and the newly available scientific and technical information.

Staff conclusions regarding support for particular quantitative analyses reflect our assessment of the degree to which updated analyses in the current review are likely to substantially add to our understanding of NO₂ exposures or health risks. These conclusions are informed by our consideration of the available health evidence and the available technical information, tools and methods. They build on the preliminary conclusions presented in the REA Planning Document (U.S. EPA, 2015), and on the CASAC's advice and public input on that document.

Based on our consideration of the above information, we have conducted updated analyses examining the occurrence of NO₂ air quality concentrations (i.e., as surrogates for potential NO₂ exposures) that may be of public health concern (see below and Appendix B). Consistent with the anticipated approach discussed in the REA Planning document (U.S. EPA, 2015a, Section 5.2), these updated analyses have been incorporated into this PA, and a separate REA will not be developed as part of the current review. The analyses discussed below and in appendix B have been informed by advice from the CASAC and input from the public on the REA Planning document (Diez Roux and Frey, 2015) and on the draft PA (Diez Roux and Sheppard, 2017).

Section 4.1 below summarizes our approach to considering potential support for updated quantitative analyses in this review. Section 4.2, along with the accompanying appendix (Appendix B), presents updated analyses comparing NO₂ air quality with health-based benchmarks. Sections 4.3 and 4.4 present our consideration of the potential support for updated exposure and risk assessments, respectively, and our conclusions that such updated assessments are not supported in the current review.

4.1 APPROACH TO CONSIDERING POTENTIAL SUPPORT FOR UPDATED QUANTITATIVE ANALYSES

In each NAAQS review, selection of the appropriate model(s) for the characterization of exposures and/or risks is influenced by the nature and strength of the evidence for the subject pollutant. Depending on the type of evidence available, analyses may include quantitative risk assessments based on dose-response, exposure-response, or ambient concentration-response relationships. Analyses may also include comparisons of health-based benchmark concentrations, drawn from controlled human exposure studies, with modeled exposure estimates or with ambient air quality concentrations (i.e., as surrogates for potential ambient exposures). The variety of approaches that have been employed in NAAQS reviews is summarized in Figure 4-1.

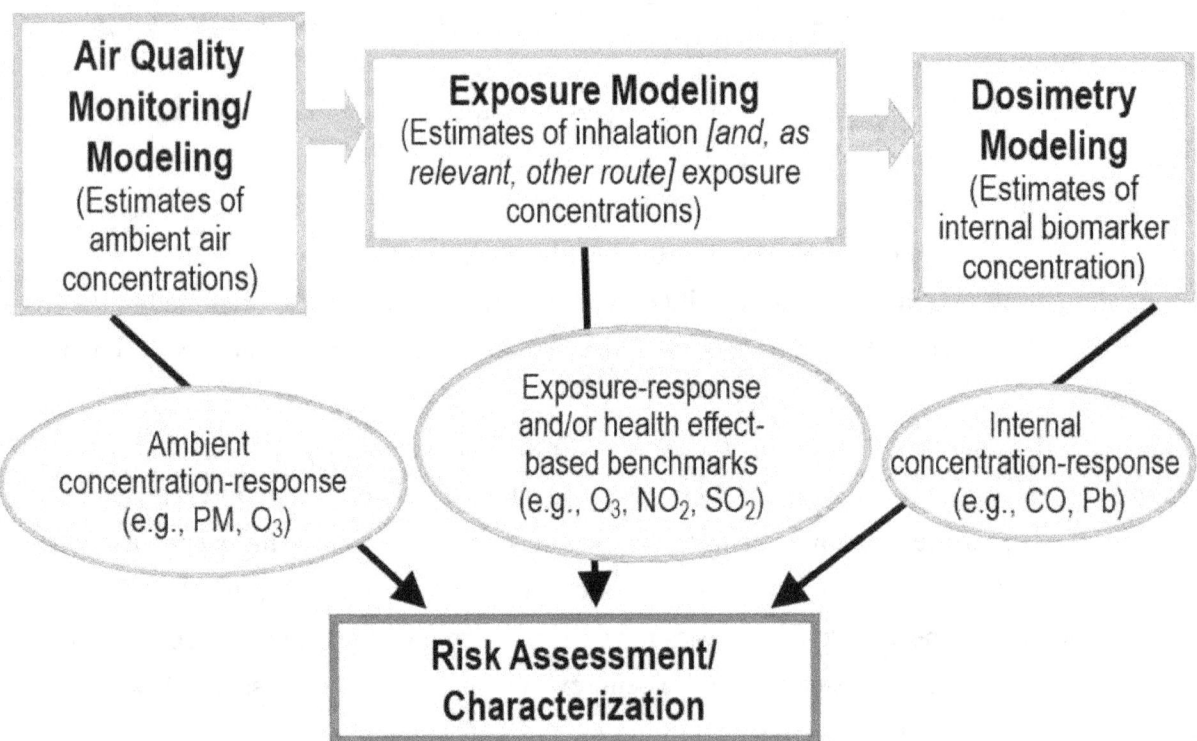

Figure 4-1. Risk characterization models employed in NAAQS Reviews.

In the last review of the primary NO_2 NAAQS, the 2008 ISA concluded that the strongest evidence supported the occurrence of respiratory effects following short-term NO_2 exposures (U.S. EPA, 2008b). Based on that evidence, the REA employed three approaches to quantify NO_2 exposures and health risks (U.S. EPA, 2008a):

1) Benchmarks were identified based on information from controlled human exposure studies of NO_2-induced increases in AR. Ambient NO_2 concentrations were compared to

these benchmarks. In urban areas across the U.S., such comparisons were made for ambient NO_2 concentrations at locations of NO_2 monitoring sites and simulated concentrations on/near roadways[93] (U.S. EPA, 2008a, Chapter 7).

2) Modeled estimates of personal NO_2 exposures were compared to benchmarks in a single urban area (Atlanta, GA), with a focus on children with asthma and people of all ages with asthma (U.S. EPA, 2008a, Chapter 8).

3) Concentration-response relationships from an epidemiologic study (Tolbert et al., 2007) were used to estimate NO_2-associated emergency department visits for respiratory causes in Atlanta, GA (U.S. EPA, 2008a, Chapter 9).

For this review, conclusions regarding the extent to which the newly available evidence and information support updated quantitative analyses are based on our consideration of a variety of factors. As noted above, these include consideration of the available health evidence and the available technical information, tools and methods. Our consideration of these factors inform judgments as to the likelihood that particular quantitative analyses will add substantially to our understanding of NO_2 exposures or health risks, beyond the insights gained from the analyses conducted in the last review. These key considerations and judgments are discussed in the REA Planning document (U.S. EPA, 2015a) and are summarized in Figure 4-2.

---

[93] Based on the available evidence, there was uncertainty regarding the locations of maximum NO_2 concentrations with respect to roadway emissions and transformation of NO to NO_2. Therefore, in the last review the EPA characterized these simulated concentrations as on- or near-road (75 FR 6474, February 9, 2010).

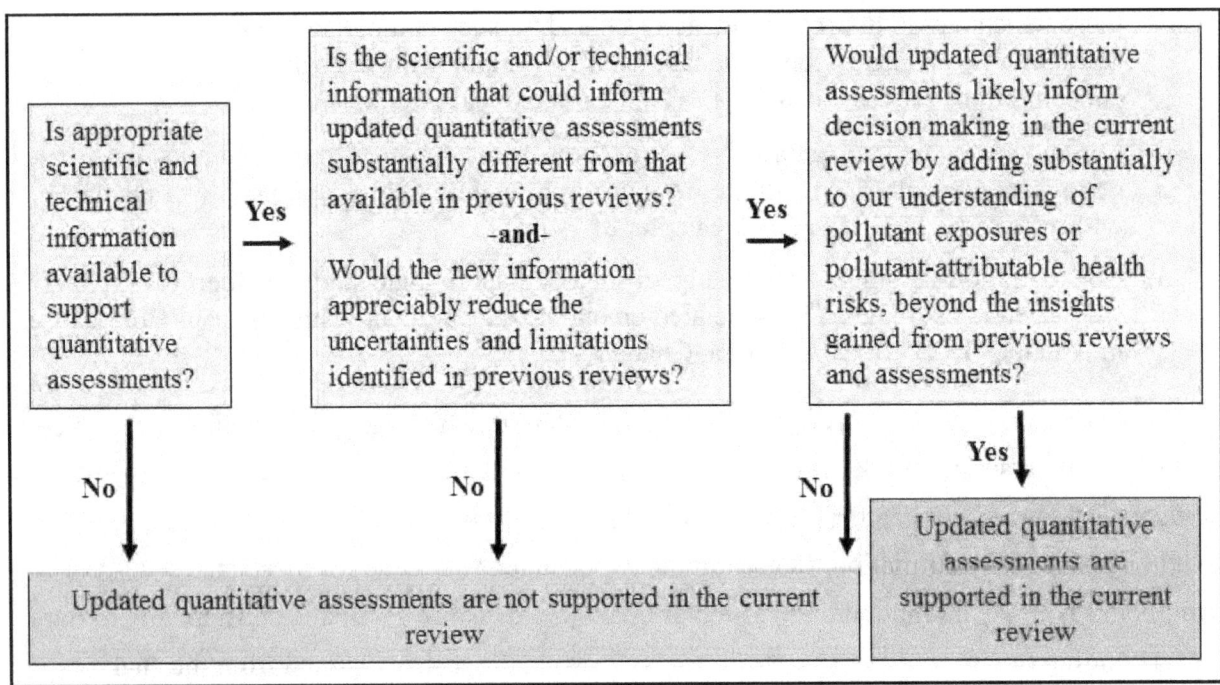

Figure 4-2. Key considerations for updated quantitative analyses.

As indicated in Figure 4-2, an initial consideration is the available health effects evidence and the foundation it may provide for updated quantitative analyses. As discussed in Chapter 3 of this PA, our evaluation of the scientific evidence is based on the assessment of the full body of evidence in the 2016 ISA (U.S. EPA, 2016). Consistent with prior reviews, in considering the evidence with regard to support for quantitative analyses, we give foremost consideration to health endpoints for which the ISA concludes the evidence supports a "causal" relationship or indicates that there is "likely to be a causal" relationship. As discussed in more detail in Chapter 3 of this PA, in the current review, the ISA reaches the following conclusions in this regard:

- "A causal relationship exists between short-term NO_2 exposure and respiratory effects based on evidence for asthma exacerbation" (U.S. EPA, 2016, p. 1-17).
- "There is likely to be a causal relationship between long-term NO_2 exposure and respiratory effects based on evidence for the development of asthma" (U.S. EPA, 2016, p. 1-20).

For all other health endpoints evaluated, the evidence was determined to be either "suggestive of, but not sufficient to infer, a causal relationship" or "inadequate to infer a causal relationship" (U.S. EPA, 2016, Table ES-1).

Given these ISA conclusions, our consideration of potential support for updated quantitative analyses in this review focuses on evidence for health outcomes related to asthma exacerbation (short-term NO_2 exposures) and the development of asthma (long-term NO_2 exposures). Our consideration of this evidence is discussed further below as it relates to the identification of NO_2 benchmarks (Sections 4.2, 4.3) and as to whether updated risk assessments are supported in the current review (Section 4.4).

4.2 COMPARISON OF NO₂ AIR QUALITY TO HEALTH-BASED BENCHMARKS

As discussed in Chapter 3, controlled human exposure studies of AR provide the strongest evidence supporting a causal relationship between short-term NO_2 exposures and respiratory effects.[94] A meta-analysis of individual-level data from these studies (Brown, 2015) supports the occurrence of increased AR in individuals with asthma following resting NO_2 exposures from 100 to 530 ppb (U.S. EPA, 2016, Section 5.2.2.1). In the last review, the 2008 REA compared NO_2 benchmarks based on information from such controlled human exposure studies with estimates of ambient NO_2 concentrations. These comparisons provided perspective on the potential for populations to experience NO_2 exposures that may be of public health concern (U.S. EPA, 2008a).

Given that mobile sources were identified in the last review as the largest contributors to U.S. NO_X emissions, and that NO_2 monitors were generally not located near heavily trafficked roads, an important uncertainty identified in the 2008 REA was the characterization of 1-hour NO_2 concentrations around major roadways. Based largely on the fact that information is newly available in this review from near-road NO_2 monitors (Chapter 2), the REA Planning document reached the preliminary conclusion that updated analyses comparing ambient NO_2 concentrations (i.e., as surrogates for potential exposure concentrations) to health-based benchmarks are supported (U.S. EPA, 2015a). In particular, the REA Planning document noted that new information from near-road monitors would be expected to provide important perspective, beyond what was available in the last review, on the extent to which NO_2 exposures around roads could have potentially important implications for public health (U.S. EPA, 2015a, Section 2.2.1). We have since conducted updated analyses comparing ambient NO_2 concentrations to benchmarks, the details of which are described in Section 4.2.1 and in Appendix B to this PA. Section 4.2.2 presents our overall conclusions based on these updated analyses.

4.2.1 Updated Analyses Comparing NO₂ Air Quality with Health-Based Benchmarks

In this PA, we have conducted updated analyses comparing NO_2 air quality to benchmarks in 23 study areas (Table 4-1). Our selection of study areas focused on CBSAs with

---

[94] Increased AR in people with asthma is the only health endpoint that has been shown to occur in controlled human exposure studies following exposures to NO_2 concentrations near those typically found in the ambient air in the U.S. (Section 3.2).

near-road monitors in operation,[95] CBSAs with the highest NO_2 design values and CBSAs with a relatively large number of NO_2 monitors overall (i.e., providing improved spatial characterization). Based on these criteria, a total of 23 CBSAs from across the U.S. were selected as study areas (See Appendix B, Figure B2-1).[96] Further evaluation indicates that these 23 study areas are among the most populated CBSAs in the U.S.; they have among the highest total NO_X emissions and mobile source NO_X emissions in the U.S.; and they include a wide range of stationary source NO_X emissions (Appendix B, Figures B2-2 to B2-8).

Air quality-benchmark comparisons were conducted in study areas with unadjusted air quality and with air quality adjusted upward to just meet the existing 1-hour standard.[97] Upward adjustment was required because all locations in the U.S. meet the current NO_2 NAAQS. These comparisons inform our consideration of the following questions:

- **To what extent are the current NO_2 standards estimated to allow ambient NO_2 concentrations that may be of public health concern? What are the important uncertainties associated with those estimates?**

In addressing these questions, an important focus is on the extent to which ambient NO_2 concentrations at or above health-based benchmarks could occur near major roadways. While data from the recently deployed near-road monitors will inform our consideration of this issue, the data available from these monitors are limited. Most near-road monitors have been in operation for only one to two years (Section 2.2.2), and true NO_2 DVs could not be calculated at

---

[95] As discussed above (Section 2.2.2), in the last review near-road monitors were required within 50 m of major roads in large urban areas that met certain criteria for population size or traffic volume. Most near-road monitors are sited within about 30 m of the road, and in some cases they are sited almost at the roadside (i.e., as close as 2 m from the road; http://www3.epa.gov/ttn/amtic/nearroad.html).

[96] Study area CBSAs are Atlanta-Sandy Springs-Roswell, GA; Baltimore-Columbia-Towson, MD; Boston-Cambridge-Newton, MA-NH; Chicago-Naperville-Elgin, IL-IN-WI; Dallas-Fort Worth-Arlington, TX; Denver-Aurora-Lakewood, CO; Detroit-Warren-Dearborn, MI; Houston-The Woodlands-Sugar Land, TX; Kansas City, MO-KS; Los Angeles-Long Beach-Anaheim, CA; Miami-Fort Lauderdale-West Palm Beach, FL; Minneapolis-St. Paul-Bloomington, MN-WI; New York-Newark-Jersey City, NY-NJ-PA; Philadelphia-Camden-Wilmington, PA-NJ-DE-MD; Phoenix-Mesa-Scottsdale, AZ; Pittsburgh, PA; Richmond, VA; Riverside-San Bernardino-Ontario, CA; Sacramento--Roseville--Arden-Arcade, CA; San Diego-Carlsbad, CA; San Francisco-Oakland-Hayward, CA; St. Louis, MO-IL; Washington-Arlington-Alexandria, DC-VA-MD-WV.

[97] In all study areas, ambient NO_2 concentrations required smaller upward adjustments to just meet the 1-hour standard than to just meet the annual standard. Therefore, when adjusting air quality to just meet the current NO_2 NAAQS, we applied the adjustment needed to just meet the 1-hour standard. Air quality was adjusted such that the three-year average of the 98th percentiles of the annual distributions of daily maximum 1-hour NO_2 concentrations equals 100 ppb. Information on the air quality adjustment approach can be found in Appendix B, Section B2.4.1.

these monitors for the one-hour standard (i.e., because 1-hour DVs are based on three years of data).[98]

In this section, we discuss our approach to identifying and interpreting the NO_2 health-based benchmarks (Section 4.2.1.1), summarize the results of the air quality-benchmark comparisons (Section 4.2.1.2) and discuss uncertainties in these analyses (Section 4.2.1.3). More detailed descriptions of the approaches used to conduct analyses, and the results of those analyses, are provided in Appendix B.

4.2.1.1   Health-Based Benchmarks

Based on the evidence from controlled human exposure studies of NO_2-induced increases in AR, the 2008 REA identified NO_2 benchmarks from 100 to 300 ppb (U.S. EPA, 2008a). As discussed further below, in the current review we have again identified benchmarks from 100 to 300 ppb for comparison to ambient NO_2 concentrations, based largely on information from the same controlled human exposure studies of AR that were available in the last review (U.S. EPA, 2016, Tables 5-1 and 5-2). This evidence indicates the potential for increased AR in some people with asthma following resting exposures to NO_2 concentrations from 100 to 530 ppb, though important uncertainties remain. In its review of the draft PA, CASAC agreed with this range of health-based benchmarks, stating that "[t]he decision to set the lowest benchmark analyses at 100 ppb NO_2 is reasonable as it reflects the lowest level, with sufficient scientific certainty, where acute NO_2 health effects have been shown to occur" (Diez Roux and Sheppard, 2017).[99]

In identifying the range of NO_2 health-based benchmarks to evaluate, and the weight to place on specific benchmarks within this range, we consider both the group mean responses reported in individual studies of AR and the results of a meta-analysis that combined individual-level data from multiple studies (Brown, 2015; U.S. EPA, 2016, Section 5.2.2.1). Group mean responses in individual studies, and the variability in those responses, can provide insight into the extent to which observed changes in AR are due to NO_2 exposures, rather than to chance alone, and have the advantage of being based on the same exposure conditions. With regard to individual studies, we consider both the direction and the statistical significance of group mean responses in AR following exposures to various NO_2 concentrations. Beyond what we can learn from individual studies, the meta-analysis by Brown (2015) can aid in identifying trends in individual-level responses across studies and can have the advantage of increased power to

---

[98] One implication of this is that near-road monitors were generally not used as the basis for adjusting air quality to just meet the current standard. As discussed below (Section 4.2.1.4), this introduces uncertainty into our air quality adjustments.

[99] Though, as noted below, CASAC also suggested consideration of sensitivity analyses based on additional benchmarks below 100 ppb. These sensitivity analyses are presented in appendix B.

detect effects, even in the absence of statistically significant effects in individual studies. With regard to the meta-analysis, we consider the fraction of people with asthma who experienced increased AR following exposures to various NO_2 concentrations, and the extent to which those fractions reflect statistically significant majorities of study volunteers.

In first considering studies conducted in resting individuals, where the data are most consistent (U.S. EPA, 2016, Section 5.2.2.1), we note that the lowest NO_2 concentration to which individuals with asthma have been exposed is 100 ppb. Of the five controlled human exposure studies conducted at 100 ppb, a statistically significant increase in AR following exposure to NO_2 was only observed in the study by Orehek et al. (1976) (N = 20; Table 3-2). Of the four studies that did not report statistically significant increases in AR following resting exposures to 100 ppb NO_2, three reported non-significant trends toward decreased AR (Ahmed et al., 1983b (N = 20); Hazucha et al., 1983 (N = 15); Tunnicliffe et al., 1994 (N = 8)), and one reported a trend towards increased AR (Ahmed et al., 1983a (N = 20)) (Table 3-2). When individual-level data from these five studies were combined in a meta-analysis, Brown (2015) reported that a marginally significant[100] majority of study participants experienced an increase in AR following exposure to 100 ppb NO_2 (i.e., 61%, $p = 0.08$; N = 76).[101] When the analysis was restricted to non-specific AR, the percentage who experienced increased AR was larger and statistically significant (i.e., 66%, p = 0.033; N = 50). In contrast, when the analysis was restricted to specific AR, study participants exposed to 100 ppb NO_2 were evenly divided between experiencing increases and decreases in AR (i.e., 50% increased and 50% decreased; N = 26) (Brown, 2015, Table 4).

Compared to 100 ppb, increased AR has been reported more consistently following exposures to higher NO_2 concentrations. In particular, most studies conducted in resting individuals report statistically significant increases in AR following exposures at or above 250 ppb NO_2 (Section 3.2.2). In addition, when resting NO_2 exposure concentrations above 100 ppb are examined in the meta-analysis, results indicate that statistically significant majorities of study

---

[100] In this study, marginal significance is defined as a P-value between 0.10 and 0.05.

[101] Brown et al. (2015) compared the number of study participants who experienced an increase in AR following NO_2 exposures to the number who experienced a decrease in AR. Study participants who experienced no change in AR were not included in comparisons. Thus, of the study participants who experienced either an increase or decrease in AR following exposure to 100 ppb NO_2, 61% experienced an increase and 39% experienced a decrease. The percentage of total study participants who experienced an increase in AR was slightly smaller than the percentage reported here and in Brown (2015) (i.e., 55% rather than 61% for 100 ppb NO_2 exposure, based on information in Table 1 of Brown (2015)).

participants experienced increased AR (Brown, 2015, Tables 3 to 5).[102] These results are largely due to non-specific AR. In contrast, specific AR was not increased in the majority of study volunteers following exposures to NO_2 concentrations at or below 300 ppb. In addition, neither specific nor non-specific AR was affected following exposures to any of the NO_2 concentrations evaluated during exercise (i.e., including exposure concentrations up to 600 ppb) (Brown, 2015, Tables 3 to 5).

In further considering these studies within the context of identifying appropriate benchmarks, we note the discussion of uncertainties in Section 3.2.2.1 above. As discussed in more detail in that section, there is no indication of a dose-response relationship between NO_2 and AR (Goodman, 2009; Brown, 2015). Though the lack of an apparent dose-response relationship does not necessarily indicate the lack of an NO_2 effect, it adds uncertainty to our interpretation of the controlled human exposure studies of AR. An additional uncertainty is the clinical relevance of the reported NO_2-induced increases in AR. While the meta-analysis by Brown (2015) has partially addressed this uncertainty by evaluating the magnitudes of responses in a subset of study participants exposed to NO_2 concentrations from 100 to 530 ppb,[103] this analysis is limited to a small subset of studies and study participants and, as noted above in section 3.2.2.1, there is uncertainty in reaching conclusions about the potential for clinically relevant effects at any particular NO_2 exposure concentration within the range evaluated.

When taken together, the results of controlled human exposure studies and of the meta-analysis by Brown (2015) support consideration of NO_2 benchmarks from 100 to 300 ppb, based largely on studies of non-specific AR in study participants exposed at rest. Benchmarks from the upper end of this range are supported by the results of individual studies, the majority of which reported statistically significant increases in AR following NO_2 exposures at or above 250 ppb, and by the results of the meta-analysis by Brown (2015). Benchmarks from the lower end of this range are supported by the results of the meta-analysis, even though individual studies do not consistently report statistically significant NO_2-induced increases in AR following exposures below 250 ppb. Given uncertainties in the evidence, including the lack of an apparent dose-response relationship and uncertainty in the potential adversity of reported increases in AR,

---

[102] Specifically, for resting exposures to concentrations of 100 ppb up to < 200 ppb, 200 ppb up to and including 300 ppb, and above 300 ppb, increased AR was reported in 63%, 65%, and 78% of study participants, respectively. The fractions of individuals who experienced increased AR following resting exposures, compared to the fraction who experienced decreased AR, reached statistical significance for all of the ranges of exposure concentrations evaluated ($p < 0.05$) (Brown, 2015, Table 5).

[103] As discussed above (Section 3.2), this analysis indicates the potential for clinically relevant increases in AR in some people with asthma exposed to NO_2 concentrations from 100 to 530 ppb.

caution is appropriate when interpreting the potential public health implications of 1-hour NO_2 concentrations at or above these benchmarks. This is particularly the case for the 100 ppb benchmark, given the less consistent results at this exposure concentration.

4.2.1.2 *Summary of Results*

We have evaluated the occurrence of 1-hour ambient NO_2 concentrations at or above the various benchmarks for as-is (i.e., unadjusted) air quality from 2010 to 2015 and for NO_2 air quality adjusted to just meet the existing 1-hour standard.[104] In considering these results, our focus is on the number of days per year that such 1-hour NO_2 concentrations could occur at each monitoring site in each study area. Detailed results of these analyses can be found in Appendix B (Section B3). In Tables 4-1 and 4-2 below, we present the number of days per year with daily maximum 1-hour NO_2 concentrations calculated to be at or above benchmarks of 100, 150,[105] and 200 ppb,[106] based on non-near-road monitors (Table 4-1)[107] and near-road monitors (Table 4-2).

---

[104] As noted above, in all study areas, ambient NO_2 concentrations required smaller upward adjustments to just meet the 1-hour standard than to just meet the annual standard. Therefore, when adjusting air quality to just meet the current NO_2 NAAQS, we applied the adjustment needed to just meet the 1-hour standard.

[105] Though no studies specifically evaluated the potential for increased AR following exposures to 150 ppb NO_2, results for the 150 ppb benchmark can provide information on the degree to which exceedances of the 100 ppb benchmark are due to ambient NO_2 concentrations closer to 100 ppb or 200 ppb.

[106] Because ambient NO_2 concentrations never reached or exceeded even the 200 ppb benchmark under the air quality scenarios that we have evaluated in this draft PA, we do not present results for the 300 ppb benchmark.

[107] Most, though not all, of these NO_2 monitors are classified as "area-wide." We use the term "area-wide" to refer to monitors sited at neighborhood, urban, and regional scales, as well as those monitors sited at either micro- or middle-scale that are representative of many such locations in the same CBSA (Section 2.2.2, above).

Table 4-1. Average and maximum number of days per year with non-near road NO₂ concentrations at or above benchmarks.

Study Area[108]	100 ppb benchmark[a,b]					150 ppb benchmark[a,b]					200 ppb benchmark[a,b]				
	As Is Air Quality	Air Quality Adjusted to Just meet the Existing 1-hr Standard[c]				As Is Air Quality	Air Quality Adjusted to Just meet the Existing 1-hr Standard				As Is Air Quality	Air Quality Adjusted to Just meet the Existing 1-hr Standard			
	2010-2015	2010-2012	2011-2013	2012-2014	2013-2015	2010-2015	2010-2012	2011-2013	2012-2014	2013-2015	2010-2014	2010-2012	2011-2013	2012-2014	2013-2015
Atlanta, GA	0 (0)	3 (15)	5 (30)	4 (16)	2 (16)	0 (0)	0 (0)	0 (0)	0 (0)	0 (0)	0 (0)	0 (0)	0 (0)	0 (0)	0 (0)
Baltimore, MD	0 (0)	4 (12)	5 (12)	4 (12)	4 (11)	0 (0)	0 (0)	0 (0)	0 (0)	0 (0)	0 (0)	0 (0)	0 (0)	0 (0)	0 (0)
Boston, MA	0 (0)	5 (15)	5 (15)	6 (18)	3 (12)	0 (0)	0.5 (1)	0.5 (1)	1 (1)	0.5 (1)	0 (0)	0 (0)	0 (0)	0 (0)	0 (0)
Chicago, IL	0.5 (1)	7 (24)	-	-	4 (12)	0 (0)	0 (0)	-	-	0 (0)	0 (0)	0 (0)	-	-	0 (0)
Dallas, TX	0 (0)	1 (9)	2 (11)	2 (16)	2 (12)	0 (0)	0 (0)	0 (0)	0 (0)	0 (0)	0 (0)	0 (0)	0 (0)	0 (0)	0 (0)
Denver, CO	0.5 (1)	-	14 (29)	5 (10)	4 (10)	0 (0)	-	0.5 (1)	1 (1)	0.5 (1)	0 (0)	-	0 (0)	0 (0)	0 (0)
Detroit, MI	0 (0)	6 (12)	9 (23)	8 (18)	4 (9)	0 (0)	0 (0)	0 (0)	0 (0)	0 (0)	0 (0)	0 (0)	0 (0)	0 (0)	0 (0)
Houston, TX	0.5 (3)	2 (12)	2 (8)	2 (15)	2 (14)	0 (0)	0.5 (2)	0 (0)	0 (0)	0 (0)	0 (0)	0 (0)	0 (0)	0 (0)	0 (0)
Kansas City, MO-KS	0 (0)	5 (9)	5 (9)	7 (11)	7 (10)	0 (0)	0 (0)	0 (0)	0 (0)	0 (0)	0 (0)	0 (0)	0 (0)	0 (0)	0 (0)
Los Angeles, CA	0.5 (1)	5 (19)	5 (22)	5 (23)	4 (27)	0 (0)	0.5 (1)	0.5 (1)	0.5 (1)	0.5 (1)	0 (0)	0 (0)	0 (0)	0 (0)	0 (0)
Miami, FL	0 (0)	5 (15)	4 (12)	5 (12)	6 (17)	0 (0)	0.5 (1)	0.5 (1)	1 (1)	0.5 (1)	0 (0)	0 (0)	0 (0)	0 (0)	0 (0)
Minneapolis, MN	0 (0)	6 (22)	4 (11)	4 (14)	2 (10)	0 (0)	0.5 (2)	0.5 (3)	0.5 (2)	0 (0)	0 (0)	0 (0)	0 (0)	0 (0)	0 (0)
New York, NY	0.5 (1)	3 (9)	3 (12)	3 (12)	4 (12)	0.5 (1)	0.5 (1)	0.5 (1)	0.5 (1)	0.5 (1)	0 (0)	0 (0)	0 (0)	0 (0)	0 (0)
Philadelphia, PA	0.5 (1)	3 (18)	3 (23)	-	2 (13)	0 (0)	0.5 (1)	0.5 (1)	-	0.5 (1)	0 (0)	0 (0)	0 (0)	-	0 (0)
Phoenix, AZ	0.5 (1)	2 (9)	4 (13)	3 (12)	3 (9)	0 (0)	0 (0)	0 (0)	0 (0)	0 (0)	0 (0)	0 (0)	0 (0)	0 (0)	0 (0)
Pittsburgh, PA	0 (0)	2 (9)	3 (22)	-	3 (14)	0 (0)	0 (0)	0 (0)	-	0 (0)	0 (0)	0 (0)	0 (0)	-	0 (0)
Richmond, VA	0 (0)	6 (17)	5 (17)	7 (15)	2 (6)	0 (0)	0.5 (1)	0.5 (1)	0.5 (1)	0 (0)	0 (0)	0 (0)	0 (0)	0 (0)	0 (0)
Riverside, CA	0.5 (7)	1 (14)	4 (25)	-	1 (5)	0 (0)	0.5 (1)	0.5 (8)	-	0 (0)	0 (0)	0 (0)	0 (0)	-	0 (0)

[108]Tables 4-1 and 4-2 identify study areas as individual cities. However, actual study areas include the entire CBSAs within which the identified cities are located. Complete study areas are defined above and in Appendix B.

Sacramento, CA	0 (0)	2 (10)	3 (14)	4 (21)	3 (17)	0 (0)	0 (0)	0 (0)	0 (0)	0 (0)	0 (0)	0 (0)	0 (0)	0 (0)	0 (0)
San Diego, CA	0.5 (1)	1 (9)	1 (12)	1 (12)	0.5 (4)	0 (0)	0 (0)	0 (0)	0 (0)	0 (0)	0 (0)	0 (0)	0 (0)	0 (0)	0 (0)
San Francisco, CA	0.5 (1)	1 (13)	1 (23)	1 (11)	2 (14)	0 (0)	0.5 (1)	0 (0)	0 (0)	0 (0)	0 (0)	0 (0)	0 (0)	0 (0)	0 (0)
St. Louis, MO	0 (0)	-	2 (8)	4 (14)	1 (5)	0 (0)	0 (0)	0 (0)	0 (0)	-	0 (0)	0 (0)	0 (0)	-	0 (0)
Washington, DC	0 (0)	4 (14)	4 (20)	5 (24)	6 (24)	0 (0)	0.5 (1)	0.5 (1)	0.5 (1)	0.5 (1)	0.5 (1)	0 (0)	0 (0)	0 (0)	0 (0)

[a] All calculated means were rounded to the nearest integer, though for mean values <0.50, rather than round downward to zero, a value of 0.5 was designated to distinguish it from instances where there were absolutely no (i.e., 0) days at or above benchmark levels during the collective years of interest.

[b] In parentheses are the maximum number of days in single year during the collective years of interest.

[d] Blank cells (indicated by "-") indicate time periods for which sufficient monitoring data were not available to generate estimates

Table 4-2. Number of days in 2014 and 2015 with near-road NO$_2$ concentrations at or above benchmarks.[109]

Study Area	Site ID	Annual Average Daily Traffic (AADT)	Distance from Road (m)	100 ppb Benchmark				150 ppb Benchmark				200 ppb Benchmark			
				As-Is Air Quality		Adjusted Air Quality		As-Is Air Quality		Adjusted Air Quality		As-Is Air Quality		Adjusted Air Quality	
				2014	2015	2014	2015	2014	2015	2014	2015	2014	2015	2014	2015
Atlanta, GA	130890003	146,000	30	-	0	-	7	-	0	-	0	-	0	-	0
	131210056	284,920	2	-	0	-	3	-	0	-	0	-	0	-	0
Baltimore, MD	240270006	186,750	16	-	0	-	7	-	0	-	0	-	0	-	0
Boston, MA	250250044	198,239	10	0	0	11	6	0	0	0	0	0	0	0	0
Chicago, IL	170313103			1	0	4	4	0	0	0	0	0	0	0	0
Dallas, TX	481131067	235,790	24	-	0	-	5	-	0	-	0	-	0	-	0
Denver, CO	080310027	249,000	9	0	0	5	3	0	0	0	0	0	0	0	0
Detroit, MI	261630093	140,500	9	0	0	9	9	0	0	0	0	0	0	0	0
	261630095	172,600	49	-	0	-	5	-	0	-	0	-	0	-	0
Houston, TX	482011066	324,119	24	0	0	4	3	0	0	0	0	0	0	0	0
Kansas City, MO-KS	290950042	114,495	20	0	0	4	3	0	0	0	0	0	0	0	0
Los Angeles, CA	060590008	272,000	9	0	0	21	1	0	0	0	0	0	0	0	0
Minneapolis, MN	270370480	87,000	30	-	0	-	3	-	0	-	0	-	0	-	0
	270530962	277,000	33	0	0	11	10	0	0	0	0	0	0	0	0
New York, NY	340030010	311,234	20		1		8	-	1	-	0	-	0	-	0
Philadelphia, PA	421010075	124,610	12	0	0	3	2	0	0	0	0	0	0	0	0
Phoenix, AZ	040134019	320,138	12	0	0	3	2	0	0	0	0	0	0	0	0
Pittsburgh, PA	420031376	87,534	18	-	0	-	11	-	0	-	0	-	0	-	0
Richmond, VA	517600025	151,000	21	-	0	-	9	-	0	-	0	-	0	-	0
Riverside, CA	060710026	245,300	50	-	0	-	8	-	0	-	0	-	0	-	0
San Francisco, CA	060010012	216,000	20	0	0	2	1	0	0	0	0	0	0	0	0
	291890016	161,338	27	-	0	-	1	-	0	-	0	-	0	-	0
St. Louis, MO	295100094	159,326	25	0	0	10	2	0	0	0	0	0	0	0	0

[109] Where monitor was in operation for 300 or more days in the year. Blank cells (indicated by "-") indicate time periods for which sufficient monitoring data were not available.

Based on the results presented in Tables 4-1 and 4-2 above, we make the following key observations for study areas when air quality was unadjusted ("as-is") and when air quality was adjusted to just meet the current 1-hour NO_2 standard:

1. For unadjusted air quality

   a. One-hour ambient NO_2 concentrations in study areas, including those near major roadways, were always below 200 ppb, and were virtually always below 150 ppb.
      i. Even in the worst-case years (i.e., the years with the largest number of days at or above benchmarks), no study areas had any days with 1-hour NO_2 concentrations at or above 200 ppb, and only one area had any days (i.e., one day) with 1-hour concentrations at or above 150 ppb.

   b. One-hour ambient NO_2 concentrations in study areas, including those near major roadways, only rarely reached or exceeded 100 ppb. On average in all study areas, 1-hour NO_2 concentrations at or above 100 ppb occurred on less than one day per year.
      i. Even in the worst-case years, most study areas had either zero or one day with 1-hour NO_2 concentrations at or above 100 ppb (7 days in the single worst-case location and worst-case year).

2. For air quality adjusted to just meet the current primary 1-hour NO_2 standard

   a. The current standard is estimated to allow no days in study areas with 1-hour ambient NO_2 concentrations at or above 200 ppb. This is true for both area-wide and near-road monitoring sites, even in the worst-case years.

   b. The current standard is estimated to allow almost no days with 1-hour ambient NO_2 concentrations at or above 150 ppb, based on both area-wide and near-road monitoring sites (i.e., zero to one day per year, on average).
      i. In the worst-case years in most study areas, the current standard is estimated to allow either zero or one day with 1-hour ambient NO_2 concentrations at or above 150 ppb. In the single worst-case year and location, the current standard is estimated to allow eight such days.

   c. At area-wide monitoring sites in most of the study areas, the current standard is estimated to allow from one to seven days per year, on average, with 1-hour ambient NO_2 concentrations at or above 100 ppb. At near-road monitoring sites in most of the study areas, the current standard is estimated to allow from about one to 10 days per year with such 1-hour concentrations.
      i. In the worst-case years in most of the study areas, the current standard is estimated to allow from about 5 to 20 days with 1-hour NO_2 concentrations at or above 100 ppb (30 days in the single worst-case location and year).

4.2.1.3 *Limitations and uncertainties*

There are a variety of limitations and uncertainties in these comparisons of NO_2 air quality with health-based benchmarks. In particular, we note uncertainties in the evidence underlying the benchmarks themselves, as discussed in Section 4.2.1.1, uncertainties in the upward adjustment of NO_2 air quality concentrations, and uncertainty in the degree to which monitored NO_2 concentrations reflect the highest potential NO_2 exposures. Each of these is discussed below.

Health-Based benchmarks

The primary goal of this analysis is to inform conclusions regarding the potential for the existing primary NO_2 standards to allow exposures to ambient NO_2 concentrations that may be of concern for public health. As discussed in detail above (Sections 3.2.2, 4.2.1.1), the meta-analysis by Brown (2015) indicates the potential for increased AR in some people with asthma following NO_2 exposures from 100 to 530 ppb. While it is possible that certain individuals could be more severely affected by NO_2 exposures than indicated by existing studies, which have generally evaluated adults with mild asthma,[110] there remains uncertainty in the degree to which the effects identified in these studies would be of public health concern. In particular, the lack of an apparent dose-response relationship between NO_2 and AR and uncertainties in the magnitude and potential adversity of the increase in AR following NO_2 exposures complicate our interpretation of comparisons between ambient NO_2 concentrations and health-based benchmarks. When considered in the context of the less consistent results observed across individual studies following exposures to 100 ppb NO_2, compared to higher exposure concentrations,[111] these uncertainties have the potential to be of particular importance for interpreting the public health implications of ambient NO_2 concentrations at or above the 100 ppb benchmark.[112]

With regard to the magnitude and clinical relevance of the NO_2-induced increase in AR in particular, we note that the meta-analysis by Brown (2015) attempts to address this uncertainty and inconsistency across individual studies. Specifically, as discussed above (Section 3.2.2), the meta-analysis evaluates the available individual-level data on the magnitude of the change in AR following resting NO_2 exposures. Brown (2015) reports that the magnitude of the increases in

---

[110] Although Brown (2015) notes that one study included in the meta-analysis (Avol et al., 1989) evaluated children aged 8 to 16 years and that disease status varied across studies, ranging from "inactive asthma up to severe asthma in a few studies" (p. 3 in published manuscript).

[111] As discussed previously, while the meta-analysis indicates that the majority of study volunteers experienced increased non-specific AR following exposures to 100 ppb NO_2, results were marginally significant when specific AR was also included in the analysis. In addition, individual studies do not consistently indicate increases in AR following exposures to 100 ppb NO_2.

[112] Sensitivity analyses included in Appendix B (Section 3.2, table B3-1) also evaluated 1-hour NO_2 benchmarks below 100 ppb (i.e., 85, 90, 95 ppb), though the available health evidence does not provide a basis for determining what exposures to such NO_2 concentrations might mean for public health.

AR observed following resting NO_2 exposures from 100 to 530 ppb were large enough to be of potential clinical relevance in about a quarter of the 72 study volunteers with available data. This is based on the fraction of exposed individuals who experienced a halving of the provocative dose of challenge agent following NO_2 exposures. This magnitude of change has been recognized by the American Thoracic Society (ATS) and the European Respiratory Society as a "potential indicator, although not a validated estimate, of clinically relevant changes in airway responsiveness" (Reddel et al., 2009) (U.S. EPA, 2016, p. 5-12). While this analysis by Brown (2015) indicates the potential for some people with asthma to experience effects of clinical relevance following resting NO_2 exposures from 100 to 530 ppb, it is based on a relatively small subset of volunteers and the interpretation of these results for any specific exposure concentration within the range of 100 to 530 ppb is uncertain (see section 3.2.2, above).

Approach to adjusting ambient NO_2 concentrations

These analyses use historical air quality relationships as the basis for adjusting ambient NO_2 concentrations to just meet the current 1-hour standard (Appendix B). The adjusted air quality is meant to illustrate a hypothetical scenario, and does not represent expectations regarding future air quality trends. If ambient NO_2 concentrations were to increase in some locations to the point of just meeting the current standards, it is not clear that the spatial and temporal relationships reflected in the historical data would persist. In particular, as discussed in Section 2.1.2 of this PA, we expect that ongoing implementation of existing regulations and the implementation of the recently revised O_3 NAAQS[113] will result in continued reductions in ambient NO_2 concentrations over much of the U.S. (i.e., reductions beyond the "unadjusted" air quality used in these analyses). Thus, if ambient NO_2 concentrations were to increase to the point of just meeting the existing 1-hour NO_2 standard in some areas, the resulting air quality patterns may not be similar to those estimated in our air quality adjustments.

There is also uncertainty in the upward adjustment of NO_2 air quality because three years of data are not yet available from most near-road monitors. In most study areas, DVs were not calculated at near-road monitors and, therefore, near-road monitors were generally not used as the basis for identifying adjustment factors for just meeting the existing standard.[114] In locations where near-road monitors measure the highest NO_2 DVs, reliance on those near-road monitors to identify air quality adjustment factors would result in smaller adjustments being applied to

---

[113] Based on analyses conducted as part of the 2015 Regulatory Impact Analysis for the O_3 NAAQS, available at https://www3.epa.gov/ttnecas1/docs/ria/naaqs-o3_ria_final_2015-09.pdf.

[114] Though in a few study locations, near-road monitors did contribute to the calculation of air quality adjustments, as described in Appendix B (Table B2-7).

monitors in the study area. Thus, monitors in such study areas would be adjusted upward by smaller increments, potentially reducing the number of days on which the current standard is estimated to allow 1-hour NO_2 concentrations at or above benchmarks. Given that near-road monitors in most areas measure higher 1-hour NO_2 concentrations than the area-wide monitors in the same CBSA (Figures 2-7 to 2-9), this uncertainty has the potential to impact results in many of our study areas. While the magnitude of the impact is unknown at present, the inclusion of additional years of near-road monitoring information in the determination of updated air quality adjustments could result in fewer estimated 1-hour NO_2 concentrations at or above benchmarks in some study areas.

Degree to which monitored NO_2 concentrations reflect the highest potential NO_2 exposures

To the extent there are unmonitored locations where ambient NO_2 concentrations exceed those measured by monitors in the current network, the potential for NO_2 exposures at or above benchmarks could be underestimated. In the last review, this uncertainty was determined to be particularly important for potential exposures around roads. The 2008 REA estimated that the large majority of modeled exposures to ambient NO_2 concentrations at or above benchmarks occurred on or near roads (U.S. EPA, 2008a, Figures 8-17 and 8-18). When characterizing ambient NO_2 concentrations, the 2008 REA attempted to address this uncertainty by estimating the elevated NO_2 concentrations that can occur on or near the road. These estimates were generated by applying literature-derived adjustment factors to NO_2 concentrations at monitoring sites located away from the road.[115]

In the current review, given that the 23 selected study areas have among the highest NO_X emissions in the U.S., and given the siting characteristics of existing NO_2 monitors, this uncertainty likely has only a limited impact on the results of the air quality-benchmark comparisons. In particular, as described above, mobile sources tend to dominate NO_X emissions within most CBSAs, and the 23 study areas evaluated have among the highest mobile source NO_X emissions in the U.S. (Appendix B, Section B2.3.2). Most study areas have near-road NO_2 monitors in operation, which are required within 50 m of the most heavily trafficked roadways in large urban areas. The majority of these near-road monitors are sited within 30 m of the road, and several are sited within 10 m (see Atlanta, Cincinnati, Denver, Detroit, Los Angeles in

---

[115] Sensitivity analyses included in Appendix B use updated data from the scientific literature (Richmond-Bryant et al., 2016) to estimate "on-road" NO_2 concentrations based on monitored concentrations around a roadway in Las Vegas (Appendix B, Section B2.4.2). However, there remains considerable uncertainty in the relationship between on-road and near-road NO_2 concentrations, and in the degree to which they may differ. Therefore, in evaluating the potential for roadway-associated NO_2 exposures, we focus on the concentrations at locations of near-road monitors.

EPA's database of metadata for near-road monitors[116]). Thus, even though the location of highest NO$_2$ concentrations around roads can vary (Section 2.1), we anticipate that the near-road NO$_2$ monitoring network, with monitors sited from 2 to 50 m away from heavily trafficked roads, effectively captures the types of locations around roads where the highest NO$_2$ concentrations can occur.[117]

This conclusion is consistent with the ISA's analysis of available data from near-road NO$_2$ monitors, which indicates that near-road monitors with target roads having the highest traffic counts also had among the highest 98th percentiles of 1-hour daily maximum NO$_2$ concentrations (U.S. EPA, 2016, Section 2.5.3.2). The ISA concludes that "[o]verall, the very highest 98th percentile 1-hour maximum concentrations were generally observed at the monitors adjacent to roads with the highest traffic counts" (U.S. EPA, 2016, p. 2-66).

It is also important to consider the degree to which air quality-benchmark comparisons appropriately characterize the potential for NO$_2$ exposures near non-roadway sources of NO$_X$ emissions. With regard to this issue, we note that the 23 selected study areas include CBSAs with large non-roadway sources of NO$_X$ emissions. This includes study areas with among the highest NO$_X$ emissions from electric power generation facilities (EGUs) and airports, the two types of non-roadway sources that emit the most NO$_X$ in the U.S. (Appendix B, Section B2.3.2). As discussed below, several study areas have non-near-road NO$_2$ monitors sited to determine the impacts of such sources.

Table 2-12 in the ISA (U.S. EPA, 2016) summarizes NO$_2$ concentrations at selected monitoring sites that are likely to be influenced by non-road sources, including nearby ports, airports, border crossings, petroleum refining, or oil and gas drilling. For example, the Los Angeles, CA CBSA includes one of the busiest ports and one of the busiest airports in the U.S. Out of 18 monitors in the Los Angeles CBSA, three of the five highest 98th percentile 1-hour maximum concentrations were observed at the near-road site, the site nearest the port, and the site adjacent to the airport. In the Chicago, IL CBSA, the highest hourly NO$_2$ concentration measured in 2014 (105 ppb) occurred at the Schiller Park, IL monitoring site, located adjacent to O'Hare International airport and very close to a major rail yard (i.e., Bedford Park Rail Yard) (U.S. EPA, 2016, Section 2.5.3.2).[118] In addition, one of the highest 1-hour daily maximum NO$_2$

---

[116] This database is found at http://www3.epa.gov/ttn/amtic/nearroad.html

[117] Though it remains possible that some areas (e.g., street canyons in urban environments) could have higher ambient NO$_2$ concentrations than indicated by near-road monitors. This issue is highlighted as a data gap and research need in section 5.4 below. Sensitivity analyses estimating the potential for on-road NO$_2$ exposures are described in Appendix B.

[118] Sections B5.1 and B5.2 of Appendix B provides data on the large sources of NO$_X$ emissions in areas where NO$_2$ monitors are located.

concentrations recorded in recent years (136 ppb) was observed at a Denver, CO non-near-road site. This concentration was observed at a monitor located one block from high-rise buildings that form the edge of the high-density central business district. This monitor is likely influenced by local traffic, as well as by commercial heating and other activities (U.S. EPA, 2016, Section 2.5.3.2).[119] Thus, beyond the NO_2 near-road monitors, some NO_2 monitors in study areas are also sited to capture high ambient NO_2 concentrations around important non-roadway sources of NO_X emissions.

Conclusions Based on Air Quality-Benchmark Comparisons

As discussed above and in the REA Planning document (U.S. EPA, 2015, Section 2.1.1), an important uncertainty identified in the 2008 REA was the characterization of 1-hour NO_2 concentrations around major roadways. In the current review, data from recently deployed near-road NO_2 monitors improves our understanding of such ambient NO_2 concentrations. In this PA we have conducted updated analyses comparing ambient NO_2 concentrations (i.e., as surrogates of potential exposures) to health-based benchmarks, with a particular focus on study areas where near-road monitors have been deployed.

As discussed in Chapter 2, recent NO_2 concentrations measured in all U.S. locations meet the existing primary NO_2 NAAQS.[120] Based on these recent (i.e., unadjusted) ambient measurements, analyses estimate almost no potential for 1-hour exposures to NO_2 concentrations at or above benchmarks, even at the lowest benchmark examined (i.e., 100 ppb).

Analyses of air quality adjusted upwards to just meet the current 1-hour standard also indicate virtually no potential for 1-hour exposures to NO_2 concentrations at or above the 200 ppb benchmark, and almost none for exposures at or above 150 ppb. This is the case for both average and worst-case years, including in study areas with near-road monitors sited within a few meters of heavily trafficked roads. With respect to the lowest benchmark evaluated, analyses estimate that the current 1-hour standard allows the potential for exposures to 1-hour NO_2

---

[119] Recent traffic counts on the nearest streets were 44,850 (in 2014) and 23,389 (in 2013) vehicles per day, respectively. Traffic counts on other streets within one block of this monitor were 22,000, 13,000, 5,000, and 2,490 vehicles per day. Together, this adds up to more than 100,000 vehicles per day on streets within one block of this non-near-road monitor (U.S. EPA, 2016, Section 2.5.3.2).

[120] As described in section 2.3.1 (above), the maximum NO_2 DVs in 2015 for the whole network were well-below the NAAQS, with the highest values being 30 ppb (annual) and 72 ppb (hourly).

concentrations at or above 100 ppb on some days (i.e., about one to 10 days per year, on average).[121,122]

These results are consistent with our expectations, given that the current 1-hour standard, with its 98th percentile form, is anticipated to limit, but not eliminate, exposures to 1-hour NO_2 concentrations at or above 100 ppb.[123] These results are similar to the results presented in the REA from the last review, based on NO_2 concentrations at the locations of area-wide ambient monitors (Appendix B, Section B5.9, Table B5-66). In contrast, compared to the on/near road simulations in the last review, these results indicate substantially less potential for 1-hour exposures to near-road NO_2 concentrations at or above benchmarks (Appendix B, Section B5.9, Table B5-66).[124]

When these results and associated uncertainties are taken together, we note that the current 1-hour NO_2 standard is expected to allow virtually no potential for exposures to the NO_2 concentrations that have been shown most consistently to increase AR in people with asthma (i.e., above 200 ppb), even under worst-case conditions across a variety of study areas with among the highest NO_X emissions in the U.S. Such NO_2 concentrations were not estimated to occur, even at monitoring sites adjacent to some of the most heavily trafficked roadways.

In addition, the current standard is expected to limit, though not eliminate, exposures to 1-hour concentrations at or above 100 ppb. Though the current standard is estimated to allow 1-hour NO_2 concentrations at or above 100 ppb on some days, the potential public health implications of exposures to these concentrations are unclear. In particular, while the meta-analysis by Brown (2015) indicates the potential for increased AR following exposure to 100 ppb NO_2 (i.e., when analyses were restricted to non-specific AR), individual studies generally do not indicate statistically significant NO_2-induced increases in AR following exposures to 100 ppb NO_2 and meta-analysis results based on all available data at 100 ppb (i.e., resting exposures, specific and non-specific AR) were only marginally significant. When combined with uncertainty due to the lack of an overall dose-response relationship and uncertainty in the degree

---

[121] Results for the 100 ppb benchmark are due primarily to 1-hour NO_2 concentrations that are closer to 100 ppb than 200 ppb, based on the results for the 150 ppb benchmark.

[122] Sensitivity analyses included in Appendix B (section 3.2) indicate a greater potential for exposures to NO_2 benchmarks below 100 ppb (i.e., 85, 90, 95 ppb). However, as noted above, the uncertainties in the available health evidence lead to uncertainties in interpreting what such NO_2 exposures might mean for public health.

[123] The 98th percentile generally corresponds to the 7th or 8th highest 1-hour concentration in a year.

[124] On-/near-road simulations in the last review estimated that a 1-hour NO_2 standard with a 98th percentile form and a 100 ppb level could allow about 20 to 70 days per year with 1-hour NO_2 concentrations at or above the 200 ppb benchmark and about 50 to 150 days per year with 1-hour concentrations at or above the 100 ppb benchmark (Appendix B, Table B5-56).

to which effects would be adverse should they occur, there is considerable uncertainty regarding the potential public health implications of exposures to 100 ppb NO_2. In limiting exposures to NO_2 concentrations at or above 100 ppb, the current standard provides protection against exposures for which the evidence of adverse NO_2-attributable effects is less certain.

In reaching an overall conclusion based on the results of analyses comparing NO_2 air quality with health-based benchmarks, we consider all of the information discussed above. Given the results of these analyses, and the uncertainties inherent in their interpretation, we conclude that there is little potential for exposures to ambient NO_2 concentrations that would be of clear public health concern in locations meeting the current 1-hour standard. Additionally, while a lower standard level (i.e., lower than 100 ppb) would be expected to further limit the potential for exposures to 100 ppb NO_2, the public health implications of such reductions are unclear, particularly given that no additional protection would be expected against exposures to NO_2 concentrations at or above the higher benchmarks (i.e., 200 ppb and above). Thus, we conclude that these analyses comparing ambient NO_2 concentrations to health-based benchmarks do not provide support for considering potential alternative standards to increase public health protection, beyond the protection provided by the current standards.

4.3   MODEL-BASED EXPOSURE ASSESSMENT

In the last review, in addition to analyses comparing NO_2 air quality to benchmarks, the REA used an exposure model to generate estimates of 1-hour personal NO_2 exposures in a single urban study area (i.e., Atlanta, GA). These modeled 1-hour personal exposures were compared to 1-hour benchmarks ranging from 100 to 300 ppb. The exposure assessment in the last review served to complement the results of the broader, less resource-intensive, analyses comparing NO_2 air quality to benchmarks (described above).

In the current review, the REA Planning document (U.S. EPA, 2015, Chapter 3) indicated that the potential utility of an updated model-based assessment of personal NO_2 exposures would depend on the results of the updated comparison of NO_2 air quality with health effect benchmarks (Section 4.2). To the extent air quality-benchmark comparisons indicate little potential for exposures to ambient NO_2 concentrations that would be of public health concern, the REA Planning document concluded that the added value of more refined estimates of personal NO_2 exposures would be limited.

As discussed in Section 4.2 above, analyses comparing NO_2 air quality with health-based benchmarks indicate little potential for exposures to ambient NO_2 concentrations that would be of clear public health concern in locations meeting the current 1-hour standard. Given the results of these analyses, more refined estimates of personal NO_2 exposures would be of limited additional value. Based on these conclusions, we have not conducted an updated assessment of modeled NO$_2$ exposures in this review. In its review of the draft PA, CASAC agreed "with the decision to not conduct any new model-based analyses" (Diez Roux and Sheppard, 2017, p.5).

4.4 EPIDEMIOLOGY-BASED RISK ASSESSMENT

Risk estimates based on epidemiologic studies have the potential to provide perspective on the most serious pollutant-associated public health risks (e.g., hospital admissions, emergency department (ED) visits, premature mortality) in populations that often include at-risk groups. However, the emphasis given to such quantitative risk estimates depends on the extent to which the underlying epidemiologic studies address key uncertainties related to NO$_2$ associations with health effects, including the potential for confounding by co-occurring pollutants. This section provides an overview of staff's considerations and conclusions regarding potential support for risk assessment based on information from epidemiologic studies of short short-term (Section 4.3.1) and a long-term (Section 4.3.2) NO$_2$ concentrations.

4.4.1 Short-Term Epidemiologic-Based Risk Assessment

In the last review of the primary NO$_2$ NAAQS, NO$_2$-associated respiratory-related ED visits in the Atlanta Metropolitan Statistical Area (MSA) were estimated for short-term ambient NO$_2$ concentrations, based on concentration-response functions from an epidemiologic study by Tolbert et al. (2007) (U.S. EPA, 2008a, Chapter 9). Specifically, the 2008 REA modeled respiratory-related ED visits (including asthma, chronic obstructive pulmonary disease (COPD), upper respiratory illness, pneumonia and bronchiolitis) for individuals of all ages based on a 3-day moving average of the daily maximum 1-hour NO$_2$ concentrations measured at a single central-site monitor. The REA reported several main findings, including the following:

1. When air quality was adjusted to simulate just meeting the existing annual standard, about 8 to 9% of respiratory-related ED visits in the Atlanta MSA were estimated to be attributable to short-term NO$_2$ exposures, based on a single-pollutant model. Risk estimates were reduced by as much as about 60% when based on co-pollutant models.[125] Ninety-five percent confidence intervals, reflecting statistical uncertainty in the NO$_2$ coefficient, included negative risk estimates when based on co-pollutant models that included O$_3$ or PM$_{10}$ (U.S. EPA, 2008a, Section 9.7).

2. When air quality was adjusted to simulate just meeting potential alternative standards with 1-hour averaging times, standards with levels of 50, 100, and 150 ppb reduced estimated NO$_2$-associated risks compared to the annual standard alone (U.S. EPA, 2008a, Tables 9-3 and 9-4).

---

[125] Risk estimates were reduced to about 3% of respiratory-related ED visits based on a co-pollutant model that included PM$_{10}$, and to about 4-5% of such visits based on a model that included O$_3$ (U.S. EPA, 2008a, Tables 9-3 and 9-4).

3. When air quality was adjusted to simulate just meeting a potential alternative standard with a 1-hour averaging time and a level of 200 ppb, estimated risks were similar to those estimated for the annual standard alone (U.S. EPA, 2008a, Tables 9-3 and 9-4).

The 2008 REA noted that a number of key uncertainties should be considered when interpreting these results with regard to decisions on the standard. These included the following (U.S. EPA, 2008a, Section 9.6):

1. Uncertainties in the estimates of NO_2 coefficients in concentration-response functions used in the assessment.

2. Uncertainties concerning the specification of the concentration-response model (including the shape of the relationships) and whether or not a population threshold exists within the range of concentrations examined in the studies.

3. Uncertainty concerning potential confounding by co-occurring pollutants.

4. Uncertainty in the adjustment of air quality distributions to simulate just meeting various standards.

Overall, the 2008 REA concluded that the risks estimated to be associated with just meeting the annual NO_2 standard (i.e., the existing standard at the time of the last review) can be judged important from a public health perspective (U.S. EPA, 2008a). In the 2010 final decision, the Agency further noted that a 1-hour standard with a level at or below 100 ppb "could substantially reduce exposures to ambient NO_2 and associated health risks (compared to just meeting the current standard)" (75 FR 6483, February 9, 2010). Upon further consideration of these results, and their associated uncertainties, the risk assessment was not used to distinguish between the support for particular standard levels at or below 100 ppb.

In considering whether to conduct an updated epidemiology-based risk assessment in the current review, in the REA Planning document we evaluated the newly available information in the context of that which was previously available. We specifically considered the extent to which the new information would be likely to reduce uncertainties and/or substantially improve our understanding of NO_2-attributable health risks, beyond the insights gained from the risk assessment conducted in the last review.

As discussed in more detail in the REA Planning document (U.S. EPA, 2015, Section 4.2.2), the evidence that has become available since the last review has not substantially changed our understanding of health effects attributable to short-term NO_2 exposures or of the populations potentially at increased risk of such effects. Updated risk estimates based on information from epidemiology studies in the current review would be subject to the same uncertainties identified in the 2008 REA.

In particular, recent studies do not provide an improved basis, compared to the last review, for quantifying NO_2-attributable risks independent of important traffic-related pollutants (e.g., CO, EC, UFP, benzene, $PM_{2.5}$, and PM_{10}) (U.S. EPA, 2016, Section 3.5). The ISA concludes that an important uncertainty in the current review continues to be the "[s]trength of inference from copollutant models about independent associations of NO_2, especially with pollutants measured at central site monitors" (U.S. EPA, 2016, Table 1-1). In particular, of the key studies supporting the determination of a causal relationship with respiratory effects (U.S. EPA, 2016, Table 5-39), two U.S. studies evaluating asthma-related hospital admissions or ED visits have become available since the last review (Strickland et al., 2010; Li et al., 2012). Neither of these studies reported NO_2 health effect associations in co-pollutant models that included other roadway-related pollutants.

Based on the above considerations, in the REA Planning document we concluded that an updated epidemiology-based risk assessment estimating respiratory-related endpoints attributable to short-term NO_2 exposures would be subject to uncertainties that are essentially the same as those identified in the 2008 REA (U.S. EPA, 2008a) (i.e., uncertainties that resulted in the risk assessment not being used to distinguish between support for particular standard levels at or below 100 ppb). We reached the preliminary conclusion that such an updated epidemiology-based risk assessment in the current review would not appreciably reduce uncertainties and limitations from the assessment conducted in the last review and would be unlikely to substantially improve our understanding of NO_2-attributable health risks or increase our confidence in risk estimates beyond the assessment from the last review. The CASAC agreed with this conclusion in its review of the REA Planning document, stating that "the CASAC concurs that an updated epidemiology-based risk assessment in the current review would be unlikely to substantially improve our understanding of NO_2-attributable health risks, or to increase our confidence in risk estimates, beyond the assessment from the last review" (Diez Roux and Frey, 2015, p. 5). Based on our consideration of the evidence, as summarized above, and CASAC advice, in this review we have not conducted an updated quantitative risk assessment of short-term NO_2 exposures based on information from epidemiology studies. In future reviews, potential support for conducting an updated risk assessment will be revisited, with consideration given to the evidence available at the time of that review and the degree to which that evidence reduces important uncertainties.

4.4.2 Long-Term Epidemiologic-Based Risk Assessment

This section discusses our consideration of potential support for a quantitative risk assessment based on information from epidemiology studies of long-term NO_2 concentrations, and presents our conclusion that such an assessment is not supported by the available evidence and information. Section 4.4.2.1 summarizes our preliminary considerations and conclusions presented in the REA Planning document (U.S. EPA, 2015). Section 4.4.2.2 summarizes the CASAC's advice on those preliminary considerations and conclusions, based on its review of the NO$_2$ REA Planning document (Diez Roux and Frey, 2015). Section 4.4.2.3 presents our additional considerations and our conclusion that a quantitative risk assessment is not supported in the current review.

4.4.2.1 Preliminary Considerations and Conclusions from the REA Planning Document

The REA Planning Document presented our preliminary considerations and conclusions regarding support for a quantitative risk assessment based on information from epidemiology studies of long-term NO$_2$ concentrations. In reaching preliminary conclusions, we considered the evidence assessed in the second draft ISA (U.S. EPA, 2015b), noting the draft ISA determination that the evidence "indicates there is likely to be a causal relationship between long-term NO$_2$ exposure and respiratory effects" (U.S. EPA, 2015b, section 1.5.1, pp. 1-19 and 1-21) and that the "strongest evidence is for effects on asthma development" (U.S. EPA, 2015b, Table 1-1). As discussed above (Section 3.3), key evidence supporting this causal determination comes from recent epidemiologic cohort studies reporting associations between long-term ambient NO$_2$ exposures (i.e., averaged over 1–10 years) and development of asthma in children, and from experimental studies for airway responsiveness and allergic responses. The REA Planning document considered the extent to which epidemiologic studies of long-term NO$_2$ exposures could support a quantitative risk assessment in the current review.

In reaching preliminary conclusions, the REA Planning document noted that, as for short-term NO$_2$ exposures, an important uncertainty in the epidemiologic studies of long-term NO$_2$ exposures and health effect associations is the extent to which effects are independently related primarily to NO$_2$ rather than one or more co-occurring pollutants. Compared to studies of short-term NO$_2$ exposures, this is an even more important issue for long-term exposures, given the higher correlations between long-term NO$_2$ concentrations and other pollutants reported in many epidemiologic studies using LUR to estimate exposures, and the lack of reported correlations in studies using monitored data (U.S. EPA, 2016, Table 6-1).[126]

The REA Planning document further noted that, of the key studies evaluating associations between long-term NO$_2$ exposures and the development of asthma (U.S. EPA, 2016,

---

[126] In studies of NO$_2$-associated asthma development, correlations with co-occurring pollutants were most often reported for PM. NO$_2$ and PM were often highly correlated in these studies, particularly in studies that used land use regression or dispersion modeling to estimate long-term NO$_2$ exposures (e.g., r values were greater than 0.9 in several studies) (U.S. EPA, 2016, Table 6-1).

Table 6-5), none evaluated associations in co-pollutant models for traffic-related pollutants. Speaking to this issue, the 2016 ISA notes uncertainty "in identifying an independent effect of NO_2 exposure from traffic-related copollutants because evidence from experimental studies for effects related to asthma development is limited, and epidemiologic analysis of confounding is lacking" (U.S. EPA, 2016, Table 1-1). In particular, the ISA states that "correlations with $PM_{2.5}$ and BC were often high (e.g., $r = 0.7-0.96$), and no studies of asthma incidence evaluated copollutant models to address copollutant confounding, making it difficult to evaluate the independent effect of NO_2" (U.S. EPA, 2016, p. 6-64). The REA Planning Document concluded that a quantitative risk assessment based on information from studies of NO_2-associated asthma development in children would be subject to considerable uncertainty due to the inability to distinguish the contributions of NO_2 from the contributions of other highly correlated pollutants (U.S. EPA, 2015a). Given this uncertainty, the REA Planning document reached the preliminary conclusion that such a risk assessment would not substantially add to our understanding of NO_2-attributable health risks and would therefore be of limited value in informing decisions in the current review.

4.4.2.2 *CASAC Advice on the REA Planning Document*

In its review of the REA Planning document, the CASAC generally agreed with staff's concerns regarding the potential for confounding by co-occurring pollutants, and the potential implications of such confounding for risk estimates. However, given the stronger evidence for NO_2-associated asthma incidence in the current review, the CASAC encouraged EPA staff to further consider the feasibility of a risk assessment based on information from epidemiology studies of long-term NO_2 exposures. Specifically, the CASAC provided the following advice (Diez Roux and Frey, 2015, pp. 5-6):

> The CASAC concurs with the assessment that a quantitative risk assessment based on the epidemiologic evidence [of long-term NO_2 exposures] would be challenged by "considerable uncertainty due to the inability to distinguish the contributions of NO_2 from the contributions of other highly correlated pollutants." Nevertheless the finding that the evidence for these relationships is likely to be causal dictates a thoughtful consideration of an updated risk assessment, even in the face of these uncertainties. The CASAC encourages the EPA to explore the feasibility of a quantitative risk assessment based on the long-term epidemiology. The agency may find that such an REA is not feasible or that it may not substantially improve the understanding of health risk attributable to long-term NO_2 exposures, in which case the CASAC would request a clear explanation for this finding.

4.4.2.3   *Staff Conclusions Regarding Potential Support for a Quantitative Risk Assessment*

In response to the CASAC's advice, we have further explored potential support for a risk assessment based on information from epidemiologic studies of long-term NO_2 exposures. In doing so, we note the determination in the final ISA that "[t]here is likely to be a causal relationship between long-term NO_2 exposure and respiratory effects based on evidence for the development of asthma" (U.S. EPA, 2016, pp. 1-20). While this causal determination provides the initial motivation for considering a potential quantitative risk assessment of asthma incidence, we also evaluate other factors important for the conduct and interpretation of such an assessment. These additional factors generally relate to the availability of required information, the degree to which that information is subject to important uncertainties and, given those uncertainties, the degree to which an assessment would be likely to improve our understanding of NO_2-associated health risks. Table 4-3 below discusses our consideration of these factors.

Table 4-3. Factors to Consider in Conducting a Risk Assessment.

Description of Factors	Staff's Consideration of Factors	Conclusions
Availability of U.S. Studies: Our consideration of epidemiologic studies for the purpose of quantitative risk assessment is focused on studies conducted in the U.S. While studies conducted outside of the U.S. form an important part of the overall evidence base supporting ISA causal determinations, when quantifying risks in the U.S. it is important to use concentration-response relationships from epidemiologic studies that reflect U.S. population demographics, air quality and exposure patterns, monitoring networks (i.e., when monitors are used as exposure surrogates), healthcare systems (i.e., particularly for outcomes based on administrative databases), and baseline incidence rates. While potential differences in such factors between the U.S. and other countries do not preclude the qualitative use of non-U.S. studies in drawing overall conclusions regarding the strength of the scientific evidence, these differences can add considerable uncertainty into quantitative risk estimates.	To identify studies that could support quantitative risk assessment, we focus primarily on the U.S. studies determined in the ISA to provide "key evidence" supporting the causal determination for long-term NO_2 exposures and respiratory effects (U.S. EPA, 2016, Table 6-5). While a number of the key epidemiologic studies identified in the ISA were not conducted within the U.S. (U.S. EPA, 2016, Table 6-5), two key U.S. studies are available (Jerrett et al., 2008; Clougherty et al., 2007). The multi-community study by Jerrett et al. (2008) reports associations with asthma development across several California communities, while the single-city study by Clougherty et al. (2007) reports an association between asthma development and long-term NO_2 exposure in Boston. Although uncertainties and limitations in the other U.S. studies of long-term NO_2 exposures resulted in the ISA placing less emphasis on them, we also consider the potential for these studies to provide an appropriate basis for a quantitative risk assessment. In particular, the multicity study by Nishimura et al. (2013) reports associations between long-term NO_2 concentrations and asthma incidence, based on five U.S. cities, and the multi-community study by McConnell et al. (2010) reports associations across several California communities (study population overlaps with Jerrett et al., 2008).	Our further consideration of potential support for a quantitative risk assessment focuses on the two key U.S. epidemiologic studies identified in the ISA (Jerrett et al., 2008; Clougherty et al., 2007). While recognizing the additional uncertainty, we also consider the U.S. studies assessed in the ISA that were not identified as key studies (McConnell et al., 2010; Nishimura et al., 2013).

Exposure Characterization: The choice of exposure surrogate in an epidemiologic study can affect the relationship estimated to exist between ambient pollutant exposures and the health outcome of interest. The ISA identifies potential exposure measurement error as an important uncertainty in interpreting epidemiologic studies. Unlike studies of short-term NO_2, in which exposure measurement error tends to bias associations toward the null, in studies of long-term NO_2, exposure measurement error could also bias reported associations away from the null (U.S. EPA, 2016, Table A-1).	U.S. epidemiologic studies use various approaches to estimate long-term NO_2 exposures. Specifically, Nishimura et al. (2013) and McConnell et al. (2010) use air quality data from central-site ambient monitors in study communities. In contrast, Jerrett et al. (2008) and Clougherty et al. (2007) placed passive NO_2 samplers in study areas. Jerrett et al. (2008) placed samplers in the yards of study volunteers for 2 weeks in the summer (mid-August) and 2 weeks in the fall–winter season (mid-November). Four week averages in each location were used to represent long-term NO_2 exposures. Clougherty et al. (2007) placed samplers in locations across study communities for one week per month, and these data were then used to build a LUR model of estimated children's residential exposures.	It is unlikely that we could recreate ambient exposure surrogates based on passive samplers at subjects' homes or other locations, as was done in the studies by Clougherty et al. (2007) and Jerrett et al. (2008). Use of concentration-response functions from these studies to estimate risks based on data from the ambient NO_2 monitoring network would introduce substantial uncertainty into risk estimates, limiting the degree to which such estimates could inform policy judgments.
Beyond consideration of the potential for exposure measurement error, when conducting a risk assessment based on a concentration-response function from an epidemiologic study, the approach used in the study to estimate pollutant exposures is an important consideration. In cases where it is not feasible to use the same, or substantially similar, approach to estimate exposures in the risk assessment, additional uncertainty is introduced into risk estimates.		Of the U.S. studies, the exposure surrogates employed by Nishimura et al. (2013) (Inverse distance weighting for the four closest monitors with a 50km buffer) and McConnell et al. (2010) (One monitor site per community) could be applied most readily to a potential quantitative risk assessment, though these surrogates based on central-site monitors may also be more prone to exposure measurement error (U.S. EPA, 2016, p. 6-18). The ambient

	monitoring data used in these studies is based on central-site monitors in study communities. Ambient data from central site monitors is publically available through the EPA's Air Quality System (AQS). Use of such data would be consistent with exposure surrogates in previous NAAQS reviews (e.g., U.S. EPA, 2011; U.S. EPA, 2014).
	Information on baseline prevalence, though not incidence, is likely available for the populations and outcomes evaluated by Jerrett et al. (2008), McConnell et al. (2010), and Nishimura et al. (2013). In addition, the use of prevalence rather than incidence would add uncertainty to quantitative risk estimates.

Neither baseline incidence nor prevalence information is available for the population evaluated by Clougherty et al. (2007), therefore, we cannot estimate risks based on information from this study. |
| **Baseline Incidence:** Epidemiology-based effect estimates used in modeling risk typically relate a unit change in the ambient concentration of a pollutant to a resulting change in the incidence of the health outcome being assessed. Thus, a critical input to a risk assessment is information on the baseline incidence of the health outcome in the study population being assessed and in the study area where the risk assessment is being conducted. | Jerrett et al. (2008), McConnell et al. (2010) and Nishimura et al. (2013) all reported associations with asthma incidence in children, though the age ranges of the children differed between studies (and Nishimura reports results up to age 21). Nishimura et al. (2013) evaluated associations specifically in Latino and African American children. National-level age-stratified asthma prevalence is available within BenMAP-CE (U.S. EPA, 2015c, Table D-10; CDC, 2008),[127] though age-stratified information on specific racial or ethnic groups is more limited. Currently, we have access to this data nationally, but not by state. We also do not have data on asthma incidence, which would likely diverge from prevalence rates, and be the more appropriate baseline metric for asthma development. |

---

[127] BenMAP-CE stands for "Environmental Benefits Mapping and Analysis Program – Community Edition." The BenMAP software and associated documentation are available for download at http://www2.epa.gov/benmap.

	Clougherty et al. (2007) reported an association with asthma incidence in children in the Boston area who are exposed to high levels of violence. Neither baseline incidence nor prevalence rates are available for this population.	Given the few studies that have evaluated this issue for NO_2 and asthma incidence, and the ambiguous results reported by those studies, there is considerable uncertainty regarding the shape of the concentration-response function for NO_2 and asthma incidence. This includes uncertainty as to whether a threshold exists below which effects are not observed.
Shape of Concentration-Response Function: An understanding of the slope of the concentration-response generated from the effect estimates is necessary to accurately understand the relationship between the outcome and exposure over a range of ambient concentrations. Uncertainty in the shape of the concentration-response function adds uncertainty to risk estimates.	In the case of NO_2 and the development of asthma, information regarding the shape of the concentration-response function is very limited. According to the ISA (U.S. EPA, 2016, p. 6-64), "[i]n limited analysis of the concentration-response relationship, results did not consistently indicate a linear relationship in the range of ambient NO_2 concentrations examined (Shima et al., 2002; Carlsten et al., 2011)." The ISA further notes that "these studies did not conduct analyses to evaluate whether there is a threshold for effects" (U.S. EPA, 2016, p. 6-64).	
Control for Potential Confounding: High correlations between multiple co-occurring pollutants complicates the interpretation of health effect associations with any individual pollutant within the mixture. The potential for copollutant confounding is of particular concern for studies of health effects associated with long-term ambient concentrations of NO_2, given the relatively high correlations between NO_2 and other traffic-related pollutants, including $PM_{2.5}$.	In considering this issue, the ISA concludes that "[e]pidemiologic studies of asthma development in children have not clearly characterized potential confounding by $PM_{2.5}$ or traffic-related pollutants [e.g., CO, BC/EC, volatile organic compounds (VOCs)]" (U.S. EPA, 2016, p. 6-64). The ISA further notes that "[i]n the longitudinal studies, correlations with $PM_{2.5}$ and BC were often high (e.g., $r = 0.7–0.96$), and no studies of asthma incidence evaluated copollutant models to address copollutant confounding, making it difficult to evaluate the independent effect of NO_2" (U.S. EPA, 2016, p. 6-64).	While studies of long-term NO_2 and asthma incidence contribute qualitatively to the ISA's causal determination, these studies do not provide a reliable basis for quantifying the magnitude of the NO_2 contribution to asthma development. Therefore, any NO_2 risk estimates developed from these studies would be subject to considerable uncertainty. If an NO_2 risk assessment were conducted based on studies of long-term NO_2, there would be particular

uncertainty regarding the extent to which NO_2 risk estimates reflect the magnitude of NO_2 health impacts rather than the health impacts of traffic-related pollutants as a whole.

After considering the factors discussed above, we conclude that a quantitative risk assessment based on epidemiologic studies of long-term NO_2 exposures is not supported in this review. This conclusion is based on our consideration of the available evidence for associations between long-term NO_2 and the development of asthma, including consideration of the uncertainties that would be inherent in any risk estimates based on that evidence. In particular, we note most of the available epidemiologic studies of long-term NO_2 and asthma incidence are not appropriate for quantitative risk assessment because they were not conducted in the U.S. Of the studies that were conducted in the U.S., most were not emphasized in the ISA's determination of a causal relationship due to important uncertainties. Additionally, most of these studies used exposure metrics that are not readily transferable to a quantitative risk assessment or they evaluated populations for which information on baseline incidence is not available (Table 4-3). The two U.S. studies that are the exception to these limitations (i.e., McConnell et al., 2010; Nishimura et al., 2013) are still subject to broader uncertainties related to the potential for co-pollutant confounding of the NO_2 association, potential bias due to exposure measurement error, and the shape of the concentration-response function.

With regard to the potential for copollutant confounding in particular, we note the high correlations between long-term NO_2 concentrations and the long-term concentrations of other traffic-related pollutants (U.S. EPA, 2016, Section 6.2.2.1, Table 6-1). Given these correlations, study authors often interpreted associations with NO_2 as reflecting associations with a marker of traffic-related pollution more broadly (e.g., Jerrett et al., 2008; McConnell et al., 2010). Based on our consideration of all of the above information, we conclude that estimates of the risk of asthma development associated with long-term NO_2 exposures, if developed, would be subject to considerable uncertainty and would likely not be informative, beyond what can be learned from our consideration of the studies themselves (Chapter 3). Thus, we have not conducted such an assessment in this PA. In its review of the draft PA, CASAC agreed "with the decision to not conduct any new…epidemiologic-based analyses" (Diez Roux and Sheppard, p. 5). In future reviews, potential support for conducting a quantitative risk assessment will be revisited, with consideration given to the evidence available at the time of that review and the degree to which that evidence reduces important uncertainties.

4.5   REFERENCES

Ahmed, T; Dougherty, R; Sackner, MA. (1983a). Effect of 0.1 ppm NO2 on pulmonary functions and non-specific bronchial reactivity of normals and asthmatics [final report]. (CR-83/11/BI). Warren, MI: General Motors Research Laboratories.

Ahmed, T; Dougherty, R; Sackner, MA. (1983b). Effect of NO2 exposure on specific bronchial reactivity in subjects with allergic bronchial asthma [final report]. (CR-83/07/BI). Warren, MI: General Motors Research Laboratories.

Brown, JS. (2015). Nitrogen dioxide exposure and airway responsiveness in individuals with asthma. Inhal Toxicol 27: 1-14. http://dx.doi.org/10.3109/08958378.2014.979960

Carlsten, C; Dybuncio, A; Becker, A; Chan-Yeung, M; Brauer, M. (2011). Traffic-related air pollution and incident asthma in a high-risk birth cohort. Occup Environ Med 68: 291-295. http://dx.doi.org/10.1136/oem.2010.055152

Centers for Disease Control and Prevention (CDC). 2008. National Center for Health Statistics. National Health Interview Survey, 1999-2008.

Clark, NA; Demers, PA; Karr, CJ; Koehoorn, M; Lencar, C; Tamburic, L; Brauer, M. (2010). Effect of early life exposure to air pollution on development of childhood asthma. Environ Health Perspect 118: 284-290. http://dx.doi.org/10.1289/ehp.0900916

Clougherty, JE; Levy, JI; Kubzansky, LD; Ryan, PB; Suglia, SF; Canner, MJ; Wright, RJ. (2007). Synergistic effects of traffic-related air pollution and exposure to violence on urban asthma etiology. Environ Health Perspect 115: 1140-1146. http://dx.doi.org/10.1289/ehp.9863

Diez Roux A and Frey H.C. (2015). Letter from Drs. Ana Diez Roux, Chair and H. Christopher Frey, Immediate Past Chair, Clean Air Scientific Advisory Committee to EPA Administrator Gina McCarthy. CASAC Review of the EPA's Review of the Primary National Ambient Air Quality Standards for Nitrogen Dioxide: Risk and Exposure Assessment Planning Document. EPA-CASAC-15-002. September 9, 2015. Available at:
https://yosemite.epa.gov/sab/sabproduct.nsf/264cb1227d55e02c85257402007446a4/A7922887D5BDD8D485257EBB0071A3AD/$File/EPA-CASAC-15-002+unsigned.pdf

Goodman, JE; Chandalia, JK; Thakali, S; Seeley, M. (2009). Meta-analysis of nitrogen dioxide exposure and airway hyper-responsiveness in asthmatics. Crit Rev Toxicol 39: 719-742. http://dx.doi.org/10.3109/10408440903283641

Hazucha, MJ; Ginsberg, JF; McDonnell, WF; Haak, ED, Jr; Pimmel, RL; Salaam, SA; House, DE; Bromberg, PA. (1983). Effects of 0.1 ppm nitrogen dioxide on airways of normal and asthmatic subjects. J Appl Physiol Respir Environ Exerc Physiol 54: 730-739.

Jerrett, M; Shankardass, K; Berhane, K; Gauderman, WJ; Künzli, N; Avol, E; Gilliland, F; Lurmann, F; Molitor, JN; Molitor, JT; Thomas, DC; Peters, J; McConnell, R. (2008). Traffic-related air pollution and asthma onset in children: A prospective cohort study with individual exposure measurement. Environ Health Perspect 116: 1433-1438. http://dx.doi.org/10.1289/ehp.10968

Li, S; Batterman, S; Wasilevich, E; Wahl, R; Wirth, J; Su, FC; Mukherjee, B. (2011). Association of daily asthma emergency department visits and hospital admissions with ambient air pollutants among the pediatric Medicaid population in Detroit: Time-series and time-stratified case-crossover analyses with threshold effects. Environ Res 111: 1137-1147. http://dx.doi.org/10.1016/j.envres.2011.06.002

McConnell, R; Islam, T; Shankardass, K; Jerrett, M; Lurmann, F; Gilliland, F; Gauderman, J; Avol, E; Künzli, N; Yao, L; Peters, J; Berhane, K. (2010). Childhood incident asthma and traffic-related air pollution at home and school. Environ Health Perspect 118: 1021-1026. http://dx.doi.org/10.1289/ehp.0901232

Nishimura, KK; Galanter, JM; Roth, LA; Oh, SS; Thakur, N; Nguyen, EA; Thyne, S; Farber, HJ; Serebrisky, D; Kumar, R; Brigino-Buenaventura, E; Davis, A; LeNoir, MA; Meade, K; Rodriguez-Cintron, W; Avila, PC; Borrell, LN; Bibbins-Domingo, K; Rodriguez-Santana, JR; Sen, S; Lurmann, F; Balmes, JR; Burchard, EG. (2013). Early-life air pollution and asthma risk in minority children: The GALA II and SAGE II studies. Am J Respir Crit Care Med 188: 309-318. http://dx.doi.org/10.1164/rccm.201302-0264OC

Orehek, J; Massari, JP; Gayrard, P; Grimaud, C; Charpin, J. (1976). Effect of short-term, low-level nitrogen dioxide exposure on bronchial sensitivity of asthmatic patients. J Clin Invest 57: 301-307. http://dx.doi.org/10.1172/JCI108281

Reddel, HK; Taylor, DR; Bateman, ED; Boulet, LP; Boushey, HA; Busse, WW; Casale, TB; Chanez, P; Enright, PL; Gibson, PG; de Jongste, JC; Kerstjens, HA; Lazarus, SC; Levy, ML; O'Byrne, PM; Partridge, MR; Pavord, ID; Sears, MR; Sterk, PJ; Stoloff, SW; Sullivan, SD; Szefler, SJ; Thomas, MD; Wenzel, SE. (2009). An official American Thoracic Society/European Respiratory Society statement: Asthma control and exacerbations: Standardizing endpoints for clinical asthma trials and clinical practice. Am J Respir Crit Care Med 180: 59-99. http://dx.doi.org/10.1164/rccm.200801-060ST

Shima, M; Nitta, Y; Ando, M; Adachi, M. (2002). Effects of air pollution on the prevalence and incidence of asthma in children. Arch Environ Health 57: 529-535. http://dx.doi.org/10.1080/00039890209602084

Strickland, MJ; Darrow, LA; Klein, M; Flanders, WD; Sarnat, JA; Waller, LA; Sarnat, SE; Mulholland, JA; Tolbert, PE. (2010). Short-term associations between ambient air pollutants and pediatric asthma emergency department visits. Am J Respir Crit Care Med 182: 307-316. http://dx.doi.org/10.1164/rccm.200908-1201OC

Tolbert, PE; Klein, M; Peel, JL; Sarnat, SE; Sarnat, JA. (2007). Multipollutant modeling issues in a study of ambient air quality and emergency department visits in Atlanta. J Expo Sci Environ Epidemiol 17: S29-S35. http://dx.doi.org/10.1038/sj.jes.7500625

U.S. EPA (2008a). Risk and Exposure Assessment to Support the Review of the NO2 Primary National Ambient Air Quality Standard. U.S. EPA, Office of Air Quality Planning and Standards. Research Triangle Park, NC. EPA 452/R-08-008a/b. November 2008. Available at: http://www.epa.gov/ttn/naaqs/standards/nox/s_nox_cr_rea.html.

U.S. EPA (2008b). Integrated Science Assessment for Oxides of Nitrogen – Health Criteria. U.S. EPA, National Center for Environmental Assessment, Research Triangle Park, NC. EPA/600/R-08/071. July 2008. Available at: http://cfpub.epa.gov/ncea/cfm/recordisplay.cfm?deid=194645.

U.S. EPA. (2014). Integrated Review Plan for the Primary National Ambient Air Quality Standards for Nitrogen Dioxide. U.S. EPA, National Center for Environmental Assessment and Office of Air Quality Planning and Standards, Research Triangle Park, NC. EPA-452/R-14-003. June 2014. Available at: http://www.epa.gov/ttn/naaqs/standards/nox/data/201406finalirpprimaryno2.pdf.

U.S. EPA (2015a). Review of the Primary National Ambient Air Quality Standards for Nitrogen Dioxide: Risk and Exposure Assessment Planning Document. U.S. EPA, Office of Air Quality Planning and Standards, Research Triangle Park, NC. EPA-452/D-15-001. May 13, 2015. Available at: https://www3.epa.gov/ttn/naaqs/standards/nox/data/20150504reaplanning.pdf

U.S. EPA. (2015b). Integrated Science Assessment for Oxides of Nitrogen – Health Criteria (Second External Review Draft). U.S. EPA, National Center for Environmental Assessment and Office, Research Triangle Park. EPA/600/R-14/006. January 2015. Available at: http://cfpub.epa.gov/ncea/isa/recordisplay.cfm?deid=288043.

U.S. EPA. (2015c). Benmap-ce User's Manual Appendices. U.S EPA, Office of Air Quality Planning and Standards, Research Triangle Park, NC. Research Triangle Park, North Carolina. Available at:

https://www.epa.gov/sites/production/files/2015-04/documents/benmap-ce_user_manual_appendices_march_2015.pdf.

U.S. EPA (2011). Policy Assessment for the Review of the Particulate Matter National Ambient Air Quality Standards. Office of Air Quality Planning and Standards, U.S. Environmental Protection Agency, Research Triangle Park, NC. EPA 452/R-11-003. April 2011. Available at: https://www3.epa.gov/ttn/naaqs/standards/pm/s_pm_2007_pa.html

U.S. EPA (2014). Policy Assessment for the Review of the Ozone National Ambient Air Quality Standards. Office of Air Quality Planning and Standards, U.S. Environmental Protection Agency, Research Triangle Park, NC. EPA 452/R-14-006. August 2014. Available at: https://www3.epa.gov/ttn/naaqs/standards/ozone/s_o3_2008_pa.html

U.S. EPA (2016). Integrated Science Assessment for Oxides of Nitrogen – Health Criteria (2016 Final Report). U.S. EPA, National Center for Environmental Assessment, Research Triangle Park, NC. EPA/600/R-15/068. January 2016. Available at: https://cfpub.epa.gov/ncea/isa/recordisplay.cfm?deid=310879.

5 CONCLUSIONS ON THE ADEQUACY OF THE CURRENT PRIMARY NO$_2$ STANDARDS

This chapter summarizes staff's consideration of the available evidence and information related to the adequacy of the current primary NO$_2$ standards, as discussed in the preceding chapters. In addition, this chapter presents staff's conclusions regarding the adequacy of those standards, including conclusions on each of the elements of the standards (i.e., indicator, averaging time, level, and form). In reaching conclusions on the adequacy of the current primary NO$_2$ standards, we revisit the following overarching policy-relevant question for this review:

- **Does the currently available scientific evidence and information from quantitative analyses support or call into question the adequacy of the public health protection afforded by the current primary NO$_2$ standards?**

As discussed in Chapter 1 of this PA, our approach to addressing this overarching question, and informing the Administrator's judgments on the primary NO$_2$ standards, builds upon the approach used in the last review of the primary NO$_2$ NAAQS. This approach is consistent with the requirements of the NAAQS provisions of the CAA and with how the EPA and the courts have historically interpreted these CAA provisions. In particular, the CAA requires primary standards that, in the judgment of the Administrator, are requisite to protect public health with an adequate margin of safety. In setting primary standards that are "requisite" to protect public health, the EPA's task is to establish standards that are neither more nor less stringent than necessary for this purpose. One intent of the requirement that primary standards provide an "adequate margin of safety" is to address uncertainties associated with inconclusive scientific and technical information. Thus, as discussed in Chapter 1, the CAA does not require that primary standards be set at a zero-risk level, but rather at a level that limits risk sufficiently so as to protect public health with an adequate margin of safety.

Section 5.1 below summarizes staff's evidence-based considerations and staff's conclusions on the extent to which the evidence supports or calls into question the basic elements of the current primary NO$_2$ standards. Section 5.2 summarizes staff's consideration of quantitative analyses comparing NO$_2$ air quality with health-based benchmarks, and staff's conclusions on the extent to which the current standards could allow NO$_2$ exposures of public health concern. Section 5.3 presents our overall conclusions regarding the adequacy of the public health protection provided by the current primary NO$_2$ standards. Section 5.4 highlights areas for additional research and data collection in order to reduce uncertainties in future reviews of the primary NO$_2$ NAAQS.

5.1 EVIDENCE-BASED CONSIDERATIONS

As discussed in Chapter 3, in considering the evidence available in the current review with regard to adequacy of the current 1-hour and annual NO_2 standards, we first consider the nature of the health effects attributable to NO_2 exposures (Sections 3.2.1, 3.3.1). In doing so, we address the following questions:

- **To what extent does the currently available scientific evidence alter or strengthen our conclusions from the last review regarding health effects attributable to ambient NO_2 exposures? Are previously identified uncertainties reduced or do important uncertainties remain? Have new uncertainties been identified?**

As described in greater detail in Sections 3.2.1 and 3.3.1, we address these questions for both short-term and long-term NO_2 exposures, with a focus on health endpoints for which the ISA concludes that the evidence indicates there is a "causal" or "likely to be a causal" relationship.

In answering the questions above with regard to short-term NO_2 exposures, section 3.2.1 notes that, as in the last review, the strongest evidence continues to come from studies examining respiratory effects. In particular, the ISA concludes that evidence indicates a "causal" relationship between short-term NO_2 exposure and respiratory effects, based on evidence related to asthma exacerbation. While this conclusion reflects a strengthening of the causal determination, compared to the last review, this strengthening is based largely on a more specific integration of the evidence related to asthma exacerbations rather than on the availability of new, stronger evidence. Though some evidence has become available since the last review, as summarized below, this evidence has not fundamentally altered our understanding of the relationship between short-term NO_2 exposures and respiratory effects.

The strongest evidence supporting this ISA conclusion comes from controlled human exposure studies demonstrating NO_2-induced increases in AR in individuals with asthma. Most of the controlled human exposure studies assessed in the ISA were available in the last review, particularly studies of non-specific AR. As in the last review, there remains uncertainty due to the lack of an apparent dose-response relationship between NO_2 exposures and AR and uncertainty in the potential adversity of NO_2-induced increases in AR. The newly available meta-analysis by Brown (2015) has partially addressed this latter uncertainty by demonstrating the potential for clinically relevant increases in AR in some asthmatics following exposures to NO_2 concentrations from 100 to 530 ppb.[128] Supporting evidence for a range of NO_2-associated

---

[128] As described in Chapter 3, consideration of clinical relevance in the ISA is based on evidence from clinical studies evaluating efficacy of inhaled corticosteroids that are used to prevent bronchoconstriction and airway

respiratory effects also comes from epidemiologic studies. While recent epidemiologic studies provide new evidence based on improved exposure characterizations and co-pollutant modeling, these studies are consistent with the evidence from the last review and do not fundamentally alter our understanding of the respiratory effects associated with ambient NO_2 exposures. Due to limitations in the available epidemiologic methods, uncertainty remains in the current review regarding the potential for confounding by traffic-related copollutants (i.e., $PM_{2.5}$, EC/BC, CO). Thus, while some new evidence is available in this review, that new evidence has not substantially altered our understanding of the respiratory effects that occur following short-term NO_2 exposures.

In answering the questions above with regard to long-term NO_2 exposures, Section 3.3.1 notes the ISA conclusion that there is "likely to be a causal relationship" between long-term NO_2 exposure and respiratory effects, based largely on the evidence for asthma development in children. New epidemiologic studies of asthma development have increasingly utilized improved exposure assessment methods (i.e., measured or modeled at or near children's homes and followed for many years), which partly reduces uncertainties from the last review related to exposure measurement error. Explicit integration of evidence for individual outcome categories (e.g. asthma incidence, respiratory infection) provides an improved characterization of biological plausibility and mode of action. This improved characterization includes the assessment of new evidence supporting a role for repeated short-term NO_2 exposures in the development of asthma. High correlations between long-term average ambient concentrations of NO_2 and long-term concentrations of other traffic-related pollutants, together with the general lack of epidemiologic studies evaluating copollutant models that include traffic-related pollutants, remains a concern in interpreting associations with asthma development. Specifically, the extent to which NO_2 may be serving primarily as a surrogate for the broader traffic-related pollutant mix remains unclear. Thus, while the evidence for respiratory effects related to long-term NO_2 exposures has become stronger since the last review, there remain important uncertainties to consider in evaluating this evidence within the context of the adequacy of the current standards.

Given the evaluation of the evidence in the ISA and the causal determinations (Sections 3.2 and 3.3), our further consideration of the evidence focuses on studies of asthma exacerbations (short-term exposures) and asthma development (long-term exposures). We next

---

responsiveness. Generally, a change of at least one doubling dose is considered to be an indication, but not validation, of clinical relevance (this represents a decline in AR as the dose to induce AR is doubled). Based on this, a halving of the provocative dose is taken in the ISA to represent an increase in AR that is an indication of clinical relevance.

consider what these bodies of evidence indicate with regard to the basic elements of the primary NO_2 standards. In particular, we consider the following question:

- **To what extent does the available evidence for respiratory effects attributable to either short- or long-term NO_2 exposures support or call into question the basic elements of the current primary NO_2 standards?**

In addressing this question, we evaluate the evidence in the context of the indicator, averaging times, levels, and forms of the current standards. Each of these elements is discussed below.

Indicator

The indicator for both the current annual and 1-hour NAAQS for oxides of nitrogen is NO_2. While the presence of gaseous species other than NO_2 has long been recognized (discussed in Section 2.1, above), no alternative to NO_2 has been advanced as being a more appropriate surrogate for ambient gaseous oxides of nitrogen. Both previous and recent controlled human exposure studies and animal toxicology studies provide specific evidence for health effects following exposure to NO_2. Similarly, the large majority of epidemiologic studies report health effect associations with NO_2, as opposed to other gaseous oxides of nitrogen, though the degree to which monitored NO_2 reflects actual NO_2 concentrations, as opposed to NO_2 plus other gaseous oxides of nitrogen, can vary (Section 2.2, above). In addition, because emissions that lead to the formation of NO_2 generally also lead to the formation of other NO_X oxidation products, measures leading to reductions in population exposures to NO_2 can generally be expected to lead to reductions in population exposures to other gaseous oxides of nitrogen. Therefore, an NO_2 standard can also be expected to provide some degree of protection against potential health effects that may be independently associated with other gaseous oxides of nitrogen even though such effects are not discernable from currently available studies. Given these considerations, we reach the conclusion that it is appropriate in the current review to consider retaining the NO_2 indicator for standards meant to protect against exposures to gaseous oxides of nitrogen. In its review of the draft PA, CASAC agreed with this conclusion (Diez Roux and Sheppard, 2017).

Averaging time

The current primary NO_2 standards are based on 1-hour and annual averaging times. Together, these standards can provide protection against short- and long-term NO_2 exposures.

In establishing the 1-hour standard in the last review, the Administrator considered evidence from both experimental and epidemiologic studies. She noted that controlled human exposure studies and animal toxicological studies provided evidence that NO_2 exposures from less than one hour up to three hours can result in respiratory effects such as increased AR and inflammation. These included five controlled human exposure studies that evaluated the potential for increased AR following 1-hour exposures to 100 ppb NO_2 in people with asthma. In addition, epidemiologic studies had reported health effect associations with both 1-hour and 24-hour NO_2 concentrations, without indicating that either of these averaging periods was more closely linked with reported effects. Thus, the available experimental evidence provided support for considering an averaging time of shorter duration than 24 hours while the epidemiologic evidence provided support for considering both 1-hour and 24-hour averaging times. Given this evidence, the Administrator concluded that, at a minimum, a primary concern with regard to averaging time was the level of protection provided against 1-hour NO_2 exposures. Based on available analyses of NO_2 air quality, she further concluded that a standard with a 1-hour averaging time could also be effective at protecting against effects associated with 24-hour NO_2 exposures (75 FR 6502, February, 9, 2010).

Based on the considerations summarized above, the Administrator judged that it was appropriate to set a new NO_2 standard with a 1-hour averaging time. She concluded that such a standard would be expected to effectively limit short-term (e.g., 1- to 24-hours) NO_2 exposures that had been linked to adverse respiratory effects. She also retained the existing annual standard to continue to provide protection against effects potentially associated with long-term exposures to oxides of nitrogen (75 FR 6502, February, 9, 2010). These decisions were consistent with CASAC advice to establish a short-term primary standard for oxides of nitrogen based on using 1-hour maximum NO_2 concentrations and to retain the current annual standard (Samet, 2008, p. 2; Samet, 2009, p. 2).

As in the last review, support for a standard with a 1-hour averaging time comes from both the experimental and epidemiologic evidence. Controlled human exposure studies evaluated in the ISA continue to provide evidence that NO_2 exposures from less than 1-hour up to three hours can result in increased AR in individuals with asthma (U.S. EPA, 2016, Tables 5-1 and 5-2). These controlled human exposure studies provide key evidence supporting the ISA's determination that "[a] causal relationship exists between short-term NO_2 exposure and respiratory effects based on evidence for asthma exacerbation" (U.S. EPA, 2016, p. 1-17). In addition, the epidemiologic literature assessed in the ISA provides support for short-term averaging times ranging from 1-hour up to 24-hours (e.g., U.S. EPA, 2016 Figures 5-3, 5-4 and Table 5-12). Consistent with the evidence in the last review, the ISA concludes that there is no indication of a stronger association for any particular short-term duration of NO_2 exposure (U.S. EPA, 2016, section 1.6.1). Thus, a 1-hour averaging time reasonably reflects the exposure durations used in the controlled human exposure studies that provide the strongest support for the

ISA's determination of a causal relationship. In addition, a standard with a 1-hour averaging time is expected to provide protection against the range of short-term exposure durations that have been associated with respiratory effects in epidemiologic studies (i.e., 1-hour to 24-hours). When taken together, we reach the conclusion that the combined evidence from experimental and epidemiologic studies continues to support an NO_2 standard with a 1-hour averaging time to protect against health effects related to short-term NO_2 exposures. In its review of the draft PA, the CASAC found that there continued to be scientific support for the 1-hour averaging time (Diez Roux and Sheppard, 2017).

With regard to protecting against long-term exposures, the evidence supports considering the overall protection provided by the combination of the annual and 1-hour standards. The current annual standard was originally promulgated in 1971 (36 FR 8186, April 30, 1971), based on epidemiologic studies reporting associations between respiratory disease and long-term exposure to NO_2. The annual standard was retained in subsequent reviews, in part to provide a margin of safety against the serious effects reported in animal studies using long-term exposures to high NO_2 concentrations (above 8,000 ppb) (U.S. EPA, 1995).

As described above, evidence newly available in the current review demonstrates associations between long-term NO_2 exposures and asthma development in children, based on NO_2 concentrations averaged over year of birth, year of diagnosis, or entire lifetime. Supporting evidence indicates that repeated short-term NO_2 exposures could contribute to this asthma development. In particular, the ISA states that "findings for short-term NO_2 exposure support an effect on asthma development by describing a potential role for repeated exposures to lead to recurrent inflammation and allergic responses," which are "identified as key early events in the proposed mode of action for asthma development" (U.S. EPA, 2016, p. 6-64 and p. 6-65). Taken together, the evidence supports the potential for recurrent short-term NO_2 exposures to contribute to the asthma development that has been reported in epidemiologic studies to be associated with long-term exposures. Thus, in establishing standards to protect against adverse health effects related to long-term NO_2 exposures, we reach the conclusion that the evidence supports the consideration of both 1-hour and annual averaging times. In its review of the draft PA, CASAC supported this approach to considering the protection provided against long-term NO_2 exposures by the combination of the annual and 1-hour NO_2 standards. CASAC specifically noted that "it is the suite of the current 1-hour and annual standards, together, that provide protection against adverse effects" (Diez Roux and Sheppard, 2017).

Level and form

In considering the extent to which evidence supports or calls into question the levels or forms of the current NO$_2$ standards, we revisit the following specific questions addressed in Chapter 3 of this PA:

- **To what extent does the evidence indicate adverse respiratory effects attributable to short- or long-term NO$_2$ exposures lower than previously identified or below the existing standards?**

In addressing this question in Chapter 3, we consider the range of NO$_2$ exposure concentrations that have been evaluated in experimental studies (controlled human exposure and animal toxicology) and the ambient NO$_2$ concentrations in locations where epidemiologic studies have reported associations with adverse outcomes.

Short-Term

Controlled human exposure studies demonstrate the potential for increased AR in some people with asthma following 30-minute to 1-hour exposures to NO$_2$ concentrations near those in the ambient air (Section 3.2.2).[129] In evaluating the NO$_2$ exposure concentrations at which increased AR has been observed, we consider both the group mean results reported in individual studies and the results from a recent meta-analysis evaluating individual-level data (Brown, 2015; U.S. EPA, 2016, Section 5.2.2.1). Group mean responses in individual studies, and the variability in those responses, can provide insight into the extent to which observed changes in AR are due to NO$_2$ exposures, rather than to chance alone, and have the advantage of being based on the same exposure conditions. The meta-analysis can aid in identifying trends in individual-level responses across studies and can have the advantage of increased power to detect effects, even in the absence of statistically significant effects in individual studies.

As discussed in more detail in Section 3.2.2.1, individual studies consistently report statistically significant NO$_2$-induced increases in AR following resting exposures to NO$_2$ concentrations at or above 250 ppb, but have generally not reported statistically significant increases in AR following resting exposures to NO$_2$ concentrations from 100 to 200 ppb. When individual-level data from these studies were combined in a meta-analysis, Brown (2015) reported that significant majorities of study participants experienced increased AR following resting exposures to NO$_2$ concentrations from 100 to 530 ppb. In some affected individuals, the

---

[129] As discussed in Chapter 3, experimental studies have not reported other respiratory effects following short-term exposures to NO$_2$ concentrations at or near those found in the ambient air. In addition, experimental studies examining asthma-related effects attributable to long-term NO$_2$ exposures are limited to exposures to NO$_2$ concentrations well-above those found in the ambient air and well-above those that could occur under the current standards (i.e. $\geq$ 1,000 ppb).

magnitudes of these increases were large enough to have potential clinical relevance. Following exposures to 100 ppb NO_2 specifically, the lowest exposure concentration evaluated, a marginally significant majority of study participants experienced increased AR.[130] Important limitations in this evidence include the lack of an apparent dose-response relationship between NO_2 and AR and uncertainty in the adversity of the reported increases in AR. These uncertainties become increasingly important at the lower NO_2 exposure concentrations (i.e., at or near 100 ppb), as the evidence for NO_2-induced increases in AR becomes less consistent across studies.

With regard to the epidemiologic evidence from U.S. and Canadian studies, as described in Sections 3.2.2.2 and 3.3.2.1, we consider the ambient NO_2 concentrations in locations where such studies have examined associations with asthma-related hospital admissions or emergency department visits (short-term) or with asthma incidence (long-term). In particular, we consider the extent to which NO_2-health effect associations are consistent, precise, statistically significant, and present for distributions of ambient NO_2 concentrations that likely would have met the current standards. To the extent NO_2-health effect associations are reported in study areas that would likely have met the current standards, the evidence supports the potential for the current standards to allow the NO_2-associated effects indicated by those studies. In the absence of studies reporting associations in locations meeting the current NO_2 standards, there is greater uncertainty regarding the potential for reported effects to be caused by NO_2 exposures that occur with air quality meeting those standards. We also note consideration of important uncertainties in the evidence, including the potential for copollutant confounding and exposure measurement error, and the extent to which near-road NO_2 concentrations are reflected in the available air quality data.

With regard to epidemiologic studies of short-term NO_2 exposures, as discussed in Section 3.2.2.2, we note the following. First, the only recent multicity study evaluated (Stieb et al., 2009), which had maximum 1-hour DVs ranging from 67 to 242 ppb, did not report a positive association between NO_2 and ED visits. In addition, of the single-city studies (see Figure 3-1) that reported positive and relatively precise associations between NO_2 and asthma hospital admissions and ED visits, most locations had NO_2 concentrations likely to have violated the current 1-hour NO_2 standard over at least part of the study period. In addition, had near-road NO_2 monitors been in place during study periods, DVs would likely have been higher. Thus, it is

---

[130] When the analysis was restricted only to non-specific AR following exposures to 100 ppb NO_2, the percentage who experienced increased AR was larger and statistically significant. In contrast, when the analysis was restricted only to specific AR following exposures to 100 ppb NO_2, the majority of study participants did not experience increased AR (U.S. EPA, 2016; Brown 2015).

likely that even the one study location with a maximum DV of 100 ppb (Atlanta) would have violated the existing 1-hour standard during study periods.[131] For the study locations with maximum DVs below 100 ppb, mixed results have been reported, with associations that are generally statistically non-significant and imprecise. As with the studies reporting more precise associations, near-road monitors were not in place during these study periods. If they had been, it is unclear whether 1-hour DVs would have been below 100 ppb. In drawing conclusions based on this epidemiologic evidence, we must also consider the potential for copollutant confounding as ambient NO_2 concentrations are often highly correlated with other pollutants. This can complicate attempts to distinguish between independent effects of NO_2 and effects of the broader pollutant mixture. While this has been addressed to some extent in available studies, uncertainty remains for the most relevant copollutants (i.e., those related to traffic such as $PM_{2.5}$, EC/BC, and CO). Taken together, we reach the conclusion that available U.S. and Canadian epidemiologic studies of hospital admissions and emergency department visits do not indicate the occurrence of NO_2-associated effects in locations and time periods with NO_2 concentrations that would clearly have met the current 1-hour NO_2 standard (i.e., with its level of 100 ppb and 98th percentile form).

In giving further consideration specifically to the form of 1-hour standard, we note that the available evidence and information is consistent with that informing consideration of form in the last review. The last review focused on the upper percentiles of the distribution of NO_2 concentrations based, in part, on evidence for health effects associated with short-term NO_2 exposures from experimental studies which provided information on specific exposure concentrations that were linked to respiratory effects. In that review, the EPA specified a 98th percentile form, rather than a 99th percentile, for the new 1-hour standard. In combination with the 1-hour averaging time and 100 ppb level, a 98th percentile form was judged to provide appropriate public health protection. In addition, compared to the 99th percentile, a 98th percentile form was expected to provide greater regulatory stability.[132] A 98th percentile form is also consistent with our consideration of uncertainties in the health effects that have the potential to occur at 100 ppb.

---

[131] Based on recent air quality information for Atlanta, 98th percentiles of daily maximum 1-hour NO_2 concentrations are higher at near-road monitors than non-near-road monitors (Figures 2-9 and 2-10, above). These differences could have been even more pronounced during study periods, when NO_X emissions from traffic sources were higher (Section 2.1.2, above).

[132] As noted in the last review, a less stable form could result in more frequent year-to-year shifts between meeting and violating the standard, potentially disrupting ongoing air quality planning without achieving public health goals (75 FR 6493, February 9, 2010).

Specifically, when combined with the 1-hour averaging time and the level of 100 ppb, the 98th percentile form limits, but does not eliminate, the potential for exposures to 100 ppb NO$_2$.[133]

Long-Term

In addressing the question posed above with regard to health effects related to long-term NO$_2$ exposures, we first consider the basis for the current annual standard. It was originally set to protect against NO$_2$-associated respiratory disease in children reported in a series of epidemiologic studies (36 FR 8186, April 30, 1973). In subsequent reviews, the EPA has retained the annual standard, judging that it provides protection with an adequate margin of safety against the serious effects that have been reported in animal studies following long-term exposures to NO$_2$ concentrations above 1,000 ppb. In the 2010 review, the EPA noted that, though some evidence supported the need to limit long-term exposures to NO$_2$, the evidence for adverse health effects attributable to long-term NO$_2$ exposures did not support changing the level of the annual standard.

In the current review, the strengthened "likely to be causal" relationship between long-term NO$_2$ exposures and respiratory effects is supported by epidemiologic studies of asthma development. While these studies strengthen the evidence for effects of long-term exposures, compared to the last review, they are subject to important uncertainties, including the potential for confounding by traffic-related copollutants. The potential for such confounding is particularly important to consider when interpreting epidemiologic studies of long-term NO$_2$ exposures given (1) the relatively high correlations observed, and modeled, between long-term ambient concentrations of NO$_2$ and long-term concentrations of other roadway-associated pollutants; (2) the general lack of information from copollutant models on the potential for NO$_2$ associations that are independent of other traffic-related pollutants or mixtures; and (3) the general lack of experimental support for effects following long-term exposures to NO$_2$ concentrations near those in the ambient air. Thus, it is unclear the degree to which the observed effects are independently related to exposure to ambient concentrations of NO$_2$. The epidemiologic evidence from several U.S. and Canadian studies is also subject to uncertainty with regard to the extent to which the studies accurately characterized exposures of the study populations, further limiting what these studies can tell us regarding the adequacy of the current standards.

While we recognize the above uncertainties, we consider what studies of long-term NO$_2$ and asthma development can tell us with regard to the adequacy of the current NO$_2$ standards. As discussed above for short-term exposures, we consider the degree to which the evidence

---

[133] The 98th percentile corresponds to about the 7th or 8th highest daily maximum 1-hour NO$_2$ concentration in a year.

indicates adverse respiratory effects associated with long-term NO_2 exposures in locations that would have met the NAAQS. As summarized in Section 3.3.1, the causal determination for long-term exposures is supported both by studies of long-term NO_2 exposures and studies indicating a potential role in asthma development for repeated short-term exposures to high NO_2 concentrations. As such, when considering the ambient NO_2 concentrations present during study periods, we consider these concentrations within the context of both the 1-hour and annual NO_2 standards. As discussed in Section 3.3.2.1, while annual DVs in study locations were often below 53 ppb, maximum 1-hour DVs in most locations were near or above 100 ppb.[134, 135] Because these study-specific DVs are based on the area-wide NO_2 monitors in place during study periods, they likely do not reflect the NO_2 concentrations near the largest roadways, which are expected to be higher in most urban areas. Had near-road monitors been in place during study periods, NO_2 DVs based on near-road concentrations likely would have been higher in many locations, and would have been more likely to exceed the level of the annual and/or 1-hour standard(s).

Overall, the evidence does not provide support for NO_2-attributable asthma development in children in locations with NO_2 concentrations that would have clearly met the current annual and 1-hour standards. The strongest evidence informing the level at which effects may occur comes from U.S. and Canadian epidemiologic studies that are subject to critical uncertainties related to copollutant confounding and exposure assessment. Even if these fundamental uncertainties were to be dismissed, our evaluation indicates that most of the locations included in epidemiologic studies of long-term NO_2 exposure and asthma incidence would likely have violated either one or both of the current NO_2 standards, over at least parts of the study periods.

Based on the information discussed above, we revisit the following question:

- **To what extent does the evidence indicate adverse respiratory effects attributable to short- or long-term NO_2 exposures lower than previously identified or below the existing standards**

    In addressing this question, we note that (1) experimental studies do not indicate adverse respiratory effects attributable to either short- or long-term NO_2 exposures lower than previously identified and that (2) epidemiologic studies do not provide support for associations between adverse effects and ambient NO_2 concentrations that would have clearly met the current standards. Taken together, we reach the conclusion that the available evidence does not support

---

[134] Mean 1-hour DVs from the study periods were also near or above 100 ppb in many study locations.

[135] As discussed in Chapter 2, analyses demonstrate that a 1-hour NO_2 DV (based on three-year averages of 98th percentiles of annual distributions of daily maximum 1-hour NO_2 concentrations) at or below 100 ppb generally corresponds to an annual DV (based on annual average NO_2 concentrations) below 35 ppb.

the need for increased protection against short- or long-term NO_2 exposures, beyond that provided by the existing standards. In its review of the draft PA, the CASAC agreed with this conclusion, stating that "[t]he CASAC concurs with the EPA that the current scientific literature does not support a revision to the primary NAAQS for nitrogen dioxide" (Diez Roux and Sheppard, 2017, p. 9). Therefore, we have not identified potential alternative standard levels or forms for consideration.

5.2   AIR QUALITY-, EXPOSURE- AND RISK-BASED CONSIDERATIONS

As described in Chapter 4, beyond our consideration of the scientific evidence, we also consider the extent to which quantitative analyses of NO_2 air quality, exposures or health risks could inform conclusions on the adequacy of the public health protection provided by the current primary NO_2 standards. Conducting such quantitative analyses, if appropriate, could inform judgments about the public health impacts of NO_2-related health effects and could help to place the evidence for specific effects into a broader public health context. To this end, in the REA Planning Document (U.S. EPA, 2015) and in this PA we have evaluated the extent to which the available evidence and information provide support for conducting new or updated analyses of NO_2 exposures and/or health risks, beyond the analyses conducted in the 2008 REA (U.S. EPA, 2008). In doing so, we carefully considered the assessments developed as part of the last review of the primary NO_2 NAAQS (U.S. EPA, 2008) and the newly available scientific and technical information, particularly considering the degree to which updated analyses in the current review are likely to substantially add to our understanding of NO_2 exposures and/or health risks. We have also considered the CASAC advice and public input received on the REA Planning Document (see Chapter 4) and on the draft PA (Diez Roux and Sheppard, 2017).

As discussed above and in the REA Planning Document (U.S. EPA, 2015, Section 2.1.1), an important uncertainty identified in the 2008 REA was the characterization of 1-hour NO_2 concentrations around major roadways. The 2008 REA estimated NO_2 concentrations on/near roads by applying literature-derived adjustment factors to NO_2 concentrations at area-wide monitoring sites. A key consideration in the current review is the extent to which newly available information could reduce uncertainties with regard to NO_2 concentrations around major roads. As discussed in Section 2.3.2, data from recently deployed near-road NO_2 monitors provide an improved understanding of such ambient NO_2 concentrations. Therefore, in this PA we have conducted updated analyses comparing ambient NO_2 concentrations (i.e., as surrogates of potential exposures) to health-based benchmarks, with a particular focus on study areas where near-road monitors have been deployed.[136]

When considering analyses comparing NO_2 air quality with health-based benchmarks, we focus on the following specific questions:

- **To what extent are ambient NO_2 concentrations that may be of public health concern estimated to occur in locations meeting the current NO_2 standards? What are the important uncertainties associated with those estimates?**

As discussed in Section 4.2, benchmarks are based on information from controlled human exposure studies of NO_2 exposures and AR. In identifying specific NO_2 benchmarks, and considering the weight to place on each, we consider both the group mean results reported in individual studies and the results of a meta-analysis that combined data from multiple studies (Brown, 2015; U.S. EPA, 2016, Section 5.2.2.1), as described above.

When taken together, the results of individual controlled human exposure studies and of the meta-analysis by Brown (2015) support consideration of NO_2 benchmarks between 100 and 300 ppb, based largely on studies of non-specific AR in people with asthma exposed at rest. As discussed in more detail in Section 4.2.1.1, benchmarks from the upper end of this range are supported by the results of individual studies, the majority of which reported statistically significant increases in AR following NO_2 exposures at or above 250 ppb, and by the results of the meta-analysis by Brown (2015). Benchmarks from the lower end of this range, including 100 ppb, are supported by the results of the meta-analysis, even though individual studies do not consistently report statistically significant NO_2-induced increases in AR at these lower concentrations. In particular, individual studies have not generally reported significant increases in AR following resting exposures to 100 ppb NO_2, but the meta-analysis indicates that a marginally significant majority of study participants experienced an increase in AR following exposures to 100 ppb NO_2 (Brown, 2015).[137] While there are a variety of factors that likely underlie the observed variability, they are not fully known and the variability remains an uncertainty in evaluating these results.

---

[136] We have not conducted more complex NO_2 exposure and risk assessments in this review. As discussed above (Sections 4.3, 4.4) and in the REA Planning document (U.S. EPA, 2015), such updated assessments would be unlikely to substantially improve our understanding of NO_2 exposures and health risks associated with the current standards, beyond what we know from the air quality-benchmark comparisons described in Chapter 4 (Section 4.2) and the risk assessment conducted in the last review.

[137] Results were statistically significant when analyses were restricted to non-specific AR, but not when analyses were restricted to specific AR.

In further considering the potential public health implications of exposures to NO_2 concentrations at or above benchmarks, we also note the discussion of uncertainties in Section 3.2.2.1. As discussed in more detail in that section, there is no indication of a dose-response relationship between NO_2 and AR in people with asthma, regardless of the challenge type (i.e., specific or non-specific) or exposure conditions (i.e., resting or exercising) (Goodman, 2009; Brown, 2015). Though the lack of an apparent dose-response relationship does not necessarily indicate the lack of an NO_2 effect, it adds uncertainty to our interpretation of the controlled human exposure studies of AR. An additional uncertainty is the clinical relevance of the reported NO_2-induced increases in AR though, as described above (section 5.1), the meta-analysis by Brown (2015) has partially addressed this uncertainty.

Thus, while we consider benchmarks from 100 to 300 ppb, uncertainties in the evidence from controlled human exposure studies suggest that caution is appropriate when interpreting the potential public health implications of 1-hour NO_2 concentrations at these benchmarks. While this is true even for the higher benchmarks (i.e., given the lack of an apparent dose-response relationship and remaining uncertainty with regard to adversity), it is particularly the case for the 100 ppb benchmark, where the results of individual studies are inconsistent.

As discussed in Section 4.2, analyses of unadjusted air quality, which meets the current standards in all locations, indicate almost no potential for 1-hour exposures to NO_2 concentrations at or above any of the benchmarks examined, including 100 ppb. Analyses of air quality adjusted upwards to just meet the current 1-hour standard[138] indicate virtually no potential for 1-hour exposures to NO_2 concentrations at or above 200 ppb (or 300 ppb), and almost none for exposures at or above 150 ppb. This is the case for both averaged estimates and estimates in worst-case years, including at near-road monitoring sites within a few meters of heavily trafficked roads. With respect to the lowest benchmark evaluated, analyses estimate that there is potential for exposures to 1-hour NO_2 concentrations at or above 100 ppb on some days (i.e., about one to 10 days per year, on average). As described above, this result is consistent with our expectations, given that the current 1-hour standard, with its 98th percentile form, is expected to limit, but not eliminate, the occurrence of 1-hour NO_2 concentrations of 100 ppb.

Thus, the current 1-hour NO_2 standard is expected to allow virtually no potential for exposures to the NO_2 concentrations that have been shown most consistently to increase AR in people with asthma, even under worst-case conditions across a variety of study areas with among

---

[138] In all study areas, ambient NO_2 concentrations required smaller upward adjustments to just meet the 1-hour standard than to just meet the annual standard. Therefore, as noted above (Section 4.2.1), when adjusting air quality to just meet the current NO_2 NAAQS, we applied the adjustment needed to just meet the 1-hour standard.

the highest NO$_X$ emissions in the U.S. Such NO$_2$ concentrations are not estimated to occur, even at monitoring sites adjacent to some of the most heavily trafficked roadways. In addition, the current standard provides protection against NO$_2$ exposures that have the potential to exacerbate asthma symptoms, but for which the evidence indicates greater uncertainty in both the occurrence of such exacerbations and in their severity, should they occur (i.e., closer to 100 ppb). Given the results of these analyses, and the uncertainties inherent in their interpretation, we conclude that there is little potential for exposures to ambient NO$_2$ concentrations that would be of public health concern in locations meeting the current 1-hour standard.

5.3 CASAC ADVICE

In addition to the evidence and quantitative information discussed above (Chapters 3 and 4), we have also considered the advice and recommendations of the CASAC, based on its review of the draft PA, and comments from the public on the draft PA (Diez Roux and Sheppard, 2017). In its comments on the draft PA, the CASAC concurred with staff's overall preliminary conclusions that it is appropriate to consider retaining the current primary NO$_2$ standards without revision, stating that, "the CASAC recommends retaining, and not changing the existing suite of standards" (Diez Roux and Sheppard, 2017). With regard to the individual elements of the standards, the CASAC stated the following:

- **Indicator and averaging time:** "[T]here is strong evidence for the selection of NO$_2$ as the indicator of oxides of nitrogen" and "for the selection of 1-hour and annual averaging times" (Diez Roux and Sheppard, 2017 pg. 9).

- **Level of the 1-hour standard:** "[T]here are notable adverse effects at levels that exceed the current standard, but not at the level of the current standard. Thus, the CASAC advises that the current 1-hour standard is protective of adverse effects and that there is not a scientific basis for a standard lower than the current 1-hour standard" (Diez Roux, and Sheppard 2017 pg. 9).

  - **Form of the 1-hour standard:** The CASAC also "recommends retaining the current form" for the 1-hour standard (Diez Roux and Sheppard, 2017).

- **Level of the annual standard:** Recognizing that the 1-hour standard can effectively contribute to limiting long-term NO$_2$ concentrations, the CASAC agreed with the EPA's decision to focus on the degree of protection provided by the suite of standards against long-term exposures, rather than the annual standard alone. In providing support for retaining the level of the existing annual standard, the CASAC specifically noted that "it is the suite of the current 1-hour and annual standards, together, that provide protection against adverse effects" (Diez Roux and Sheppard, 2017, p. 9). As noted above, "the CASAC recommends retaining, and not changing the existing suite of standards" (Diez Roux and Sheppard, 2017).

5.4 STAFF CONCLUSION ON THE ADEQUACY OF THE CURRENT STANDARDS

The overarching question guiding our consideration of the available evidence and information for the current review is:

- **Does the available scientific evidence and information support or call into question the adequacy of the public health protection afforded by the current primary NO₂ standards?**

Staff has reached the conclusion that the current body of evidence, in combination with the available information from quantitative analyses, supports the adequacy of the public health protection provided by the current primary NO_2 standards and does not call into question any of the elements of the current standards. Staff further reaches the conclusion that it is appropriate to consider retaining the current standards, without revision, in this review. In reaching these conclusions we particularly note the following:

- The strongest evidence for NO_2-related effects comes from controlled human exposure studies, and a meta-analysis of individual-level data from those studies, demonstrating the potential for people with asthma to experience NO_2-induced increases in AR following exposures under resting conditions from 100 to 530 ppb. While increases in AR are considered to be a hallmark of asthma and can lead to poorer control of symptoms, the potential public health implications of these results are not clear due to the lack of an apparent dose-response relationship and uncertainty in the potential adversity of the reported changes in AR. There is additional uncertainty at the lower end of this range because, while the meta-analysis indicates that the majority of study volunteers with asthma experienced increased AR following resting exposures to 100 ppb NO_2, individual studies do not consistently report NO_2-induced increases in AR at this exposure concentration.

- While epidemiologic studies provide consistent evidence for associations with asthma-related effects, studies conducted in the U.S. and Canada do not provide support for associations of asthma-related hospital admissions or emergency department visits with exposure to short-term NO_2 concentrations in locations that would have clearly met the current standards. This is particularly the case given that NO_2 concentrations near the most heavily-trafficked roadways are not likely reflected by the NO_2 concentrations measured at monitors in operation during study years. We additionally note that there is potential for copollutant confounding contributing to some uncertainty regarding the extent to which the observed effects can be attributed independently to NO_2 exposure.

- While epidemiologic studies report associations between long-term NO_2 exposures and asthma development in children, these studies are subject to important uncertainties that limit the extent to which they provide insight into the adequacy of the public health protection provided by the current standards. These uncertainties include the potential for copollutant confounding, given the high correlations between long-term averages of NO_2 and other traffic-related pollutants, and the potential for exposure measurement error. Even if these uncertainties were to be dismissed, epidemiologic studies conducted in the U.S. and Canada do not indicate associations of asthma incidence with exposures to long-term NO_2 in locations that would have clearly met the current standards. This is particularly the case given that NO_2 concentrations near the most heavily-trafficked roadways are not likely reflected by monitors in operation during study years.

- The current 1-hour NO_2 standard is expected to allow virtually no potential for exposures to the NO_2 concentrations that have been shown most consistently to increase AR in people with asthma (i.e., above 200 ppb), even under worst-case conditions across a variety of study areas with among the highest NO_X emissions in the U.S. Such NO_2 concentrations were not estimated to occur, even at monitoring sites adjacent to some of the most heavily trafficked roadways.

- The current 1-hour standard is expected to limit, though not eliminate, the potential for exposures to 1-hour concentrations at or above 100 ppb. Thus, the current standard provides protection against NO_2 exposures that have the potential to exacerbate asthma symptoms, but for which the evidence indicates uncertainty in the occurrence of such exacerbations and in their severity, should they occur.

As noted in Chapter 1 above, in establishing primary standards that, in the Administrator's judgment, are requisite to protect public health with an adequate margin of safety, the Administrator seeks to establish standards that are neither more nor less stringent than necessary for this purpose. The Act does not require that primary standards be set at a zero-risk level, but rather at a level that reduces risk sufficiently so as to protect public health with an adequate margin of safety. We additionally note that different public health policy judgments could lead to different conclusions regarding the extent to which the current standards protect the public health. Such judgments include those related to the appropriate degree of public health protection that should be afforded as well as the appropriate weight to be given to various aspects of the evidence and information, including how to consider uncertainties.

In this context, we recognize that the uncertainties and limitations associated with the many aspects of the estimated relationships between NO_2 exposures and potentially adverse respiratory effects are amplified with consideration of increasingly lower NO_2 concentrations. In staff's view, there is appreciable uncertainty in the extent to which reductions in asthma exacerbations or asthma incidence would result from revising the current NO_2 standards. The basis for any consideration of alternative standards would reflect different public health policy judgments as to the appropriate approach for weighing uncertainties in the evidence.

Based on all of the above considerations, and on CASAC advice and public input, we reach the conclusion that it is appropriate to consider retaining the current standards, without revision, in this review. The available evidence and information do not support the identification of potential alternative standards that provide a different degree of public health protection. In light of this conclusion, we have not identified potential alternative standards for consideration.

5.5 AREAS FOR FUTURE RESEARCH AND DATA COLLECTION

The uncertainties and limitations that remain in the review of the primary NO_2 standards are largely related to understanding the range of ambient concentrations over which we have confidence in the occurrence of NO_2-attributable adverse health effects, as indicated by available epidemiologic, controlled human exposure, and animal toxicological studies. We encourage continued investigation to further reduce these uncertainties, as described below, including additional efforts to characterize the adversity and clinical-significance of reported effects. Looking across the literature, we further encourage the synthesis of evidence into meta-analyses, such as those recently conducted by Goodman et al. (2009) and Brown (2015). Meta-analyses can facilitate an integrated assessment of the available evidence, and can provide additional power to detect effects, beyond that found in many individual studies.

In this section, we highlight areas for future health-related research, model development, and data collection activities to address the uncertainties and limitations in the current body of scientific evidence for NO_2. These future research areas reflect advice from the CASAC (Diez Roux and Sheppard, 2017). If undertaken, research focused on the uncertainties highlighted in this section could provide important evidence for informing future reviews of the primary NO_2 NAAQS.[139]

Epidemiologic Evidence:

In the current review, epidemiologic studies provide the strongest evidence for effects that can clearly be considered adverse. However, the degree to which these studies provide clarity regarding the NO_2 exposure concentrations at which potentially adverse effects are likely to occur is limited by inherent uncertainties, including the potential for copollutant confounding from other traffic-related pollutants, uncaptured NO_2 exposure gradients, differential population exposure, and unexplored effect measure modification. We encourage research to improve our understanding of these issues, as described below.

---

[139] In some cases, research in these areas can go beyond aiding standard setting to also inform the development of more efficient and effective control strategies.

- Understanding Potential Confounding: Within the current body of scientific literature, there are uncertainties related to the extent of potential confounding by traffic-related copollutants. Given the relatively high correlations between concentrations of NO_2 and several co-occurring pollutants, this issue particularly impacts our consideration of the epidemiologic evidence for associations between long-term NO_2 exposures and asthma development. As stated in section 3.3.2, the emphasis that is placed on epidemiology studies reflects, in part, the degree to which the evidence indicates that copollutant confounding may occur. Thus, additional research into this issue could reduce an important uncertainty in the existing evidence and could further inform decisions in future reviews of the primary NO_2 standards. In its advice to the Administrator, CASAC has reiterated the need for additional research on this issue (Diez Roux and Frey, 2015b).

- Seasonal and Indoor/Outdoor exposure gradients: Issues of seasonal differences in NO_2 exposures and distinguishing between ambient and indoor exposures need to be better addressed to improve inferences of health effects (e.g., see Diez Roux and Frey, 2015a). As noted in the CASAC response to charge questions regarding the second draft of the ISA, "[t]here can be more interpretation from studies of indoor exposure and for studies undertaken in different seasons," and "[t]he indoor exposure studies can be informative because they do not have the same mix of co-pollutants as the outdoor exposure studies." (Diez Roux and Frey, 2015a).

- Differential Population Exposure: There is a need to better address issues of equity and environmental justice related to the distributions of NO_2 exposures among and between communities of varying socioeconomic status. Such distributions may be related to the identification of groups at higher risk for adverse effects as a result of combinations of exposure scenarios, populations, lifestages, and socioeconomic factors. As noted by the CASAC (Diez Roux and Frey, 2015a), "[t]here is substantial evidence that groups in poverty or who are non-white experience higher exposures to NO_2, but the epidemiological evidence is still lacking. It is important to clearly show how the exposure differences follow socioeconomic status (SES) or racial gradients, because for those that are considered causal or likely to be causal, there is high potential for larger health effects even if the epidemiological evidence of a direct effect modification is lacking." Related to this, it is important to better characterize the locations where peak NO_2 exposures occur (e.g., on-road in vehicles, roadside as pedestrians, in urban street canyons, near other non-road facilities such as rail yards or industrial facilities) to potentially identify where higher population exposures may be occurring. Such improved characterization of peak NO_2 exposures may improve our understanding of NO_2-related health effects in at-risk populations.

- Characterizing At-Risk Populations and Effect Measure Modification: The degree to which people with asthma are more responsive to NO_2 exposures could vary depending on the disease phenotype (i.e., atopic versus non-atopic). In addition, sensitivity to NO_2

exposures may be enhanced for people who have other conditions, such as diabetes or cardiovascular disease. These and other factors that could confer sensitivity to NO_2 exposures (e.g., psychosocial stress, copollutant exposures) should be further investigated.

Mode of Action

There is an ongoing scientific need to improve our mechanistic understanding related to effects of exposure to NO_2, including effects on respiratory endpoints (e.g., related to asthma) and on endpoints related to cardiovascular disease and premature mortality. Research needs within this category are discussed below.

- Temporal exposure patterns: Research evaluating the latency period for the development of new asthma, and the NO_2 temporal exposure patterns that can contribute to the disease, would be beneficial to future reviews. For example, an important uncertainty in the current review is the role of single or repeated short-term NO_2 exposures versus persistent long-term exposures in the development of asthma. This type of work could inform future consideration of the protection provided by the 1-hour and annual standards.

- Worsening Asthma Symptoms: In the current review, evidence from controlled human exposure studies supports the occurrence of increased AR in people with asthma following resting exposures to NO_2 concentrations from 100 to 530 ppb. This increase in AR suggests the potential for NO_2 exposures to worsen asthma symptoms. However, results of these studies are not always consistent, potentially due to differences in the endpoints examined, the challenge agents used, or the exposure conditions (exercise versus rest). In addition, a dose-response relationship is not apparent from the available data, contributing to uncertainty in our interpretation of these studies. To address these uncertainties, we encourage future research into the occurrence of increased AR, or other effects, following exposures to NO_2 concentrations found in the ambient air. We further encourage additional efforts to characterize the potential for exposures to such ambient NO_2 concentrations to result in effects that are adverse or that are clinically relevant. Such research could help to inform future consideration of the health protection provided by the NO_2 NAAQS.

- Better characterization of non-respiratory endpoints: While the body of epidemiologic studies reporting associations between NO_2 and cardiovascular disease, cardio-metabolic disease, birth outcomes, and cancer is rapidly growing, there is uncertainty in the degree to which such effects are specifically associated with NO_2. Controlled human NO_2 exposure studies would be most informative in elucidating these potential relationships, but are unlikely to be feasible for many outcomes. Additionally, controlled animal exposure studies and other mechanistic studies may be particularly informative for the next NO_2 NAAQS review, as they may foster greater understanding of mode of action for a wider range of endpoints.

Spatial and Temporal Gradients in Ambient NO₂ Concentrations

An improved understanding of ambient NO_2 concentrations around important sources of NO_X emissions would help to inform consideration of potential exposures in future reviews.

- Near-road monitor data: Since the last review, NO_2 monitors have been deployed near major roads in large urban areas across the U.S. Future research should use the data from these monitors to better understand spatial and temporal gradients in ambient NO_2 concentrations. Additionally, characterizations of near road NO_2 concentrations should include measurements of other traffic-related pollutants such as ultra-fine particles (number concentration), black carbon, PM, and CO.

- Other near-source NO_2 concentrations: In addition, we encourage research to better characterize other near-source environments where NO_2 exposures may be of importance. These environments could include urban street canyons and areas near large sources of NO_X emissions, such as airports or rail yards. Widespread passive NO_2 sampling may be a practical way to identify hotspots, and to characterize NO_2 concentration gradients in such locations.

- Factors that affect ambient NO_2 concentrations: Further collection and refinement of information to better characterize the factors contributing to variability in ambient NO_2 concentrations is also an ongoing need, including information on air quality monitor site characteristics, available traffic counts, fleet mix data, and historical emissions information and trends.

Quantification of Risk and Exposure Estimates

Research into better understanding the uncertainties surrounding risk estimates, as they relate to risk quantification, as well as research into new methods of risk quantification, may be helpful to future reviews, as detailed below.

- Quantification of Uncertainties: There is a need for improved methods to quantify key uncertainties in epidemiology-based risk estimates. In this review, key uncertainties included those related to copollutant confounding, exposure characterization, baseline incidence, and the shape of the concentration-response function. Taken together, these uncertainties contributed to the staff conclusion not to conduct an updated epidemiology-based risk assessment (as described in section 4.4.2).

- Risk Quantification Methods: Although in this review there was not sufficient new scientific information to support an updated risk assessment, the development of exposure quantification methods, models, and data, or new interpretations of existing information, may inform risk assessments in future reviews.

- Health Benchmarks: There is an ongoing need for scientific information to support the identification of health-based benchmarks. In the current review, we focus on benchmarks from 100 to 300 ppb, based on information from controlled human exposure studies of AR. We encourage additional research into the occurrence of AR, or other effects, following exposures to NO_2 concentrations at and below 100 ppb. Such research could provide a more robust body of evidence for reaching conclusions on the occurrence of increased AR or other effects, and on the potential adversity of those effects, following exposures to NO_2 concentrations near those found in the ambient air in the U.S. In general, studies are most useful if they characterize the effects of exposure to NO_2 itself, independent of co-occurring pollutants.

5.6   REFERENCES

Brown, JS. (2015). Nitrogen dioxide exposure and airway responsiveness in individuals with asthma. Inhal Toxicol 27: 1-14. http://dx.doi.org/10.3109/08958378.2014.979960

Diez Roux A and Frey H.C. (2015a). Letter from Drs. Ana Diez Roux, Chair and H. Christopher Frey, Immediate Past Chair, Clean Air Scientific Advisory Committee to EPA Administrator Gina McCarthy. CASAC Review of the EPA's Integrated Science Assessment for Oxides of Nitrogen- Health Criteria (Second External Review Draft). EPA-CASAC-15-001. September 9, 2015. Available at: https://yosemite.epa.gov/sab/sabproduct.nsf/6612DAF24438687B85257EBB0070369C/$File/EPA-CASAC-15-001+unsigned.pdf

Diez Roux A and Frey H.C. (2015b). Letter from Drs. Ana Diez Roux, Chair and H. Christopher Frey, Immediate Past Chair, Clean Air Scientific Advisory Committee to EPA Administrator Gina McCarthy. CASAC Review of the EPA's Review of the Primary National Ambient Air Quality Standards for Nitrogen Dioxide: Risk and Exposure Assessment Planning Document. EPA-CASAC-15-002. September 9, 2015. Available at: https://yosemite.epa.gov/sab/sabproduct.nsf/264cb1227d55e02c85257402007446a4/A7922887D5BDD8D485257EBB0071A3AD/$File/EPA-CASAC-15-002+unsigned.pdf

Goodman, JE; Chandalia, JK; Thakali, S; Seeley, M. (2009). Meta-analysis of nitrogen dioxide exposure and airway hyper-responsiveness in asthmatics. Crit Rev Toxicol 39: 719-742. http://dx.doi.org/10.3109/10408440903283641

Reddel, HK; Taylor, DR; Bateman, ED; Boulet, LP; Boushey, HA; Busse, WW; Casale, TB; Chanez, P; Enright, PL; Gibson, PG; de Jongste, JC; Kerstjens, HA; Lazarus, SC; Levy, ML; O'Byrne, PM; Partridge, MR; Pavord, ID; Sears, MR; Sterk, PJ; Stoloff, SW; Sullivan, SD; Szefler, SJ; Thomas, MD; Wenzel, SE. (2009). An official American Thoracic Society/European Respiratory Society statement: Asthma control and exacerbations: Standardizing endpoints for clinical asthma trials and clinical practice. Am J Respir Crit Care Med 180: 59-99. http://dx.doi.org/10.1164/rccm.200801-060ST

Sheppard, E (2017). Letter form Dr. Elizabeth A. (Lianne) Sheppard, Chair, Clean Air Scientific Advisory Committee to EPA Administrator E. Scott Pruitt. CASAC Review of the EPA's Policy Assessment for the Review of the Primary National Ambient Air Quality Standards for Nitrogen Dioxide (External Review Draft- September 2016). EPA-CASAC-17-001. March 7th, 2017. Available at: https://yosemite.epa.gov/sab/sabproduct.nsf/LookupWebProjectsCurrentCASAC/7C2807D0D9BB4CC8852580DD004EBC32/$File/EPA-CASAC-17-001.pdf

Stieb, DM; Szyszkowicz, M; Rowe, BH; Leech, JA. (2009). Air pollution and emergency department visits for cardiac and respiratory conditions: A multi-city time-series analysis. Environ Health 8. http://dx.doi.org/10.1186/1476-069X-8-25

U.S. EPA (1995). Review of the National Ambient Air Quality Standards for Nitrogen Dioxide: Assessment of Scientific and Technical Information. U.S. EPA, Office of Air Quality Planning and Standards, Research Triangle Park, NC. EPA-452/R-95-005. Available at: https://www3.epa.gov/ttn/naaqs/standards/nox/data/noxsp1995.pdf

U.S. EPA (2008). Risk and Exposure Assessment to Support the Review of the NO_2 Primary National Ambient Air Quality Standard. U.S. EPA, Office of Air Quality Planning and Standards. Research Triangle Park, NC. EPA 452/R-08-008a/b. November 2008. Available at: http://www.epa.gov/ttn/naaqs/standards/nox/s_nox_cr_rea.html.

U.S. EPA (2015). Review of the Primary National Ambient Air Quality Standards for Nitrogen Dioxide: Risk and Exposure Assessment Planning Document. U.S. EPA, Office of Air Quality Planning and Standards, Research Triangle Park, NC. EPA-452/D-15-001. May 13, 2015. Available at: https://www3.epa.gov/ttn/naaqs/standards/nox/data/20150504reaplanning.pdf

U.S. EPA (2016). Integrated Science Assessment for Oxides of Nitrogen – Health Criteria (2016 Final Report). U.S. EPA, National Center for Environmental Assessment, Research Triangle Park, NC. EPA/600/R-15/068. January 2016. Available at: https://cfpub.epa.gov/ncea/isa/recordisplay.cfm?deid=310879.

APPENDIX A

NITROGEN DIOXIDE AIR QUALITY

Table of Contents

LIST OF TABLES

LIST OF FIGURES

1. CALCULATION OF ANNUAL AND 1-HOUR NO₂ DESIGN VALUES

The following procedures were used to calculate annual and 1-hour design values (DVs) for NO₂.

- Raw hourly NO₂ data was downloaded from the following sources:
    - EPA's AQS database (parameter code 42602)
    - Canadian NAPs network website (http://maps-cartes.ec.gc.ca)
    - US SEARCH network ftp site (ftp://mail.atmospheric-research.com)

- Two types of DVs were calculated for each site in each of the 3 networks.
- Annual DV:
    - For each site and year, the annual DV is the mean hourly concentration.
    - 75% of hours in the year must be present for the annual DV to be valid.
- Hourly DV:
    - Daily max was identified for each sampling day for each site
    - Two different methods were used to calculate 98th percentile values:
    - Using days with 18 or more hourly samples
    - Using days with 1 or more hourly samples

- The final 98th percentile reported is the maximum of the 2 methods for each site and year.

- DV for each 3-year window was calculated by averaging the annual final 98ᵗʰ percentile values.

- Hourly DV were considered valid if they meet the following criteria:
    - Each day must have samples for >= 75% of hours to be valid.
    - 75% of days in a quarter must be valid for the quarter to be valid.
    - A year must have had 4 complete quarters for it to be valid.
    - A DV must have had 3 valid years to be valid.

2. ANNUAL AND HOURLY DESIGN VALUES FOR SELECT EPIDEMIOLOGIC STUDIES

Design values (DVs) have been calculated for locations of select epidemiologic studies examining respiratory effects associated with NO₂ exposures. These studies were identified from the Integrated Science Assessment for Oxides of Nitrogen – Health Criteria. Calculations were based on methods outlined above.

The DVs reported in the tables below are the highest DVs in the specified location for each year as calculated according to the methods above. The respective completeness of the hourly and annual DVs are also reported. For annual DVs, this is reported as the percentage of complete days in the year. For hourly DVs, this is reported as the number of complete quarters (75% of hours in a day, 75% of days in a quarter) for the 3-year period (the specified year and two years before).

The DVs reported in this technical memo are limited to the locations where key epidemiologic studies in the current review of the Primary NO_2 NAAQS have been conducted. For each city/CBSA, the respective study(ies) and relevant study details are reported, including the monitor IDs that were used in determining DVs

2.1 Short-Term Epidemiologic Studies For Asthma-Related Emergency Department Visits And Hospital Admissions

- Atlanta, GA   Strickland et al. (2010)

  - Pediatric asthma ED visits (5-17 yr) 1993-2004 (entire years)
  - Exposure assignment: "Daily concentrations of ambient 1-hour maximum [NO_2]…were obtained from several networks of ambient monitors…We used population-weighting to combine daily pollutant measurements across monitors."
  - Design values: We used 5 specific AQS monitors and a SEARCH monitor for which we were able to obtain data to compute design values based on personal communication with author. (131210001, 131210047, 131210048, 132230003, 162470001, SEARCH monitor on Jefferson Street)

  Strickland et al. (2011)

  - Pediatric asthma ED visits (5-17 yr) 1993-2004 (May-Oct)
  - Exposure assignment: NO_2 concentrations obtained from 3 networks of stationary monitors. Three exposure metrics used: (1) one downtown monitor was selected as central site, (2) all monitors used to calculate unweighted average of pollutant concentrations for all monitors, and (3) population-weighted average concentration.
  - Design values: We used the same monitors indicated in the Strickland et al. (2010) paper. (131210001, 131210047, 131210048, 132230003, 162470001, SEARCH monitor on Jefferson Street)

  Peel et al. (2005)

  - Asthma ED visits across all ages and in children (2-18 yr) (January 1993- August 2000)
  - Exposure assignment: Average of NO_2 concentrations from monitors for several monitoring networks

- Design values: We used the 5 AQS monitors and the SEARCH monitor that represent air quality in metro Atlanta. (131210001, 131210047, 131210048, 132230003, 162470001, SEARCH monitor on Jefferson Street)

Table A-1. Hourly and Annual Design Values for Atlanta, GA from 1993-2004.

Hourly			Annual		
Year	Max DV (Monitor ID)[a]	Number of Complete Quarters[a]	Year	Max DV (Monitor ID)[a]	Completeness (%)[c]
1993	-		1993	25 (131210048)	57.3
1994	-		1994	23 (131210048)	83.2
1995	76 (131210048)	9	1995	19 (131210048)	96.3
1996	83 (131210048)	11	1996	27 (131210048)	98.4
1997	88 (131210048)	11	1997	25 (131210048)	87.5
1998	99 (JST)	2	1998	26 (JST)	36.5
1999	100 (JST)	5	1999	26 (JST)	80
2000	95 (JST)	9	2000	23 (JST)	95.1
2001	86 (131210048)	12	2001	23 (JST)	97.5
2002	86 (131210048)	12	2002	19 (JST)	97.9
2003	81 (131210048)	12	2003	20 (JST)	89
2004	75 (131210048)	12	2004	20 (JST)	92

[a]JST refers to the SEARCH network monitor in Atlanta, GA located on Jefferson Street
[b]In the respective 3-year period (12 quarters) for the hourly DV; 9 or more quarters (75%) are required for a DV to be valid
[c]Valid annual DVs require 75% completeness

- Detroit, MI   Li et al. (2011)   - Pediatric asthma ED visits (2-18 yr), 2004-2006 (entire years)
- Exposure assignment: Average NO_2 concentrations across two monitors
- Design values: We used the two monitors indicated in the study. These were the only monitors with valid data during the study period. (261630016, 26160019)

Table A-2. Hourly and Annual Design Values for Detroit, MI from 2004-2006.

Hourly			Annual		
Year	Max DV (Monitor ID)	Number of Complete Quarters[a]	Year	Max DV (Monitor ID)	Completeness (%)[b]
2004	-	12	2004	19 (261630016)	77.2
2005	-	12	2005	20 (261630016)	95.4
2006	55 (261630016)	12	2006	16 (261630016)	98.9

[a]In the respective 3-year period (12 quarters) for the hourly DV; 9 or more quarters (75%) are required for a DV to be valid
[b]Valid annual DVs require 75% completeness

- Los Angeles, CA Linn et al. (2000) - Hospital admissions, all ages, 1992-1995 (entire years)
- Exposure assignment: Average NO_2 concentration over all monitors; Study indicated AQS monitors by map
- Design values: We identified AQS monitors by approximation of location by map and operation during study period: 060370002, 060371002, 060371103, 060590001, 060658001, 060371201, 0603716002, 060370113, 060375001, 060371301, 060372005, 060371601, 060370206, 060591003, 160659001, 060658001, 060711004, 060719004, 060371701

Table A-3. Hourly and Annual Design Values for Los Angeles, CA from 1992-1995.

Hourly			Annual		
Year	Max DV (Monitor ID)	Number of Complete Quarters[a]	Year	Max DV (Monitor ID)	Completeness (%)[b]
1992	-	12	1992	51 (060371301)	93.6
1993	-	12	1993	50	92.8
1994	171 (060371103)	12	1994	50	95.4
1995	168 (060371103)	12	1995	46	94.1

[a]In the respective 3-year period (12 quarters) for the hourly DV; 9 or more quarters (75%) are required for a DV to be valid
[b]Valid annual DVs require 75% completeness

- Cleveland/ Cincinnati, OH Jaffe et al. (2003) -Asthma ED visits, 5-34 yr, July 1991- June 1996 (summers, June-August)
-Exposure assignment: Monitor with highest 24-h avg concentration
- Design values: We used all operating monitors in Cleveland and Cincinnati for the study period (390350033, 390350043, 390350060, 390610035, 390610037)

Table A-4. Hourly and Annual Design Values for Cleveland and Cincinnati, OH from 1991-1996.

Hourly			Annual		
Year	Max DV (Monitor ID)	Number of Complete Quarters[a]	Year	Max DV (Monitor ID)	Completeness (%)[b]
Cleveland					
1991	-		1991	30 (390610035)	83.6

Year	Max DV (Monitor ID)	Number of Complete Quarters	Year	Max DV (Monitor ID)	Completeness
1992	-		1992	29 (390350033)	81.3
1993	89 (390350033)	8	1993	28 (390350060)	90.1
1994	92 (390350033)	4	1994	28 (390350060)	92.6
1995	83 (390350060)	12	1995	27 (390350060)	93.7
1996	83 (390610035)	4	1996	29 (390610037)	98.9
Cincinnati					
1991	-		1991	17 (210371001)	94.2
1992	-		1992	15 (210371001)	94.4
1993	77 (390610035)	10	1993	18 (210371001)	89.5
1994	76 (390610035)	12	1994	20 (210371001)	92.6
1995	80 (390610037)	4	1995	20 (210371001)	94.2
1996	83 (390610035)	4	1996	19 (210371001)	94.2

[a]In the respective 3-year period (12 quarters) for the hourly DV; 9 or more quarters (75%) are required for a DV to be valid
[b]Valid annual DVs require 75% completeness

- New York City, NY   Ito et al. (2007)   - Asthma ED visits, all ages, 1999-2002 (entire years)
  - Exposure assignment: Average NO_2 concentrations from 15 monitors
  - Design values: We used the 15 monitors closest to city center that were in operation during study years (360050073, 360050080, 360050083, 360050110, 360610010, 360610056, 360810097, 360810098, 360810124, 340130011, 340130016, 340131003, 340170006, 340230011, 340390004)

- Bronx/Manhattan, NY   ATSDR 2006   - Asthma ED visits, all ages, Bronx: January 1999-November 2000, Manhattan: September 1999-November 2000
  - Exposure assignment: NO_2 concentrations from a monitor in the Bronx and a monitor in Manhattan
  - Design values: We used the monitors specified by location in the study (360050073, 360610010)
  - Design values for these encompass years before the study period.

Table A-5. Hourly and Annual Design Values for New York, NY from 1999-2002.

Hourly			Annual		
Year	Max DV (Monitor ID)	Number of Complete Quarters[a]	Year	Max DV (Monitor ID)	Completeness (%)[b]
New York City (including Newark, NJ)					
1999	-		1999	42 (340390004)	97.8
2000	-		2000	41 (340390004)	98
2001	102 (340390004)	12	2001	40 (340390004)	97.1

2002	101 (340390004)	12	2002	40 (340390004)	92.6
Bronx					
1999	94 (360050073)	1	1999	32 (360050073)	25.4
2000	94 (360050073)	1	2000	N/A	
Manhattan					
1999	86 (360610010)	11	1999	36 (360610010)	89
2000	86 (360610010)	11	2000	36 (360610010)	92

[a]In the respective 3-year period (12 quarters) for the hourly DV; 9 or more quarters (75%) are required for a DV to be valid
[b]Valid annual DVs require 75% completeness

- Edmonton, Canada   Villeneuve et al. (2007)   -Asthma emergency department visits, all ages > 2 yr, 1992-2002 (entire years)
-Exposure assignment: Average NO_2 concentration across three monitoring stations
-Design values: We used the three NAPS monitors in Edmonton (90121, 90122, 90130)

Table A-6. Hourly and Annual Design Values for Edmonton, Canada from 1992-2002.

Hourly			Annual		
Year	Max DV (Monitor ID)	Number of Complete Quarters[a]	Year	Max DV (Monitor ID)	Completeness (%)[b]
1992	-		1992	36 (90122)	99.5
1993	-		1993	27 (90130)	95.2
1994	242 (90122)	12	1994	27 (90130)	97.6
1995	87 (90122)	12	1995	27 (90130)	97.1
1996	96 (90122)	12	1996	25 (90130)	97.2
1997	96 (90122)	12	1997	26 (90130)	93.3
1998	100 (90122)	12	1998	27 (90130)	98
1999	86 (90122)	12	1999	24 (90130)	99.4
2000	74 (90122)	12	2000	25 (90130)	98.9
2001	70 (90122)	12	2001	25 (90130)	98
2002	76 (90122)	12	2002	25 (90130)	98.3

[a]In the respective 3-year period (12 quarters) for the hourly DV; 9 or more quarters (75%) are required for a DV to be valid
[b]Valid annual DVs require 75% completeness

- Toronto, Canada   Burnett et al. (1999)   -Hospital admissions, all ages, 1980-1994 (entire years)
-Exposure assignment: Average NO_2 concentration across four monitoring stations that are not likely influence by a local source (site-specific) (reference Burnett et al. JAWMA 1998 (48))
-Design values: We identified study monitors using the map provided in the reference study (60403, 60410, 60413, 60418). We also used other NAPS monitors in

metro Toronto as they were also representative of potential population exposures (60401, 60402, 60412, 60414, 60417, 60419, 60420, 60421, 60422, 60423, 60424, 60425)

Table A-7. Hourly and Annual Design Values for Toronto, Canada from 1980-1994.

Hourly			Annual		
Year	Max DV (Monitor ID)	Number of Complete Quarters[a]	Year	Max DV (Monitor ID)	Completeness (%)[b]
1980	-		1980	36 (60403)	95.4
1981	-		1981	36 (60403)	95
1982	127 (60412)	8	9182	34 (60412)	89.4
1983	120 (60412)	10	1983	30 (60414)	97.5
1984	123 (60412)	12	1984	36 (60412)	95.6
1985	113 (60412)	10	1985	38 (60412)	38.7
1986	115 (60412)	6	1986	34 (60418)	97
1987	110 (60412)	2	1987	32 (60403)	94.6
1988	130 (60412)	3	1988	38 (60422)	56.9
1989	120 (60412)	7	1989	33 (60403)	97.9
1990	117 (60403)	12	1990	30 (60403)	81.2
1991	110 (60403)	12	1991	30 (60422)	91.7
1992	220 (60403)	12	1992	45 (60422)	95.6
1993	227 (60403)	12	1993	31 (60420)	28.7
1994	223 (60403)	12	1994	30 (60424)	99.4

[a]In the respective 3-year period (12 quarters) for the hourly DV; 9 or more quarters (75%) are required for a DV to be valid
[b]Valid annual DVs require 75% completeness

- Multicity (Montreal, Ottawa, Edmonton, St. John, Halifax, Toronto, Vancouver) Stieb et al. (2009) - Asthma ED visits, all ages, 1992-2003 (entire years)
- Exposure assignment: Average NO_2 concentrations from all monitors in each city.
- Design values: All monitors in the NAPS database for each city were used.

Table A-8. Hourly and Annual Design Values for 7 Canadian cities from 1992-2003.

Hourly			Annual		
Year	Max DV (Monitor ID)	Number of Complete Quarters[a]	Year	Max DV (Monitor ID)	Completeness (%)[b]
Montreal 1997-2002					

1997	-		1997	31 (50115)	96.8
1998	-		1998	28 (50115)	96.7
1999	85 (50109)	12	1999	29 (50109)	98.9
2000	83 (50109)	12	2000	26 (50109)	92
2001	78 (50109)	10	2001	30 (50109)	55
2002	76 (50109)	8	2002	26 (50109)	77.5
Ottawa (60101, 60104) 1992-2000					
1992	-		1992	72 (60101)	14.3
1993	-		1993	24 (60101)	98.3
1994	198 (60101)	9	1994	25 (60101)	98.9
1995	78 (60101)	12	1995	25 (60101)	95.8
1996	74 (60101)	10	1996	23 (60101)	66.5
1997	70 (60101)	7	1997	25 (60101)	15.8
1998	73 (60101)	7	1998	25 (60101)	93.9
1999	83 (60101)	7	1999	33 (60101)	59.5
2000	82 (60101)	10	2000	22 (60101)	90.1
Edmonton 1992-2002					
1992	-		1992	36 (90122)	99.5
1993	-		1993	27 (90130)	95.2
1994	242 (90122)	12	1994	27 (90130)	97.6
1995	87 (90122)	12	1995	27 (90130)	97.1
1996	96 (90122)	12	1996	25 (90130)	97.2
1997	96 (90122)	11	1997	26 (90122)	93.3
1998	100 (90122)	11	1998	27 (90130)	98
1999	86 (90122)	11	1999	24 (90130)	99.4
2000	74 (90122)	12	2000	25 (90130)	98.9
2001	70 (90122)	12	2001	25 (90130)	98
2002	76 (90122)	12	2002	25 (90122)	95.7
St. John 1992-1996					
1992	-		1992	14 (10102)	88.7
1993	-		1993	15 (10102)	83.9
1994	96 (10102)	7	1994	15 (10102)	21
1995	50 (10102)	4	1995	21 (10102)	1.7
1996	48 (10102)	1	1996	#N/A	
Halifax 1999-2002					
1999	-		1999	#N/A	
2000	-		2000	18 (30118)	83.5
2001	67 (30118)	6	2001	17 (30118)	76
2002	61(30118)	9	2002	17 (30118)	80
Toronto 1999-2003					
1999	-		1999	28 (60430)	79.7
2000	-		2000	30 (60430)	33.1
2001	94 (60423)	7	2001	31 (60403)	11.4
2002	98 (60423)	4	2002	26 (60429)	81.3
2003	84 (60429)	9	2003	27 (60429)	92.3

Vancouver 1999-2003					
1999	-		1999	32 (100120)	3.4
2000	-		2000	27 (100112)	97.9
2001	86 (100120)	0	2001	26 (100112)	97.7
2002	56 (100121)	12	2002	25 (100112)	97.3
2003	56 (100121)	12	2003	25 (100112)	98

[a]In the respective 3-year period (12 quarters) for the hourly DV; 9 or more quarters (75%) are required for a DV to be valid
[b]Valid annual DVs require 75% completeness

2.2 Design Values For Epidemiologic Studies Of Asthma-Related Respiratory Symptoms

- New Haven, CT   Gent et al. (2009)   -Wheeze in asthmatic children (4-12 yr), Aug 2000-Feb 2004
  -NO_2 central site
  -NO_2 effect estimated from multipollutant model with source apportionment factor of EC, zinc, lead, copper, and selenium

Table A-9. Hourly and Annual Design Values for New Haven, CT for 2000-2004.

Hourly			Annual		
Year	Max DV (Monitor ID)	Number of Complete Quarters[a]	Year	Max DV (Monitor ID)	Completeness (%)[b]
2000	-		2000	25 (90091123)	95.8
2001	-		2001	27 (90091123)	98.3
2002	75 (090091123)	12	2002	25 (90091123)	98.2
2003	73 (090091123)	12	2003	25 (90091123)	97.6
2004	70 (090090027)	4	2004	23 (90091123)	11

[a]In the respective 3-year period (12 quarters) for the hourly DV; 9 or more quarters (75%) are required for a DV to be valid
[b]Valid annual DVs require 75% completeness

- Bronx, NY   Patel et al. (2010)   -Wheeze and chest tightness in asthmatic adolescents (14-20 yr), 2003 - 2005
  - NO_2 central site (PS52, MONITOR ID 360050110)

Table A-10. Hourly and Annual Design Values for Bronx, NY for 2003-2005.

Hourly	Annual

Year	Max DV (Monitor ID)	Number of Complete Quarters[a]	Year	Max DV (Monitor ID)	Completeness (%)[b]
2003	82 (360050110)	9	2003	30 (360050110)	78.4
2004	79 (360050110)	11	2004	30 (360050110)	97.4
2005	78 (360050110)	11	2005	29 (360050110)	89.1

[a]In the respective 3-year period (12 quarters) for the hourly DV; 9 or more quarters (75%) are required for a DV to be valid
[b]Valid annual DVs require 75% completeness

- Fresno/Clovis, CA   Mann et al. (2010)   - Wheeze in asthmatic children (6-11 yr), Nov 2000- April 2005 (subgroup analyses for boys with intermittent asthma and atopy)
  - NO_2 central site (CARB monitoring stations used in study; AQS data used for DVs)

Table A-11. Hourly and Annual Design Values for Fresno, CA from 2000-2005.

Hourly			Annual		
Year	Max DV (Monitor ID)	Number of Complete Quarters[a]	Year	Max DV (Monitor ID)	Completeness (%)[b]
2000	-		2000	21 (60190008)	93.8
2001	-		2001	21 (60190008)	94.8
2002	74 (60190008)	12	2002	20 (60190008)	94.4
2003	77 (60190008)	12	2003	19 (60190008)	94.6
2004	72 (60190008)	12	2004	17 (60190008)	94.7
2005	71 (60190008)	12	2005	17 (60190008)	90.2

[a]In the respective 3-year period (12 quarters) for the hourly DV; 9 or more quarters (75%) are required for a DV to be valid
[b]Valid annual DVs require 75% completeness

- Multicity   O'Connor et al. (2008)   - Wheeze/cough, slow play, and missed school in children with asthma (5-12 yr), Aug 1998 – July 2001
  - NO_2 central site (monitors near residences)

Table A-12. Hourly and Annual Design Values for 7 U.S. cities from 1998-2001.

Hourly			Annual		
Year	Max DV (Monitor ID)	Number of	Year	Max DV (Monitor ID)	Completeness (%)[b]

		Complete Quarters[a]			
Boston, MA					
1998	83 (250250021)	12	1998	31 (250250002)	94.1
1999	81 (250250002)	12	1999	30 (250250002)	93.2
2000	76 (250250002)	11	2000	43 (250250002)	0.1
2001	73 (250250002)	11	2001	30 (250250002)	93.1
Bronx, NY					
1998	96 (36050080)	11	1998	36 (360050080)	97.8
1999	95 (36050080)	11	1999	33 (360050080)	96.7
2000	94 (360050073)	1	2000	33 (360050080)	32.3
2001	94 (360050073)	1	2001	32 (360050110)	56.9
Chicago, IL					
1998	-		1998	32 (170310063)	95.3
1999	-		1999	31.5 (170310063)	98.4
2000	87 (170310063)	12	2000	32 (170310063)	94.7
2001	86 (170310063)	12	2001	32 (170310063)	98.7
Dallas, TX					
1998	-	12	1998	20 (481130069)	98.3
1999	-	12	1999	21 (481130069)	88.3
2000	75 (481130069)	11	2000	19 (481130069)	59.5
2001	73 (481130069)	11	2001	19 (481130069)	90
New York, NY					
1998	-		1998	42 (340390004)	98.9
1999	-		1999	42 (340390004)	97.8
2000	102 (340390004)	12	2000	41 (340390004)	98
2001	102 (340390004)	12	2001	40 (340390004)	97.1
Seattle, WA					
1998	-		1998	20 (530330020)	67
1999	-		1999	22 (530330020)	37.7
2000	65 (530330032)	4	2000	21 (530330032)	24.1
2001	65 (530330032)	8	2001	22 (530330032)	32.3
Tucson, AZ					
1998	-		1998	17 (40191011)	98.3
1999	-		1999	19 (40191028)	91.7
2000	58 (40191011)	12	2000	17 (40191011)	97.1
2001	58 (40191011)	12	2001	17 (40191028)	98.8

[a]In the respective 3-year period (12 quarters) for the hourly DV; 9 or more quarters (75%) are required for a DV to be valid
[b]Valid annual DVs require 75% completeness

- Atlanta, GA   Darrow et al. (2010)   - Respiratory ED visits, all ages, 1993-2004
  - Average daily concentration across monitors with population weighting (5 AQS monitors +SEARCH monitor, operational 1998)

Tolbert et al. (2007)
- Respiratory ED visits, all ages, March 1994 – Dec 2004
- Average of NO_2 concentrations across monitors (5 AQS monitors + SEARCH, operational 1998)

Peel et al. (2005)
- Respiratory ED visits, all ages, Jan 1993 - Aug 2000)
- Average of NO_2 concentrations across monitors (5 AQS monitors + SEARCH, operational 1998))

Table A-13. Hourly and Annual Design Values for Atlanta, GA for 1993-2004

Hourly			Annual		
Year	Max DV (Monitor ID)[a]	Number of Complete Quarters[b]	Year	Max DV (Monitor ID)[a]	Completeness (%)[c]
1993	-		1993	25 (131210048)	57.3
1994	-		1994	23 (131210048)	83.2
1995	76 (131210048)	9	1995	19 (131210048)	96.3
1996	83 (131210048)	11	1996	27 (131210048)	98.4
1997	88 (131210048)	11	1997	25 (131210048)	87.5
1998	99 (JST)	2	1998	26 (JST)	36.5
1999	100 (JST)	5	1999	26 (JST)	80
2000	95 (JST)	9	2000	23 (JST)	95.1
2001	86 (131210048)	12	2001	23 (JST)	97.5
2002	86 (131210048)	12	2002	19 (JST)	97.9
2003	81 (131210048)	12	2003	20 (JST)	89
2004	75 (131210048)	12	2004	20 (JST)	92

[a]JST refers to the SEARCH network monitor in Atlanta, GA located on Jefferson Street
[b]In the respective 3-year period (12 quarters) for the hourly DV; 9 or more quarters (75%) are required for a DV to be valid
[c]Valid annual DVs require 75% completeness

- Toronto, Canada

Burnett et al. (2001)
- Pediatric respiratory hospital admissions (< 2 yr), 1980-1994
- Average NO_2 concentrations across 4 monitors with continuous data and not influence by any local source (16 monitors from NAPS database; some monitors with high DV likely influenced by traffic)

Table A-14. Hourly and Annual Design Values for Toronto, Canada from 1980-1994.

Hourly			Annual		
Year	Max DV (Monitor ID)	Number of Complete Quarters[a]	Year	Max DV (Monitor ID)	Completeness (%)
1980	-		1980	36 (60403)	95.4

1981	-		1981	36 (60403)	95
1982	127 (60412)	8	1982	34 (60412)	89.4
1983	120 (60412)	10	1983	30 (60403)	93.8
1983	120 (60412)	10	1983	30 (60403)	97.5
1984	123 (60412)	12	1984	36 (60412)	95.6
1985	113 (60412)	10	1985	38 (60412)	38.7
1986	115 (60412)	6	1986	34 (60418)	97
1987	110 (60412)	2	1987	32 (60403)	94.6
1988	130 (60422)	3	1988	38 (60422)	56.9
1989	120 (60422)	7	1989	33 (60403)	97.9
1990	117 (60403)	12	1990	30 (60403)	81.2
1991	110 (60403)	12	1991	30 (60422)	91.7
1992	220 (60423)	12	1992	45 (60422)	95.6
1993	227 (60423)	12	1993	31 (60420)	28.7
1994	223 (60423)	12	1994	30 (60424)	99.4

[a]In the respective 3-year period (12 quarters) for the hourly DV; 9 or more quarters (75%) are required for a DV to be valid
[b]Valid annual DVs require 75% completeness

- Vancouver, Canada   Yang et al. (2003)   - Respiratory hospital admissions (< 3 yr and ≥ 65 yr), 1986-1998
  - Average NO_2 concentrations from 30 monitors from the British Columbia network (DVs from 7 NAPS monitors)

Table A-15. Hourly and Annual Design Values for Toronto, Canada from 1986-1998.

Hourly			Annual		
Year	Max DV (Monitor ID)	Number of Complete Quarters[a]	Year	Max DV (Monitor ID)	Completeness (%)
1986	-		1986	37 (100118)	11.3
1987	-		1987	29 (100112)	93.8
1988	104	8 (100120)	1988	33 (100108)	23.7
1989	100	12 (100110)	1989	31 (100112)	94
1990	90	12 (100110)	1990	31 (100112)	92.1
1991	85	12 (100110)	1991	33 (100112)	92.4
1992	230	11 (100108)	1992	60 (100112)	89.8
1993	241	8 (100108)	1993	31 (100108)	23.6
1994	322	4 (100108)	1994	24 (100112)	97.6
1995	104	1 (100108)	1995	26 (100112)	97.7
1996	66	12 (100121)	1996	27 (100112)	97.8
1997	68	12 (1001120)	1997	29 (100112)	97.5
1998	67	12 (1001120)	1998	29 (100112)	97.4

[a]In the respective 3-year period (12 quarters) for the hourly DV; 9 or more quarters (75%) are required for a DV to be valid
[b]Valid annual DVs require 75% completeness

Fung et al. (2006)   - Respiratory hospital admissions, 1995-1999
- Average of NO_2 concentrations across monitors in city
-% increase per 30 ppb increment of NO_2: 8.6 (4.2, 13.3)

Table A-16. Hourly and Annual Design Values for Toronto, Canada from 1995-1999.

Hourly			Annual		
Year	Max DV (Monitor ID)	Number of Complete Quarters[a]	Year	Max DV (Monitor ID)	Completeness (%)
1995	-		1995	26 (100112)	97.7
1996	-		1996	27 (100112)	97.8
1997	68 (100112)	12	1997	29 (100112)	97.5
1998	67 (100112)	12	1998	29 (100112)	97.4
1999	66 (100120)	5	1999	32 (100120)	3.4

[a]In the respective 3-year period (12 quarters) for the hourly DV; 9 or more quarters (75%) are required for a DV to be valid
[b]Valid annual DVs require 75% completeness

- Multicity   Dales et al. (2009)   - Pediatric respiratory hospital admissions (0-24 days), 1986-2000
- NO_2 concentrations across all monitors in cities

Table A-17. Hourly and Annual Design Values for 11 Canadian cities from 1986-2000.

Hourly			Annual		
Year	Max DV (Monitor ID)	Number of Complete Quarters[a]	Year	Max DV (Monitor ID)	Completeness (%)
Calgary					
1986	-		1986	34 (90227)	97.6
1987	-		1987	34 (90227)	98.9
1988	110 (90227)	12	1988	35 (90227)	98.8
1989	105 (90227)	12	1989	35 (90227)	98.9
1990	103 (90227)	12	1990	34 (90227)	94
1991	95 (90227)	12	1991	37 (90227)	98.6
1992	338 (90218)	8	1992	49 (90218)	99
1993	250 (90218)	12	1993	31 (90227)	99
1994	254 (90218)	12	1994	29 (90227)	98.9
1995	86 (90227)	12	1995	28 (90227)	98.9
1996	93 (90218)	12	1996	29 (90227)	98.7
1997	98 (90218)	12	1997	30 (90227)	99.4

1998	106 (90218)	12	1998	31 (90227)	98.5
1999	95 (90218)	12	1999	28 (90227)	97.9
2000	86 (90218)	12	2000	28 (90227)	99.2
Edmonton					
1986	-		1986	30 (90130)	95.6
1987	-		1987	31 (90130)	99.5
1988	86 (90130)	12	1988	28 (90130)	98.2
1989	78 (90130)	12	1989	26 (90130)	94.5
1990	80 (90122)	12	1990	27 (90130)	97.9
1991	100 (90121)	4	1991	29 (90130)	98.7
1992	243 (90122)	12	1992	36 (90122)	99.5
1993	239 (90122)	12	1993	27 (90130)	95.2
1994	242 (90122)	12	1994	27 (90130)	97.6
1995	87 (90122)	12	1995	27 (90130)	97.1
1996	96 (90122)	12	1996	25 (90130)	97.2
1997	96 (90122)	11	1997	26 (90122)	93.3
1998	100 (90122)	11	1998	27 (90130)	98
1999	86 (90122)	11	1999	24 (90130)	99.4
2000	74 (90122)	12	2000	25 (90130)	98.9
Halifax					
1986	-		1986	5 (30115)	97.6
1987	-		1987	19 (30117)	72.6
1988	62 (30117)	6	1988	21 (30117)	65.1
1989	60 (30117)	9	1989	17 (30117)	88
1990	58 (30177)	6	1990	21 (30118)	7.8
1991	62 (30118)	4	1991	22 (30118)	90.8
1992	148 (30118)	8	1992	40 (30118)	93.4
1993	154 (30118)	11	1993	21 (30118)	91.4
1994	148 (30118)	10	1994	18 (30118)	86
1995	61 (30118)	10	1995	19 (30118)	82.4
1996	55 (30118)	9	1996	18 (30118)	86.7
1997	58 (30118)	7	1997	#N/A	
1998	59 (30118)	7	1998	21 (30118)	96.3
1999	64 (30118)	4	1999	#N/A	
2000	59 (30118)	7	2000	18 (30118)	83.5
London					
1986	-		1986	26 (60901)	86.1
1987	-		1987	21 (60901)	94.8
1988	73 (60901)	12	1988	20 60901)	97.6
1989	70 (60901)	12	1989	22 (60901)	98.6
1990	70 (60901)	12	1990	21 (60901)	97.9
1991	70 (60901)	12	1991	19 (60901)	98.9
1992	113 (60901)	12	1992	13 (60901)	99.2
1993	113 (60901)	12	1993	20 (60901)	97.9
1994	113 (60901)	12	1994	23 (60901)	98.7

1995	70 (60901)	10	1995	22 (60901)	46.5
1996	70 (60901)	6	1996	18 (60903)	98.2
1997	70 (60901)	2	1997	18 (60903)	90.7
1998	59 (60903)	10	1998	18 (60903)	99.4
1999	63 (60903)	10	1999	19 (60903)	96.7
2000	65 (60903)	11	2000	17 (60903)	86.7
Ottawa					
1986	-		1986	36 (60101)	96
1987	-		1987	37 (60101)	97.8
1988	93 (60101)	12	1988	34 (60101)	95.4
1989	93 (60101)	12	1989	38 (60101)	97.8
1990	92 (60101)	12	1990	31 (60101)	99
1991	96 (60101)	8	1991	20 (60104)	48.3
1992	263 (60101)	5	1992	72 (60101)	14.3
1993	258 (60101)	5	1993	24 (60101)	98.3
1994	198 (60101)	9	1994	25 (60101)	98.9
1995	78 (60101)	12	1995	25 (60101)	95.8
1996	74 (60101)	10	1996	23 (60101)	66.5
1997	70 (60101)	7	1997	25 (60101)	15.8
1998	73 (60101)	7	1998	25 (60101)	93.9
1999	83 (60101)	7	1999	33 (60101)	59.5
2000	82 (60101)	10	2000	22 (60101)	90.1
St. John					
1986	N/A		1986	N/A	
1987	N/A		1987	N/A	
1988	N/A		1988	N/A	
1989	N/A		1989	N/A	
1990	57 (10102)	3	1990	15 (10102)	93.1
1991	56 (10102)	6	1991	14 (10102)	92.9
1992	99 (10102)	9	1992	14 (10102)	88.7
1993	98 (10102)	9	1993	15 (10102)	83.9
1994	96 (10102)	7	1994	15 (10102)	21
1995	50 (10102)	4	1995	21 (10102)	1.7
1996	48 (10102)	1	1996	N/A	
1997	43 (10102)	2	1997	8 (10102)	40.3
1998	32 (10102)	6	1998	5 (10102)	93.4
1999	35 (10102)	9	1999	7 (10102)	88.1
2000	36 (10102)	10	2000	8 (10102)	92.8
Toronto					
1986	-		1986	34 (60418)	97
1987	-		1987	32 (60403)	94.6
1988	130 (60422)	3	1988	38 (60422)	56.9
1989	120 (60422)	7	1989	33 (60403)	97.9
1990	117 (60403)	12	1990	30 (60403)	81.2
1991	110 (60403)	12	1991	30 (60422)	91.7

1992	220 (60423)	12	1992	45 (60422)	95.6
1993	227 (60423)	12	1993	31 (60420)	28.7
1994	223 (60423)	12	1994	30 (60424)	99.4
1995	120 (60420)	1	1995	32 (60425)	13.8
1996	108 (60423)	12	1996	34 (60425)	99.3
1997	103 (60423)	11	1997	35 (60425)	59.3
1998	87 (60423)	11	1998	30 (60403)	97.5
1999	91 (60425)	2	1999	28 (60403)	79.7
2000	89 (60423)	11	2000	30 (60430)	33.1
Vancouver					
1986	-		1986	37 (100118)	11.3
1987	-		1987	30 (100112)	93.8
1988	104 (100120)	8	1988	33 (100108)	23.7
1989	100 (100110)	12	1989	31 (100112)	94
1990	90 (100110)	12	1990	31 (100112)	92.1
1991	85 (100110)	12	1991	33 (100112)	92.4
1992	230 (100108)	11	1992	60 (100112)	89.8
1993	241 (100108)	8	1993	31 (100108)	23.6
1994	322 (100108)	4	1994	24 (100112)	97.6
1995	104 (100108)	1	1995	26 (100112)	97.7
1996	66 (100121)	12	1996	27 (100112)	97.8
1997	68 (100112)	12	1997	29 (100112)	97.5
1998	67 (100112)	12	1998	29 (100112)	97.4
1999	66 (100120)	5	1999	32 (100120)	3.4
2000	67 (100120)	1	2000	27 (100112)	97.9
Winnipeg					
1986	-		1986	20 (70119)	84.7
1987	-		1987	20 (70119)	87.2
1988	66 (70119)	11	1988	19 (70119)	93.5
1989	60 (70119)	12	1989	19 (70119)	93.9
1990	59 (70119)	12	1990	15 (70119)	90.9
1991	57.3 (70118)	12	1991	17 (70119)	93.9
1992	132 (70119)	11	1992	14 (70119)	81.7
1993	-		1993	17 (70119)	93
1994	-		1994	17 (70119)	89.4
1995	57 (70119)	12	1995	18 (70119)	94.3
1996	55 (70119)	12	1996	18 (70119)	94
1997	60 (70119)	12	1997	18 (70119)	94.3
1998	62 (70119)	12	1998	17 (70119)	93.7
1999	65 (70119)	12	1999	18 (70119)	94
2000	61 (70119)	12	2000	16 (70119)	93.7
Windsor					
1986	-		1986	25 (60204)	92.9
1987	-		1987	27 (60204)	88.4
1988	97 (60204)	11	1988	30 (60204)	29.6

Year	Max DV (Monitor ID)		Year	Max DV (Monitor ID)	Completeness
1989	90 (60204)	11	1989	28 (60204)	82.2
1990	87 (60204)	12	1990	25 (60204)	95
1991	83 (60204)	12	1991	25 (60204)	96.4
1992	123 (60204)	12	1992	18 (60204)	98.2
1993	127 (60204)	12	1993	26 (60204)	98.7
1994	130 (60204)	12	1994	28 (60204)	99.5
1995	87 (60204)	12	1995	25 (60204)	98.3
1996	79 (60204)	12	1996	26 (60204)	99.3
1997	71 (60204)	12	1997	24 (60204)	94.6
1998	66 (60204)	12	1998	24 (60204)	98.7
1999	67 (60204)	12	1999	23 (60204)	98.2
2000	67 (60204)	12	2000	22 (60204)	97.4
Hamilton					
1986	N/A		1986	27 (60501)	91.8
1987	N/A		1987	30 (60501)	14.6
1988	80 (60515)	11	1988	25 (60511)	95.4
1989	77 (60515)	11	1989	26(60512)	98.9
1990	73 (60515)	12	1990	22 (60512)	98.6
1991	70 (60515)	12	1991	22 (60512)	98.8
1992	160 (60515)	12	1992	28 (60515)	98.7
1993	163 (60515)	11	1993	23 (60511)	98.2
1994	163 (60515)	11	1994	23 (60511)	94.2
1995	73 (60515)	10	1995	24 (60511)	98.6
1996	67 (60515)	11	1996	22 (60511)	89.8
1997	66 (60515)	11	1997	19 (60511)	77
1998	68 (60515)	12	1998	23 (60515)	79.6
1999	78 (60511)	8	1999	28 (60511)	74.5
2000	80 (60511)	9	2000	23 (60511)	90

[a]In the respective 3-year period (12 quarters) for the hourly DV; 9 or more quarters (75%) are required for a DV to be valid
[b]Valid annual DVs require 75% completeness

- Multicity   Cakmak et al. 2011   - Respiratory hospital admissions, April 1993- March 2000
  - NO_2 concentrations across all monitors in cities

Table A-18. Hourly and Annual Design Values for 10 Canadian Cities from 1993-2000.

Year			Max DV (Monitor ID)		
Year	Max DV (Monitor ID)	Number of Complete Quarters[a]	Year	Max DV (Monitor ID)	Completeness (%)
Calgary					
1993	-		1993	31 (90227)	99

1994	-		1994	29 (90227)	98.9
1995	86 (90227)	12	1995	28 (90227)	98.9
1996	93 (90218)	12	1996	29 (90227)	98.7
1997	98 (90218)	12	1997	30 (90227)	99.4
1998	106 (90218)	12	1998	31 (90227)	98.5
1999	95 (90218)	12	1999	28 (90227)	97.9
2000	86 (90218)	12	2000	28 (90227)	99.2
Edmonton					
1993	-		1993	27 (90130)	95.2
1994	-		1994	27 (90130)	97.6
1995	87 (90122)	12	1995	27 (90130)	97.1
1996	96 (90122)	12	1996	25 (90130)	97.2
1997	96 (90122)	11	1997	26 (90130)	93.3
1998	100 (90122)	11	1998	27 (90130)	98
1999	86 (90122)	11	1999	24 (90130)	99.4
2000	74 (90122)	12	2000	25 (90130)	98.9
Halifax					
1993	-		1993	21 (30118)	91.4
1994	-		1994	18 (30118)	86
1995	61 (30118)	10	1995	19 (30118)	82.4
1996	55 (30118)	9	1996	18 (30118)	86.7
1997	58 (30118)	7	1997	N/A	#N/A
1998	59 (30118)	7	1998	21 (30118)	96.3
1999	64 (30118)	4	1999	N/A	#N/A
2000	59 (30118)	7	2000	18 (30118)	83.5
London					
1993	-		1993	20 (60901)	97.9
1994	-		1994	23 (60901)	98.7
1995	70 (60901)	10	1995	22 (60901)	46.5
1996	70 (60901)	6	1996	18 (60903)	98.2
1997	70 (60901)	2	1997	18 (60903)	90.7
1998	59 (60903)	10	1998	18 (60903)	99.4
1999	63 (60903)	10	1999	20 (60903)	96.7
2000	65 (60903)	11	2000	17 (60903)	86.7
Ottawa					
1993	-		1993	24 (60101)	98.3
1994	-		1994	25 (60101)	98.9
1995	78 (60101)	12	1995	25 (60101)	95.8
1996	74 (60101)	10	1996	23 (60101)	66.5
1997	70 (60101)	7	1997	25 (60101)	15.8
1998	73 (60101)	7	1998	25 (60101)	93.9
1999	83 (60101)	7	1999	33 (60101)	59.5
2000	82 (60101)	10	2000	22 (60101)	90.1
St. John					
1993	-		1993	15.2 (10102)	83.9

1994	-		1994	15.3 (10102)	21
1995	50 (10102)	4	1995	21.3 (10102)	1.7
1996	48 (10102)	1	1996	N/A	N/A
1997	43 (10102)	2	1997	8.2 (10102)	40.3
1998	32 (10102)	6	1998	4.8 (10102)	93.4
1999	35 (10102)	9	1999	6.6 (10102)	88.1
2000	36 (10102)	10	2000	8.2 (10102)	92.8
Toronto					
1993	-		1993	31 (60420)	28.7
1994	-		1994	30 (60424)	99.4
1995	120 (60420)	1	1995	32 (60425)	13.8
1996	108 (60423)	12	1996	34 (60425)	99.3
1997	103 (60423)	11	1997	35 (60425)	59.3
1998	87 (60423)	11	1998	30 (60403)	97.5
1999	91 (60425)	2	1999	28 (60403)	79.7
2000	89 (60423)	11	2000	30 (60430)	33.1
Vancouver					
1993	-		1993	31 (100108)	23.6
1994	-		1994	24 (100112)	97.6
1995	104 (100108)	1	1995	26 (100112)	97.7
1996	66 (100121)	12	1996	27 (100112)	97.8
1997	68 (100112)	12	1997	29 (100112)	97.5
1998	67 (100112)	12	1998	29 (100112)	97.4
1999	66 (100120)	5	1999	32 (100120)	3.4
2000	67(100120)	1	2000	27 (100112)	97.9
Winnipeg					
1993	-		1993	17 (70119)	93
1994	-		1994	17 (70119)	89.4
1995	57 (70119)	12	1995	18 (70119)	94.3
1996	55 (70119)	12	1996	18 (70119)	94
1997	60 (70119)	12	1997	187 (70119)	94.3
1998	62 (70119)	12	1998	17 (70119)	93.7
1999	65 (70119)	12	1999	18 (70119)	94
2000	61 (70119)	12	2000	16 (70119)	93.7
Windsor					
1993	-		1993	26 (60204)	98.7
1994	-		1994	28 (60204)	99.5
1995	87 (60204)	12	1995	25 (60204)	98.3
1996	79 (60204)	12	1996	26 (60204)	99.3
1997	71 (60204)	12	1997	24 (60204)	96.4
1998	66 (60204)	12	1998	24 (60204)	98.7
199	68 (60204)	12	1999	23 (60204)	98.2
2000	67 (60204)	12	2000	22 (60204)	97.4

2.3 Design Values In Locations For Long-Term Epidemiologic Studies Of Asthma Incidence In Children

- Vancouver, British Columbia

  Carlsten et al. 2011

  - LUR used to estimate annual concentrations at birth residential address (birth year exposure) for each subject; Air pollution estimates for 1995 generated from 2003 annual averages; Asthma assessed at 7 yrs
  - The 1-hour design value for this study encompasses years before the study period (i.e. 1993 and 1994)

Table A-19. Hourly and Annual Design Values for Vancouver, British Columbia in 1995.

Hourly			Annual		
Year	Max DV (Monitor ID)	Number of Complete Quarters[a]	Year	Max DV (Monitor ID)[a]	Completeness (%)[b]
1995	104 (100108)	1	1995	26 (100112)	97.7

[a]In the respective 3-year period (12 quarters) for the hourly DV; 9 or more quarters (75%) are required for a DV to be valid
[b]Valid annual DVs require 75% completeness

- British Columbia

  Clark et al. (2010)

  - LUR, central site monitors, and IDW used for postal code level exposure assignment for duration of pregnancy and first year of life (1999-2000); Asthma assessed at 36-59 mos
  - Methods from manuscript indicated that exposure measures were collected from 22 monitors for NO and NO_2, but only 7 monitors from the NAPS Canadian database had data for use in our approach.
  - The 1-hour design values for this study encompass years before the study period.

Table A-20. Hourly and Annual Design Values for Vancouver, British Columbia from 1999-2000.

Hourly			Annual		
Year	Max DV (Monitor ID)	Number of Complete Quarters[a]	Year	Max DV (Monitor ID)	Completeness (%)[b]
1999	66 (100120)	5	1999	32 (100120)	3.7
2000	67 (100120)	1	2000	27 (100112)	97.9

[a]In the respective 3-year period (12 quarters) for the hourly DV; 9 or more quarters (75%) are required for a DV to be valid
[b]Valid annual DVs require 75% completeness

- Boston   Clougherty et al. (2007)
  - LUR model based on 1 week of monitoring per month from 1987-2004 at 28 sampling sites was used to assign residential exposure for exposure in year of asthma diagnosis; Mean age of asthma diagnosis 6.8 yrs
  - Cohort from East Boston; AQS monitor 250250021 was used as representative based on study map

Table A-21. Hourly and Annual Design Values for East Boston from 1987-1994.

Hourly			Annual		
Year	Max DV (Monitor ID)	Number of Complete Quarters[a]	Year	Max DV (Monitor ID)[a]	Completeness (%)[b]
1987	-		1987	37 (250250021)	92.1
1988	-		1988	33 (250250021)	94.2
1989	91 (250250021)	12	1989	32 (250250021)	98.6
1990	90 (250250021)	12	1990	32 (250250021)	97.8
1991	84 (250250021)	12	1991	32 (250250021)	97.2
1992	81 (250250021)	12	1992	30 (250250021)	96.6
1993	79 (250250021)	12	1993	32 (250250021)	97.2
1994	79 (250250021)	11	1994	30 (250250021)	87.5
1995	76 (250250021)	11	1995	27 (250250021)	96.7
1996	83 (250250021)	11	1996	28 (250250021)	97
1997	82 (250250021)	12	1997	27 (250250021)	97.1
1998	83 (250250021)	12	1998	28 (250250021)	95.6
1999	75 (250250021)	11	1999	27 (250250021)	85.5
2000	72 (250250021)	11	2000	22 (250250021)	86.2
2001	64 (250250021)	10	2001	21 (250250021)	89
2002	60 (250250021)	10	2002	23 (250250021)	82.8
2003	54 (250250021)	6	2003	32 (250250021)	0
2004	54 (250250021)	3	2004	N/A	

[a]In the respective 3-year period (12 quarters) for the hourly DV; 9 or more quarters (75%) are required for a DV to be valid
[b]Valid annual DVs require 75% completeness

- Multicity (San Francisco, Houston, Chicago, Bronx, Puerto Rico)   Nishimura et al. (2013)
  - IDW with 4 closest monitors within 50 km of residence used for exposure assignment
  - AQS study monitors identified by visual approximation from study maps

Table A-22. Hourly and Annual Design Values for 5 U.S. Cities from 1986-2008.

Hourly			Annual		
Year	Max DV (Monitor ID)	Number of Complete Quarters[a]	Year	Max DV (Monitor ID)	Completeness (%)[b]
Chicago, IL					
1986	-		1986	42 (170310040)	95.6
1987	-		1987	43 (170310040)	76.6
1988	131 (170310053)	0	1988	34 (170310063)	38.9
1989	180 (170310053)	0	1989	34 (170310039)	96.6
1990	94 (170310039)	12	1990	31 (170310039)	98.5
1991	88 (170310039)	12	1991	32 (170310039)	98.4
1992	84 (170310039)	12	1992	30 (170310039)	89.3
1993	82 (170310039)	12	1993	31 (170310063)	89.2
1994	82 (170310063)	11	1994	34 (170310039)	62.8
1995	83 (170310063)	12	1995	32 (170310063)	98.4
1996	87 (170310039)	3	1996	31 (170310063)	95.7
1997	88 (170310063)	12	1997	34 (170310063)	97.5
1998	89 (170310037)	3	1998	32 (170310063)	95.3
1999	88 (170310063)	12	1999	32 (170310063)	98.4
2000	87 (170310063)	12	2000	32 (170310063)	94.7
2001	86.3 (170310063)	12	2001	32 (170310063)	98.7
2002	90 (170310063)	12	2002	32 (170310063)	98.8
2003	87 (170310063)	12	2003	31 (170310063)	96.9
2004	86 (170310063)	12	2004	29 (170310063)	96.6
2005	83 (170310063)	12	2005	30 (170310063)	95.7
Houston, TX					
1986	-		1986	28 (482011037)	57
1987	-		1987	30 (482011037)	93
1988	110 (482011037)	10	1988	28 (482011037)	91
1989	110 (482011037)	12	1989	28 (482011037)	84.3
1990	107 (482011037)	11	1990	29 (482011037)	71.3
1991	103 (482011037)	11	1991	28 (482011037)	85.6
1992	97 (482011037)	11	1992	28 (482011037)	80.3
1993	-		1993	24 (482011037)	81.6
1994	-		1994	28 (482011037)	82.6
1995	86 (482011037)	10	1995	26 (482011037)	76.4
1996	81 (482011037)	10	1996	23 (482011037)	87.2
1997	79 (482010047)	11	1997	25 (482011037)	81.3
1998	79 (482010047)	11	1998	23 (482011035)	87.4
1999	80 (482010047)	10	1999	24 (482010047)	83.2
2000	76 (482010047)	10	2000	21 (482011037)	93.2
2001	79 (482011037)	9	2001	29 (482011037)	22.1
2002	82 (482011037)	5	2002	18 (482010047)	88.2
2003	86 (482011037)	1	2003	18 (482010047)	96

Year			Year		
2004	62 (482010047)	12	2004	19 (482011034)	96.7
2005	62 (482010047)	12	2005	18 (482010047)	92.9
New York, NY					
1986	-		1986	49 (360610056)	68.6
1987	-		1987	43 (360610056)	40.7
1988	129 (360610056)	6	1988	44 (360610056)	44.5
1989	131 (360610056)	7	1989	49 (360610056)	94.3
1990	128 (360610056)	10	1990	46 (360610056)	96
1991	121 (360610056)	10	1991	47 (360610056)	96
1992	113 (360610056)	7	1992	36 (360610056)	59.4
1993	110 (360610056)	6	1993	43 (360610056)	87.1
1994	114 (360610056)	7	1994	46 (360610056)	86.6
1995	113 (360610056)	10	1995	42 (360610056)	94.4
1996	110 (360610056)	11	1996	42 (360610056)	86.1
1997	99 (360610056)	12	1997	40 (360610056)	94.6
1998	96 (360050080)	11	1998	40 (360610056)	91.2
1999	95 (360050080)	11	1999	41 (360610056)	97
2000	94 (360050073)	1	2000	38 (360610056)	95.5
2001	94 (360050073)	1	2001	38 (360610056)	39.1
2002	96 (360610010)	5	2002	38 (360610010)	95.5
2003	105 (360610010)	1	2003	38 (360610056)	58.5
2004	79 (360050110)	11	2004	35 (360610056)	62.6
2005	78 (360050110)	11	2005	37 (360610056)	97.3
San Francisco, CA					
1986	-		1986	24 (060410001)	98.5
1987	-		1987	24 (060750005)	97.3
1988	90 (060750005)	12	1988	26 (060750005)	96.5
1989	93 (060750005)	10	1989	26 (060750005)	87.9
1990	87 (060750005)	10	1990	21 (060750005)	88.3
1991	87 (060750005)	10	1991	24 (060750005)	99.1
1992	83 (060750005)	11	1992	22 (060750005)	97.5
1993	77 (060750005)	12	1993	24 (060750005)	93.1
1994	74 (060750005)	12	1994	22 (060750005)	96.8
1995	70 (060750005)	12	1995	21 (060750005)	97.2
1996	70 (060750005)	12	1996	22 (060750005)	96.8
1997	65 (060750005)	12	1997	20 (060750005)	93.6
1998	63 (060750005)	12	1998	20 (060750005)	95.1
1999	63 (060750005)	12	1999	21 (060750005)	94.6
2000	64 (060750005)	12	2000	20 (060750005)	95
2001	65 (060750005)	12	2001	19 (060750005)	95
2002	61 (060750005)	12	2002	19 (060750005)	93.3
2003	60 (060750005)	12	2003	18 (060750005)	94.3
2004	57 (060750005)	12	2004	17 (060750005)	92.3
2005	55 (060750005)	12	2005	16 (060750005)	94.9
Puerto Rico					

1986	N/A		1986	N/A	
1987	N/A		1987	N/A	
1988	N/A		1988	N/A	
1989	N/A		1989	N/A	
1990	N/A		1990	N/A	
1991	N/A		1991	N/A	
1992	N/A		1992	N/A	
1993	N/A		1993	N/A	
1994	N/A		1994	N/A	
1995	N/A		1995	N/A	
1996	N/A		1996	N/A	
1997	83 (720330006)	0	1997	20 (720330006)	16.4
1998	83 (720330006)	1	1998	12 (720330006)	64.8
1999	67 (720330006)	2	1999	7.3 (720330006)	66.6
2000	80 (720330006)	4	2000	18 (720330006)	75.5
2001	71 (720330006)	6	2001	9 (721270009)	87.5
2002	68 (720330006)	8	2002	7 (721270009)	92.2
2003	41 (720330006)	6	2003	7 (721270009)	60.2
2004	29 (720330008)	0	2004	11 (720330008)	18.1
2005	32 (720330008)	2	2005	9 (721270009)	72.2

[b]In the respective 3-year period (12 quarters) for the hourly DV; 9 or more quarters (75%) are required for a DV to be valid
[c]Valid annual DVs require 75% completeness

- Southern CA   McConnell et al. (2010)

- Community central site monitors and line source dispersion models used for residential and school NOx; Kindergarten and 1st grade children (4.8-9 yrs) enrolled in 2002-2003 and followed for 3 yrs (2002-2006)
- Models include TRP show independent, significant association for TRP and attenuation of NO_2 effect
-DVs were identified for Santa Maria, Santa Barbara, Alpine, Los Angeles CBSA, and Riverside CBSA. Several study communities were located in the Los Angeles CBSA (Long Beach, Anaheim, Mira Loma, and Glendora) and several others in the Riverside CBSA (San Dimas, Upland, Lake Arrowhead, San Bernardino, Riverside, and Lake Elsinore). The high DVs for each year were taken for these CBSAs rather than the study communities due to their close proximity and inability to determine specific community monitors.

Table A-23. Hourly and Annual Design Values for Southern California Communities for 2002-2006.

Hourly			Annual		
Year	Max DV (Monitor ID)	Number of Complete Quarters[a]	Year	Max DV (Monitor ID)[a]	Completeness (%)[b]
Santa Barbara, CA					
2002	-		2002	N/A	
2003	-		2003	15 (60830011)	36.1
2004	52 (60830011)	6	2004	13 (60830011)	92.8
2005	50 (60830011)	10	2005	12 (60830011)	95.1
2006	48 (60830011)	11	2006	11 (60830011)	55.1
Santa Maria, CA					
2002	-		2002	11 (60831008)	43.2
2003	-		2003	11 (60831008)	92.7
2004	42 (60831008)	10	2004	10 (60831008)	85.2
2005	41 (60831008)	12	2005	10 (60831008)	94.5
2006	37 (60831008)	10	2006	8 (60831008)	55.1
Alpine, CA					
2002	-		2002	13 (60731006)	93.5
2003	-		2003	14 (60731006)	87.1
2004	54 (060731006)	11	2004	12 (60731006)	94.1
2005	51 (060731006)	11	2005	11 (60731006)	92.2
2006	46 (060731006)	12	2006	10 (60731006)	94
Los Angeles-Long Beach, CA					
2002	-		2002	40 (60371002)	93.2
2003	-		2003	35 (60371002)	90.7
2004	121 (60370030)	2	2004	34 (60371103)	85.4
2005	101 (60371103)	11	2005	31 (60371301)	93.5
2006	92 (60371602)	2	2006	31 (60371301)	94.1
Riverside, CA					
2002	-		2002	36 (60711004)	94.5
2003	-		2003	34 (60711004)	94.8
2004	95 (60711004)	12)	2004	31 (60711004)	94.8
2005	92 (60711004)	12	2005	31 (60711004)	94.7
2006	98 (60711004)	12	2006	31 (60711004)	88

[a]In the respective 3-year period (12 quarters) for the hourly DV; 9 or more quarters (75%) are required for a DV to be valid
[b]Valid annual DVs require 75% completeness

Jerrett et al. (2008)   - LUR model based on Palmes tubes outside homes for 2 weeks in summer and winter used for exposure assessment; children were enrolled at age 10 in 1993 or 1996 and followed for 8 yrs or until high school graduation
-DVs were identified for Santa Maria, Lompoc, Santa Barbara, Alpine, Atascadero, Los Angeles CBSA, and Riverside CBSA. Several study communities were located in the Los Angeles CBSA

(Long Beach, Anaheim, Mira Loma, and Glendora) and several others in the Riverside CBSA (San Dimas, Upland, Lake Arrowhead, San Bernardino, Riverside, and Lake Elsinore). The high DVs for each year were taken for these CBSAs rather than the study communities due to their close proximity and inability to determine specific community monitors.

Table A-24. Hourly and Annual Design Values for Southern California Communities from 1993-2004.

Hourly			Annual		
Year	Max DV (Monitor ID)[a]	Number of Complete Quarters[b]	Year	Max DV (Monitor ID)[a]	Completeness (%)[c]
Santa Maria					
1993	-		1993	7 (060831010)	93.3
1994	-		1994	7 (060831010)	93.3
1995	45 (060831007)	4	1995	12 (060831007)	92.6
1996	45 (060831007)	7	1996	12 (060831007)	88.5
1997	45 (060831007)	11	1997	13 (060831007)	95.1
1998	43 (060831007)	11	1998	12 (060831007)	93.6
1999	44 (060831007)	11	1999	11 (060831007)	89.7
2000	42 (060831007)	8	2000	13 (060831007)	14.6
2001	43 (060831007)	4	2001	15 (060831008)	0.1
2002	37 (060831008)	6	2002	11 (060831008)	43.2
2003	38 (060831008)	6	2003	11 (060831008)	92.7
2004	42 (060831008)	10	2004	10 (060831008)	85.2
Lompoc, CA					
1993	-	12	1993	9 (060832004)	92
1994	-	12	1994	9 (060832004)	92.3
1995	39 (060831013)	12	1995	7 (060832004)	93.5
1996	36 (060831013)	12	1996	7 (060832004)	91.8
1997	36 (060831013)	12	1997	7 (060832004)	90.7
1998	35 (060831013)	12	1998	7 (060832004)	92.5
1999	35 (060831013)	12	1999	7 (060832004)	93.4
2000	33 (060831013)	12	2000	6 (060832004)	92.1
2001	33 (060831013)	12	2001	6 (060832004)	93.7
2002	31 (060831013)	11	2002	4 (060832004)	79.8
2003	32 (060831013)	11	2003	6 (060832004)	92.9
2004	31 (060831013)	11	2004	6 (060832004)	93.3
Riverside, CA					
1993	-		1993	42 (060711004)	94.9
1994	-		1994	41 (060711004)	95
1995	140 (060711004)	12	1995	46 (060711004)	95.1
1996	137 (060711004)	11	1996	38 (060711004)	77.1
1997	127 (060711004)	11	1997	36 (060712002)	90.6

1998	112 (060711004)	11	1998	36 (060712002)	95.5
1999	111 (060650012)	11	1999	39 (060711004)	95.8
2000	134 (060650012)	12	2000	38 (060711004)	95.9
2001	142 (060650012)	12	2001	37 (060711004)	95.7
2002	135 (060650012)	12	2002	36 (060711004)	94.5
2003	107 (060650012)	12	2003	34 (060711004)	94.8
2004	95 (060711004)	12	2004	31 (060711004)	94.8
Los Angeles-Long Beach, CA					
1993	-		1993	50 (060371701)	92.8
1994	-		1994	48 (060371701)	94
1995	149 (060371701)	12	1995	46 (060711004)	95.1
1996	145 (060374002)	12	1996	42 (060371701)	94.2
1997	140 (060374002)	12	1997	43 (060371701)	95.4
1998	129 (060374002)	12	1998	43 (060371701)	94.9
1999	127 (060371701)	12	1999	51 (060371701)	95.4
2000	125 (060371701)	12	2000	44 (060371701)	95.8
2001	116 (060371701)	12	2001	37 (060371701)	95.7
2002	106 (060374002)	12	2002	36 (060371701)	95
2003	104 (060374002)	12	2003	35 (060371701)	95.2
2004	105 (060374002)	12	2004	31 (060371701)	94.2
Alpine, CA (San Diego CBSA)					
1993	-		1993	14 (060731006)	97.7
1994	-		1994	13 (060731006)	96.2
1995	61 (060731006)	12	1995	13 (060731006)	97.8
1996	61 (060731006)	12	1996	12 (060731006)	94.5
1997	60 (060731006)	11	1997	11 (060731006)	61.9
1998	54 (060731006)	11	1998	12 (060731006)	93.2
1999	57 (060731006)	11	1999	15 (060731006)	91.3
2000	59 (060731006)	12	2000	15 (060731006)	83.4
2001	59 (060731006)	12	2001	14 (060731006)	94.8
2002	56 (060731006)	12	2002	13 (060731006)	93.5
2003	55 (060731006)	11	2003	14 (060731006)	87.1
2004	54 (060731006)	11	2004	12 (060731006)	94.1
Atascadero, CA (San Luis Obispo CBSA)					
1993	-		1993	14 (060798001)	95.1
1994	-		1994	14 (060798001)	93
1995	55 (060798001)	12	1995	12 (060798001)	94.4
1996	52 (060798001)	12	1996	12 (060798001)	94.5
1997	52 (060798001)	12	1997	12 (060798001)	92.3
1998	51 (060798001)	12	1998	11 (060798001)	94.9
1999	56 (060798001)	12	1999	14 (060798001)	93.8
2000	55 (060798001)	12	2000	12 (060798001)	93.6
2001	54 (060798001)	12	2001	11 (060798001)	87.5
2002	51 (060798001)	12	2002	11 (060798001)	92.2
2003	50 (060798001)	12	2003	9 (060798001)	95.2

| 2004 | 48 (060798001) | 12 | 2004 | 8 (060798001) | 95.1 |

[a]In the respective 3-year period (12 quarters) for the hourly DV; 9 or more quarters (75%) are required for a DV to be valid
[b]Valid annual DVs require 75% completeness

3. DISTRIBUTIONS OF DAILY MAXIMUM 1-HOUR NO₂ CONCENTRATIONS FOR LOCATIONS OF SHORT-TERM EPIDEMIOLOGIC STUDIES

Table A-25: Distribution of daily maximum 1-hour concentrations of NO₂ (ppb) for locations of U.S. epidemiologic studies of short-term hospital admissions and emergency department visits.

	U.S. Study Locations (years)							
	Atlanta (1993-2000)	Atlanta (1994-2004)	Atlanta (1993-2004) (May-Oct.)	Bronx (1999-2000)	Manhattan (1999-2000)	Detroit (2004-2006)	Los Angeles (1992-1995)	New York (1999-2002)
5th	22	22	21	25	32	16	30	38
10th	26	26	25	28	35	18	36	42
50th	44	43	42	41	47	33	60	59
75th	57	55	54	50	56	40	90	72
90th	70	68	64	62	66	47	120	85
98th	92	90	85	80	82	57	177	108
99th	100	99	92	85	90	60	197	114

Shaded columns represent study locations/years for studies in Figure 3-1 with DVs ≤ 100 ppb.

4. TRENDS IN 1-HOUR AND ANNUAL NO₂ DESIGN VALUES

Figures 2-4 and 2-5 present different properties of trends of annual and hourly DVs from 1980-2015. For both of these graphics, only DVs considered "valid" by the completeness criteria in the CFR were used. In Figure 2-4, five percentiles (5, 25, 50, 75, and 95) of the year-specific distributions of DVs across the U.S. were calculated and plotted. In Figure 2-5, the correlation between valid DVs and year are calculated and categorized into one of three bins: increasing, decreasing, or insignificant. This calcuation used the cor.test() function included in the R statistical package to return the Spearman correlation ("r") and associated p-value for each NO2 monitoring site and DV type. If the p-value was greater than 0.05, then the trend was deemed insignificant. If the p-value was less than or equal to 0.05, then the sign of the r-value was used to determine the direction of the trend: positive for increasing and negative for decreasing.

5. EVALUATION OF ROADWAY GRADIENTS OF NO₂ CONCENTRATIONS FROM 1980-2015

Figures 2-6was generated by plotting boxplots of NO₂ DVs from urban areas as a function of bins of distances in meters from the nearest major road. A monitor was considered "Urban" if it resided inside the boundary of a core-based statistical area (CBSA) as defined by the U.S. Census Bureau in 2014 (https://www.census.gov/geo/maps-data/data/tiger-line.html). These distances were determined using 2012 data from the Highway Performance Monitoring System (HPMS) created by the Federal Highway Administration (http://www.fhwa.dot.gov/policyinformation/hpms/shapefiles.cfm). The R statistical program (https://www.r-project.org/) and the gDistance() function in the rgeos package (https://CRAN.R-project.org/package=rgeos) were used calculate the distances between monitor long/lat points and the nearest road in the HPMS shapefile. (sentence about the HPMS file is all "major" roads). The distances and DV datasets were then merged together by monitors and input to graphical commands to produce the boxplots in Figure 2-6. All DVs, both valid and invalid according the CFR completeness criteria, were included in the boxplots to more robustly explore the physical relationship between NO2 concentrations and distance from vehicular sources.

6. COMPARISON OF DISTRIBUTION OF NO₂ CONCENTRATIONS FROM NEAR-ROAD MONITORS AND NON-NEAR-ROAD MONITORS FOR 2013-2015

Figures 2-7, 2-8, 2-9 and 2-10 were created by calculating the 10th, 25th, 50th, 75th, 90th, 98th, and 99th percentiles of annual NO₂ concentrations at each monitor in a CBSA that contained a monitor in the EPA Near Road Network. No consideration was given to data completeness for these calculations. If more than one monitor of each given type (i.e. near road or non-near road) was present in a CBSA, only data from the monitor with the highest 98th percentile of that type was included. The above percentiles were then graphed as the boxplots specific to each CBSA, road type and year as shown in Figures 2-7 through 2-10.

7. RELATIONSHIP BETWEEN 1-HOUR AND ANNUAL NO₂ DESIGN VALUES

Figure 2-11 was generated by plotting DVs from monitors and years where both the annual and hourly DV was valid according to the CFR completeness criteria. Regression statistics (slope, intercept and R2) were calculated using the lm() function included in the R statistical package.

8. RELATIONSHIP BETWEEN ANNUAL AVERAGES OF THE NEAR-ROAD AND NON-NEAR-ROAD MAXIMUM 1-HR DAILY NO₂ CONCENTRATIONS FROM 2013-2016

In Figure A-1, we examine how the relationship between hourly and annual NO_2 concentrations, based on 1-hour and annual DVs, has changed since 1980.

Figure A-1. Averages of the near-road and non-near-road maximum 1-hr daily NO₂ concentrations from 2013-2016. Each point represents the medians of the distributions displayed in Figures 2-7 through 2-10 in the main body of the PA.

APPENDIX B

NITROGEN DIOXIDE AIR QUALITY CHARACTERIZATION

Table of Contents

List of Figures

This page intentionally left blank

List of Tables

B1. INTRODUCTION

This document provides the detailed results of an air quality characterization (AQC) performed as part of the primary NO_2 NAAQS review. Ambient concentration measurements (1980-2015), along with adjusted and simulated ambient concentrations, were evaluated using approaches described in Chapter 2 of the NO_2 REA planning document (REA PD; US EPA, 2015). The approaches and data sets used in this characterization were also informed by review by the CASAC (Diez-Roux and Frey, 2015) and public comments, with appropriate modifications noted here.

As indicated in the REA PD (section 2.1.1), there is a substantially improved body of information available in the current review to inform an updated characterization of 1-hour ambient NO_2 concentrations. In particular, data from recently deployed NO_2 monitors near major roads, combined with new information from monitoring and modeling studies of NO_2 concentration gradients around roads, add to our understanding of ambient NO_2 concentrations in near-road and on-road environments. This new information, combined with recent information on NO_X emissions provides important perspective, beyond what was available from the last review, on the extent to which NO_2 exposures could have potentially important implications for public health.

We evaluated ambient NO_2 concentrations and compared them with health-based benchmarks, with a particular focus on updating analyses of concentrations occurring near roads. The following sections describe our technical approach used to conduct these analyses (Section B2) including a representativeness evaluation of the study areas selected for focused analysis. Then, the number of days per year having daily maximum 1-hour (DM1H) concentrations at or above the selected benchmarks was evaluated using historical (1980-2015) unadjusted ambient monitoring data and several adjusted and simulated air quality scenarios using recent (2010-2015) ambient monitoring data in selected urban study areas (Section B3). Section B4 follows in identifying reference material used in developing this document. And finally, detailed supplemental data are provided in Section B5 to support the analyses presented in Sections B2 and B3.

B2. APPROACH

This air quality characterization (AQC) requires benchmark concentrations of interest, the identification of a study area(s) of interest, and the characterization of respective air quality. Further, the overall representativeness of the characterization is informed by ambient monitoring physical attributes and local NO_X emission source information. Each of these fundamental components of the AQC, the data and approaches used, and an overview of the air quality benchmark summary metrics are described in the following sections. SAS version 9.4 (SAS/STAT 13.2; SAS, 2015) was used to process all ambient monitor data files and to perform all mathematical and statistical analyses of NO_2 concentration data.

2.1 AIR QUALITY BENCHMARK LEVELS

The primary goal of this NO_2 AQC is to inform policy decisions regarding the likelihood that the existing or potential alternative standards would allow for exposures to ambient NO_2 concentrations that could be of concern for public health. To facilitate such an analysis, we evaluated the daily maximum 1-hour (DM1H) ambient NO_2 concentrations adjusted to just meet the existing NO_2 standards at varying air quality benchmark levels. We identified air quality benchmark levels based on the range of data from controlled human exposure studies of non-specific airway responsiveness in people with asthma.[1] Because there are few instances where 1-hour ambient concentrations could go above 200 ppb when considering concentrations that just meet the existing standards, we focused on the lower end of this range and selected three air quality benchmark levels of 100, 150 and 200 ppb. Instances when ambient concentrations in selected study areas are at or above these levels are counted and summarized, as detailed in section 2.4.3.

2.2 AMBIENT MONITORING DATA

All of the existing ambient NO_2 monitoring data from 1980-2015 are considered in this AQC. Hourly ambient concentration data were obtained from the U.S. EPA's AirData and Air Quality System (AQS) Data Mart websites.[2] Any replicate NO_2 measures occurring at the same

---

1 The majority of study volunteers experienced increased airway responsiveness following exposures to NO_2 concentrations of 100-300 ppb (or higher) for 30 minutes to 2 hours.

2 Data for 1990-2015 obtained from http://aqsdr1.epa.gov/aqsweb/aqstmp/airdata/download_files.html#Raw. Data for 1980-1989 downloaded from https://aqs.epa.gov/aqsweb/documents/data_mart_welcome.html.

monitoring site (though having multiple parameter occurrence codes or POCs) were averaged. Except for the newly designated near-road monitors, only monitors having a complete year of data were used in our analyses. Ascertaining a complete year of monitoring data is a multi-step process. First, valid days are defined as those having at least 18 hours of measurements. Next, a valid quarter is identified as having at least 75% of valid days within a three-month calendar period (68-70 days). Finally, where all four quarters in a calendar year are valid, the year of monitoring data was considered complete. The near-road monitoring data were used as reported for any year available, regardless of how many hours or days observations were collected.

2.3 STUDY AREAS

We have conducted updated analyses comparing NO_2 air quality to benchmarks in several study areas. Our selection of study areas focused on CBSAs having newly designated near-road monitors, CBSAs having the highest NO_2 design values (thus requiring the smallest adjustment to just meet the existing 1-hour standard) and CBSAs with a relatively large number of NO_2 monitors overall (i.e., providing improved spatial characterization). Additional considerations for evaluating representativeness of selected study areas include CBSA population, ranking of total NO_X emissions and mobile source NO_X emissions, and that they include a wide range of other non-mobile sources having relatively high NO_X emissions. The specific steps taken to select the study areas and how they were evaluated for representativeness are described below.

2.3.1 Selection approach

While all of the existing ambient monitoring data from 1980-2015 are considered in this AQC, particular adjusted air quality scenarios (i.e., air quality adjusted to just meet the existing standards) and simulated microenvironmental evaluations (i.e., on-road NO_2 concentrations) warranted the defining of a specific geographic domain, or study area. For added context regarding the potential exposed population, the core-based statistical area (CBSA) was chosen as the fundamental geographic area. The following are the approach steps used to identify CBSAs to consider evaluating in the study area focused portion of this AQC, followed by the application of this approach to select study areas.

1. CBSAs were defined at a county level using delineation files available from the U.S. Census Bureau.[3]

2. All available ambient NO_2 monitor design values[4] were linked with CBSA definitions using the state and county level identifiers contained in the first five digits of the ambient monitor IDs.[5] CBSAs having at least one monitor design value were retained for further selection steps.

3. The single year 98th percentile DM1H was calculated for all near-road monitors in any CBSA for 2014 and 2015.[6] This near-road monitor summary statistic was then merged with the CBSA-monitor design value data, again using the appropriate state and county level identifiers.

4. This combined CBSA-design value-near road data were then grouped into one of two study area selection pools:

Near-road CBSAs. CBSAs having a newly designated near-road monitor and reporting NO_2 concentrations were given preference for inclusion in an initial study area selection pool. Because most of the near-road monitors began operating in 2014, all near-road NO_2 concentration data were considered usable regardless as to whether or not a complete year was available to maximize the amount of near-road data available for analysis. Fifty CBSAs were identified as part of this initial selection pool (

---

[3] The counties comprising each CBSA are listed at http://www.census.gov/population/metro/data/def.html, as defined by the Office of Management and Budget (OMB) (http://www.whitehouse.gov/sites/default/files/omb/bulletins/2013/b-13-01.pdf). Instances where only a portion of the county was identified as part of a given CBSA by the OMB memo, it was assumed in this analysis that the entire county was part of the CBSA in an effort to be more inclusive in developing the study area selection pools.

[4] A spreadsheet file containing annual average and 98th percentile daily maximum 1-hour (DM1H) design values was obtained from http://www.epa.gov/airtrends/values.html. The file obtained contains design values for 2005-2014, though only years 2010-2014 were used from this file in identifying CBSAs as part of the study area selection pool. Design values for 2015 were calculated using complete year ambient monitor data available as part of this analysis.

[5] Each ambient monitor ID has 9 characters (XXYYYZZZZ) as follows: 'XX' indicates the U.S. State code, YYY is used to identify the county code, and 'ZZZZ' is used for the monitor number.

[6] The near-road monitor IDs and attributes were obtained from http://www3.epa.gov/ttnamti1/nearroad.html.

a. Table B2-1).

b. *Other CBSAs.* CBSAs not having a newly designated near-road monitor,[7] or CBSAs not having reported concentrations at a newly designated near-road monitor though having at least one design value were included in this group. One-hundred twenty-seven CBSAs were identified as part of this group (not shown).

5. The top twenty[8] near-road CBSAs were identified from the "near-road CBSA" group using a rank value, generated using the following quantitative information:

a. For each CBSA, the maximum design value and the total number of monitors reporting a design value was obtained for all analysis years considered (i.e., 2010-2015) and the 98th percentile DM1H near-road monitor concentration for 2014 and 2015 were retained. This yielded upwards to twenty-two variables[9] describing each CBSA.

b. A global mean value for each of these twenty-two variables was calculated. Each individual CBSA variable value was then normalized by the global mean variable value.

c. A rank value for each CBSA was calculated as an equally weighted sum of the mean normalized variable values, though relative to unity (i.e., the average CBSA rank value would equal 1). The rank values were then sorted in

---

[7] There are a few CBSAs that have an existing monitor sited in close proximity to a major roadway, though not necessarily meeting the new near-road monitor designation requirements. For example, a Chicago, IL monitor (ID 170313103) is about 20 meters from a major road (see US EPA, 2008a; 2008 REA Appendix A, Table A-7). It is possible NO_2 concentrations from this monitor (and other monitors in close proximity to a major road) can be used in a manner similar to that of the formally designated near-road monitors and supplement our characterization of the near-road and on-road microenvironments.

[8] This number of "identified CBSA" (i.e., 20) was subjectively chosen for analysis tractability while also retaining a generally geographically diverse and high NO_2 concentration representative collection of CBSAs from the near-road CBSAs group.

[9] Considered here are the six annual average design values, four 3-year average 98th percentile DM1H design values, number of monitors in each of six years, number of monitors in four 3-year periods, and the 98th percentile DM1H values for the near-road monitor in 2014 and 2015.

descending order with the top twenty near-road CBSA[10] presented in Table B2-2.

6. Twenty additional CBSAs were then identified using a rank value generated using the same variables and approach listed in step 5, though considering the following modifications:

   a. Not included in the calculation of the rank value was the 98th percentile DM1H concentration from a newly designated near-road monitor, thus there were upwards to twenty variables describing each CBSA.

   b. CBSAs evaluated in this step were those in the "other CBSA" group (step 4b) and any "near-road CBSA" not identified in step 5 (i.e., totaling 157 CBSAs).

   c. The rank values were sorted in descending order with the top twenty "additional CBSA" provided in Table B2-3.

We prioritized the order by which the above identified CBSAs could be evaluated for the benchmark analysis, rather than proceed with performing analyses using all 40 CBSAs. First, all twenty "near-road CBSAs" in Table B2-2 were selected as study areas, again, based on their having the newly sited near-road monitors and associated NO_2 concentration measurements. Next, because Detroit, MI and Richmond, VA have new near-road measurement data and add to the overall U.S. geographic representation, these CBSAs were selected as a study area from the "additional CBSA" top twenty list. We then selected one CBSA (Chicago, IL) from the list of additional CBSAs, largely based on it further enhancing overall U.S. geographic representation, and having the highest population of the remaining CBSAs identified for analysis (see below). In addition, while Chicago does not have any new near-road measurement data available, we were interested in evaluating results that could be generated using historically sited monitors that are situated near a major road. Thus, a total of twenty-three study areas were selected for the focused analysis in this AQC (Figure B2-1). If other study areas are determined as useful in characterizing air quality for future assessments as part of this review, the 17 CBSAs remaining

---

[10] The use of unweighted variables in calculating the rank value places emphasis on within-CBSA spatial representation and areas measuring the highest NO_2 concentrations. Justification for this emphasis would include, 1) when properly sited regarding the most important direct source or precursor emissions, the greater the number of monitors in a CBSA would better represent ambient concentrations than in CBSAs having fewer monitors, 2) monitors having the highest concentrations would require the smallest adjustment upwards to just meet the existing standard, possibly limiting uncertainty in generated results, and 3) the risk associated with highest concentrations (even considering unadjusted concentrations) is by definition of greatest importance when performing an assessment that uses health effect benchmark concentrations.

from the list of "additional CBSAs" above, along with any new near-road CBSAs (e.g., perhaps those having newly available 2016 near-road NO_2 measurements) would likely be considered first in selecting additional study areas.

Table B2-1. CBSAs having NO$_2$ measurement data from the newly designated near-road monitoring sites (2014-2015).

CBSA Name	Abbr.	CBSA Name	Abbr.
Atlanta-Sandy Springs-Roswell, GA	ATLA	Milwaukee-Waukesha-West Allis, WI	MILW
Austin-Round Rock, TX	AUST	Minneapolis-St. Paul-Bloomington, MN-WI	MINE
Baltimore-Columbia-Towson, MD	BALT	Nashville-Davidson--Murfreesboro--Franklin, TN	NASH
Birmingham-Hoover, AL	BIRM	New Orleans-Metairie, LA	NORL
Boise City, ID	BOIS	New York-Newark-Jersey City, NY-NJ-PA	NYNY
Boston-Cambridge-Newton, MA-NH	BOST	Oklahoma City, OK	OKLA
Buffalo-Cheektowaga-Niagara Falls, NY	BUFF	Philadelphia-Camden-Wilmington, PA-NJ-DE-MD	PHIL
Charlotte-Concord-Gastonia, NC-SC	CHAR	Phoenix-Mesa-Scottsdale, AZ	PHOE
Cincinnati, OH-KY-IN	CINC	Pittsburgh, PA	PITT
Cleveland-Elyria, OH	CLEV	Portland-Vancouver-Hillsboro, OR-WA	PORT
Columbus, OH	COLO	Providence-Warwick, RI-MA	PROV
Dallas-Fort Worth-Arlington, TX	DALL	Raleigh, NC	RALE
Denver-Aurora-Lakewood, CO	DENV	Richmond, VA	RICH
Des Moines-West Des Moines, IA	DESM	Riverside-San Bernardino-Ontario, CA	RIVR
Detroit-Warren-Dearborn, MI	DETR	Rochester, NY	RONY
Hartford-West Hartford-East Hartford, CT	HART	Sacramento--Roseville--Arden-Arcade, CA	SACR
Houston-The Woodlands-Sugar Land, TX	HOUS	San Antonio-New Braunfels, TX	SANA
Indianapolis-Carmel-Anderson, IN	INDI	San Diego-Carlsbad, CA	SAND
Jacksonville, FL	JACK	San Francisco-Oakland-Hayward, CA	SANF
Kansas City, MO-KS	KANS	San Jose-Sunnyvale-Santa Clara, CA	SANJ
Las Vegas-Henderson-Paradise, NV	LASV	San Juan-Carolina-Caguas, PR	SJPR
Los Angeles-Long Beach-Anaheim, CA	LOSA	Seattle-Tacoma-Bellevue, WA	SEAT
Louisville/Jefferson County, KY-IN	LOUI	St. Louis, MO-IL	STLO
Memphis, TN-MS-AR	MEMP	Tampa-St. Petersburg-Clearwater, FL	TAMP
Miami-Fort Lauderdale-West Palm Beach, FL	MIAM	Washington-Arlington-Alexandria, DC-VA-MD-WV	WASH

Table B2-2. The top twenty near-road CBSAs identified for analysis based on 2010-2015 design values, number of monitors in operation, and near-road NO2 concentrations.

CBSA/ Study Area[1]	Number of Ambient Monitors in Area with Valid Data										Max Annual Average NO2						Max 98th pct 1-hr daily max NO2				2014 near-road monitor		2015 near-road monitor		Study area rank value[3]
	2010	2011	2012	2013	2014	2015	2010-2012	2011-2013	2012-2014	2013-2015	2010	2011	2012	2013	2014	2015	2010-2012	2011-2013	2012-2014	2013-2015	days (n)[2]	98th pct DM1H NO2	days (n)[2]	98th pct DM1H NO2	
HOUS	16	16	16	16	16	13	9	9	11	13	15	14	15	13	13	11	60	59	56	52	344	48.8	261	59.2	3.16
LOSA	16	14	9	13	16	15	4	4	5	7	26	25	21	23	22	22	67	64	63	62	361	66	255	74.8	2.74
DALL	11	14	11	11	11	10	9	9	8	8	13	13	12	12	10	10	56	53	49	47	274	40	365	44.9	2.48
SANF	10	10	11	10	9	9	8	8	8	8	16	16	15	17	14	14	74	68	61	57	334	51.7	365	48.9	2.45
NYNY	10	7	8	8	9	8	5	6	7	7	22	25	22	22	22	22	70	67	66	66	184	90	365	67	2.37
RIVR	12	9	11	11	13	11	6	3	2	1	23	21	22	21	20	19	72	62	53	50	0	0	153	77.2	2.15
SAND	8	8	8	8	6	6	6	5	4	4	21	20	20	19	13	14	73	73	57	56	0	0	269	52	1.89
PHOE	6	5	5	5	5	6	4	4	4	4	25	25	26	25	25	22	66	64	64	63	322	59	121	64	1.84
SACR	8	8	8	6	7	7	7	4	4	5	12	13	12	10	11	11	51	50	48	50	0	0	80	52	1.71
BOST	5	5	6	5	5	6	5	4	3	4	19	20	19	18	17	17	51	50	49	51	365	53	362	50	1.64
WASH	7	6	6	7	6	5	5	4	2	3	18	16	17	13	11	13	55	51	48	48	0	0	212	47.1	1.52
PHIL	5	4	5	5	5	4	3	3	1	3	23	20	18	17	18	18	65	61	39	49	358	51.4	357	47.8	1.45
PITT	5	5	5	6	4	3	4	4	1	2	15	16	14	11	10	10	53	49	36	42	122	39.7	365	44.8	1.27
DENV	2	2	2	2	2	3	0	1	2	2	28	24	25	24	23	22	0	62	72	72	361	69.6	92	73.15	1.27
ATLA	3	3	3	3	3	3	3	3	3	3	14	13	12	9	11	10	56	51	49	48	200	49.95	363	54.1	1.16
MINE	3	3	3	3	3	3	2	3	3	3	10	9	11	9	9	8	46	44	45	46	365	48	362	48	1.06
BALT	2	2	2	2	2	1	2	2	2	1	18	18	16	15	16	11	57	52	52	46	275	50.6	352	49.8	1.03
KANS	3	3	3	3	2	2	3	1	1	1	15	15	14	13	13	13	53	52	51	51	358	45.6	352	44.5	1.02
STLO	3	3	2	3	4	2	0	2	2	2	13	13	14	11	12	11	0	53	49	46	365	58.2	364	47.4	0.97
MIAM	5	5	3	4	3	2	2	2	2	2	10	8	8	8	9	5	47	46	46	44	0	0	56	40	0.95

[1] abbreviated CBSA name to allow for extended number of columns (see Table B2-1 for complete CBSA title).

[2] not used in calculating the rank value, included for information purposes only. Also, where more than one near road monitor was operating in a study area, the number of days corresponds to the monitor having the greatest 98th pct DM1H (i.e., it is possible that the other near road monitor had a greater number of days where measurements were collected).

[3] CBSA's are sorted by descending rank value.

Table B2-3. The top twenty additional CBSAs identified for analysis based on 2010-2015 design values and number of monitors in operation.

CBSA Title	Number of Ambient Monitors in Area with Valid Data										Max Annual Average NO$_2$						Max 98th pct 1-hr daily max NO$_2$				Study area rank value[1]
	2010	2011	2012	2013	2014	2015	2010-2012	2011-2013	2012-2014	2013-2015	2010	2011	2012	2013	2014	2015	2010-2012	2011-2013	2012-2014	2013-2015	
Santa Maria-Santa Barbara, CA	11	11	11	11	11	10	11	10	10	9	9	9	10	10	8	8	43	36	36	36	6.57
Baton Rouge, LA	10	8	8	8	8	6	7	7	7	5	13	12	11	10	11	9	54	52	48	41	4.88
Chicago-Naperville-Elgin, IL-IN-WI	6	5	6	5	6	5	2	0	1	3	25	23	22	21	21	18	62	0	47	63	3.07
Beaumont-Port Arthur, TX	4	4	4	4	4	4	4	4	4	4	8	7	6	6	6	5	37	35	33	33	2.75
Springfield, MA	3	3	3	3	3	3	3	3	3	3	15	16	14	14	13	13	47	46	42	46	2.55
Farmington, NM	4	3	3	3	3	3	2	3	3	3	12	13	13	12	11	11	38	41	37	36	2.39
Bakersfield, CA	4	3	3	4	3	4	3	2	1	1	14	15	15	14	13	12	58	46	48	45	2.32
El Paso, TX	3	2	2	3	3	3	1	1	2	3	17	17	16	14	14	14	61	59	57	58	2.20
Detroit-Warren-Dearborn, MI	1	1	3	3	3	4	1	1	2	3	12	12	19	18	16	18	48	44	45	50	2.10
San Luis Obispo-Paso Robles-Arroyo Grande, CA	3	3	3	3	3	2	3	3	3	2	6	6	7	7	6	3	38	38	39	31	2.05
Richmond, VA	3	3	3	2	2	3	3	2	1	1	12	10	10	8	8	14	52	47	43	41	1.94
Gillette, WY	3	3	4	4	4	3	1	1	1	3	7	6	8	9	10	7	32	32	35	49	1.93
Oklahoma City, OK	2	2	2	2	3	2	2	2	2	1	9	10	9	9	7	7	54	54	51	45	1.75
Oxnard-Thousand Oaks-Ventura, CA	2	2	2	2	2	2	2	2	2	2	10	9	10	9	9	8	38	37	39	38	1.73
Stockton-Lodi, CA	2	2	2	2	2	2	1	1	1	1	14	15	14	16	13	12	51	53	55	53	1.71
San Jose-Sunnyvale-Santa Clara, CA	1	2	2	2	1	2	1	1	1	1	14	15	13	15	13	18	50	51	53	50	1.71
El Centro, CA	2	2	2	2	2	2	1	1	1	1	14	14	14	13	12	11	62	64	47	44	1.67
Allentown-Bethlehem-Easton, PA-NJ	1	2	2	2	2	1	1	2	2	1	11	14	13	13	13	12	40	45	45	48	1.66
Cincinnati, OH-KY-IN	2	2	1	2	3	3	1	1	3	1	15	13	4	12	23	22	32	30	31	32	1.66
Providence-Warwick, RI-MA	1	1	2	2	2	3	1	1	1	2	10	11	10	10	10	22	42	43	42	46	1.61

[1] CBSA's are sorted by descending rank value.

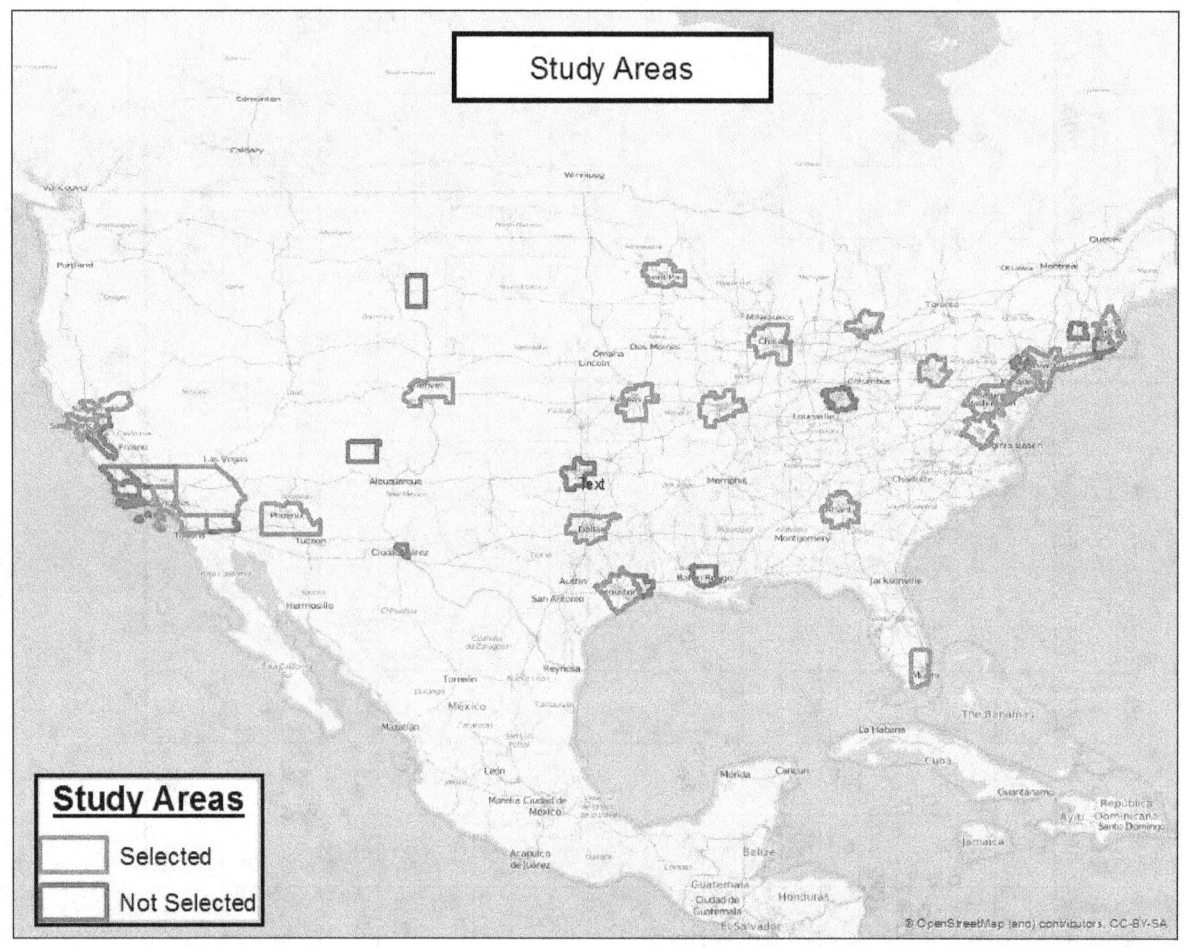

Figure B2-1. The 40 CBSAs identified as potentially useful to inform the air quality characterization, including the 23 selected study areas for focused analyses.

2.3.2 Representativeness evaluation

The geographical locations of the study areas should adequately represent areas across the U.S. having seasonal, atmospheric, or other influential factors that could contribute to variability in NO₂ concentrations and potential exposures. Figure B2-1 provides perspective to such an assessment showing reasonable geographic representation by the 23 selected study areas for the continental U.S., except for the upper northwestern U.S.,[11] even when considering the pool of "other CBSAs" not having a newly designated near-road monitor. Obviously, the NO_2 design values and number of ambient monitors operating in any northwestern CBSAs did not rise to a level necessitating their consideration as a study area of particular interest. Regarding the study area selection pool of 50 "near-road CBSAs", the three available northwestern CBSAs (Portland-Vancouver-Hillsboro, OR-WA; Seattle-Tacoma-Bellevue, WA; Boise City, ID) had the 32nd, 45th and 47th highest rank values using the above selection criteria, respectively. When considering the rank value of available northwestern CBSAs in the "other CBSAs" data set, only Gillette, WY (12th highest) was ranked within the top 50.

The selected study areas should also capture areas where large portions of the U.S. population reside, as this would better represent potential risks to populations at a local, urban, and national scales as well as increasing the likelihood for appropriately representing important study groups (e.g., children with asthma). To evaluate population representativeness, data were obtained from the U.S. Census Bureau to characterize the 2013 population in the 23 selected study areas relative to that of other CBSAs.[12] In total, the 23 selected study areas include just over 124 million people or nearly 40% of total U.S. population. When considering the 17 additional CBSAs identified as potentially useful to inform the AQC, another 14 million people could be included to the total population considered, thus comprising approximately 44% of the total U.S. population.

The individual CBSA population values were ranked by descending population, with the top 100 CBSAs retained and plotted in Figure B2-2. For perspective, all 23 selected study areas are named and highlighted in the figure. Also named in the figure are CBSAs within the top 20 CBSAs for population (if any other than the 23 selected study areas). This analysis indicates the selected study areas are representative of the most populated CBSAs in the U.S., having 18 of the top 20 populated CBSAs in the U.S.

---

[11] Considered here are the states of Washington, Oregon, Idaho, Wyoming, Montana. As an example of the limitations to number of monitors having complete year data, the Portland-Vancouver-Hillsboro, OR-WA CBSA had 1 monitor available in each of the years evaluated. Annual average and 3-year 98th percentile DM1H NO_2 concentrations were about 9 ppb and 35 ppb, respectively.

[12] Data were downloaded from http://www.census.gov/popest/data/metro/totals/2013/CSA-EST2013-alldata.html. Two files were used and processed to generate population data that included all study areas: CSA-EST2013-alldata_USCENSUS.txt and CBSA-EST2013-alldata_USCENSUS.txt. Population data were available for a total of 382 CBSAs.

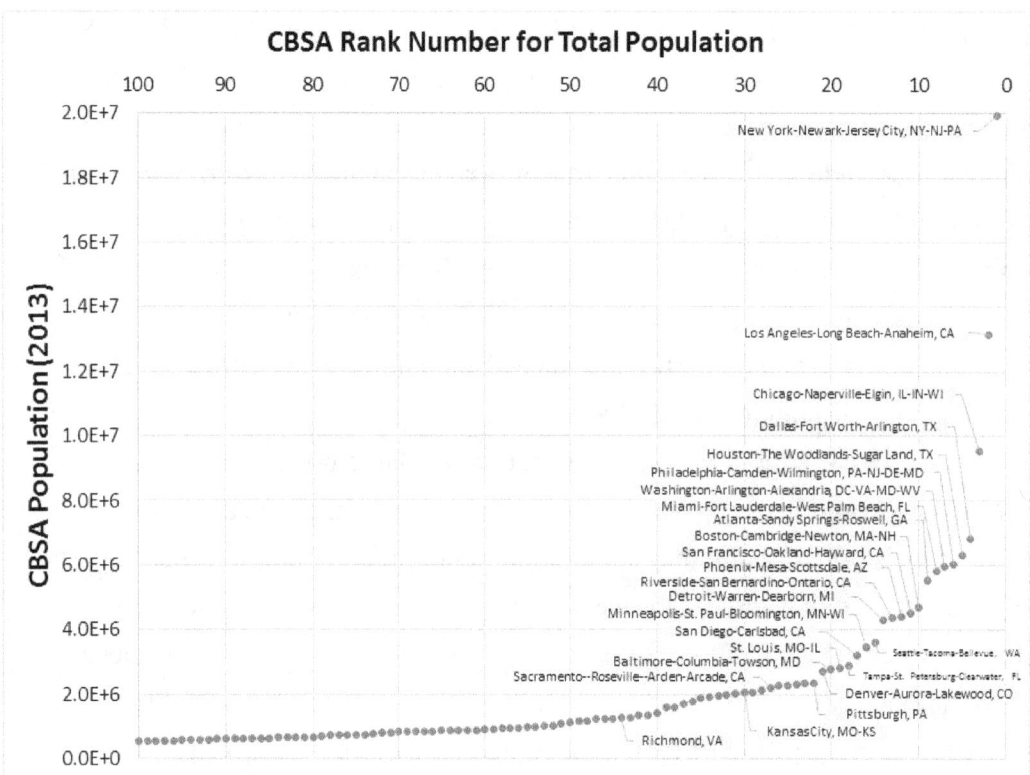

Figure B2-2. The top 100 CBSAs ranked by 2013 population. Highlighted and named are the 23 selected study areas; also named are those CBSAs ranked in the top 20.

Information on NO_X emissions[13] was used to inform our characterization of the important sources potentially contributing to monitored NO_2 concentrations. The recent EPA's National Emissions Inventory (NEI), the 2011 NEI[14] is the most comprehensive source of emissions data to identify, categorize, and evaluate NO_X emissions and emission source types. At a national level, anthropogenic sources account for more than 90% of NO_X emissions (US EPA, 2016, section 2.3.1, Figure 2-3). Motor vehicles are the largest source, with on-road and off-road mobile sources contributing nearly 60% of the total NO_X emissions nationally. Other important sources include fuel combustion-utilities (14% of total), fuel combustion-other (11% of total), and biogenics and wildfires (8% of total). Compared to the national averages, urban areas have

---

[13] Oxidized nitrogen compounds are emitted to the atmosphere primarily as NO, with NO converting to NO_2 following its reaction with O_3. Collectively, NO and NO_2 are referred to as NO_X (U.S. EPA, 2008b, section 2.2).

[14] The NEI is a national compilation of emissions estimates from all source sectors, collected from state, local, and tribal air agencies as well as those developed by EPA. The NEI is developed on a tri-annual basis, with 2011 being the most recent base year currently available and referred to as 2011 NEI. The next NEI base year will be 2014 and will be available in late 2016. For information on the NEI, see http://www.epa.gov/ttn/chief/eiinformation.html.

greater contributions to total NO$_X$ emissions from both on-road and off-road mobile sources and smaller contributions from other sources (US EPA, 2016, Figure 2-4, Table 2-1). For example, in the 21 largest CBSAs in the U.S., more than half of the urban NO$_X$ emissions are from on-road mobile sources, and when combined with off-road mobile sources, account for more than three quarters of total emissions in these large CBSAs (US EPA, 2016, section 2.3.2).

While this emissions summary at a national level is useful, the most important emission sources can vary substantially across smaller spatial scales. We evaluated the NO$_X$ emissions in all 177 CBSAs having ambient NO$_2$ concentration measurements (and emissions data were reported), including the 23 selected study areas to determine the total emissions and emission sources represented by the CBSAs that were selected as study areas. Two data sets were used to accomplish this, generated using the 2011 National Emissions Inventory (NEI)[15] as follows:

1. *County level emissions.* This data set contains NO$_X$ emissions (in tons per year) aggregated to the county level. Over 50 sectors are identified, including various mobile (e.g., "Mobile - Non-Road Equipment – Diesel") and industrial emission sources (e.g., Industrial Processes - Pulp & Paper). Emissions were summed based on the counties comprising each CBSA that had at least one ambient monitor in operation during 2010-2015.[16] Using this data set we calculated total NO$_X$ emissions, total mobile source NO$_X$ emissions,[17] and percent mobile source NO$_X$ emissions relative to total for each CBSA. Results for the 177 CBSAs were ranked for each of these three emissions variables and plotted in Figure B2-3 to Figure B2-5. In each of these

---

[15] Data downloaded from: http://www.epa.gov/air-emissions-inventories/2011-national-emissions-inventory-nei-data. Additional processing was performed here to characterize facilities that did not have entries for the 'facility type' field. Of the approximately 23,000 NO$_X$ facilities, nearly 13,000 were missing a facility type. Information from the available facility name and the NAICS codes were used to generate the facility type where missing, using the existing list of facility types. The majority of updates, designated as important primarily based on high total facility emissions, regarded the characterization of Mines/Quarries, Compressor Stations, Gas Plants, and Chemical Plants. Details regarding these updates to the data file used are available upon request.

[16] We felt that by using emissions from only those counties that had ambient monitors would best represent the emission levels and source types that could influence available ambient monitoring concentrations. It is possible that ambient monitors could be proximal to one or more other counties in a CBSA, resulting in potentially mischaracterizing or underestimating emissions and the influence of particular source types, however, we felt that including emissions from all counties in each CBSA would more likely result in a greater mischaracterization, and clearly overestimating emission levels and source types that might influence monitor concentrations. For the purposes of this broad study area evaluation, this assumption was judged as reasonable.

[17] All sectors having a description beginning with the word "Mobile" were included in the mobile source NO$_X$ emission sum. Recognizing this includes both on-road and non-road sources, analyses were also performed using only the "Mobile – On-Road" data. Results for the Mobile – On-Road emissionswere effectively the same as presented here for all mobile sources, i.e., the rank order of the study areas/CBSA were the same, only differed in that the total emissions and percent values were less.

figures, all 23 of the selected study areas are named and highlighted to indicate where these areas are ranked relative to the other 154 CBSAs. Any CBSA within the top 20 ranked emissions are also named (approximating an upper 90th percentile of CBSAs).

Clearly, the study areas selected for focused analyses in this AQC are among those CBSAs having the highest total and mobile source NO_X emissions (Figure B2-3 and Figure B2-4). The top 15 CBSAs having the greatest NO_X emissions were selected as study areas and nearly all of the study areas are within the 80th percentile or above for both total and mobile source emissions. The percent of total NO_X contributed by mobile sources considering the 23 study areas varied (Figure B2-5), and for the most part, did so in a similar fashion as that exhibited by most CBSAs, ranging from about 65-85% of total NO_X emissions. This range in mobile source contribution indicates the 23 selected study areas may contain a variety of important NO_X emission source types, albeit relatively smaller in magnitude when compared with mobile source emissions alone. Three selected study areas (Atlanta, Phoenix, and San Diego CBSAs) were ranked within the top 10 CBSAs providing reasonable representation of CBSAs where nearly all NO_X emissions are from mobile sources. About half of the selected study areas had a mobile source emissions contribution of 65-75%, again, providing representation of most CBSAs, while representation of lesser mobile source influenced areas are provided by the Denver (61%), Kansas City (57%), and Pittsburgh (45%) study areas. While there is a collection of CBSAs having greater relative contributions from non-mobile sources, their lack of representation is of limited importance considering the overall low total emissions in these CBSAs.

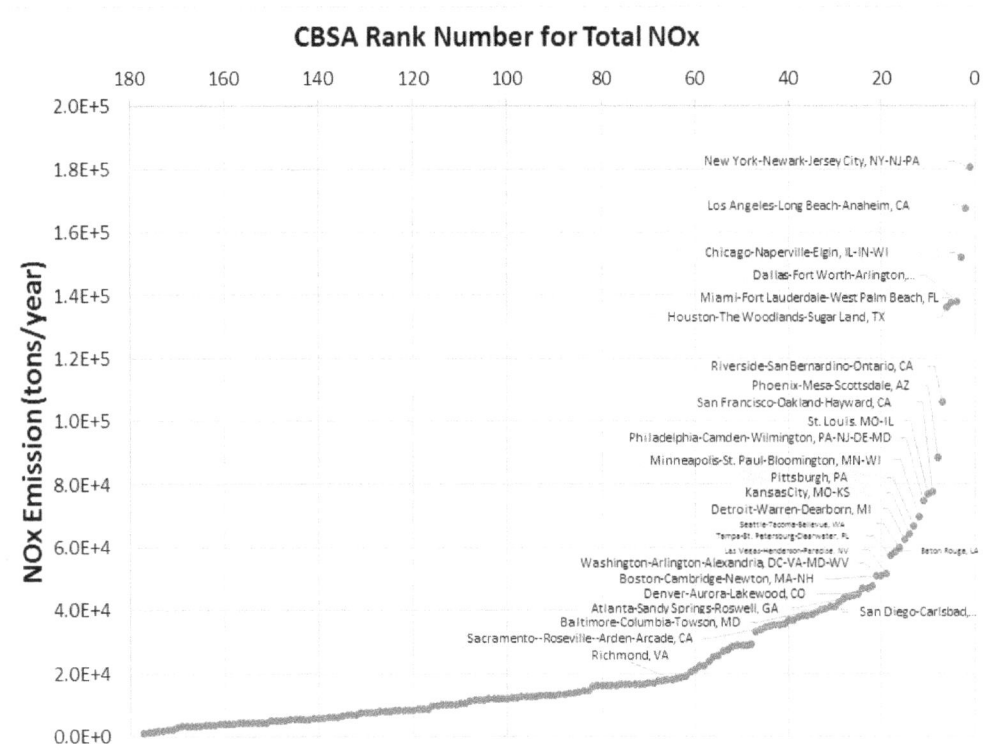

Figure B2-3. CBSAs ranked by total NOₓ emissions. Highlighted and named are the 23 selected study areas; also named are those CBSA ranked in the top 20.

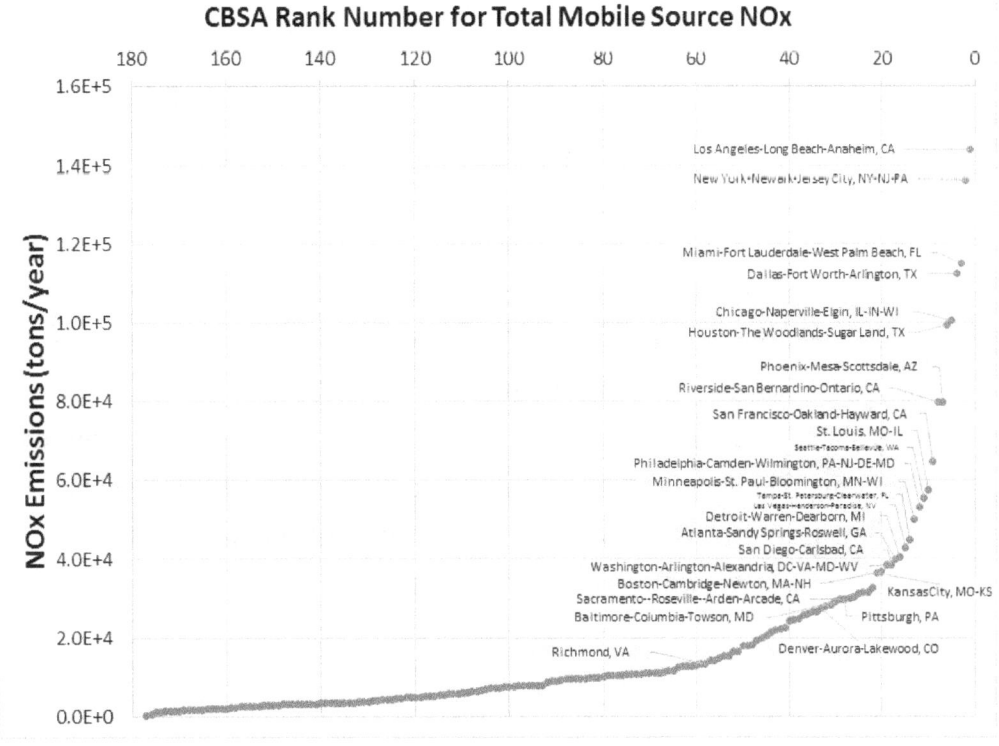

Figure B2-4. CBSAs ranked by mobile source NOₓ emissions. Highlighted and named are the 23 selected study areas; also named are those CBSA ranked in the top 20.

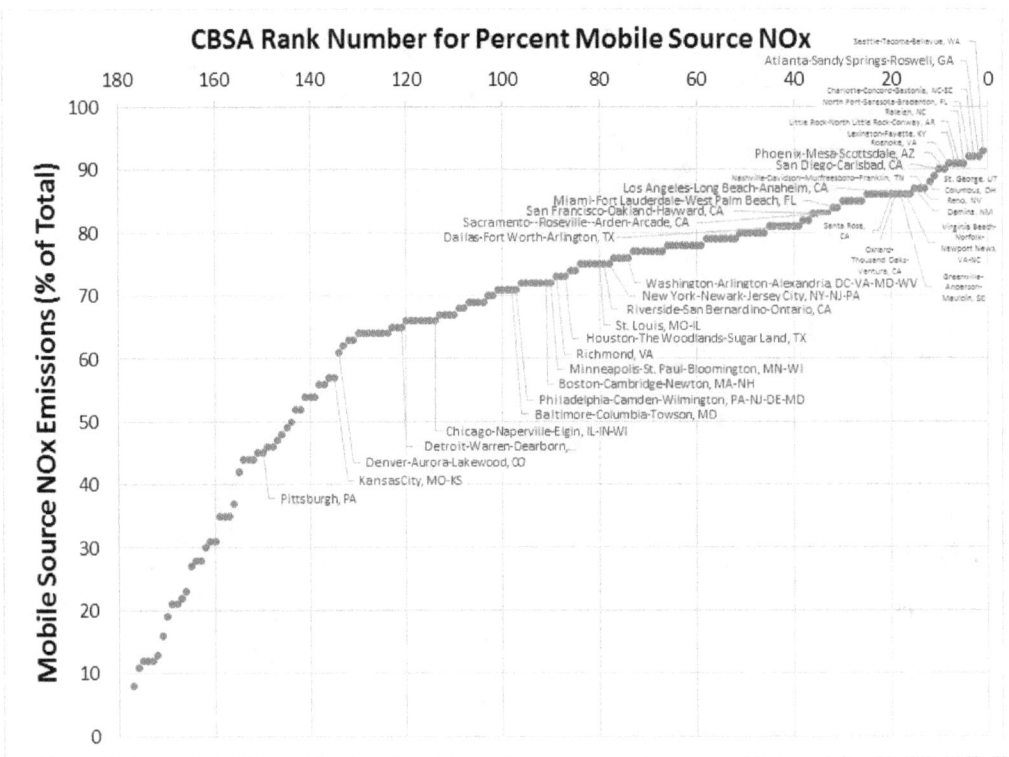

Figure B2-5. CBSAs ranked by percent all mobile source NO$_X$ emissions. Highlighted and named are the 23 selected study areas; also named are those CBSA ranked in the top 20.

2. *Facility level emissions.* This data set contains NO$_X$ emissions (in tons per year) at the individual facility level, also using only those emissions in counties that had at least one ambient monitor in operation during 2010-2015. A total of 72 facility types were identified here based on general source categories such as 'electricity generation via combustion' (EGUs), 'chemical plant' and 'compressor station'. Facilities without a source description were aggregated into a 'not characterized' facility type. To identify the most important sources in each CBSA, first the NO$_X$ emissions (tons/year) were summed by facility type. In addition, the maximum individual emission (tons/year) was retained for each facility type. The top 20 facility types were identified considering first the total NO$_X$ emissions, then considering the maximum NO$_X$ emissions for each facility types (**Error! Reference source not found.**). Eighteen of the same facility types were within the top 20 emission sources considering both metrics.

Table B2-4. The top twenty facility types ranked by total NOx emissions.

Facility Type	Facilities (n)	Total NOx (tpy)	Facility Maximum NOx (tpy)
Electricity Generation via Combustion	670	399292	17104
Airport	4323	96060	5485
Chemical Plant	372	52784	9113
Petroleum Refinery	82	52568	3655
Portland Cement Manufacturing	34	35230	2542
Mines/Quarries	124	34778	11726
Rail Yard	307	33745	1328
Compressor Station	512	33148	1583
not characterized	10804	30897	1688
Municipal Waste Combustor	49	28102	1617
Steel Mill	71	23197	4813
Gas Plant	73	15642	2268
Mineral Processing Plant	29	13929	2942
Pulp and Paper Plant	54	12415	2036
Institutional (school, hospital, prison, etc.)[1]	843	9705	739
Glass Plant	36	9393	2364
Chlor-alkali Plant	7	6936	6194
Fertilizer Plant	34	6815	2857
Coke Battery	9	6814	3075
Plastic, Resin, or Rubber Products Plant[2]	195	6215	1221

[1] Ranked 26th for facility maximum.

[2] Ranked 21st for facility maximum.

By far, EGUs contribute the most NOx emissions compared to any other facility type, emitting over four times that of the 2nd greatest emission facility type (i.e., airports). Facilities having emissions originating from a 'chemical plant' contributed to the 3rd greatest amount of NOx emissions, though constituting approximately half of those by 'airports'. The NOx emissions were evaluated considering each of the 177 CBSAs, considering each of the top 20 facility types (see supplemental data in Section B5), with Figures provided here of the top three facility types (Figure B2-6 to Figure B2-8). Reasonable representation is given by the selected study areas regarding each of the most important facility types, particularly considering the relative importance of these facility emissions with respect to mobile source contributions (i.e., being significantly lower in terms of emissions). For instance, the Houston CBSA has the greatest NOx emissions contributed by chemical plants (~13,000 tpy, Figure B2-8), though this amount is small when compared with total mobile source NOx emissions in the CBSA (~100,000 tpy, Figure B2-4). A few facility types were not well represented by the selected study areas: 'mines/quarries', 'gas plant' and 'mineral processing', facility types on frequent occasion located in the upper northwestern U.S. (see Table B5-3, Table B5-6, and Table B5-7). Considering the amount of NO$_X$ emissions from these particular facilities relative to the most important emissions nationwide, lack of representation by the selected study areas was considered as having limited influence to our overall study objectives. Furthermore, on occasion it may appear peculiar that a certain emission source is not apparent in selected study area, but this is a function of the data used for this this analysis. For example, while the Hartsfield-Jackson Atlanta International Airport NO$_X$ emissions are greater than 4,000 tpy, the data are not included in this assessment because the facility is actually located in Clayton County, a county that does not have an ambient monitor.

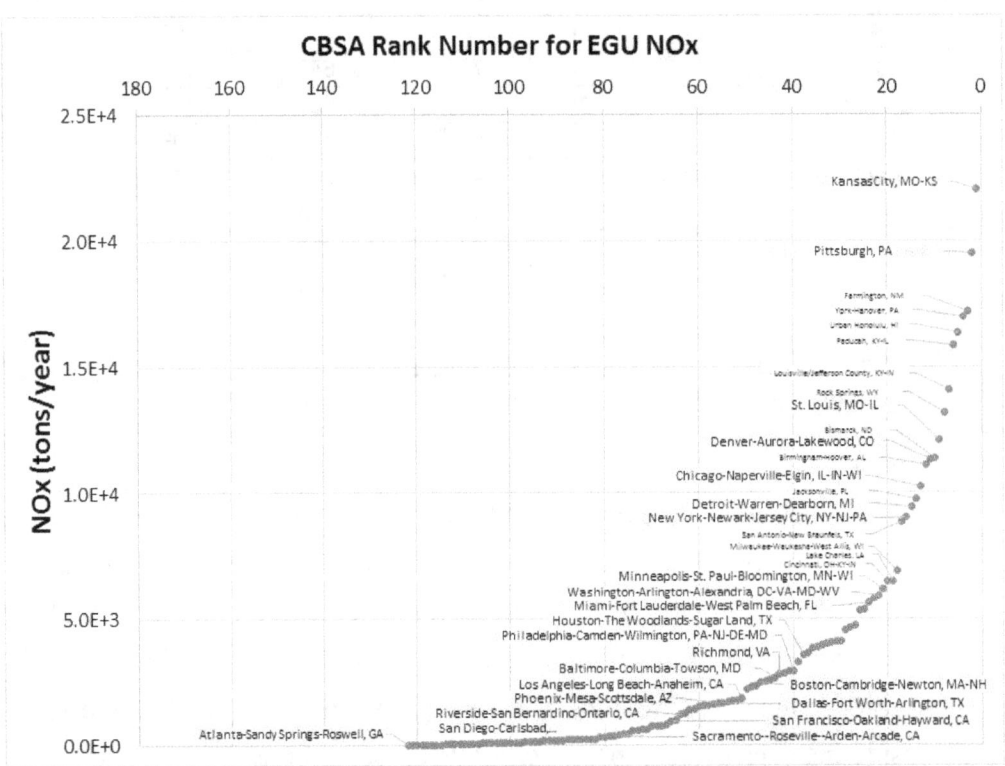

Figure B2-6. CBSAs ranked by NO$_X$ emissions from electricity generation via combustion. Highlighted and named are the 23 selected study areas; also named are CBSA ranked in the top 20.

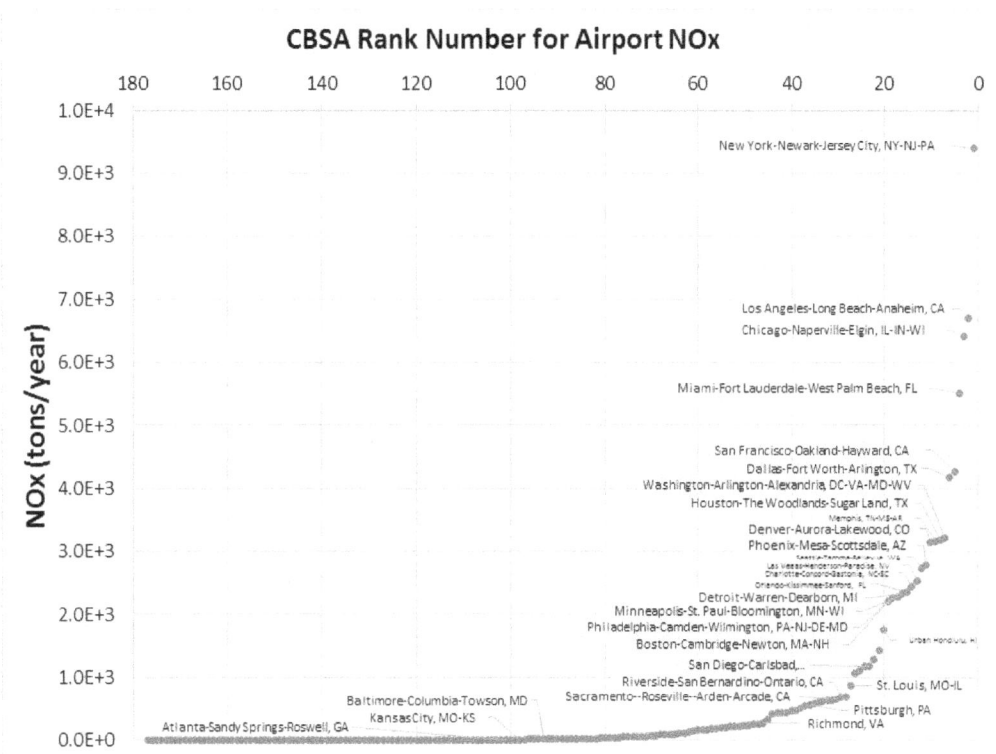

Figure B2-7. CBSAs ranked by NOₓ emissions from airports. Highlighted and named are the 23 selected study areas; also named are CBSA ranked in the top 20.

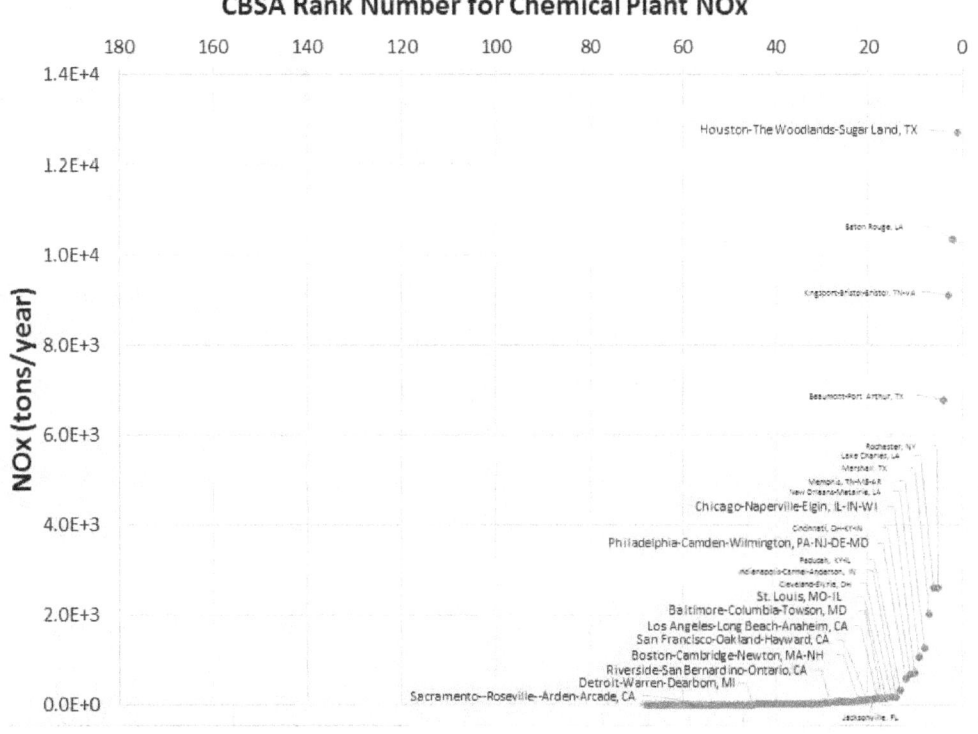

Figure B2-8. CBSAs ranked by NOₓ emissions from chemical plants. Highlighted and named are 11 of the 23 selected study areas (all other study areas had no emissions for this facility type); also named are CBSA ranked in the top 20.

2.3.3 Ambient monitor attributes

We evaluated the attributes of all individual ambient monitors in each of the 23 study areas. For added context and historical perspective, data for monitors in operation currently (2010-2015) and for those operating at any time since 1990 are provided. The attributes for each set of ambient monitors are summarized in Section 5.2 (monitors used for focused analysis) and Section 5.3 (monitors not used for focused study area analysis). For monitors used in the focused analysis, figures are provided showing a satellite view of each the design and near-road monitors.

2.4 ESTIMATED AMBIENT NO$_2$ CONCENTRATIONS

This section describes the approaches used to extend the information provided by the ambient monitor measurement data alone by developing additional air quality standard scenarios and addressing selected high-concentration environments that may not necessarily be captured by the existing monitoring network. The approach describing how air quality would be adjusted to just meet the existing is provided in section 2.4.1. A discussion of how we plan to use the ambient monitors along with factors to characterize NO$_2$ concentrations across urban study areas, including those occurring near-roads and on-roads follows (section 2.4.2). A final section (2.4.3) discusses the calculation of the air quality benchmarks and summary output metrics of interest.

2.4.1 Adjusting air quality to just meet the existing standards

Unadjusted air quality represents ambient conditions as they are at the time of measurement. While unadjusted air quality presents perspective regarding existing conditions, it does not necessarily provide the specific effect that just meeting a specific standard has on ambient concentrations, exposures, and health risk. To evaluate the ability of a specific air quality standard to protect public health, ambient NO$_2$ concentrations need to be adjusted such that they simulate levels of NO$_2$ that would just meet the existing standards (i.e., 100 ppb, 98th percentile DM1H averaged across 3-years; 53 ppb, annual average) or potential alternative standards. Such adjustments allow for comparisons of the level of public health protection that could be associated with just meeting the existing and potential alternative standards.

All areas of the United States currently have ambient NO$_2$ levels below the existing standards, albeit to varying degrees. Therefore, to simulate just meeting the existing NO$_2$ standards, NO$_2$ air quality levels in all study areas must be adjusted upward. Based on evaluating changes in the distribution in air quality over time, a two-step adjustment approach was used to adjust the recent ambient concentrations in each study area to just meet the existing standards. For this two-step adjustment approach, a proportional approach was used, as described in the REA Planning document (and used in the 2008 REA), though here the proportional adjustment was only applied to concentrations up to and including the 98th percentile DM1H (adjustment step 1). This was based on the observation that concentration changes over time (historical, high concentration years compared with recent, low concentration years) occur in a largely proportional manner (e.g., Figure B2-9 and Figure B2-10).

An additional modification to address observed deviations from linearity that could occur at the upper percentile concentrations was used for concentrations above the 98th percentile (adjustment step 2). In this way, this two-step approach utilizes the simplicity of the proportional approach used in the 2008 REA but addresses more fully, the observed changes in peak concentration distributions to better capture the distribution of high NO_2 concentrations when adjusting air quality to meet the existing standards.

To calculate the proportional adjustment factors used for step 1 to estimate concentrations up to and including the 98th percentile DM1H:

1. Using design values for each monitor having recent (2010-2015) and complete air quality and considering both the 1-hour and annual standards,[18] calculate the proportional adjustment factor needed to just meet the existing standards by dividing the standard level by the appropriate monitor design value. Then, identify the monitor in each study area having lowest adjustment factor for each year/3-year period (i.e., indicating the area design value monitor and the controlling standard).[19]

2. Calculate 98th percentile DM1H concentrations for the near-road monitor (where data are available in a single year)[20] and for the highest concentration monitor in each study area for the same years available, using the full concentration distribution and for where simultaneous measurements exist between the two monitor types.

Select the set of proportional adjustment factors based on overall consistency in area design value monitor identity, consistency and reasonableness in the factor level, and whether the limited new near road monitor data available indicate the potential for a dramatically different adjustment factor.

---

[18] See http://www.epa.gov/airtrends/values.html. For this draft AQC, 2014 was the most recent complete year data available. The period 2010-2014 includes five annual design values and three 3-year averaged hourly design values. Design values for 2015 were calculated here using complete year ambient monitor data available.

[19] Often times, only one of the two existing standards (1-hour or annual) would be the controlling standard in a particular area, and is identified by the monitor design value that is arithmetically closest to the particular standard. Based on the level and form of the existing standard DM1H standard, it is expected to be the controlling standard.

[20] Only two near-road monitors had three years of continuous monitoring available for this review (one each in Detroit and St. Louis). Therefore, in most stuidy areas design values DM1H standard cannot be calculated using the near-road monitor data. It remains informative to calculate similar averaging time metrics for each monitor (near-road and area-wide) using years where comparable statistics can be calculated.

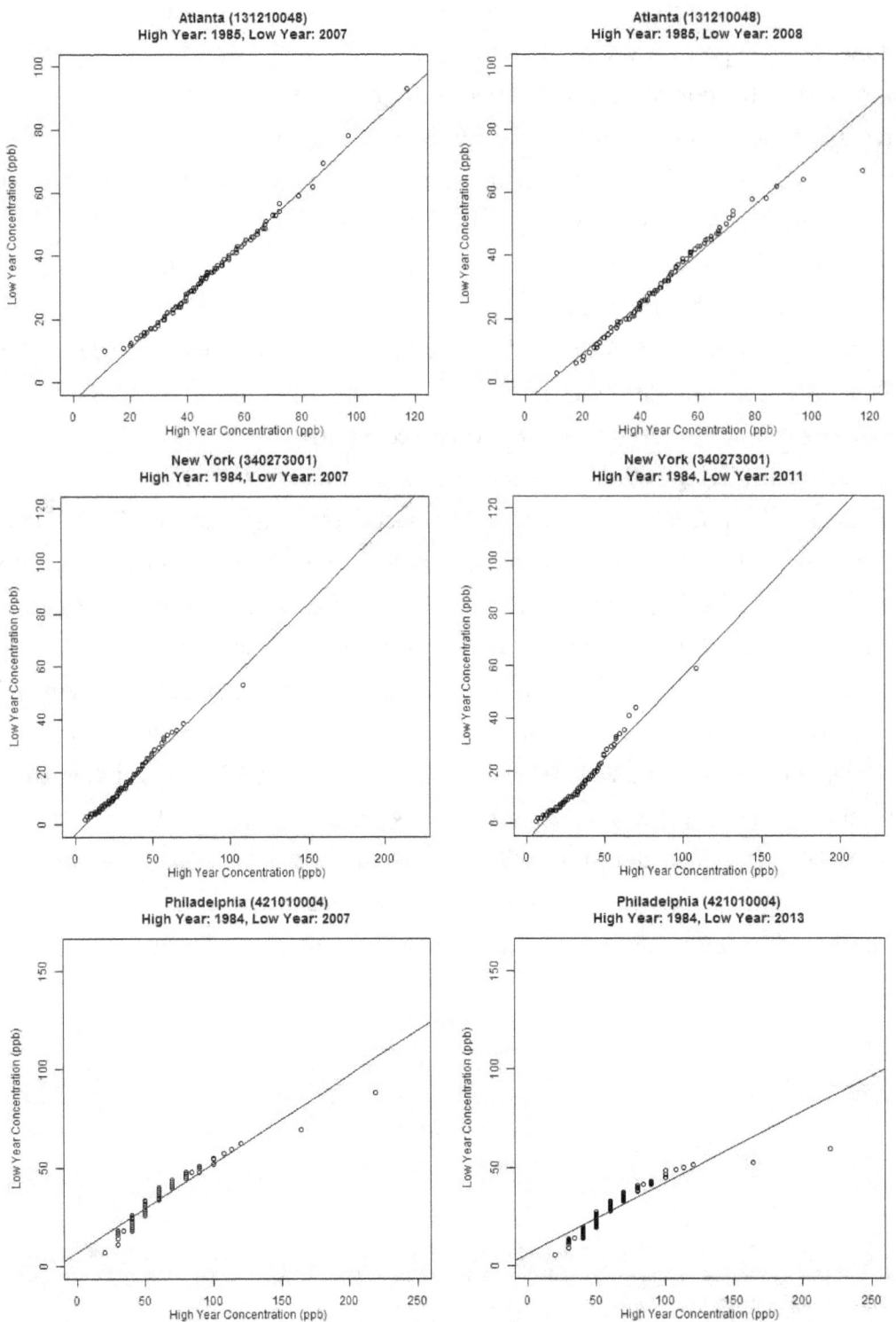

Figure B2-9. Distribution of DM1H NO₂ concentrations (0 – 100th percentile) for a high-concentration year (1980s) versus a low-concentration year (2000s) adapted from Rizzo (2008) (left panel) and updated comparison with a recent low-concentration year (right panel). Atlanta (top panel), New York/New Jersey (middle panel), and Philadelphia (bottom panel) study areas.

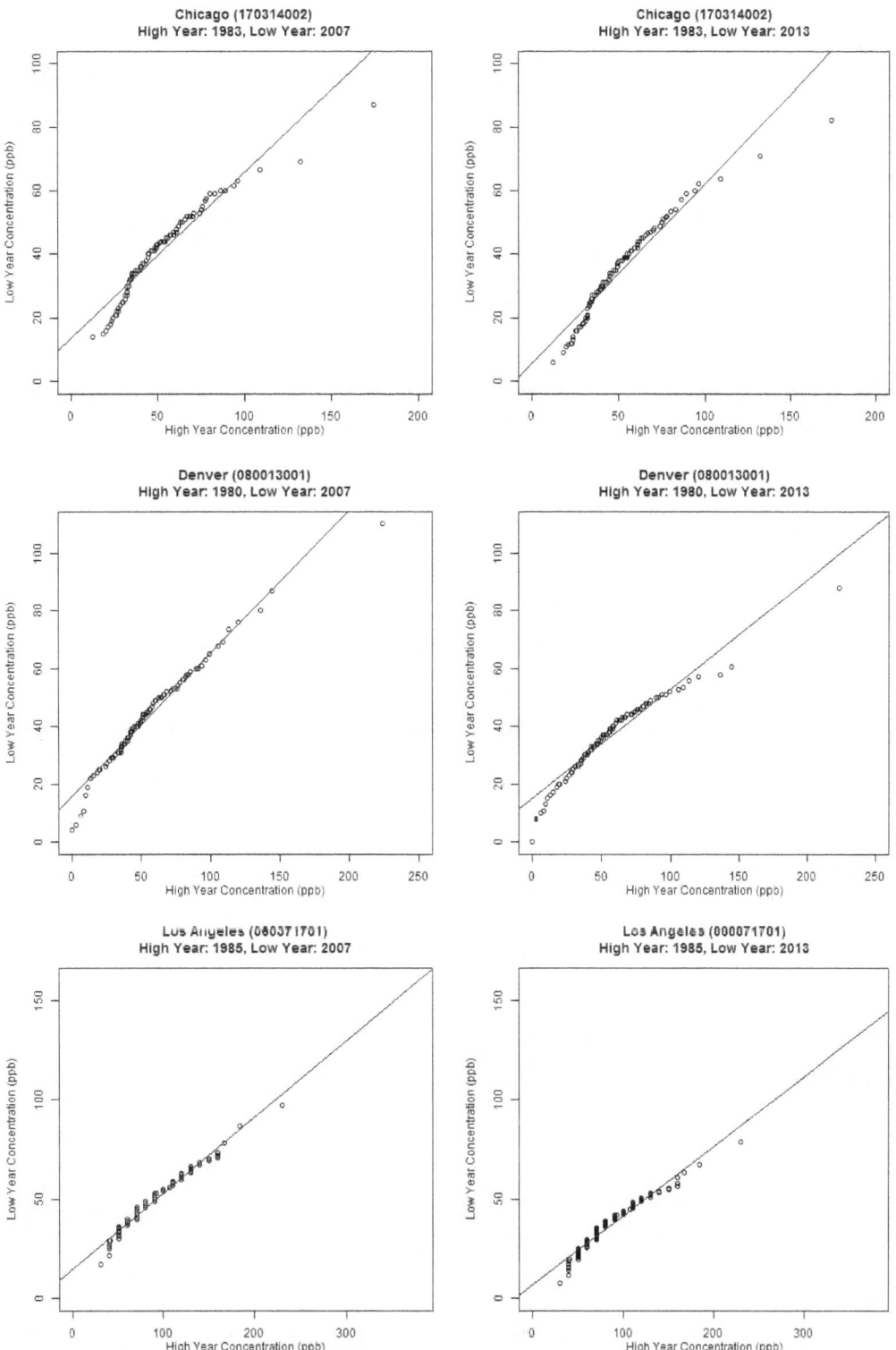

Figure B2-10. Distribution of DM1H NO₂ concentrations (0 – 100ᵗʰ percentile) for a high-concentration year (1980s) versus a low-concentration year (2000s) adapted from Rizzo (2008) (left panel) and updated comparison with a recent low-concentration year (right panel). Chicago (top panel), Denver (middle panel), and Los Angeles (bottom panel) study areas.

The collection of all proportional adjustment factors calculated and those used for each study area are provided in two tables; those using the 3-year average design value data (Table B2-5) and/or the single year near-road/highest monitor data (Table B2-6) for adjusting individual years. Table B2-7 indicates the decision process and selection reasoning, where judgements were made beyond simply choosing any available design value. In all instances where design values were calculated, the DM1H concentrations resulted in the lowest adjustment factors, thus indicating the hourly standard is the controlling standard in each of the 23 selected study areas.

To calculate the second set of factors used for adjusting the ambient concentrations above the 98th percentile DM1H to just meet the existing standard, the following approach was used:

1. For each monitor, using recent, complete year ambient monitor concentrations (2003-2015)[21], divide the DM1H concentrations that are above the 98th percentile DM1H[22] by the 98th percentile DM1H, considering each year separately.

2. Generate the mean ratio for each DM1H across all years available[23] for each individual monitor by averaging the values calculated in step 1.

3. Where ratios cannot be reasonably calculated for an individual monitor (i.e., the monitor is newly sited, such as the near-road monitors), ratios from the area design value monitor are used.

There is general consistency in the adjustment factor across monitors upwards to about the 4th highest maximum, but beyond this point up to the maximum DM1H adjustment factor there is increasing variability in the factor across the monitors (Table B2-8). The maximum adjustment factor by far exhibited the widest range of values within each study area and across study areas. When comparing the overall distribution of the area design value monitor ratios to the other area-wide monitor ratios within each study area, the area design value monitor had the highest ratios in 7 of study areas (Baltimore, Detroit, Denver, Miami, Richmond, St. Louis, and Washington DC). In 7 study areas the area design value monitor had some of the lowest ratios (Atlanta, Boston, Houston, Kansas, Los Angeles, Pittsburgh, and Sacramento), while the remaining 8 study areas had their design value monitor ratios falling within the middle of the collection of

---

[21] The collection of years used to calculate the upper percentile concentraiton ratios (2003-2015) was based on the "NOx SIP CALL" (63 FR 57354) finalized October 27, 1998 that required reduction measures to be in place for most of the U.S. by 2003.

[22] For a full year of ambient data, there would be 365 or 366 DM1H concentrations. Therefore, upwards to seven unique ratios could be calculated using these seven days having DM1H concentrations above the 98th percentile DM1H.

[23] Thus, an adjustment factor for each monitor and DM1H was calculated by averaging across the, at most, 13 ratio values (years).

area-wide monitors. Given the observed range of these adjustment factors, it is possible that, if the variability in upper percentile concentrations at the area design value monitors is not appropriately reflecting that of the near-road monitors, there could be an under-/over-estimation in the number of days per year at or above benchmark levels when considering the near-road and on-road results.

Table B2-5. Proportional adjustment factors calculated from ambient monitor design values in each of the 23 selected study areas.

CBSA Title	Adjustment Factors Calculated Using Annual Design Values						Adjustment Factors Calculated Using DM1H Design Values			
	2010	2011	2012	2013	2014	2015	2010-12	2011-13	2012-14	2013-15
Atlanta-Sandy Springs-Roswell, GA	3.786	4.077	4.417	5.889	4.818	2.650	1.786	1.961	2.041	2.083
Baltimore-Columbia-Towson, MD	2.944	2.944	3.313	3.533	2.944	2.944	1.754	1.923	1.923	2.174
Boston-Cambridge-Newton, MA-NH	2.789	2.650	2.789	2.944	3.118	3.118	1.961	2.000	2.041	1.961
Chicago-Naperville-Elgin, IL-IN-WI	2.120	2.304	2.409	2.524	2.524	2.944	1.613		2.128	1.587
Dallas-Fort Worth-Arlington, TX	4.077	4.077	4.417	4.417	5.300	5.300	1.786	1.887	2.041	2.128
Denver-Aurora-Lakewood, CO	1.893	2.208	2.120	2.208	2.120	1.963		1.613	1.389	1.389
Detroit-Warren-Dearborn, MI	4.417	4.417	2.789	2.944	3.313	2.944	2.083	2.273	2.222	2.000
Houston-The Woodlands-Sugar Land, TX	3.533	3.786	3.533	4.077	4.077	4.077	1.667	1.695	1.786	1.923
Kansas City, MO-KS	3.533	3.533	3.786	4.077	4.077	4.077	1.887	1.923	1.961	1.961
Los Angeles-Long Beach-Anaheim, CA	2.038	2.120	2.524	2.304	1.963	2.120	1.493	1.563	1.587	1.613
Miami-Fort Lauderdale-West Palm Beach, FL	5.300	6.625	6.625	6.625	5.889	10.600	2.128	2.174	2.174	2.273
Minneapolis-St. Paul-Bloomington, MN-WI	5.300	5.889	4.818	5.889	3.313	3.786	2.174	2.273	2.222	2.174
New York-Newark-Jersey City, NY-NJ-PA	2.409	2.120	2.409	2.409	2.409	2.409	1.429	1.493	1.515	1.515
Philadelphia-Camden-Wilmington, PA-NJ-DE-MD	2.304	2.650	2.944	3.118	2.944	2.944	1.538	1.639	2.564	2.041
Phoenix-Mesa-Scottsdale, AZ	2.120	2.120	2.038	2.120	2.120	2.409	1.515	1.563	1.563	1.587
Pittsburgh, PA	3.533	3.313	3.786	4.818	5.300	4.077	1.887	2.041	2.778	2.381
Richmond, VA	4.417	5.300	5.300	6.625	6.625	3.786	1.923	2.128	2.326	2.439
Riverside-San Bernardino-Ontario, CA	2.304	2.524	2.409	2.524	2.650	1.767	1.389	1.613	1.887	2.000
Sacramento--Roseville--Arden-Arcade, CA	4.417	4.077	4.417	5.300	4.818	4.818	1.961	2.000	2.083	2.000
San Diego-Carlsbad, CA	2.524	2.650	2.650	2.789	4.077	3.786	1.370	1.370	1.754	1.786
San Francisco-Oakland-Hayward, CA	3.313	3.313	3.533	3.118	3.118	2.944	1.351	1.471	1.639	1.754
St. Louis, MO-IL	4.077	4.077	3.786	3.786	3.786	4.077	1.887	1.887	2.041	1.923
Washington-Arlington-Alexandria, DC-VA-MD-WV	2.944	3.313	3.118	4.077	4.818	4.077	1.818	1.961	2.083	2.083

Highlighted are the factors used to adjust ambient NO$_2$ concentrations to just meet the existing (controlling) standard, up to and including the 98th percentile DM1H concentrations. Where additional data were available to inform the calculation of factors missing or not used from this table (not highlighted), those factors can be found highlighted in Table B2-6.

Table B2-6. Proportional adjustment factors calculated from the single year 98th percentile DM1H NO_2 at the highest near-road (NR) and highest concentration area wide (AW) monitor in each of the 23 selected study areas.

Column key: **S** = Based on Simultaneous 1-hr Measurements — Adjustment Factors; **n** = Based on All Available 1-hr Measurements — Number of Samples (n); **AF²** = Based on All Available 1-hr Measurements — Adjustment Factors[2].

CBSA Title	S 2013 AW	S 2013 NR	S 2014 AW	S 2014 NR	S 2015[1] AW	S 2015[1] NR	n 2013 AW	n 2013 NR	n 2014 AW	n 2014 NR	n 2015 AW	n 2015 NR	AF² 2013 AW	AF² 2013 NR	AF² 2014 AW	AF² 2014 NR	AF² 2015 AW	AF² 2015 NR
Atlanta-Sandy Springs-Roswell, GA			2.203	1.988	2.083	2.028	361		358	200	355	363	2.331		1.901	2.002	2.083	1.848
Baltimore-Columbia-Towson, MD			2.070	1.976	1.848	1.965	359		353	275	341	352	2.141		1.942	1.976	1.848	2.008
Boston-Cambridge-Newton, MA-NH	2.041	2.222	1.613	1.786	1.852	2.000	356	202	181	365	365	362	2.000	2.273	1.613	1.887	1.852	2.000
Chicago-Naperville-Elgin, IL-IN-WI							356		359		365		1.563		1.493		1.603	
Dallas-Fort Worth-Arlington, TX			2.119	2.500	2.203	2.227	363		364	274	361	365	2.028		2.105	2.500	2.203	2.227
Denver-Aurora-Lakewood, CO	1.479	1.650	1.305	1.445	1.433	1.567	365	213	365	361	365	359	1.479	1.621	1.305	1.437	1.397	1.567
Detroit-Warren-Dearborn, MI	2.326	2.083	1.961	2.000	2.062	2.381	365	360	364	364	61	361	2.326	2.083	1.961	1.961	2.062	2.000
Houston-The Woodlands-Sugar Land, TX			1.949	2.049	1.852	1.869	359		365	344	349	261	1.739		1.923	2.049	1.852	1.689
Kansas City, MO-KS	2.262	2.519	1.957	2.193	1.912	2.247	354	171	365	358	365	352	2.079	2.457	1.898	2.193	1.912	2.247
Los Angeles-Long Beach-Anaheim, CA			1.179	1.502	1.582	1.629	336		340	361	353	255	1.404		1.179	1.515	1.553	1.337
Miami-Fort Lauderdale-West Palm Beach					2.941	2.500	344		364		341	56	2.273		2.000		2.326	2.500
Minneapolis-St. Paul-Bloomington, MN-WI	2.632	2.222	2.000	2.083	2.294	2.083	363	271	365	365	361	362	2.326	2.222	2.000	2.083	2.294	2.083
New York-Newark-Jersey City, NY-NJ-PA			1.695	1.111	1.479	1.493	365		363	184	365	365	1.613		1.429	1.111	1.479	1.493
Philadelphia-Camden-Wilmington, PA-NJ-DE-MD			1.695	1.946	1.587	2.053	355		310	358	345	357	1.942		1.695	1.946	1.587	2.092
Phoenix-Mesa-Scottsdale, AZ			1.587	1.695	1.539	1.887	362		365	322	363	365	1.587		1.563	1.695	1.639	1.887
Pittsburgh, PA			2.564	2.519	2.273	2.232	365		42	122	357	365	2.488		2.222	2.519	2.273	2.232
Richmond, VA	2.725	2.174	2.320	2.237	2.538	2.141	350	92	347	269	352	353	2.497	2.174	2.299	2.237	2.625	2.160
Riverside-San Bernardino-Ontario, CA					1.513	1.370	334		330		358	360	1.580		1.572		1.513	1.370
Sacramento--Roseville--Arden-Arcade, CA					2.083	1.923	363		351		365	80	1.916		1.901		2.198	1.923
San Diego-Carlsbad, CA					1.961	1.923	357		220		365	269	1.333		1.429		1.887	1.923
San Francisco-Oakland-Hayward, CA			1.976	1.934	1.380	2.045	365		365	334	365	365	1.678		1.727	1.934	1.880	2.045
St. Louis, MO-IL	2.041	1.951	2.198	1.718	2.203	2.110	365	365	361	365	352	364	1.980	1.951	2.198	1.718	2.203	2.110
Washington-Arlington-Alexandria, DC-VA-MD-WV					1.961	2.132	365		72		316	212	1.570		1.587		1.730	2.123

[1] Used data from the first near road monitor in operation in the area where data were collected from more than one near road monitor in 2015 or where the value was used in adjustment as an alternative to the factor generated using the design value.

[2] Used data from either the near-road monitor having the greatest number of days monitored in 2015 or the monitor having the lowest adjustment factor when the number of days monitored was similar.

Highlighted are the factors used to adjust ambient NO_2 concentrations to just meet the existing (controlling) standard, up to and including the 98th percentile DM1H concentrations.

Table B2-7. Information supporting the selection of proportional factors used to adjust ambient concentrations (up to and including the 98th percentile DM1H) to just meet the existing standard.

CBSA Title	Adjustment Factor (AF) Selection Reasoning	Comparison of near-road (NR) to area wide (AW) 98th pct derived adjustment factors	
		simultaneous measurement hours	all measurement hours
Atlanta-Sandy Springs-Roswell, GA	Same monitor (13089002) had valid design values for all four periods. NR AFs are lower though generally similar in value. Preliminary results show on-road estimation (2015) unusually higher than expected, likely a function of the DV based AFs. Will use the 2013-15 DV for 2013-14, for 2015 use the lower value (130890003) generated from the two NR monitors.	NR1<AW (2014-15), NR2<AW (2015)	AW<NR1 (2014), NR1<AW (2015), NR2<AW (2015)
Baltimore-Columbia-Towson, MD	One monitor (245100040) had valid design values for three averaging periods 2010-14 (2015 incomplete but had ~7,900 1-hr samples), monitor (240053001) for 2013-2015 period had higher AF. Used avg. of single year calculated AFs for 2013-15 period (245100040).	NR<AW (2014); AW<NR (2015)	AW<NR (2014-15)
Boston-Cambridge-Newton, MA-NH	Same monitor (250250002) had valid design values for first 3 averaging periods. Similar AF value for 2013-15 period (250250042).	AW<NR (2013-15)	AW<NR (2013-15)
Chicago-Naperville-Elgin, IL-IN-WI	No design value for 2011-13. Adjustment factor for 2012-14 (different monitor 17034201) unusually high compared to single year 98th pct values for highest monitors in those years (170314002, 170310076 which are similar to adjustment for 2010-12 period). DV for 2013-15 similar to 2010-12. Only evaluating two 3-yr periods, 10-12 and 13-15.		
Dallas-Fort Worth-Arlington, TX	One monitor (484391002) had valid design value for first three averaging periods across 2010-14, used monitor 481130069 DV for 2013-15 (also had same DV for 2012-14 as 484391002).	AW<NR1 (2014-15), AW<NR2 (2015)	AW<NR1 (2014-15), AW<NR2 (2015)
Denver-Aurora-Lakewood, CO	No design value for 2010-12; for 2011-13 (080013001) AF somewhat higher than 2012-14 and 2013-15 (080310002).	AW<NR1 (2013-15), AW<NR2 (2015)	AW<NR1 (2013-15), NR2<AW (2015)
Detroit-Warren-Dearborn, MI	Same area design value monitor for first three averaging periods across 2010-14 (261630019), NR monitor (261630093) has complete year data for 2013-15 period and used for AF.	NR1<AW (2013, AW<NR1(2014-15), AW<NR2 (2015)	NR1<AW (2013-15), AW<NR2 (2015)
Houston-The Woodlands-Sugar Land, TX	Same monitor (482010075) had valid design values for first three averaging periods across 2010-14. DV for 482010416 (and 482011015) used for 2013-14 (though somewhat higher than prior years), lowest near-road AF (4882011052) used for 2015.	AW<NR1 (2014-15), NR2<AW (2015)	AW<NR1 (2014-15), NR2<AW (2015)
Kansas City, MO-KS	Same monitor (290950034) had valid design values for all four periods.	AW<NR (2013-15)	AW<NR (2013-15)
Los Angeles-Long Beach-Anaheim, CA	Same monitor (060371701) had valid design values for all four periods. Near road 060374008 adjustment was quite lower though for 2015 and used for that single year adjustment.	AW<NR1 (2014-15); NR2<AW (2015)	AW<NR1 (2014-15); NR2<AW (2015)
Miami-Fort Lauderdale-West Palm Beach, FL	Same monitor (120864002) had valid design values for first three averaging periods across 2010-14. Other monitor (120118002) similar value for 2013-15.	NR<AW (2015)	AW<NR (2015)

| CBSA Title | Adjustment Factor (AF) Selection Reasoning | Comparison of near-road (NR) to area wide (AW) 98th pct derived adjustment factors | |
		simultaneous measurement hours	all measurement hours
Minneapolis-St. Paul-Bloomington, MN-WI	Valid, similar design values for all averaging periods. For 2011-13 (270370020); 2012-14 and 2013-15 (270031002); both had exact same adjustment used for 2010-12. NR (270530962) full distribution quite lower in 2015. Averaging all 2013-2015 NR data for adjustment using both NR monitors (270370480 and 270530962).	NR1<AW (2013), AW<NR1 (2014) NR1<AW (2015), AW<NR2 (2015)	NR1<AW (2013), AW<NR1 (2014-15), NR2<AW (2015)
New York-Newark-Jersey City, NY-NJ-PA	Same monitor (340390004) had valid design values for all four periods.	NR<AW (2014) but unusual conc on 1 day, AW<NR (2015)	NR<AW (2014) but unusual conc on 1 day, AW<NR (2015)
Philadelphia-Camden-Wilmington, PA-NJ-DE-MD	Calculated AF for 2012-14 (420170012) and 2013-15 (340070002) much higher than 2010-12 and 2011-13 (421010004). Single year AF for max monitor in 2013 (also unusually high, though is 421010004). Using single year adjustments for 2013-2015 (421010004). Not evaluating 2012-14 period.	AW<NR1 (2014-15), AW<NR2 (2015)	AW<NR1 (2014-15), AW<NR2 (2015)
Phoenix-Mesa-Scottsdale, AZ	Same monitor (040133010) had valid design values for all four periods.	AW<NR1 (2014-15) NR2<AW (2015)	AW<NR1 (2014-15) NR2<AW (2015)
Pittsburgh, PA	2012-14 AF (421250005) unusually high compared with 2010-12 and 2011-13 (420030010); 2013-15 (420070014) is closer though still higher than 2010-12 and 2011-13. Max mon in 2014 (420031005) has smaller AF than 2012-14 value and that calculated using NR monitor, though only has 42 days. Thus, 3 periods will be evaluated (2010-12, 2011-13, 2013-15).	NR<AW (2014-15), but close in value	AW<NR (2014), NR<AW (2015)
Richmond, VA	Same monitor (510360002) had valid design values for all four periods, though NR (517600025) consistently lower for each year of 2013-2015. Used avg. of these three years to generate AF for 2013-15 period.	NR<AW (2013-15)	NR<AW (2013-15)
Riverside-San Bernardino-Ontario, CA	Four different monitors initially used for AFs, with each progressively increasing in value from about 1.4 to 2.0. Use of 2.0 resulted in unusually high concentrations at NR monitors for 2015, also unusually high concentrations for the area wide monitors in 2012 using the 12_14 value of 1.887. For 12_14, averaging the 11_13 (060712002) and 10_12 values (060710001). For 2013_15, combining single year values from highest monitor in first two years (060712002, 2014; 060710001, 2013) and averaging. For 2015, using factor derived from the near-road monitor (060710026) having the greatest number of days with measurements. Note, AF used is highly influential in on-road estimations as well.	NR<AW both NR mons (2015) - NR2 1/2 year	NR<AW both NR mons (2015) - NR2 1/2 year
Sacramento--Roseville--Arden-Arcade, CA	Two monitors (060670006 and 060670010) have valid design values over at least two periods each, similar in value. Used to evaluate all four periods.	NR<AW (2015), similar though NR only 80 days	AW<NR (2015)
San Diego-Carlsbad, CA	Two monitors (060732007, 2010-12 and 2011-13; 060731010, 2012-14 and 2013-15) have valid design values over period though some difference in value, particularly when compared with the 2013-14 single year values for 060732007. Using the mean of 2013-14 060732007 value for 2013-14 (in 2012-14 period), using 2013-15 design value from 060731010 to simulate 2015 only for 2013-15 period.	NR<AW (2015) though similar	AW<NR (2015) though similar

CBSA Title	Adjustment Factor (AF) Selection Reasoning	Comparison of near-road (NR) to area wide (AW) 98th pct derived adjustment factors	
		simultaneous measurement hours	all measurement hours
San Francisco-Oakland-Hayward, CA	Same monitor (060750005) had valid design values for all four periods	NR<AW (2014), AW<NR (2015) but close in value	AW<NR (2014-15)
St. Louis, MO-IL	No design value for 2010-12. Valid design values for 2011-13 and 2012-14 for same monitor (2951000086). Near road monitor (295100094) is area design value monitor for 2013-15, similar values to earlier years.	NR1<AW (2013-15), AW<NR2 (2015)	NR1<AW (2013-15), AW<NR2 (2015)
Washington-Arlington-Alexandria, DC-VA-MD-WV	Two monitors (110010041, 2010-12 and 2011-13; 110010043, 2012-14 and 2013-15) have valid design values over period with generally small differences.	AW<NR (2015)	AW<NR (2015)

Table B2-8. Individual monitor-based factors calculated for use in adjusting DM1H ambient NO$_2$ concentrations above the 98th percentile DM1H in the 23 selected study areas.

CBSA	Site ID	Adjustment factor derived from ratio of DM1H concentrations to 98th percentile DM1H, averaged across years 2003-2015						
		1st DM1H	2nd DM1H	3rd DM1H	4th DM1H	5th DM1H	6th DM1H	7th DM1H
Atlanta-Sandy Springs-Roswell, GA	130890003[a]	1.258	1.167	1.123	1.09	1.061	1.03	1.018
	131210056[a]	1.258	1.167	1.123	1.09	1.061	1.03	1.018
	130890002[b]	1.258	1.167	1.123	1.09	1.061	1.03	1.018
	132230003	1.529	1.333	1.268	1.202	1.141	1.058	1.028
	132470001	1.361	1.25	1.179	1.135	1.113	1.04	1.022
Baltimore-Columbia-Towson, MD	240270006[a]	1.253	1.156	1.120	1.076	1.054	1.026	1.016
	240053001	1.232	1.130	1.096	1.057	1.033	1.019	1.006
	245100040[b]	1.253	1.156	1.120	1.076	1.054	1.026	1.016
Boston-Cambridge-Newton, MA-NH	250250044[a]	1.242	1.149	1.1	1.076	1.056	1.037	1.013
	250092006	1.259	1.179	1.137	1.092	1.064	1.049	1.024
	250094005	1.407	1.367	1.301	1.158	1.144	1.117	1.079
	250095005	1.179	1.132	1.093	1.070	1.047	1.031	1.016
	250213003	1.387	1.369	1.205	1.189	1.118	1.053	1.019
	250250002[b]	1.242	1.149	1.100	1.076	1.056	1.037	1.013
	250250040	1.754	1.266	1.178	1.124	1.064	1.034	1.023
	250250042	1.278	1.199	1.142	1.093	1.077	1.04	1.018
	330150018	1.670	1.300	1.185	1.163	1.053	1.026	1.018
Chicago-Naperville-Elgin, IL-IN-WI	170310063	1.278	1.179	1.104	1.067	1.046	1.028	1.014
	170310076	1.325	1.212	1.146	1.099	1.058	1.032	1.014
	170313103	1.295	1.163	1.104	1.077	1.047	1.032	1.009
	170314002	1.249	1.178	1.147	1.117	1.070	1.029	1.017
	170314201	1.258	1.165	1.118	1.098	1.055	1.034	1.019
	180890022	1.308	1.231	1.126	1.083	1.040	1.018	1.010
Dallas-Fort Worth-Arlington, TX	484391053[a]	1.249	1.158	1.124	1.092	1.049	1.032	1.011
	481131067[a]	1.249	1.158	1.124	1.092	1.049	1.032	1.011
	481130069	1.247	1.167	1.111	1.068	1.058	1.033	1.025
	481130075	1.198	1.143	1.101	1.062	1.045	1.025	1.008
	481130087	1.279	1.138	1.095	1.072	1.047	1.024	1.012
	481210034	1.304	1.191	1.142	1.099	1.058	1.037	1.016
	481390016	1.295	1.158	1.084	1.048	1.039	1.022	1.015
	481391044	1.209	1.155	1.101	1.072	1.046	1.025	1.018

CBSA	Site ID	Adjustment factor derived from ratio of DM1H concentrations to 98th percentile DM1H, averaged across years 2003-2015						
		1st DM1H	2nd DM1H	3rd DM1H	4th DM1H	5th DM1H	6th DM1H	7th DM1H
	482311006	1.244	1.146	1.125	1.100	1.071	1.046	1.026
	482570005	1.292	1.208	1.151	1.101	1.061	1.050	1.023
	483670081	1.718	1.224	1.154	1.143	1.112	1.046	1.027
	484390075	1.170	1.163	1.100	1.036	1.019	1.012	1.012
	484391002[b]	1.249	1.158	1.124	1.092	1.049	1.032	1.011
	484392003	1.235	1.096	1.090	1.070	1.052	1.034	1.027
	484393009	1.215	1.162	1.107	1.058	1.039	1.028	1.012
	484393011	1.211	1.127	1.094	1.072	1.037	1.028	1.017
Denver-Aurora-Lakewood, CO	080310028[a]	1.428	1.163	1.134	1.103	1.069	1.033	1.021
	080310027[a]	1.428	1.163	1.134	1.103	1.069	1.033	1.021
	080013001	1.305	1.188	1.134	1.097	1.063	1.030	1.012
	080310002[b]	1.428	1.163	1.134	1.103	1.069	1.033	1.021
	080310026	1.136	1.108	1.072	1.054	1.034	1.030	1.008
Detroit-Warren-Dearborn, MI	261630095[a]	1.292	1.160	1.109	1.076	1.053	1.030	1.018
	261630019[b]	1.292	1.160	1.109	1.076	1.053	1.030	1.018
	261630093	1.302	1.119	1.109	1.050	1.020	1.015	1.013
	261630094	1.194	1.119	1.087	1.071	1.048	1.027	1.012
Houston-The Woodlands-Sugar Land, TX	482011052[a]	1.310	1.178	1.129	1.087	1.049	1.031	1.018
	482011066[a]	1.310	1.178	1.129	1.087	1.049	1.031	1.018
	480391004	1.292	1.231	1.145	1.096	1.072	1.041	1.026
	480391016	1.341	1.182	1.125	1.105	1.071	1.042	1.037
	481671034	1.302	1.216	1.149	1.113	1.083	1.069	1.019
	482010024	1.300	1.166	1.099	1.067	1.035	1.016	1.009
	482010026	1.564	1.370	1.267	1.180	1.109	1.064	1.019
	482010029	1.545	1.293	1.195	1.137	1.093	1.037	1.013
	482010047	1.366	1.217	1.143	1.089	1.051	1.035	1.013
	482010055	1.219	1.148	1.114	1.081	1.05	1.028	1.009
	482010075[b]	1.310	1.178	1.129	1.087	1.049	1.031	1.018
	482010416	1.260	1.133	1.098	1.082	1.047	1.034	1.018
	482011015	1.476	1.173	1.135	1.086	1.056	1.029	1.012
	482011034	1.441	1.221	1.151	1.100	1.055	1.043	1.016
	482011035	1.343	1.172	1.122	1.081	1.056	1.030	1.008
	482011039	1.406	1.211	1.144	1.091	1.049	1.032	1.011
	482011050	1.605	1.401	1.220	1.149	1.109	1.045	1.025
	483390078	1.478	1.287	1.186	1.140	1.104	1.051	1.032
Kansas City, MO-KS	290950042[a]	1.417	1.223	1.163	1.135	1.077	1.039	1.021
	201070002	1.682	1.495	1.313	1.233	1.171	1.100	1.023
	202090021	1.396	1.181	1.099	1.056	1.036	1.020	1.013
	290950034[b]	1.417	1.223	1.163	1.135	1.077	1.039	1.021
Los Angeles-Long Beach-Anaheim, CA	060374008[a]	1.242	1.146	1.099	1.076	1.042	1.029	1.013
	060590008[a]	1.242	1.146	1.099	1.076	1.042	1.029	1.013
	060370002	1.248	1.166	1.141	1.096	1.07	1.047	1.018
	060370016	1.272	1.184	1.134	1.097	1.071	1.039	1.020
	060370113	1.290	1.159	1.124	1.093	1.074	1.035	1.019
	060371103	1.357	1.214	1.160	1.117	1.060	1.032	1.017
	060371201	1.343	1.185	1.116	1.060	1.033	1.022	1.009
	060371302	1.197	1.151	1.122	1.082	1.066	1.030	1.011
	060371602	1.237	1.172	1.104	1.058	1.035	1.023	1.018
	060371701[b]	1.242	1.146	1.099	1.076	1.042	1.029	1.013
	060372005	1.415	1.267	1.168	1.130	1.099	1.062	1.036
	060374002	1.404	1.256	1.136	1.100	1.062	1.038	1.006
	060374006	1.399	1.156	1.128	1.063	1.042	1.027	1.017
	060375005	1.354	1.189	1.140	1.105	1.069	1.037	1.017

CBSA	Site ID	Adjustment factor derived from ratio of DM1H concentrations to 98th percentile DM1H, averaged across years 2003-2015						
		1st DM1H	2nd DM1H	3rd DM1H	4th DM1H	5th DM1H	6th DM1H	7th DM1H
	060376012	1.339	1.201	1.163	1.109	1.064	1.034	1.018
	060379033	1.219	1.146	1.081	1.051	1.027	1.014	1.009
	060590007	1.295	1.186	1.143	1.092	1.062	1.033	1.018
	060591003	1.297	1.212	1.150	1.105	1.062	1.040	1.02
	060595001	1.280	1.151	1.104	1.075	1.049	1.045	1.015
Minneapolis-St. Paul-Bloomington, MN-WI	270370480[a]	1.352	1.231	1.163	1.111	1.080	1.031	1.025
	270530962[a]	1.352	1.231	1.163	1.111	1.080	1.031	1.025
	270031002	1.202	1.151	1.093	1.064	1.053	1.030	1.010
	270370020[b]	1.352	1.231	1.163	1.111	1.080	1.031	1.025
	270370423	1.561	1.205	1.121	1.097	1.069	1.044	1.015
Miami-Fort Lauderdale-West Palm Beach, FL	120110035[a]	1.808	1.235	1.100	1.079	1.067	1.034	1.022
	120110031	1.569	1.375	1.177	1.114	1.072	1.045	1.010
	120118002	1.566	1.210	1.142	1.103	1.077	1.049	1.028
	120860027	1.290	1.178	1.134	1.089	1.062	1.046	1.026
	120864002[b]	1.808	1.235	1.100	1.079	1.067	1.034	1.022
	120990020	1.263	1.194	1.169	1.153	1.067	1.025	1.000
New York-Newark-Jersey City, NY-NJ-PA	340030010[a]	1.442	1.270	1.154	1.097	1.060	1.038	1.020
	340030006	1.218	1.157	1.116	1.066	1.046	1.015	1.015
	340130003	1.290	1.225	1.178	1.116	1.077	1.038	1.024
	340131003	1.377	1.240	1.172	1.115	1.077	1.038	1.020
	340170006	1.731	1.280	1.225	1.133	1.083	1.048	1.015
	340230011	1.326	1.175	1.118	1.088	1.067	1.038	1.014
	340273001	1.437	1.255	1.175	1.148	1.101	1.064	1.042
	340390004[b]	1.442	1.270	1.154	1.097	1.060	1.038	1.020
	360050110	1.402	1.207	1.158	1.109	1.078	1.051	1.019
	360050133	1.207	1.165	1.112	1.069	1.040	1.023	1.013
	360590005	1.486	1.325	1.203	1.141	1.094	1.045	1.021
	360810124	1.420	1.253	1.149	1.072	1.039	1.023	1.009
Philadelphia-Camden-Wilmington, PA-NJ-DE-MD	421010076[a]	1.377	1.205	1.142	1.084	1.057	1.035	1.013
	421010075[a]	1.377	1.205	1.142	1.084	1.057	1.035	1.013
	100032004	1.575	1.228	1.147	1.109	1.072	1.023	
	340070002[c]	1.238	1.130	1.097	1.059	1.049	1.043	1.013
	420170012	1.309	1.179	1.102	1.075	1.037	1.029	1.011
	420450002	1.485	1.308	1.140	1.090	1.048	1.030	1.023
	421010004[b]	1.377	1.205	1.142	1.084	1.057	1.035	1.013
	421010047	1.546	1.354	1.195	1.141	1.091	1.041	1.013
Phoenix-Mesa-Scottsdale, AZ	040134020[a]	1.300	1.194	1.102	1.066	1.053	1.036	1.017
	040134019[a]	1.300	1.194	1.102	1.066	1.053	1.036	1.017
	040130019	1.250	1.152	1.112	1.078	1.054	1.028	1.016
	040133002	1.147	1.084	1.064	1.050	1.038	1.029	1.014
	040133003	1.177	1.116	1.071	1.051	1.029	1.024	1.003
	040133010[b]	1.300	1.194	1.102	1.066	1.053	1.036	1.017
	040134011	1.384	1.252	1.155	1.119	1.073	1.046	1.021
	040139997	1.156	1.095	1.055	1.045	1.028	1.019	1.017
Pittsburgh, PA	420031376[a]	1.199	1.144	1.088	1.067	1.043	1.021	1.015
	420030008	1.230	1.132	1.097	1.066	1.050	1.037	1.020
	420030010[b]	1.199	1.144	1.088	1.067	1.043	1.021	1.015
	420031005	1.454	1.243	1.179	1.123	1.088	1.050	1.028
	420031008	1.101	1.043	1.043	1.043	1.043	1.014	1.014
	420070014	1.265	1.189	1.142	1.106	1.066	1.036	1.022
	421250005	1.368	1.170	1.124	1.103	1.061	1.037	1.012
	421255200	1.185	1.161	1.161	1.093	1.069	1.047	1.023
Richmond, VA	517600025[a]	1.349	1.214	1.178	1.125	1.087	1.063	1.033

CBSA	Site ID	Adjustment factor derived from ratio of DM1H concentrations to 98th percentile DM1H, averaged across years 2003-2015						
		1st DM1H	2nd DM1H	3rd DM1H	4th DM1H	5th DM1H	6th DM1H	7th DM1H
	510360002[b]	1.349	1.214	1.178	1.125	1.087	1.063	1.033
	510870014	1.194	1.125	1.087	1.066	1.049	1.035	1.013
	517600024	1.357	1.162	1.095	1.062	1.050	1.025	1.015
Riverside-San Bernardino-Ontario, CA	060710027[a]	1.334	1.194	1.122	1.079	1.053	1.035	1.017
	060710026[a]	1.334	1.194	1.122	1.079	1.053	1.035	1.017
	060650009	1.251	1.161	1.117	1.084	1.058	1.034	1.012
	060650012	1.250	1.112	1.066	1.045	1.033	1.025	1.014
	060651003	1.212	1.104	1.099	1.072	1.032	1.019	1.007
	060651016	1.862	1.451	1.379	1.219	1.210	1.027	1.018
	060655001	1.238	1.156	1.074	1.058	1.037	1.020	1.005
	060658001	1.220	1.158	1.104	1.060	1.037	1.019	1.011
	060658005	1.286	1.127	1.096	1.056	1.037	1.030	1.022
	060659001	1.239	1.169	1.120	1.097	1.060	1.041	1.018
	060710001	1.189	1.120	1.080	1.062	1.040	1.029	1.011
	060710306	1.245	1.187	1.091	1.055	1.029	1.015	1.008
	060711004	1.223	1.156	1.123	1.089	1.053	1.036	1.017
	060711234	1.246	1.180	1.137	1.081	1.054	1.033	1.023
	060712002[b]	1.334	1.194	1.122	1.079	1.053	1.035	1.017
	060719004	1.349	1.188	1.127	1.093	1.053	1.031	1.009
Sacramento--Roseville--Arden-Arcade, CA	060670015[a]	1.207	1.124	1.092	1.072	1.054	1.023	1.010
	060610006	1.203	1.143	1.103	1.074	1.067	1.042	1.017
	060670002	1.504	1.210	1.125	1.073	1.051	1.030	1.012
	060670006[b]	1.207	1.124	1.092	1.072	1.054	1.023	1.010
	060670010	1.225	1.131	1.098	1.070	1.049	1.038	1.014
	060670011	1.391	1.223	1.158	1.107	1.068	1.039	1.019
	060670012	1.288	1.192	1.131	1.098	1.066	1.036	1.025
	060670014	1.328	1.182	1.144	1.079	1.051	1.035	1.029
	061130004	1.268	1.188	1.126	1.097	1.070	1.053	1.030
San Diego-Carlsbad, CA	060731017[a]	1.281	1.213	1.129	1.078	1.060	1.043	1.018
	060730001	1.197	1.118	1.089	1.066	1.047	1.033	1.012
	060730003	1.259	1.106	1.082	1.057	1.035	1.017	1.003
	060730006	1.348	1.182	1.154	1.105	1.064	1.046	1.019
	060731002	1.281	1.152	1.092	1.069	1.042	1.021	1.010
	060731006	1.411	1.232	1.150	1.117	1.067	1.042	1.009
	060731008	1.312	1.224	1.137	1.095	1.055	1.028	1.010
	060731010	1.245	1.182	1.130	1.091	1.053	1.019	1.008
	060731014	1.196	1.118	1.078	1.039	1.020	1.000	1.000
	060731016	1.239	1.153	1.095	1.080	1.058	1.037	1.000
	060732007[b]	1.281	1.213	1.129	1.078	1.06	1.043	1.018
San Francisco-Oakland-Hayward, CA	060010012[a]	1.325	1.150	1.092	1.077	1.056	1.035	1.016
	060010007	1.245	1.165	1.103	1.079	1.052	1.028	1.010
	060010009	1.274	1.193	1.145	1.083	1.063	1.032	1.025
	060010011	1.233	1.160	1.097	1.073	1.051	1.034	1.014
	060012004	1.200	1.168	1.136	1.099	1.053	1.030	1.020
	060012005	2.050	1.262	1.204	1.158	1.090	1.086	1.009
	060130002	1.250	1.159	1.116	1.082	1.056	1.031	1.019
	060131002	1.233	1.176	1.133	1.097	1.058	1.043	1.017
	060131004	1.311	1.183	1.131	1.107	1.056	1.034	1.012
	060132007	1.335	1.107	1.082	1.067	1.061	1.034	1.027
	060410001	1.227	1.138	1.101	1.073	1.052	1.035	1.012
	060750005[b]	1.325	1.150	1.092	1.077	1.056	1.035	1.016
	060811001	1.299	1.189	1.137	1.082	1.066	1.049	1.023
St. Louis, MO-IL	291890016[a]	1.332	1.158	1.095	1.061	1.034	1.016	1.011

CBSA	Site ID	Adjustment factor derived from ratio of DM1H concentrations to 98th percentile DM1H, averaged across years 2003-2015						
		1st DM1H	2nd DM1H	3rd DM1H	4th DM1H	5th DM1H	6th DM1H	7th DM1H
	295100094[a]	1.332	1.158	1.095	1.061	1.034	1.016	1.011
	171630010	1.317	1.151	1.113	1.065	1.044	1.022	1.017
	171630900	1.179	1.143	1.119	1.096	1.068	1.044	1.029
	295100085	1.185	1.141	1.125	1.105	1.068	1.064	1.020
	295100086[b]	1.332	1.158	1.095	1.061	1.034	1.016	1.011
Washington-Arlington-Alexandria, DC-VA-MD-WV	110010051[a]	1.490	1.334	1.209	1.140	1.096	1.060	1.032
	110010025	1.269	1.135	1.103	1.057	1.040	1.022	1.007
	110010041[b]	1.490	1.334	1.209	1.140	1.096	1.060	1.032
	110010043	1.347	1.219	1.138	1.078	1.059	1.033	1.018
	110010050	1.211	1.139	1.052	1.020	1.020	1.010	1.000
	240330030	1.236	1.154	1.081	1.059	1.048	1.009	1.002
	510130020	1.211	1.132	1.089	1.066	1.050	1.038	1.020
	511071005	1.220	1.141	1.090	1.057	1.028	1.019	1.008
	511530009	1.333	1.255	1.124	1.078	1.053	1.034	1.012
	515100009	1.190	1.136	1.113	1.069	1.038	1.031	1.020
	515100021	1.166	1.129	1.086	1.066	1.050	1.031	1.013

[a] The near-road monitor used the ratios derived from the monitor having the highest design value.
[b] The area design value monitor ratios used for adjustments made to upper percentile concentrations at the near-road monitor.
[c] Monitor ID 340070003 (operating during 2003-2008) is sited in close proximity to newly sited monitor ID 340070002 (operating during 2012-2015). The data from both monitors were combined to calculate ratios.

Regarding the application of these factors used to adjust air quality to just meet the existing 1-hr standard, the following two steps were applied.

1. Adjust all DM1H concentrations proportionally, up to and including the 98th percentile DM1H, at all monitors and for each single year in each study area using the appropriate study area and year adjustment factors derived from Table B2-5 and Table B2-6 multiplied by the DM1H concentrations. Thus, for all of the proposed study areas selected in this air quality assessment, a single proportional factor, derived from one monitor in a study area, is used to adjust concentrations (up to and including the DM1H 98th percentile) measured at all of the other monitors in that study area. The monitor having the highest design value will have adjusted concentrations that just meet the existing hourly standard (a 3-year average DM1H 98th percentile of 100 ppb), while all other monitors will have adjusted hourly

concentrations/design values less than that value.[24] Because there is overlap in the 3-year averaging periods considered here, there will be instances where multiple values are generated for a single year of data. For example, for each monitor in operation during 2012, three separate estimates of adjusted air quality could be calculated for that year, using the unique adjustment factors derived from each of the three 3-year periods (2010-2012, 2011-2013, and 2012-2014), where available.

2. For all DM1H concentrations above the 98th percentile DM1H at each individual monitor, the adjusted 98th percentile DM1H concentration from step 1 is multiplied by the suite of mean ratios from Table B2-8 to develop the series of days having DM1H concentrations above the 98th percentile DM1H. As described above, there are up to seven DM1H ratios, thus upward to seven adjusted concentrations are estimated above the 98th percentile DM1H for each monitoring year.

As an example, the results of applying these two adjustments are illustrated in Figure B2-11 for ambient concentrations measured at the monitor (ID 421010004) across a three-year standard averaging period (2011-2013). Plotted in this figure are the unadjusted DM1H concentrations, concentrations adjusted to just meet the existing hourly standard using a proportional factor alone (the approach used in the 2008 REA), and concentrations adjusted to just meet the existing hourly standard using a proportional factor and additional factors for estimating concentrations above the 98th percentile DM1H. In this figure, concentrations at or above the DM1H 80th percentile are plotted to highlight the upper percentiles of the distribution.

We selected a single year of ambient concentrations (2011) to demonstrate the calculations, though the same approach applies for all three years in that averaging period. Using the proportional adjustment alone (an increase of 63.9%, Table B2-5, based on the 2011-2013 averaging period) appropriately increases the 80th and 100th DM1H unadjusted concentrations of 44 and 88 ppb (top panel, gray line, Figure B2-11) to 73 and 144 ppb (top panel, red line, Figure B2-11). When addressing deviations from linearity above the 98th percentile, the suite of ratios

---

[24] Ideally this is how the standard would work but this is not always the case considering the suite of monitors in operation at any one time in the study areas. The design value for the hourly standard requires three consecutive years of ambient concentraitons meeting the completeness criteria. For instance, a monitor may not have met completeness criteria for one year, but during two of the three years it could have concentrations (the full distribution including the 98th percentile) greater than the monitor used to calculate the highest design value. Effectively when adjusting concentrations to just meet the standard, the area design value monitor will have a DM1H 98th percentile average concentration of 100 ppb while this other monitor in this hypothetical situation will have a DM1H 98th percentile average concentration greater than 100 ppb. It is also possible each of the two individual years within that three year averaging period have a DM1H 98th percentile greater than 100 ppb.

used from Table B2-8 extend the upper percentile concentrations to somewhat higher concentrations (top panel, blue line, Figure B2-11) when compared with that using a proportional factor alone to adjust all concentrations. For example, the proportionally adjusted 98th percentile DM1H concentration of 124 ppb is used with the maximum DM1H adjustment factor of 1.377 to estimate a maximum DM1H concentration of about 172 ppb for this new air quality distribution. Note also, because of the three-year averaging period and that each monitoring year has a unique, variable concentration distribution, there will be years where the 98th percentile DM1H is above 100 ppb, and years where the 98th percentile DM1H is below 100 ppb (though still the 3-year average, [124.4+92.4+84.4]/3, is equivalent to 100 ppb). It also follows that there will be variable numbers of days per year where DM1H concentration are at or above the selected benchmarks in each year.

While on average, the expected number of days per year would fall somewhere around 8 for the three-year period (i.e., the 98th percentile DM1H value would be the 8th highest value in a 365-day monitor period), because of this year-to-year variation in ambient concentrations, the actual range benchmark exceedances when considering the individual years will likely extend above and below this average value. For instance, Figure B2-11 (top panel) indicates that 2011 has the highest adjusted 98th percentile DM1H of 124.4 ppb in the three-year period. Based on this we know, at a minimum, there would be at least 8 days in that year having DM1H concentrations $\geq$ 100 ppb. Further, the figure indicates the DM1H concentration of 100 ppb falls somewhere between the 93rd and 94th percentile of the 2011 adjusted ambient concentration distribution, indicating that the number of days per year with DM1H concentrations $\geq$ 100 ppb is likely between 21-25 days/year for this year of air quality (i.e., the actual number of benchmark exceedances for 2011 is calculated as 23 days/year). However, when considering the year having the lowest adjusted 98th percentile DM1H in that same three-year period (i.e., Figure B2-11, bottom panel, the 98th pct DM1H for 2013 is 84.4 ppb), there are likely far fewer days/year having DM1H concentrations $\geq$ 100 ppb. This is because, for this lowest concentration year within the standard 3-year averaging period, the DM1H concentration of 100 ppb falls between the 99th and 100th percentile of the adjusted ambient concentration distribution (i.e., the actual number of benchmark exceedances for 2013 is calculated as 2 days/year).

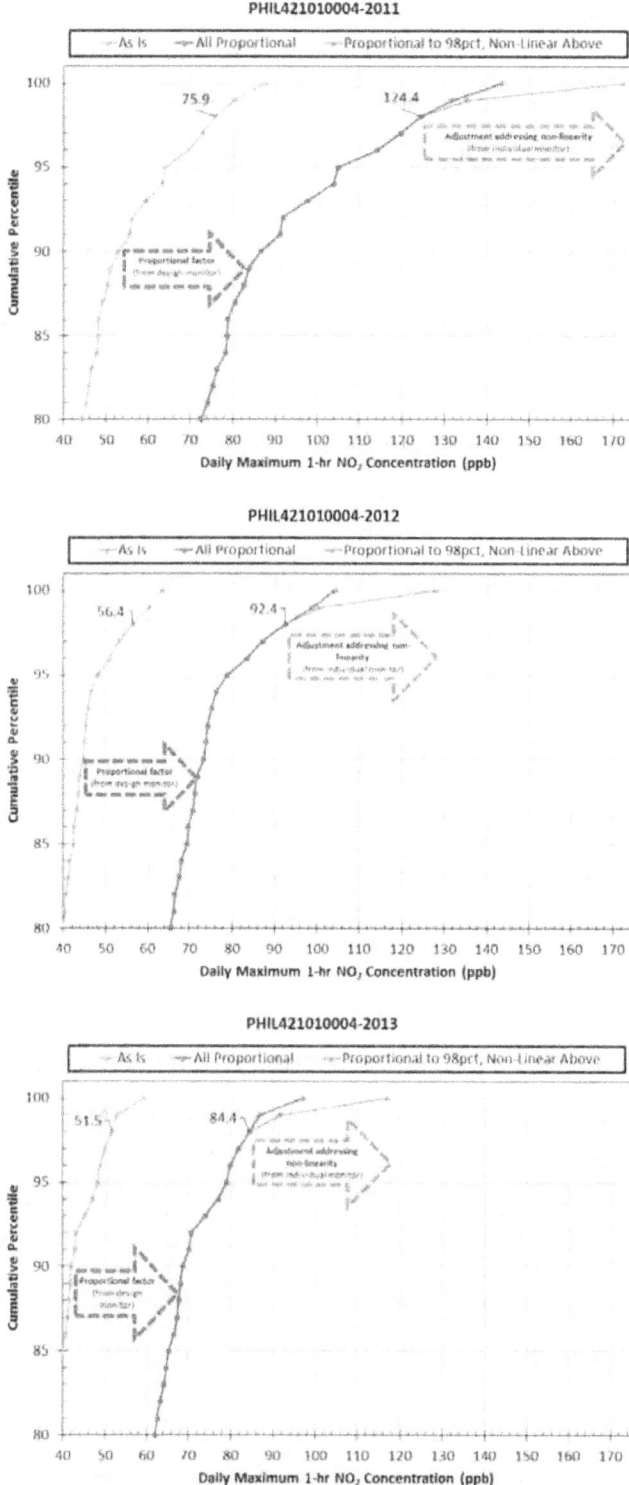

Figure B2-11. Distribution of unadjusted (as is) ambient NO₂ concentrations, that adjusted using a proportional factor alone (all proportional), and that adjusted using a combined proportional factor and ratio approach (proportional to 98th percentile, non-linear above) in the Philadelphia study area at monitor ID 421010004 across a three-year averaging period, 2011 (top panel), 2012 (middle panel) and 2013 (bottom panel).

2.4.2 Simulating on-road concentrations

The newly designated near-road monitors would best characterize NO_2 concentrations occurring around roadways, when compared to the monitoring information available in the last review. Based on this newly available data, as well as overall consistency in the information obtained from the types of monitoring and modeling analyses described in the REA Planning document, we have concluded that the near-road data would also serve as the basis for estimating on-road concentrations, generally expected to be similar to or perhaps greater than the near-road concentrations.

We developed a series of simulation factors to estimate on-road ambient NO_2 concentrations from the near-road monitor data, based on measurement data and a statistical model developed from a near-road study. The simulation factors were derived from a statistical model similar to that developed in the 2008 REA, though using newly available near-road monitoring data collected in Las Vegas, NV. Details of the statistical model development and on-road simulation factor output is fully described in Richmond-Bryant et al. (2017), while information regarding the measurement data collection are found in Kimbrough et al. (2013). Fundamentally, based on their forms and the data used to construct these types of statistical models, the simulation factors derived will generate on-road concentrations that will be greater than that of the ambient monitor concentrations used to approximate them. That said, uncertainties remain for where and when the maximum concentrations could occur at fine temporal and spatial scales based on influential factors such as local meteorological conditions (e.g., wind speed and direction) and NO_X conversion rates.

Briefly, near-road measurements of air quality, traffic, and meteorology were collected at a study area located adjacent to Interstate-15 (I-15) in Las Vegas NV during Dec. 2008 to Jan. 2010. Downwind sampling sites were located approximately 20 m, 100 m, and 300 m east of the interstate. Logit-ln functions were developed, considering the influence of local meteorological conditions (e.g., wind direction and approximate mixing heights).[25] The logit-ln functions were

---

[25] We evaluated five model scenarios considering atmospheric conditions and wind direction: 1) all wind and atmospheric stability conditions combined, 2) winds from the west (210º-330º, where the monitors were downwind of the highway), 3) winds

then used to estimate on-road NO_2 concentrations and concentrations predicted at varying distances from the road (i.e., 5, 10, 20, and 30 meters).[26] Using each hourly prediction, statistically modeled on-road NO_2 concentrations were compared to the modeled near-road concentrations to calculate the percent increase in on-road concentrations. As a reminder, while modeled and observed concentration agreement was reasonable (Figure B2-12), there were no on-road measurement collected as part of this study, thus an evaluation of the modeled on-road concentrations could not be performed and remains as an uncertainty.

To a limited extent (and as discussed in the ISA Section 2.5.3.1 that describes near-road concentration gradients), concentration level can affect the value of the on-road simulation factor generated using the Las Vegas study data. In general, lower concentration quintiles have greater percent differences between on-road and away-from-road concentrations than higher concentration quintiles. Therefore, the distributions of percent increases were stratified by the near-road concentration distribution quintiles. Because upper percentile concentrations are those that will most likely lead to the majority of concentrations at or above benchmark levels, here we selected the factors derived from the upper quantiles (i.e., when near-road concentrations at or above the 80th percentile). Meteorological conditions also affect the value of the estimated simulation factor; study conditions where winds were predominantly from the west (downwind from the road) combined with the presence of atmospheric inversions had a lower percent difference between on-road and away-from-road concentrations than when winds were perpendicular to the road combined with conditions associated with greater mixing heights. In considering these differing meteorological conditions, and seeking a generally conservative though simple approach using these results to simulate on-road NO_2 concentrations here, we chose three values for estimating on-road concentrations for each distance, based on the two lowest and highest median values, and the average of these two median values. These on-road simulation factors are provided in

Table B2-9.

_____

from the east (30°-150°, where the monitors were upwind of the highway), 4) inversion conditions (convective mixing height less than 300 m), and 5) non-inversion conditions (convective mixing height greater than 300 m).

[26] It is possible to generate on-road simulation factors to use for any monitor distance (e.g., 80 m, 200 m from the road), though the principal objective here was to develop factors to apply to the new near-road monitor concentration data.

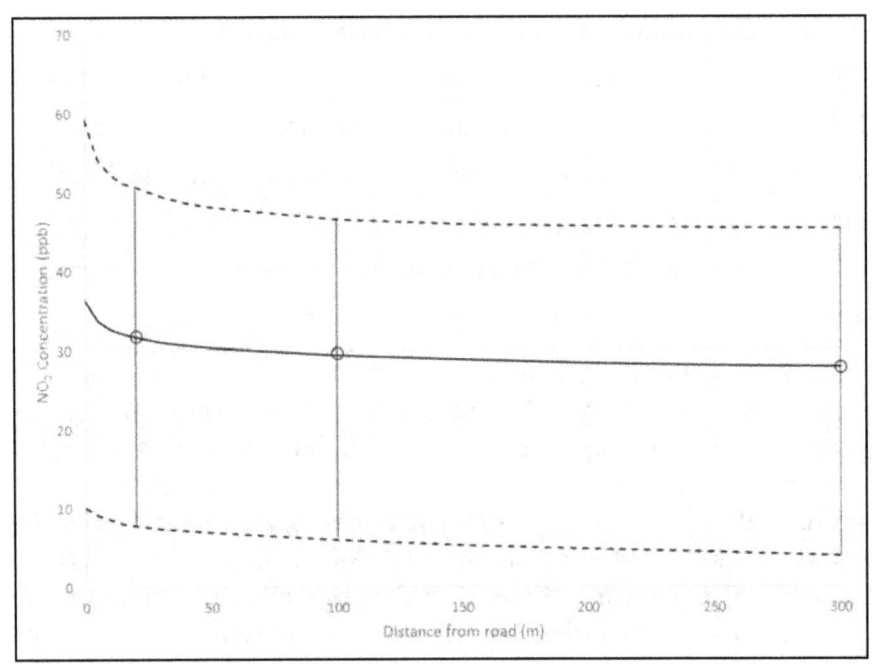

Figure B2-12. Predicted and observed NO₂ concentrations for winds from the west using based on data from a Las Vegas NV near-road measurement study. Predicted median (solid), predicted 98ᵗʰ and 2ⁿᵈ percentile (dotted), observed median (circles), and observed 98ᵗʰ and 2ⁿᵈ percentiles (error bars) are shown.

Table B2-9. Factors used to simulate on-road NO₂ concentrations from near-road monitors sited at varying distance from a major road.

Distance from Road (meters)	Average upwards adjustment from near-road concentrations given selected simulation factor used[1]		
	Low	**Mid**	**High**
5	7%	9%	11%
10	9%	12%	15%
20	12%	16%	19%
30	13%	17%	21%

[1] based on Las Vegas, NV near-road measurement study data (see Appendix A of NO₂ REA PD).

2.4.3 Calculating number of days at or above benchmark levels

As was discussed above in section 2.4.1, we have identified 1-hour concentrations ranging from 100, 150, and 200 ppb as the air quality benchmark levels to consider in this air quality assessment. The complete distribution of DM1H concentrations are used to calculate the number of days per year concentrations are at or above these benchmark levels at each monitor, for all air quality scenarios, and within each study area. Specifically, for this draft AQC we generated NO₂ air quality benchmark analysis results for 22 study areas having new near-road monitor data and one study area (Chicago) not yet having a newly designated near-road monitor. The mean and maximum number of days per year having concentrations at or above benchmark levels were calculated using 2010-2015 air quality for two ambient monitor types (i.e., area wide, near road) and for where on-road concentrations were simulated.[27] The five air quality scenarios include *as is* ambient concentrations (unadjusted concentrations, for all years where measured/estimated), CS1012 (2010-2012 air quality adjusted upwards to just meet the existing standard across the three-year period), CS1113 (2011-2013, similarly adjusted), CS1214 (2012-2014, similarly adjusted), and CS1315 (2013-2015, similarly adjusted).

The number of days per year a DM1H exceeds a particular benchmark level are counted for each monitor in operation. Then these results were summarized using two metrics as follows:

1) *Site-Year Metric.* Consistent with the NO_2 and SO_2 REAs (US EPA, 2009; US EPA, 2010, respectively) the first summary result metric is based on monitor site-years of data, whereas daily exceedances are calculated as such (and considers all individual site-years in the period of interest, neither weighted by site or year). Because the means are calculated using data from all monitors in a study area (not just the area design value monitor), the mean value for this metric would very likely be less than 7 days per year.[28] This mean metric can represent, on-average, the number of days per year at any one site in the study area we could have an exceedance over that averaging period. The maximum is the single monitor worst air quality year for that study area. These results are presented using the phrase '*site-year*' with calculations performed for the area wide and near-road monitors separately. Because the on-road data were mainly limited to measurements during 2014-15 and were mostly from a single monitor in each study area, these data were presented for those individual years (means and maximums are not calculated).

For example, the number of days per year having DM1H concentrations above benchmarks in the St. Louis study area is presented in

---

[27] The number of monitoring days per year can vary, even considering the use of monitors designated as having a valid year of data. The number of days per year having concentrations at or above benchmark levels was not adjusted considering this fact, that is to say, if fewer than 365 days were monitored (e.g., only 344 days were available), the number of benchmark exceedances was not increased by a factor of 1.06. It was assumed that the valid monitoring year reasonably represented what would be considered a complete year. The same approach was used for the near-road/on-road calculations, that is to say, if only 183 days had valid measurement data, the number of days per year above benchmarks was not increased by a factor of 1.99 to reflect a full 365-day period. When presented alone, the near-road/on-road DM1H concentrations at or above benchmarks are generaly presented as "number of days" rather than "number of days per year".

[28] Based on the form of the standard, a value of 7 days per year would expected considering the adjusted area design value monitor concentrations alone and that all three years of data within the 3-year averaging period meet completeness criteria.

Table B2-10. Means were rounded to the nearest integer, though for values <0.50, rather than round to zero, a value of 0.50 was designated to distinguish it from instances where there were no (or 0) days where the DM1H was at or above the benchmarks during the period of interest. Presented in the summary figures that follow in Section B3 are the mean and maximum values for each of the air quality scenarios, in each study area, for the area-wide and near-road monitor data. Because there were three simulation factors (low, mid, high) used in calculating the on-road concentrations, the results are presented in a summary figure for each study area, considering 2014 and 2015 individually (the years where ambient monitors had the greatest number of measurements within the 2013-2015 adjusted air quality scenario).

An important feature to note is that there can be variability in the concentrations within CBSAs, based on both the year to year variability in concentrations and the adjustment factor used. For example, when considering monitor 295100086, within the 3-year averaging period of 2012-14 and using the same series of adjustment factors for that period, the number of days per year having concentrations at or above 100 ppb is 14, 9, and 2 for years 2012, 2013, and 2014, respectively (

Table B2-10), resulting in a coefficient of variation (COV) of about 72%. When considering benchmark exceedances at the same monitor and within the same year, though across the three different averaging periods encompassing that year (e.g., 2013 – benchmark exceedances of 4, 9, and 5), the resulting COV is about 44%. In this latter instance, the three proportional adjustment factors of 1.887, 2.041, and 1.923 (Table B2-5) exhibited much less variability (COV ~ 5%), though when applied to the same exact ambient concentrations, the slightly higher adjustment factor increased the estimated concentrations above the benchmark levels by a factor of about 2 when compared with results from the other averaging periods.

Table B2-10. Example site-year metric results in the St. Louis study area: number of days per year 1-hour NO₂ concentrations are at or above 100 ppb, by monitor, year, and air quality scenario along with summary statistics.

Site ID	year	Number of Days	Number of days per year 1-hour NO$_2$ concentration ≥ 100 ppb			
			As Is	CS1113	CS1214	CS1315
Area-wide						
171630010	2010	363	0			
	2011	346	0	0		
	2013	351	0	1	1	1
	2014	365	0		2	1
	2015	356	0			0
171630900	2011	361	0	0		
	2012	365	0	0	0	
	2013	365	0	0	0	0
	2014	359	0		0	0
295100085	2014	361	0		4	1
295100086	2011	349	0	8		
	2012	350	0	6	14	
	2013	365	0	4	9	5
	2014	365	0		2	1
	2015	352	0			2
mean			0	2	4	2
max			0	8	14	5
Near-road						
295100094	2013	365	0	5	11	6
	2014	365	0		21	10
	2015	364	0			2
291890016	2015	357	0			1
mean			0	5	16	5
max			0	5	21	10

On-road						
Monitor used	year		scenario	low	mid	high
295100094	2013		As is	0	0	0
	2013		CS1315	21	26	37
	2014		As is	0	0	0
	2014		CS1315	30	37	39
	2015		As is	0	0	0
	2015		CS1315	13	17	26
291890016	2015		As is	0	0	0
	2015		CS1315	4	5	8

Highlighted are the summary statistics provided in the results figures that follow in Section B3.

2) *CBSA-wide Metric.* A new frame of reference was developed for this draft AQC to account for all days having concentrations at or above benchmarks in each study area. Rather than consider the monitors singularly (as in the above site-year calculation), the calendar year for the study area as a whole was considered, that is, there could be days where more than one monitor has a DM1H at or above a benchmark, though at times, the exceedances occur on different days of the year. Further, there could be days where more than one monitor exceeded a given benchmark on the same day. To analyze the data for each year and study area, the days of the year having a DM1H concentration at or above a benchmark were retained for each monitor, then combined to yield a CBSA-wide calendar record for all unique days where a benchmark was exceeded (i.e., exceedances occurring at any monitor in the CBSA, including the near-road monitor). The mean value is calculated here as the average of the 3 years considered in the 3-year averaging period for each study area, the maximum is the single worst air quality year in the study area. Of course, because it is a calendar record and includes instances where benchmark exceedances are not restricted to a single monitoring site (but now the days per year is CBSA-wide), the number of days per year at or above a benchmark calculated using this CBSA-wide metric is greater than that generated using the site-year metric, and the means are typically at or above that would be expected to occur at an area design value monitor when concentrations are adjusted to just meet the existing standard (i.e., greater than 7 days per year). This mean metric represents 'on-average' the number of days per year somewhere across the study area the DM1H was at or above a benchmark level across the 3-year period. Again, the maximum number of days is the worst air quality year for that study area. These same two statistics (mean and maximum) were calculated for the number of days in the study area having two monitors having DM1H concentrations at or above the benchmarks, though occurring on the same day. And finally, this CBSA-wide metric was expanded to also include all three concentration locations (area-wide, near-road and on-road). Because most locations only had the 2014-15 near road data (and hence 2014-15 on-road estimates), the results are presented for the 2013-2015 period. Also, only the on-road concentrations simulated using the 'mid' factor were used for the CBSA-wide metric.

An example is provided below, continuing with the St. Louis study area results. The summary data are first presented in

Table B2-11

Table B2-11. These summary data are what would be presented in figures in Section B3, though would also include results for the other study areas. Results are shown for the100 ppb benchmark only, because as in most study areas, exceedances of the higher benchmarks occurred much less frequently, if at all. In this study area, there were no simulated on-road concentrations at or above any benchmarks when considering the unadjusted as is air quality, thus this data is not shown. The data from which the summary means and maximums were calculated are derived from

Table B2-12, which in turn were generated based on the individual monitor DM1H concentrations presented in Table B2-13. The CBSA-wide metric calculation including the simulated on-road concentrations is calculated in a similar manner, though in addition to the concentrations at all monitors also includes the concentrations on days when the simulated on-road concentrations are at or above the benchmark levels (not shown). As expected given that the on-road concentration simulation is based on the near-road monitor data, at a minimum the benchmark exceedances occur on the same days the near-road monitor had a DM1H concentration at or above a benchmark level. Though most often, the number of days per year DM1H on-road concentrations are at or above benchmarks is greater than that estimated for the near-road monitor.

Table B2-11. Example CBSA-wide metric results in the St. Louis study area: summary statistics for the number of days per year where NO$_2$ concentration ≥ 100 ppb - anytime in the study area and for instances when it occurred on the same day at two monitoring locations.

| CBSA-wide metric | Number of days per year where DM1H NO$_2$ concentration ≥ 100 ppb | | | | | | | | | |
| | mean | | | | | maximum | | | | |
	As is	CS1012	CS1113	CS1214	CS1315	As is	CS1012	CS1113	CS1214	CS1315
Any monitor in CBSA	0	No DV	7	17	8	0	No DV	8	23	13
At two monitors (2 x 100 ppb)	0	No DV	1	3	1	0	No DV	3	5	4

Presented are the summary statistics provided in the results figures that follow in Section B3.

Table B2-12. Example CBSA-wide metric results in the St. Louis study area: number of days per year where NO$_2$ concentrations ≥ 100 ppb - data stratified by averaging period and year.

| Year | Number of days per year where DM1H NO$_2$ concentration ≥ 100 ppb | | Number of Monitors (n) | 3-year averaging period |
	Any monitor in study area	At two monitors on same day		
2011	8	0	3	1113
2012	6	0	2	1113
2013	6	3	4	1113

2012	14	0	2	1214
2013	15	5	4	1214
2014	23	5	5	1214
2013	7	4	4	1315
2014	13	0	5	1315
2015	5	0	4	1315

Table B2-13. Example CBSA-wide metric results in the St. Louis study area: NO₂ concentrations stratified by averaging period, year, month, day, and monitor on days where NO₂ concentrations ≥ 100 ppb.

year	month	day	Daily maximum 1-hour NO$_2$ on days when concentrations ≥ 100 ppb					
			mon171630010	mon171630900	mon295100086	mon295100094	mon295100085	mon291890016
2011-2013 (CS1113) averaging period								
2011	1	2	45.3	15.9	114.7			
2011	1	3	37.7	50.9	139.5			
2011	1	5	47.2	56.8	104.7			
2011	1	25	80.4	63.0	101.1			
2011	2	3	72.9	67.0	111.1			
2011	2	4	67.9	61.5	108.3			
2011	10	4	69.8	78.8	121.3			
2011	10	22	39.6	83.0	106.4			
2012	1	6		48.7	100.3			
2012	3	28		59.6	114.3			
2012	4	1		11.3	108.1			
2012	6	13		54.0	102.1			
2012	8	23		46.2	104.7			
2012	9	4		32.6	131.4			
2013	2	23	91.2	63.2	110.4	128.8		
2013	2	24	104.4	71.4	126.9	112.0		
2013	3	15	73.6	42.5	101.1	105.9		
2013	3	22	24.5	14.7	43.8	100.0		
2013	3	28	34.0	18.3	57.0	102.6		
2013	4	4	88.2	62.3	104.4	98.3		
2012-2014 (CS1214) averaging period								
2012	1	6		52.7	108.5			
2012	2	17		52.0	107.9			
2012	3	28		64.5	123.6			
2012	4	1		12.2	116.9			
2012	4	8		45.1	101.8			
2012	5	12		45.1	103.1			
2012	5	14		45.3	106.5			
2012	5	22		61.6	100.4			
2012	6	9		20.2	103.3			
2012	6	13		58.4	110.4			

year	month	day	mon171630010	mon171630900	mon295100086	mon295100094	mon295100085	mon291890016
					Daily maximum 1-hour NO$_2$ on days when concentrations ≥ 100 ppb			
2012	8	23		50.0	113.3			
2012	9	4		35.3	142.2			
2012	11	15		67.9	100.6			
2012	11	16		65.7	101.8			
2013	1	15	55.1	42.5	103.1	56.1		
2013	2	16	75.5	38.2	88.4	105.7		
2013	2	23	98.7	68.4	119.4	139.3		
2013	2	24	112.9	77.2	137.3	121.1		
2013	3	15	79.6	45.9	109.4	114.6		
2013	3	22	26.5	15.9	47.4	108.2		
2013	3	27	73.5	74.9	106.6	100.2		
2013	3	28	36.7	19.8	61.6	111.0		
2013	3	29	91.3	40.4	104.7	69.4		
2013	4	2	87.2	37.8	104.2	90.7		
2013	4	3	81.6	23.9	100.0	89.1		
2013	4	4	95.4	67.4	112.9	106.3		
2013	4	5	79.6	55.3	94.1	101.3		
2013	9	6		34.9	58.8	100.2		
2013	10	28	81.6	58.2	91.6	104.6		
2014	1	28	71.4	56.3	58.2	108.1	77.1	
2014	2	7	115.6	75.2	58.8	86.0	50.2	
2014	2	11	83.7	53.3	59.0	130.1	89.2	
2014	2	12	97.7	79.2	75.9	120.7	105.9	
2014	2	13	101.0	71.7	82.5	137.6	99.2	
2014	2	14	91.6	61.8	54.3	120.1	84.1	
2014	3	1	81.6	44.9	68.8	100.9	80.0	
2014	3	4	93.4	65.7	66.9	103.0	110.0	
2014	3	6	87.8	73.6	75.3	126.0	91.0	
2014	3	7	87.8	69.1	79.8	102.8	80.8	
2014	3	10	81.6	37.4	48.4	118.8	79.2	
2014	3	13	63.3	54.9	58.8	103.9	80.4	
2014	3	15	81.6	44.9	64.3	100.2	85.7	
2014	3	20	87.8	33.1	70.0	103.0	78.0	
2014	4	11	65.3	41.2	70.8	100.9	102.7	
2014	4	20	57.1	42.2	88.2	101.9	80.0	
2014	4	21	46.9	26.9	85.5	110.3	94.7	
2014	7	11	34.7	25.1	32.0	100.0	51.2	
2014	9	25	44.9	47.8	70.2	122.9	84.9	
2014	9	26	49.0	51.6	117.7	97.2	92.9	
2014	9	27	51.0	63.7	102.3	106.1	104.5	
2014	11	24	30.6	14.1	25.1	158.2	32.7	

year	month	day	Daily maximum 1-hour NO$_2$ on days when concentrations ≥ 100 ppb					
			mon171630010	mon171630900	mon295100086	mon295100094	mon295100085	mon291890016
2014	12	12	44.7	34.9	57.6	104.0	57.1	
2013-2015 (CS1315) averaging period								
2013	2	23	93.0	64.4	112.5	131.2		
2013	2	24	106.3	72.8	129.3	114.1		
2013	3	15	75.0	43.3	103.0	107.9		
2013	3	22	25.0	15.0	44.6	101.9		
2013	3	27	69.2	70.5	100.5	94.4		
2013	3	28	34.6	18.7	58.1	104.6		
2013	4	4	89.9	63.5	106.4	100.2		
2014	1	28	67.3	53.1	54.8	101.8	72.7	
2014	2	7	108.9	70.8	55.4	81.1	47.3	
2014	2	11	78.8	50.2	55.6	122.6	84.0	
2014	2	12	92.0	74.6	71.5	113.7	99.8	
2014	2	13	95.2	67.6	77.7	129.6	93.5	
2014	2	14	86.3	58.3	51.2	113.1	79.2	
2014	3	4	88.0	61.9	63.1	97.0	103.6	
2014	3	6	82.7	69.4	71.0	118.7	85.8	
2014	3	10	76.9	35.2	45.6	111.9	74.6	
2014	4	21	44.2	25.4	80.6	103.9	89.2	
2014	9	25	42.3	45.0	66.2	115.8	80.0	
2014	9	26	46.2	48.7	110.9	91.5	87.5	
2014	11	24	28.8	13.3	23.7	149.0	30.8	
2015	3	2			88.2	105.6		71.3
2015	3	7	80.3		101.1	89.8		96.9
2015	3	31	72.9		87.1	99.8		111.4
2015	4	1	58.7		79.0	121.4		79.2
2015	9	25	69.2		116.3	55.8		61.5

Highlighted are the DM1H NO$_2$ concentrations that were used in counting the number of days per year at or above benchmark levels summarized in Table B2-11 and Table B2-12.

B3. RESULTS

This section contains the analysis results for the AQC. An overview of the two distinct analyses is presented in this section followed by sections detailing the analytical results along with a brief discussion.

For the first analysis, we used the historical (1980-2015) unadjusted (as is) air quality to provide context for the benchmark exceedance calculations, focusing here on instances where ambient concentrations have been at or just around the level of the existing 1-hour standard. It is worth noting that the historical measurement data are representative of real air quality scenarios that existed at the time the monitoring took place and that changes in emissions control and atmospheric conditions that have occurred since that time would preclude us from drawing complete conclusions about the number of exceedances associated with a given 98th percentile DM1H NO_2 concentration, if attempting to use such information as a prediction for future air quality. Nevertheless, using these unadjusted ambient concentration data remain informative given the general consistencies in the overall concentration distribution over time at each monitor and what would be expected regarding the number of exceedances given the form of the existing 1-hour standard (i.e., for a complete year of data, on average, there would be about 7 days having concentrations at or above the 98th percentile DM1H value of 100 ppb) and the approach used to adjust concentrations to just meet the existing 1-hour standard.

For the second analysis, i.e., the core results for the AQC in the selected study areas, we used the most recently available (2010-2015) ambient NO_2 monitoring data. Exceedances of benchmark levels are calculated for the five air quality scenarios (unadjusted as is ambient concentrations and three sets of concentrations adjusted to just meeting the existing 1-hour standard) and for the three distinct ambient concentration types (area-wide, near-road, and simulated on-road).

3.1 ANALYSIS OF HISTORICAL (1980-2015) AIR QUALITY

For this analysis, we first calculated all of the rolling 3-year average 98th percentile DM1H values for all individual monitors in operation from 1980-2015 that met the completeness criteria described above (section 2.2). Historical data were used to ensure that concentrations would be at or around the level of the existing hourly standard. Counted first were the number of days per year NO_2 concentrations were at or above the 1-hour benchmark levels (i.e., 100, 150, and 200, if any) for the individual years within each 3-year averaging period. Then the mean number of days per year was calculated (thus, the average of the observed number of days for each monitor across the 3-year averaging period). Also identified were the maximum number of days per year NO$_2$ concentrations were at or above the 1-hour air quality benchmark levels (thus, the highest observed number of exceedances at each monitor for a single year in the averaging period) given the 3-year average 98th percentile DM1H for that monitor. Results of this analysis are presented in Figure B3-1.

Based on the analysis of all available historical NO$_2$ ambient monitored concentrations and considering the form, level, and averaging time of the existing 1-hour standard, on average across a 3-year period, the number of days/year where the DM1H was $\geq$ 100 ppb ranged from about 6 to 13 days (Figure B3-1, top left panel). This mean number of days per year, on average, corresponds well with general expectations described above (i.e., on average there could be about 8 days/year at or above the 100 ppb benchmark given the form of the standard). The maximum number of days in a single year that the DM1H was $\geq$ 100 ppb ranged from about 10 to 20 days (Figure B3-1, top right panel). When considering the 150 ppb benchmark, there are far fewer days per year exceeding that level when considering ambient concentration at the existing standard. In most instances, the mean number of days per year where the DM1H was $\geq$ 150 was less than two (Figure B3-1, middle left panel), while the maximum number of days per year at or above the same benchmark was less than five (Figure B3-1, middle right panel). Furthermore, and according to this analysis of all available historical ambient measurement data, exceeding a 1-hour benchmark level of 200 ppb is a rare occurrence when considering the form and level of the existing 1-hour standard. For example, of the 23 instances a monitor had a 3-year average 98th percentile DM1H of 100 ppb, there were no exceedances of the 200 ppb benchmark on 19 of these occasions (Figure B3-1, bottom right panel). When averaging across the 3-year period, the mean number of days per year having DM1H NO$_2$ concentrations at or above 200 ppb drops to 1 or less, again with most monitors recording no concentrations at or above the 200 ppb 1-hour benchmark.

It should be noted that for this analysis of historical ambient NO$_2$ concentrations, monitors in the California CBSAs (Los Angeles, San Francisco, etc.) constitute the bulk of the data where the 3-year average 98th percentile DM1H concentrations were at or above 100 ppb. However, the results of this analysis when excluding these areas are similar (data not shown), albeit with a generally tighter range of values than when including the monitoring data in California (e.g., the mean number of days per year having DM1H $\geq$ 100 ppb ranged from about 6 to 9). When considering data only from the 23 selected study areas (not shown), the number of days per year having DM1H at or above benchmarks and given a particular 3-year average 98th percentile DM1H is also consistent with that presented in Figure B3-1.

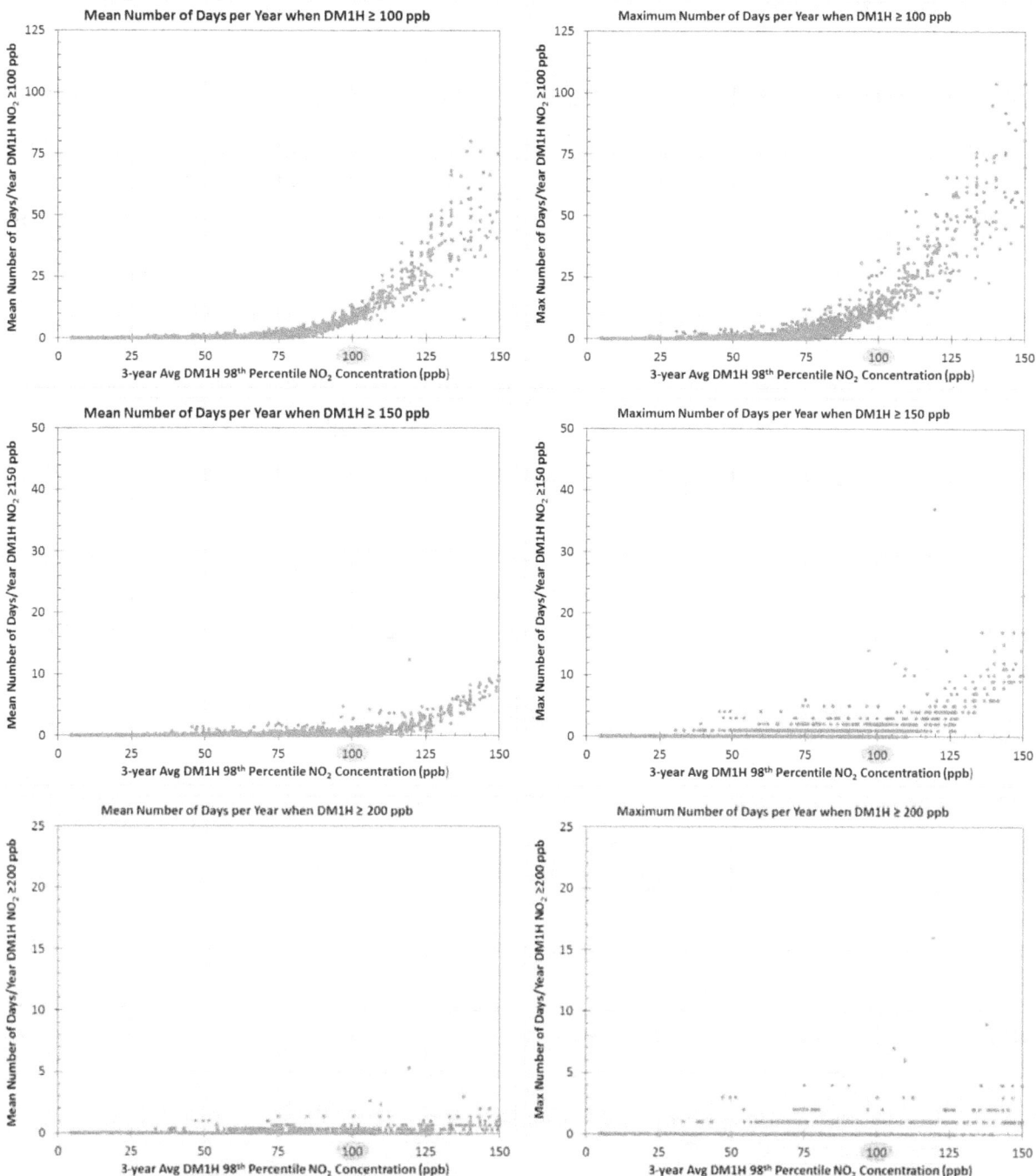

Figure B3-1. The mean (left panel) and maximum (right panel) number of days per year where DM1H NO₂ concentration was ≥ 100 ppb (top panel), ≥ 150 ppb (middle panel), and ≥ 200 ppb (bottom panel) associated with 3-year average 98th percentile DM1H NO₂ concentrations, using valid-year 1980-2015 ambient monitor data.

To provide additional context for upper percentile daily maximum concentrations below the lowest benchmark level of 100 ppb, we also used the historical (1980-2015) ambient monitoring data to evaluate the number of days per year where the DM1H was at or above three other concentration levels (i.e., 85, 90, and 95 ppb). To do so, we first isolated ambient monitor data where the 3-year average 98th percentile DM1H NO_2 concentrations were at or between 95 and 105 ppb. Then, as was done above for the selected benchmark levels, counted for each of the years within a 3-year averaging period were the number of days per year each monitor had an observed concentration at or above 85, 90, and 95 ppb. Table B3-1 provides similar results as shown in Figure B3-1, though focusing primarily where the 3-year average 98th percentile DM1H NO_2 concentrations were close to the existing NO_2 NAAQS. For tracking purposes and to conceptually link the tabular results to what had been described previously for the 100 ppb benchmark results depicted in Figure B3-1, the results for the 100 ppb benchmark are provided in Table B3-1.

Table B3-1. The number of days per year where DM1H NO_2 concentration was ≥ 85 ppb, ≥ 90 ppb, ≥ 95 ppb, and ≥ 100 ppb associated with 3-year average 98th percentile DM1H NO_2 concentrations at or near the existing NO_2 NAAQS, using valid-year 1980-2015 ambient monitor data.

Selected DM1H Level	Mean number of days per year where DM1H NO_2 concentration at or above selected level		
	min	mean	max
85	7	20	44
90	7	16	30
95	6	10	22
100	4	8	15
Min: the minimum value of any 3-year average, i.e. the smallest mean number of days per year occurring at one monitor and averaged over a 3-year period.			
Mean: the mean of all means, i.e., the mean of the mean number of days per year occurring at each monitor and averaged over a 3-year period.			
Max: the maximum value of any 3-year average, i.e. the greatest mean number of days per year occurring at one monitor and averaged over a 3-year period.			

3.2 ANALYSIS OF RECENT AIR QUALITY (2010-2015) IN SELECTED STUDY AREAS

A total of five air quality scenarios were evaluated in each of the 23 selected study areas (i.e., unadjusted as is ambient NO_2 concentrations and NO_2 concentrations adjusted to just meet the existing hourly standard for four 3-year averaging periods) considering the area-wide and near-road monitoring data available for 2010-2015. We also simulated on-road NO_2

concentrations based on the available near-road monitoring data.[29] Counted were the number of days per year the DM1H ambient concentrations exceeded the 1-hour benchmark levels of 100, 150, and 200 ppb, considering the area-wide, near-road and on-road concentrations. Because most study areas had very few days where concentrations were at or above the 150 ppb benchmark level, the results presented in this section are for the 100 ppb benchmark level only. Complete results for the 100 ppb benchmark and the other benchmarks are provided in tables in the Section B5. Presented in each figure in this section are the mean and maximum number of days per year considering the study area data on a site-year basis, and CBSA-wide basis, as was described in Section 2.4.3.

Figure B3-2 presents the results for the site-year metric, considering the area-wide monitors only. Note that where a value does not appear in this Figure for a particular study area, in all instances, that is because it is equal to zero (no benchmark exceedances). Further, any non-zero mean values that were less than 1 though not rounding upwards to 1 are presented as equal to 0.5. When considering the unadjusted (as is) air quality, very few exceedances of the 100 ppb benchmark were observed during the recent area-wide monitor data analyzed. The maximum observed values in a single year was seven (Riverside CA) and three (Houston), while most study areas had a one or no days per year DM1H concentrations $\geq$ 100 ppb.

When considering air quality adjusted to just meet the existing standard, the results were generally similar for the four different 3-year averaging periods. In general, the average number of days per year having 1-hr NO_2 concentrations at or above 100 ppb ranged from about 2 to 5 in the majority of study areas, while the maximum number of days above this same benchmark was generally about 10 days per year, consistent with the expected number of days per year when considering the complete suite of monitors in each study area, the elements of the 1-hr standard, and the procedure used to adjust air quality, as discussed above.

Figure B3-3 presents the results for the site-year metric, considering the near-road monitors only. Note that where a value does not appear in this figure for a particular study area

---

[29] A few study areas had 2013 data, many more had 2014 data, while all had but one (Chicago) had 2015 data. Where available, 2013 data were included in the analysis and generation of results though the focus of the near-road and on-road analysis is on 2014-15.

and considering the unadjusted (as is) air quality, in all instances, that is because it is equal to zero (i.e., no benchmark exceedances) because all study areas had measured concentrations at some time during the 2010-2015 data analysis period. Where a value does not appear in this figure for a particular study area and considering the adjusted air quality scenarios, in all but one instance,[30] that is because there were no near-road data available in that 3-year averaging period. Therefore, in considering these results, compared with the results generated using the area-wide monitors (Figure B3-2), inferences should be limited to results generated using the unadjusted concentrations, and the 2013-2015 3-year averaging period.

There were very few simulated on-road concentrations at or above the 1-hour 100 ppb benchmark considering the unadjusted as is air quality (results not shown). Only the Chicago, Denver, Los Angeles, Miami, New York/Jersey, and Riverside study areas were estimated to have at least one day in the monitored year, with the New York/Jersey study area also having one day per year at or above the 150 and 200 ppb benchmark levels (data shown in Table B5-58). Figure B3-4 and Figure B3-5 show the number of days where the simulated on-road DM1H NO$_2$ concentration was at or above the 100 ppb benchmark considering 2014 and 2015 air quality adjusted to just meet the existing 1-hour standard, respectively. In general, most study areas had between 10 to 20 days in the year ≥100 ppb and there was limited variability in the range of values estimated using the three simulation factors. Of the few study areas having > 20 days in the year, Baltimore, Minneapolis, Pittsburgh, and Richmond required some of the greatest upward adjustments to simulate just meeting the existing standard, i.e., adjustment factors used in these study areas were ≥ 2 (Table B2-5), while most other study area adjustment factors were generally within 1.5 to 1.9. In addition, the Minneapolis study area used the highest on-road simulation factors because both monitors were sited about 30 meters from the road (

Table B2-9).

---

[30] The Chicago study area had no (zero) days per year with a DM1H ≥ 100 ppb during the 2011-2013 3-year averaging period.

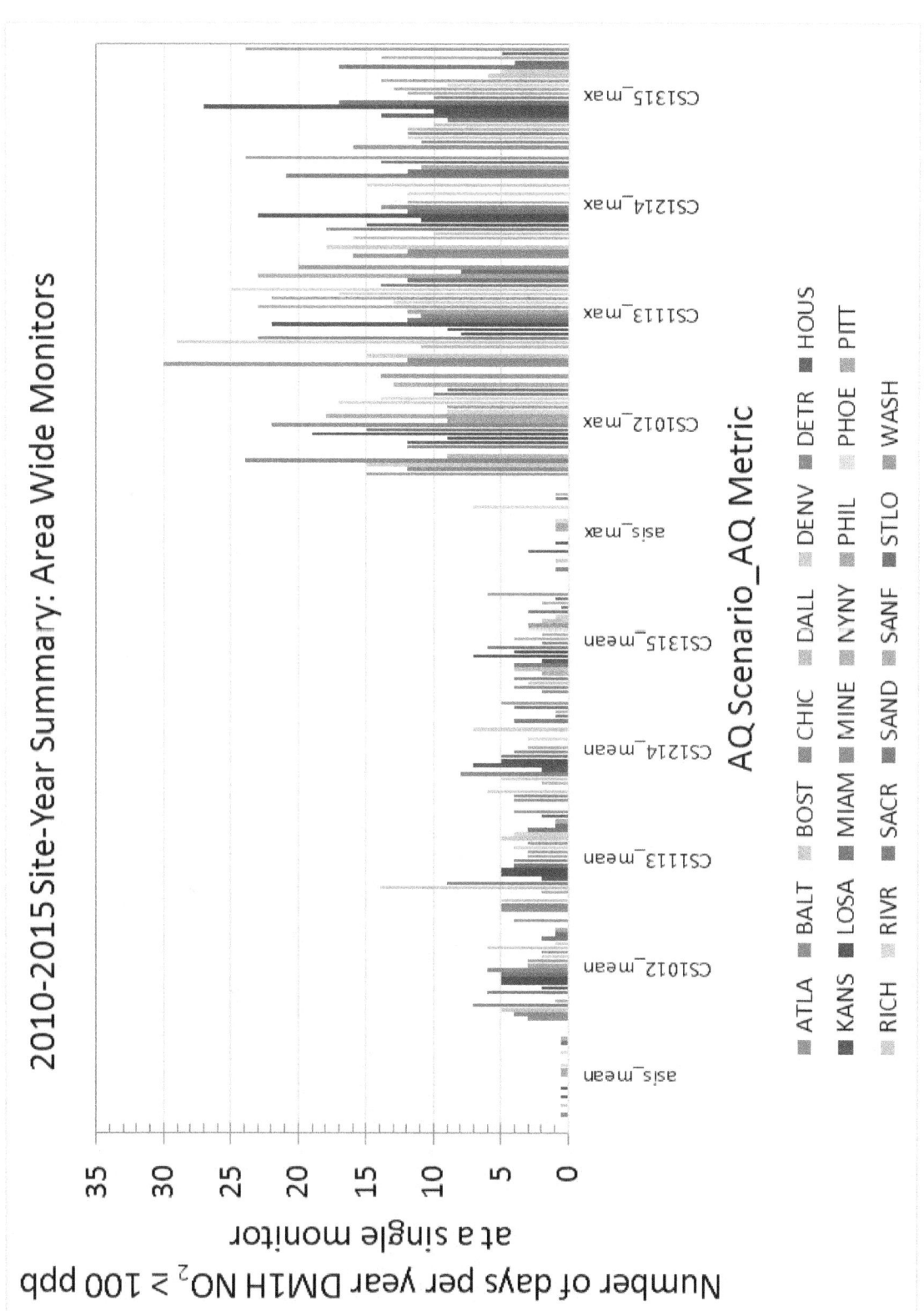

Figure B3-2. Mean and maximum number of days per year where DM1H NO$_2$ concentration at or above 100 ppb: 2010-2015 area-wide monitor site-year summary.

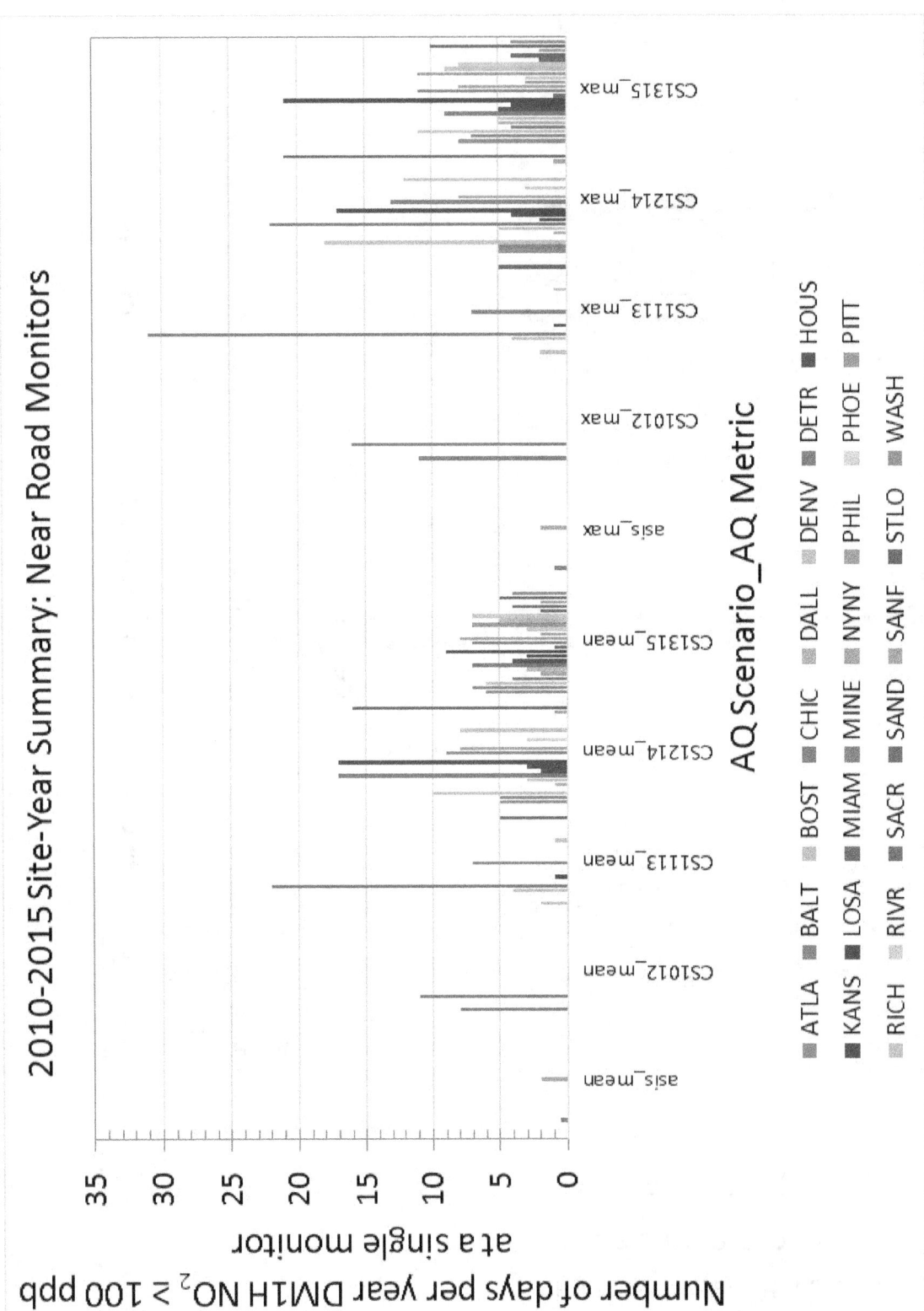

Figure B3-3. Mean and maximum number of days per year where DM1H NO₂ concentration at or above 100 ppb: 2010-2015 near-road monitor site-year summary.

B3-8

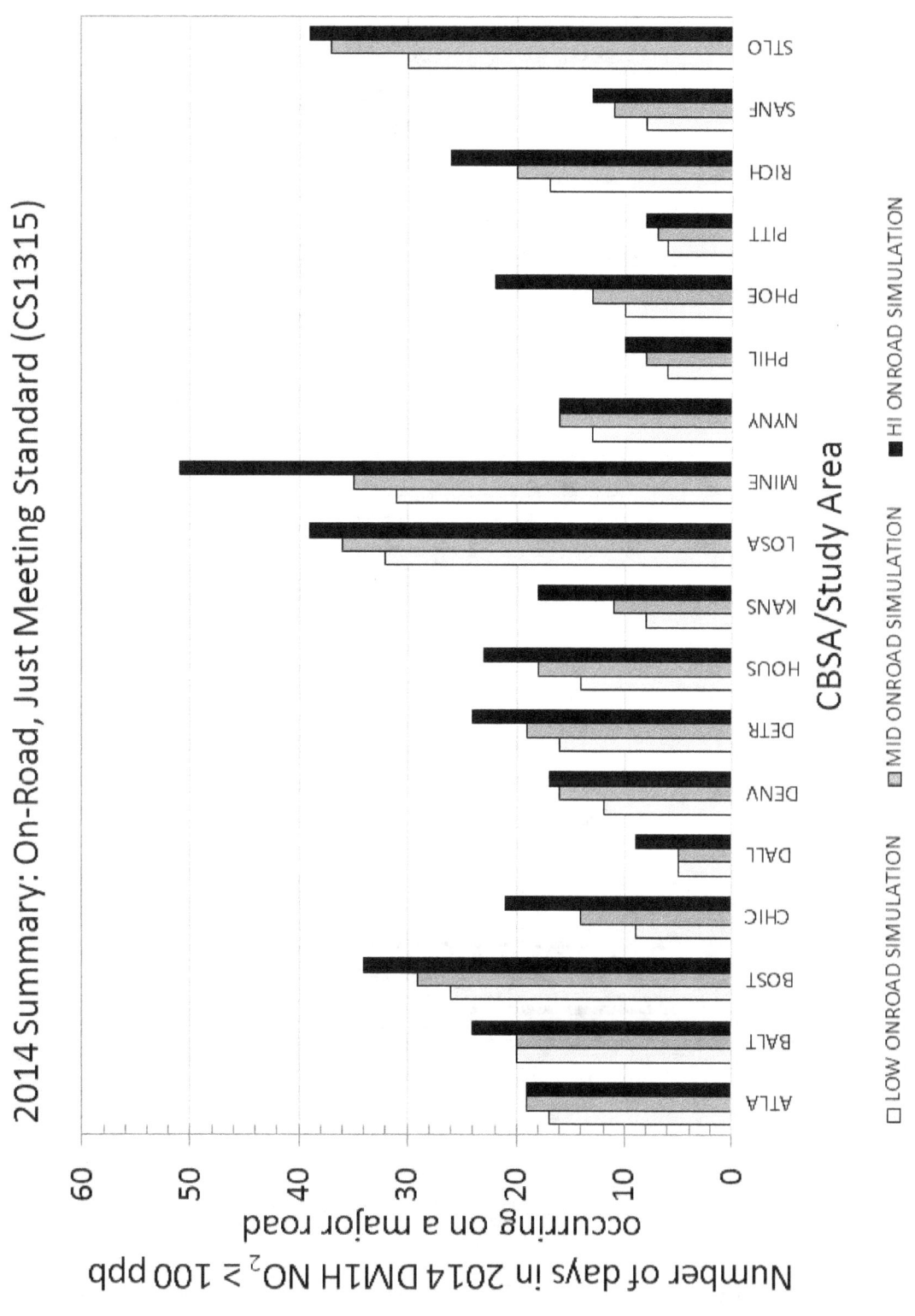

Figure B3-4. Number of days in the year where DM1H NO$_2$ concentration at or above 100 ppb: simulated on-road concentrations (2014) based on near-road monitor data and using three simulation factors.

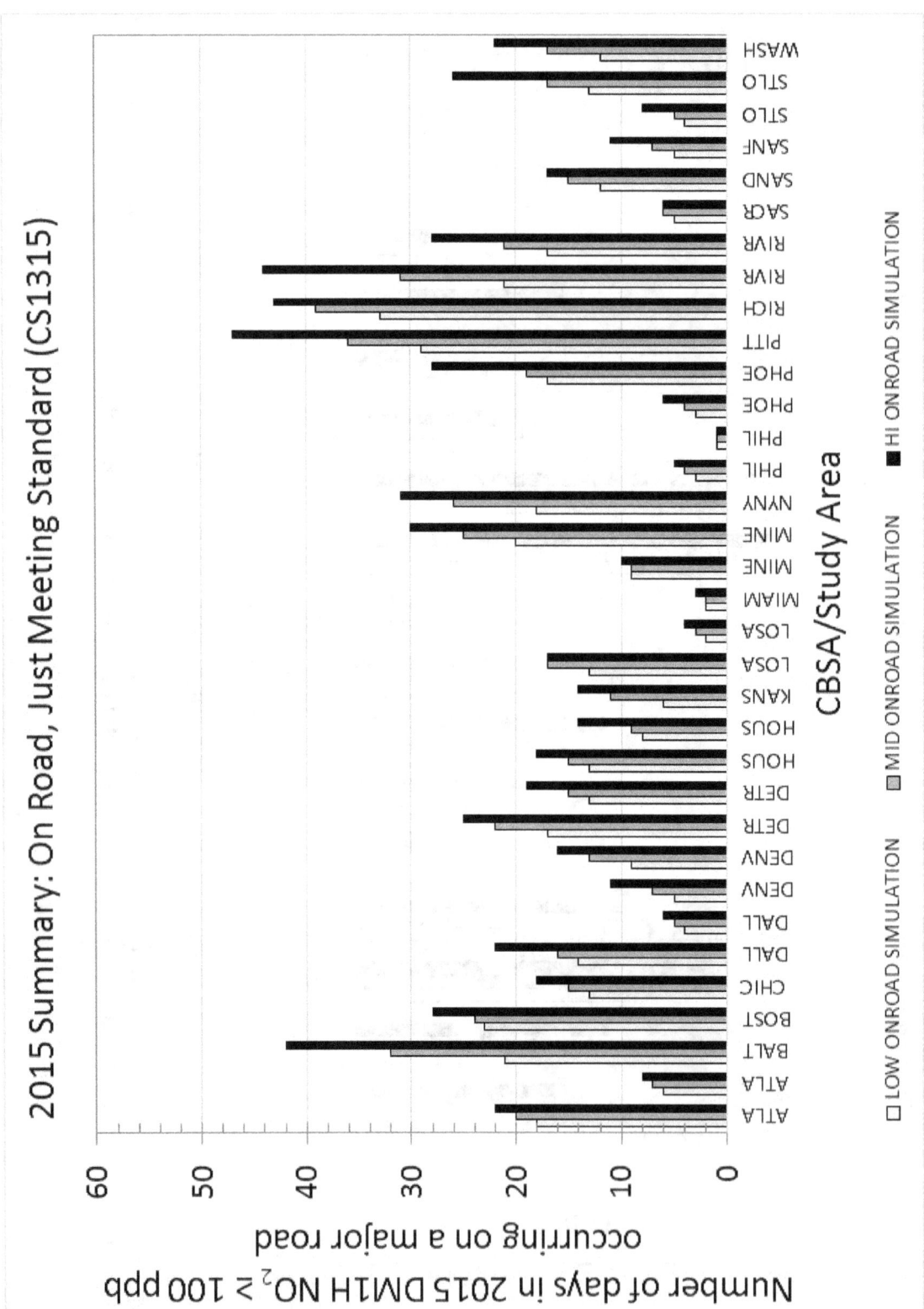

Figure B3-5. Number of days in the year where DM1H NO$_2$ concentration at or above 100 ppb: simulated on-road concentrations (2015) based on near-road monitor data and using three simulation factors.

As expected, there were a greater number of benchmark exceedances considering the CBSA-wide metric (Figure B3-6) compared to results generated using the site-year metric (Figure B3-2 and Figure B3-3). This is because the CBSA-wide metric accounts for the total number of days in the study area when DM1H NO_2 concentrations go above benchmark levels, not just those estimated to occur at a single monitor at a time but considering all monitors simultaneously operating in the study area. The difference between the two result metrics is an indicator of how the peak NO_2 concentrations vary by day across the study area, and in most instances, the variability is driven by the highest concentrations occurring at the area design value monitor and/or the near-road monitor within a study area. On average, most study areas had about 10 to 15 days per year at or above the 100 ppb benchmark, while the maximum number of days per year was about 20 when considering the CBSA-wide metric, generally about a factor or two greater than that estimated using the site-year metric. The number days per year where DM1H NO_2 concentrations were at or above benchmark levels and occurring at more than one site on the same day was much less; on average 3 days per year DM1H NO_2 concentrations were at or above 100 ppb, while the maximum number of days per year was about 5 (Figure B3-7), thus indicating simultaneous-site exceedances about a third or less of the time there was at least one day at or above 100 ppb.

The CBSA-wide metric was also expanded to include the benchmark exceedances resulting from the simulated on-road concentrations (Figure B3-8) in addition to consideration of the near-road and area-wide CBSA-wide results discussed above. Because the majority of near-road data were from 2014-2015, only the results for the 2013-2015 3-year averaging period are shown. CBSA-wide results considering the area-wide and near-road monitors (Figure B3-8, top panels) are identical to that shown in Figure B3-6 and Figure B3-7 for that 3-year averaging period. They were reproduced in this figure for comparison with the CBSA-wide results that also include exceedances from all three concentration types (i.e., area-wide, near-road, and simulated on-road) (Figure B3-8, bottom panels). For most study areas, there were about 5 to 10 additional days per year in the CBSA (on average and for the maximum) having concentrations at or above the 100 ppb benchmark level, when including the simulated on-road concentrations to the CBSA-wide metric (visually compare the top left to bottom left panels Figure B3-8). Again, there fewer days per year where concentrations are at or above the 100 ppb benchmark level and occurring at two or more locations on the same day (e.g., at both an area-wide and near-road monitor), though when including the simulated on-road concentrations in the CBSA-wide metric (Figure B3-8, bottom right panel), the number of occurrences is about twice that of the results generated using the area-wide and near-road data alone (Figure B3-8, top right panel)

3.3 UNCERTAINTY CHARACTERIZATION

Overall, there were a number of assumptions made in this AQC, potentially leading to uncertainty in the results. Assumptions were well informed and controlled to reduce uncertainties to the maximum extent possible such as, 1) selection of study areas having the greatest number of monitors and highest concentrations, thereby reducing the uncertainty in the spatial representation of ambient concentrations and adjustments made to meet the existing standards, 2) a systematic selection of the most appropriate factors used to adjust concentrations to just meet the existing standard considering both the varying degree of completeness of the area-wide monitors and the limited availability of near-road monitor concentrations, and 3) the use of newly designated near-road monitor data to best characterize near-road and on-road concentrations.

The direction of influence these assumptions and modeling approaches could have on the results of this AQC would vary, and be marked as either under- or over-estimations in concentrations and number of days at or above benchmark levels, while also potentially having varied magnitudes. For instance, it is a worthwhile reminder that in some study areas, the near-road monitor did not have a full monitoring year of data (e.g., in 2014 the NYNJ CBSA near-road monitor had concentrations for 184 days). Depending on the time of year and hour of day, it is possible that having a full year of near-road data could lead to a different number of benchmark exceedances for each of the summary calculations presented here. For example, maximum concentrations typically occur during the fall/winter early morning weekday commute hours, thus presence or absence of monitoring during these time could greatly influence the calculated number of days having benchmark exceedances. In the absence of having a robust data near-road dataset describing more specifically when or where peak concentrations would occur (the particular days or periods of the year), we felt at this early stage of the monitor implementation it was best to characterize the near-road data and simulated on-road concentrations simply using the actual number of monitored days in the year rather than including predicted concentrations for the days the near-road monitor was not in operation.

Another important uncertainty regards the design values used to generate factors for adjusting concentrations to just meet the existing standard. Individual results presented for St. Louis (section 2.4.3) indicated that small variation in adjustment factors used (COV of ~5%) could have a much greater effect (i.e., a factor of two) in the estimated number of days at or above the 100 ppb benchmark, even considering the same distribution of ambient concentrations. While appropriate steps were taken to account for instances where the highest monitor in operation was included in the calculation of these adjustment factors, having a very limited number of years (or even days in a year) available for the near-road monitors in most instances led to the use of area-wide monitor design values to calculate the adjustment factors. It is possible that having complete year data for the near-road monitors in each of the study areas could result their use as the area design value monitor, such as what has been realized during the 2013-15 averaging period in the Detroit and St. Louis study areas. By not having several complete years of near-road data available for this analysis, it is possible that both near-road and on-road concentrations are overestimated, particularly in instances where available comparisons indicate the area design value monitor concentrations are less than that at the near-road monitor. Further contributing to the uncertainty in adjusted concentrations were inconsistencies in the area design value monitor in some study area (e.g., Philadelphia) due to absence of meeting completeness criteria across several averaging periods. Often times, pseudo single-year design values were calculated from the highest available monitor (sometime near-road other time near-road) and used to estimate the adjustment factors to overcome this uncertainty, other times the adjustment factor was not used and the period was not simulated.

It is also possible that the upper (>98th) percentile DM1H ratios developed from the area design value monitor and used to adjust upper percentile concentrations for the near-road monitors (and hence, on-road concentrations) could lead to undesired variability in near-road and on-road concentrations. However, when comparing the ratio values derived from the area design value monitor to ratios observed at the other area-wide monitors within each study area, the majority fell within the range exhibited by the area-wide monitors, suggesting the magnitude of influence is limited. Even when ratios derived from the area design value monitor were at the maximum range of values, they are considered reasonable approximations, because in most areas, the observed variability in the upper percentile concentrations at the near-road monitors is expected to be similar to that observed at the highest monitors in the study areas.

And finally, as mentioned earlier in describing the approach to simulating on-road concentrations using factors derived from a statistical model along with the near-road monitored concentrations, in some instances could lead to overestimations in the estimated number of days at or above the benchmark levels occurring on-roads. This is in part because the model-derived factors used assume concentrations on-roads are always greater that those occurring away from roads. It is possible that for some near-road monitors sited in close proximity to the road (i.e., within 10 meters), that measured near-road concentrations reasonably approximate concentrations occurring on-roads. Further, while a range of on-road simulation factors were used to represent potentially important influential conditions (e.g., atmospheric stability and wind direction), the application of the factors was candid, not accounting for the complexities associated with these conditions at each monitor for every hour measurements were recorded. This along with not accounting for the variability in the estimated factors (the mean of each the low, mid and high conditions was used rather than the full distribution of values) would also contribute to uncertainty in the estimated concentrations, likely resulting in both under and over estimations.

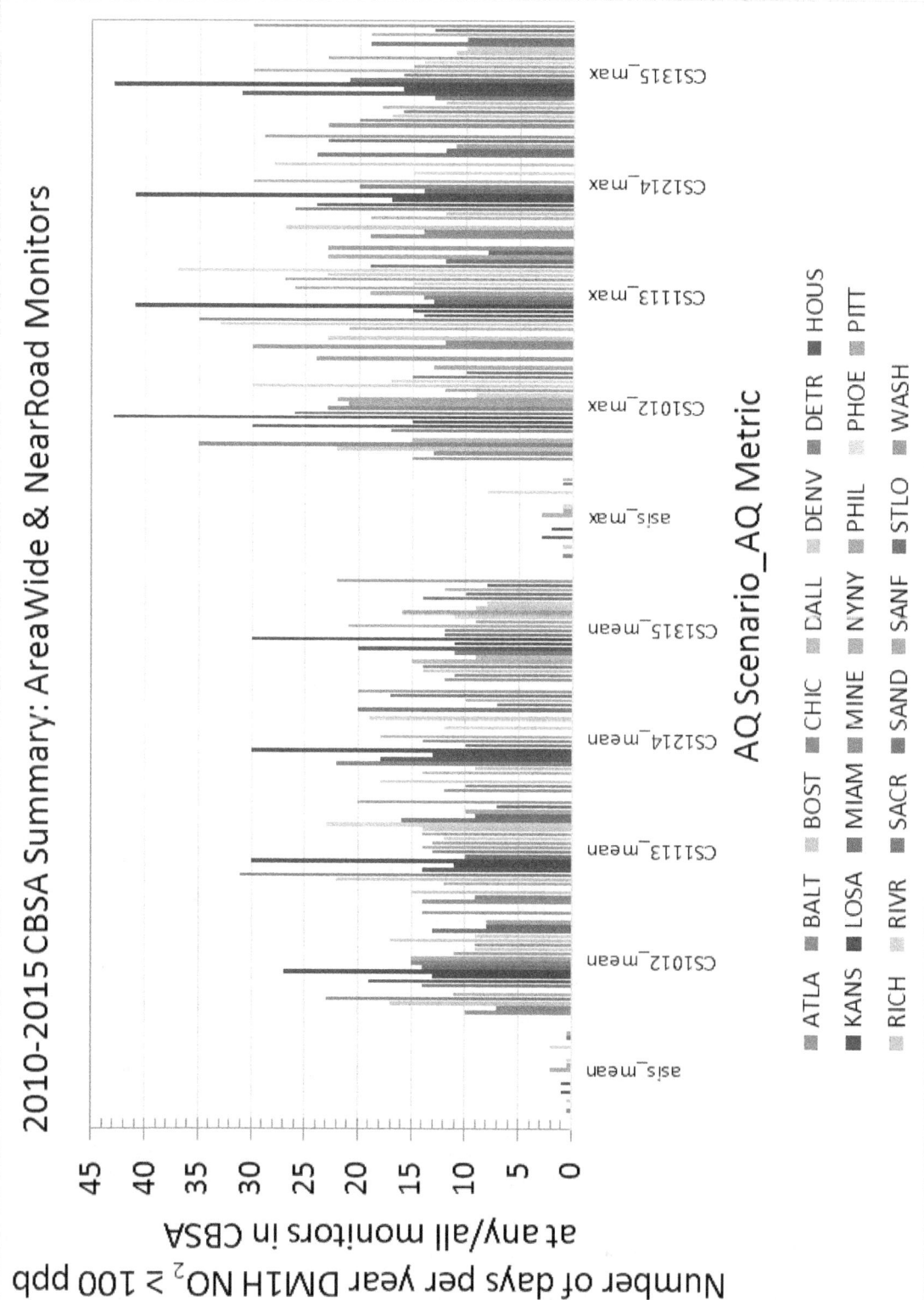

Figure B3-6. Mean and maximum number of days per year where DM1H NO$_2$ concentration at or above 100 ppb at any site: 2010-2015 CBSA-wide summary using area-wide and near-road monitors.

B3-15

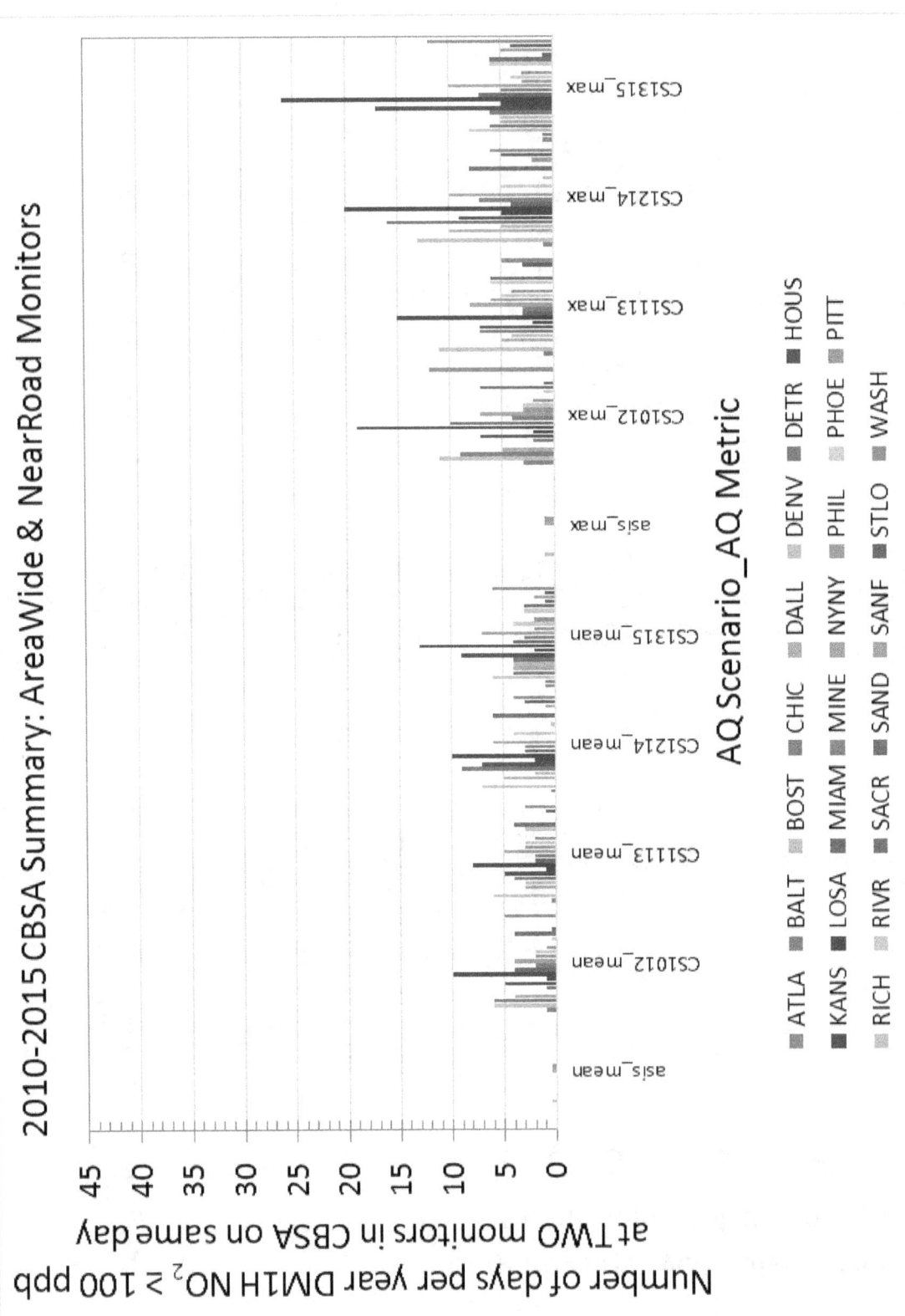

Figure B3-7. Mean and maximum number of days per year where DM1H NO$_2$ concentration at or above 100 ppb at more than one location on the same day: 2010-2015 CBSA-wide summary using area-wide and near-road monitors.

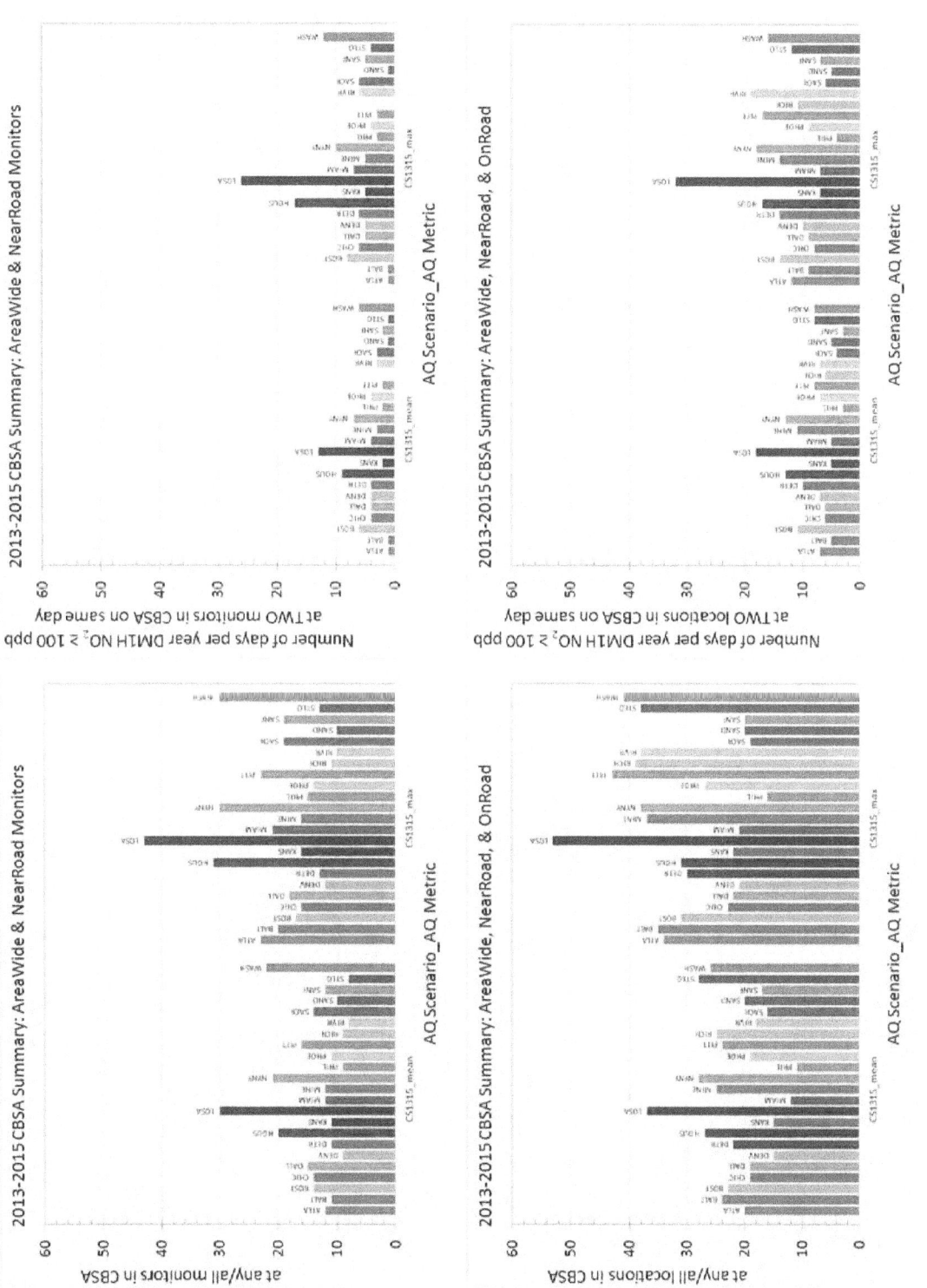

Figure B3-8. Mean and maximum number of days per year where DM1H NO$_2$ concentration at or above 100 ppb: CBSA-wide summary using area-wide and near-road monitors (top panels), using area-wide, near-road, and on-road (bottom panels), at any location (left panels), at more than one location on the same day (right panels).

B4. REFERENCES

Diez-Roux, AD and HC Frey. (2015). CASAC Review of the EPA's Review of the Primary National Ambient Air Quality Standards for Nitrogen Dioxide: Risk and Exposure Assessment Planning Document. Letter to EPA Administrator Gina McCarthy. EPA-CASAC-15-002. Available at: https://yosemite.epa.gov/sab/sabproduct.nsf/LookupWebReportsLastMonthCASAC/A7922887D5BDD8D485257EBB0071A3AD/$File/EPA-CASAC-15-002+unsigned.pdf

Kimbrough ES, Baldauf RW, Watkins N. (2013). Seasonal and diurnal analysis of NO_2 concentrations from a long-duration study conducted in Las Vegas, Nevada. *J Air Waste Manag Assoc.* 63:934-942.

Richmond-Bryant J, Owen RC, Graham S, Snyder MG, McDow S, Oakes M, Kimbrough S. (2017). Estimation of on-road NO_2 concentrations, NO_2/NO_X ratios, and related roadway gradients from near-road monitoring data. *Air Quality, Atmosphere and Health.* doi:10.1007/s11869-016-0455-7.

Rizzo M. (2008). Investigation of how distributions of hourly nitrogen dioxide concentrations have changed over time in six cities. Nitrogen Dioxide NAAQS Review Docket (EPA–HQ–OAR–2006–0922). Available at http://www.epa.gov/ttn/naaqs/standards/nox/s_nox_cr_rea.html

SAS Institute Inc. (2015). Base SAS® 9.4 Procedures Guide, Fourth Edition. Cary, NC: SAS Institute Inc.

U.S. EPA. (2008a). Risk and Exposure Assessment to Support the Review of the NO_2 Primary National Ambient Air Quality Standard. U.S. EPA, Office of Air Quality Planning and Standards. Research Triangle Park, NC. EPA 452/R-08-008a/b. November 2008. Available at: http://www.epa.gov/ttn/naaqs/standards/nox/s_nox_cr_rea.html

U.S. EPA. (2008b). Integrated Science Assessment for Oxides of Nitrogen – Health Criteria. U.S. EPA, National Center for Environmental Assessment and Office, Research Triangle Park, NC. EPA/600/R-08/071. July 2008. Available at: http://cfpub.epa.gov/ncea/cfm/recordisplay.cfm?deid=194645

U.S. EPA. (2009). Risk and Exposure Assessment to Support the Review of the SO_2 Primary National Ambient Air Quality Standard. U.S. EPA, Office of Air Quality Planning and Standards. Research Triangle Park, NC. July 2009. EPA- EPA-452/R-09-007. Available at: http://www.epa.gov/ttn/naaqs/standards/so2/data/200908SO2REAFinalReport.pdf

U.S. EPA. (2010). Quantitative Health Risk Assessment for Particulate Matter. U.S. EPA, Office of Air Quality Planning and Standards. Research Triangle Park, NC. EPA-452/R-10-005. June 2010. Available at: http://www.epa.gov/ttn/naaqs/standards/pm/data/PM_RA_FINAL_June_2010.pdf

U.S. EPA. (2014a). Integrated Review Plan for the Primary National Ambient Air Quality Standards for Nitrogen Dioxide. U.S. EPA, National Center for Environmental Assessment and Office of Air Quality Planning and Standards, Research Triangle Park, NC. EPA-452/R-14-003. June 2014. Available at: http://www.epa.gov/ttn/naaqs/standards/nox/data/201406finalirpprimaryno2.pdf

U.S. EPA. (2014b). Health Risk and Exposure Assessment for Ozone, Final Report. U.S. EPA, Office of Air Quality Planning and Standards. Research Triangle Park, NC. EPA-452/R-14-004a. August 2014. Available at: http://www.epa.gov/ttn/naaqs/standards/ozone/data/20140829healthrea.pdf

U.S. EPA. (2015). Review of the Primary National Ambient Air Quality Standards for Nitrogen Dioxide: Risk and Exposure Assessment Planning Document. U.S. EPA, Office of Air Quality Planning and Standards. Research Triangle Park, NC. EPA-452/D-15-001. May 2015. Available at: https://www3.epa.gov/ttn/naaqs/standards/nox/data/20150504reaplanning.pdf

U.S. EPA. (2016). Integrated Science Assessment for Oxides of Nitrogen – Health Criteria (2016 Final Report). U.S. Environmental Protection Agency, Washington, DC, EPA/600/R-15/068, 2016. Available at: https://cfpub.epa.gov/ncea/isa/recordisplay.cfm?deid=310879

B5. SUPPLEMENTAL DATA

5.1 CBSA RANKED NO$_X$ EMISSION TABLES FOR TOP 20 FACILITY TYPES

Table B5-1. CBSA ranked NO$_X$ emissions for top 20 facility types: electricity via combustion and airports.

1. Electricity Generation via Combustion		2. Airports	
CBSA	**NO$_X$ (tpy)**	**CBSA**	**NO$_X$ (tpy)**
Kansas City, MO-KS	22053	New York-Newark-Jersey City, NY-NJ-PA	9396
Pittsburgh, PA	19492	Los Angeles-Long Beach-Anaheim, CA	6713
Farmington, NM	17204	Chicago-Naperville-Elgin, IL-IN-WI	6424
York-Hanover, PA	16965	Miami-Fort Lauderdale-West Palm Beach, FL	5518
Urban Honolulu, HI	16330	San Francisco-Oakland-Hayward, CA	4266
Paducah, KY-IL	15825	Dallas-Fort Worth-Arlington, TX	4171
Louisville/Jefferson County, KY-IN	14108	Washington-Arlington-Alexandria, DC-VA-MD-WV	3214
Rock Springs, WY	13179	Houston-The Woodlands-Sugar Land, TX	3178
St. Louis, MO-IL	12118	Memphis, TN-MS-AR	3173
Bismarck, ND	11366	Denver-Aurora-Lakewood, CO	3153
Denver-Aurora-Lakewood, CO	11310	Phoenix-Mesa-Scottsdale, AZ	2789
Birmingham-Hoover, AL	11091	Seattle-Tacoma-Bellevue, WA	2733
Chicago-Naperville-Elgin, IL-IN-WI	10251	Las Vegas-Henderson-Paradise, NV	2541
Jacksonville, FL	9757	Charlotte-Concord-Gastonia, NC-SC	2435
Detroit-Warren-Dearborn, MI	9431	Orlando-Kissimmee-Sanford, FL	2352
New York-Newark-Jersey City, NY-NJ-PA	9007	Detroit-Warren-Dearborn, MI	2326
San Antonio-New Braunfels, TX	8845	Minneapolis-St. Paul-Bloomington, MN-WI	2287
Milwaukee-Waukesha-West Allis, WI	6906	Philadelphia-Camden-Wilmington, PA-NJ-DE-MD	2260
Lake Charles, LA	6505	Boston-Cambridge-Newton, MA-NH	2216
Cincinnati, OH-KY-IN	6490	Urban Honolulu, HI	1758

Highlighted are those selected as study areas.

Table B5-2. CBSA ranked NOx emissions for top 20 facility types: chemical plants and petroleum refineries.

3. Chemical Plant		4. Petroleum Refineries	
CBSA	NOx (tpy)	CBSA	NOx (tpy)
Houston-The Woodlands-Sugar Land, TX	12719	Houston-The Woodlands-Sugar Land, TX	7971
Baton Rouge, LA	10345	Beaumont-Port Arthur, TX	5554
Kingsport-Bristol-Bristol, TN-VA	9113	Lake Charles, LA	4726
Beaumont-Port Arthur, TX	6778	Philadelphia-Camden-Wilmington, PA-NJ-DE-MD	4533
Rochester, NY	2610	Los Angeles-Long Beach-Anaheim, CA	3918
Lake Charles, LA	2590	San Francisco-Oakland-Hayward, CA	3204
Marshall, TX	2031	Mount Vernon-Anacortes, WA	2983
Memphis, TN-MS-AR	1267	Chicago-Naperville-Elgin, IL-IN-WI	2548
New Orleans-Metairie, LA	1053	Baton Rouge, LA	2253
Chicago-Naperville-Elgin, IL-IN-WI	695	Gulfport-Biloxi-Pascagoula, MS	2177
Cincinnati, OH-KY-IN	688	Vallejo-Fairfield, CA	1969
Philadelphia-Camden-Wilmington, PA-NJ-DE-MD	570	Minneapolis-St. Paul-Bloomington, MN-WI	1296
Paducah, KY-IL	336	Urban Honolulu, HI	1272
Indianapolis-Carmel-Anderson, IN	175	New York-Newark-Jersey City, NY-NJ-PA	1143
Cleveland-Elyria, OH	165	Huntington-Ashland, WV-KY-OH	1023
St. Louis, MO-IL	158	Tulsa, OK	987
Baltimore-Columbia-Towson, MD	150	Salt Lake City, UT	703
Los Angeles-Long Beach-Anaheim, CA	136	Denver-Aurora-Lakewood, CO	670
Jacksonville, FL	114	El Paso, TX	628
San Francisco-Oakland-Hayward, CA	111	Memphis, TN-MS-AR	532

Highlighted are those selected as study areas.

Table B5-3. CBSA ranked NO$_X$ emissions for top 20 facility types: Portland cement manufacturing and mines/quarries.

5. Portland Cement Manufacturing		6. Mines/Quarries	
CBSA	**NO$_X$ (tpy)**	**CBSA**	**NO$_X$ (tpy)**
Riverside-San Bernardino-Ontario, CA	4831	Gillette, WY	28560
Allentown-Bethlehem-Easton, PA-NJ	3621	Salt Lake City, UT	4478
San Antonio-New Braunfels, TX	3439	Rock Springs, WY	446
Bakersfield, CA	3390	Birmingham-Hoover, AL	386
Dallas-Fort Worth-Arlington, TX	3077	Cheyenne, WY	154
Austin-Round Rock, TX	2245	Riverside-San Bernardino-Ontario, CA	125
Birmingham-Hoover, AL	1819	Las Vegas-Henderson-Paradise, NV	118
San Jose-Sunnyvale-Santa Clara, CA	1755	Riverton, WY	92
Miami-Fort Lauderdale-West Palm Beach, FL	1663	Urban Honolulu, HI	69
Tucson, AZ	1634	Cleveland-Elyria, OH	68
Davenport-Moline-Rock Island, IA-IL	1605	Glenwood Springs, CO	54
Reading, PA	1225	Nashville-Davidson--Murfreesboro--Franklin, TN	25
Louisville/Jefferson County, KY-IN	1097	Carlsbad-Artesia, NM	21
Seattle-Tacoma-Bellevue, WA	918	Minneapolis-St. Paul-Bloomington, MN-WI	20
Rapid City, SD	900	Casper, WY	19
Albuquerque, NM	718	Worcester, MA-CT	16
Kansas City, MO-KS	647	Baltimore-Columbia-Towson, MD	14
Waco, TX	465	Philadelphia-Camden-Wilmington, PA-NJ-DE-MD	11
York-Hanover, PA	158	Kansas City, MO-KS	10
Bozeman, MT	13	San Diego-Carlsbad, CA	9

Highlighted are those selected as study areas.

Table B5-4. CBSA ranked NO$_X$ emissions for top 20 facility types: rail yards and compressor stations.

7. Rail Yard		8. Compressor Station	
CBSA	**NO$_X$ (tpy)**	**CBSA**	**NO$_X$ (tpy)**
Chicago-Naperville-Elgin, IL-IN-WI	3005	Hobbs, NM	5208
Memphis, TN-MS-AR	2022	Glenwood Springs, CO	4516
Houston-The Woodlands-Sugar Land, TX	1950	Farmington, NM	3105
Dallas-Fort Worth-Arlington, TX	1714	Gillette, WY	2358
Kansas City, MO-KS	1714	Oklahoma City, OK	2135
Riverside-San Bernardino-Ontario, CA	1264	Rock Springs, WY	1725
Austin-Round Rock, TX	1197	Riverside-San Bernardino-Ontario, CA	1648
Salt Lake City, UT	1180	Indianapolis-Carmel-Anderson, IN	1583
St. Louis, MO-IL	867	Dallas-Fort Worth-Arlington, TX	1198
Atlanta-Sandy Springs-Roswell, GA	856	Denver-Aurora-Lakewood, CO	659
Buffalo-Cheektowaga-Niagara Falls, NY	757	Carlsbad-Artesia, NM	639
Denver-Aurora-Lakewood, CO	662	Baton Rouge, LA	610
Birmingham-Hoover, AL	661	Birmingham-Hoover, AL	509
Cincinnati, OH-KY-IN	636	Harrisburg-Carlisle, PA	502
Portland-Vancouver-Hillsboro, OR-WA	595	Athens, OH	501
Stockton-Lodi, CA	569	Chicago-Naperville-Elgin, IL-IN-WI	482
Sacramento--Roseville--Arden-Arcade, CA	557	Marshall, TX	481
Little Rock-North Little Rock-Conway, AR	550	Reading, PA	444
Pittsburgh, PA	547	York-Hanover, PA	444
Beaumont-Port Arthur, TX	540	Riverton, WY	392

Highlighted are those selected as study areas.

Table B5-5. CBSA ranked NO$_X$ emissions for top 20 facility types: sources not characterized and municipal waste combustors.

9. Not Characterized		10. Municipal Waste Combustor	
CBSA	NO$_X$ (tpy)		NO$_X$ (tpy)
Chicago-Naperville-Elgin, IL-IN-WI	3827	Miami-Fort Lauderdale-West Palm Beach, FL	4994
Riverside-San Bernardino-Ontario, CA	1539	New York-Newark-Jersey City, NY-NJ-PA	3058
New York-Newark-Jersey City, NY-NJ-PA	1464	Tampa-St. Petersburg-Clearwater, FL	2552
Baltimore-Columbia-Towson, MD	1456	Boston-Cambridge-Newton, MA-NH	2494
Pensacola-Ferry Pass-Brent, FL	1365	Philadelphia-Camden-Wilmington, PA-NJ-DE-MD	2308
Minneapolis-St. Paul-Bloomington, MN-WI	1004	Detroit-Warren-Dearborn, MI	1617
Los Angeles-Long Beach-Anaheim, CA	891	Bridgeport-Stamford-Norwalk, CT	1220
Santa Maria-Santa Barbara, CA	838	Baltimore-Columbia-Towson, MD	1184
Bakersfield, CA	693	Indianapolis-Carmel-Anderson, IN	1076
Austin-Round Rock, TX	667	Hartford-West Hartford-East Hartford, CT	1064
San Francisco-Oakland-Hayward, CA	614	Urban Honolulu, HI	1043
Denver-Aurora-Lakewood, CO	588	Worcester, MA-CT	865
Dallas-Fort Worth-Arlington, TX	567	Minneapolis-St. Paul-Bloomington, MN-WI	594
Miami-Fort Lauderdale-West Palm Beach, FL	548	Lancaster, PA	577
Philadelphia-Camden-Wilmington, PA-NJ-DE-MD	534	Tulsa, OK	564
Houston-The Woodlands-Sugar Land, TX	529	York-Hanover, PA	498
Las Vegas-Henderson-Paradise, NV	528	Washington-Arlington-Alexandria, DC-VA-MD-WV	471
Baton Rouge, LA	497	Los Angeles-Long Beach-Anaheim, CA	424
San Diego-Carlsbad, CA	492	Modesto, CA	317
Rock Springs, WY	491	Ogden-Clearfield, UT	261

Highlighted are those selected as study areas.

Table B5-6. CBSA ranked NOx emissions for top 20 facility types: steel mills and gas plants.

11. Steel Mill		12. Gas Plant	
CBSA	NOₓ (tpy)	CBSA	NOₓ (tpy)
Chicago-Naperville-Elgin, IL-IN-WI	10939	Hobbs, NM	5051
Detroit-Warren-Dearborn, MI	2776	Farmington, NM	4053
Pittsburgh, PA	1886	Rock Springs, WY	1404
Cleveland-Elyria, OH	1291	Gillette, WY	1041
Huntington-Ashland, WV-KY-OH	1243	Longview, TX	571
Baltimore-Columbia-Towson, MD	1166	Carlsbad-Artesia, NM	556
Birmingham-Hoover, AL	1095	Evanston, WY	392
Harrisburg-Carlisle, PA	310	Bakersfield, CA	372
Dallas-Fort Worth-Arlington, TX	298	Riverton, WY	359
Philadelphia-Camden-Wilmington, PA-NJ-DE-MD	261	Baton Rouge, LA	274
Reading, PA	239	Beaumont-Port Arthur, TX	259
El Paso, TX	234	Marshall, TX	212
Beaumont-Port Arthur, TX	230	San Antonio-New Braunfels, TX	198
Seattle-Tacoma-Bellevue, WA	205	Kingsport-Bristol-Bristol, TN-VA	162
Portland-Vancouver-Hillsboro, OR-WA	193	Dallas-Fort Worth-Arlington, TX	160
Fort Wayne, IN	170	Casper, WY	145
Memphis, TN-MS-AR	125	Oklahoma City, OK	128
Longview, TX	113	Cheyenne, WY	98
Jacksonville, FL	78	Houston-The Woodlands-Sugar Land, TX	88
Charlotte-Concord-Gastonia, NC-SC	53	Minneapolis-St. Paul-Bloomington, MN-WI	63

Highlighted are those selected as study areas.

Table B5-7. CBSA ranked NO$_X$ emissions for top 20 facility types: mineral processing plants and pulp and paper plants.

13. Mineral Processing Plant		14. Pulp and Paper Plant	
CBSA	**NO$_X$ (tpy)**	**CBSA**	**NO$_X$ (tpy)**
Rock Springs, WY	7689	Columbia, SC	2036
Riverside-San Bernardino-Ontario, CA	1872	York-Hanover, PA	1721
Las Vegas-Henderson-Paradise, NV	1200	Pensacola-Ferry Pass-Brent, FL	1654
State College, PA	940	Duluth, MN-WI	1255
Corsicana, TX	705	Baton Rouge, LA	1132
Detroit-Warren-Dearborn, MI	547	Charleston-North Charleston, SC	952
Manitowoc, WI	389	Kingsport-Bristol-Bristol, TN-VA	818
Fort Smith, AR-OK	221	Beaumont-Port Arthur, TX	752
Bakersfield, CA	159	Portland-South Portland, ME	463
Cincinnati, OH-KY-IN	60	Philadelphia-Camden-Wilmington, PA-NJ-DE-MD	459
Little Rock-North Little Rock-Conway, AR	44	Riverside-San Bernardino-Ontario, CA	140
Baltimore-Columbia-Towson, MD	26	Memphis, TN-MS-AR	125
Salinas, CA	22	Cincinnati, OH-KY-IN	117
Boston-Cambridge-Newton, MA-NH	18	Dallas-Fort Worth-Arlington, TX	74
Chicago-Naperville-Elgin, IL-IN-WI	16	San Jose-Sunnyvale-Santa Clara, CA	68
Minneapolis-St. Paul-Bloomington, MN-WI	5	Reading, PA	61
Tampa-St. Petersburg-Clearwater, FL	5	New York-Newark-Jersey City, NY-NJ-PA	59
Los Angeles-Long Beach-Anaheim, CA	4	Owensboro, KY	59
Oxnard-Thousand Oaks-Ventura, CA	4	Oxnard-Thousand Oaks-Ventura, CA	55
Glenwood Springs, CO	2	Tulsa, OK	52

Highlighted are those selected as study areas.

Table B5-8. CBSA ranked NO$_x$ emissions for top 20 facility types: institutional (e.g., schools, hospitals, prisons) and gas plants.

15. Institutional (school, hospital, prison, etc.)		16. Glass Plant	
CBSA	NO$_x$ (tpy)	CBSA	NO$_x$ (tpy)
Chicago-Naperville-Elgin, IL-IN-WI	797	Corsicana, TX	2364
New York-Newark-Jersey City, NY-NJ-PA	769	Pittsburgh, PA	1128
Lansing-East Lansing, MI	739	Nashville-Davidson--Murfreesboro--Franklin, TN	850
Nashville-Davidson--Murfreesboro--Franklin, TN	602	Waco, TX	781
South Bend-Mishawaka, IN-MI	580	Fresno, CA	619
Worcester, MA-CT	347	Jacksonville, FL	484
Fairbanks, AK	339	Chicago-Naperville-Elgin, IL-IN-WI	464
Los Angeles-Long Beach-Anaheim, CA	294	Portland-Vancouver-Hillsboro, OR-WA	407
Philadelphia-Camden-Wilmington, PA-NJ-DE-MD	287	Stockton-Lodi, CA	404
Boston-Cambridge-Newton, MA-NH	279	Atlanta-Sandy Springs-Roswell, GA	350
Minneapolis-St. Paul-Bloomington, MN-WI	269	San Francisco-Oakland-Hayward, CA	328
State College, PA	258	Madera, CA	327
Lexington-Fayette, KY	251	Seattle-Tacoma-Bellevue, WA	299
Denver-Aurora-Lakewood, CO	231	Worcester, MA-CT	230
Cincinnati, OH-KY-IN	198	Modesto, CA	195
Portland-South Portland, ME	197	Los Angeles-Long Beach-Anaheim, CA	70
San Francisco-Oakland-Hayward, CA	175	Houston-The Woodlands-Sugar Land, TX	42
Washington-Arlington-Alexandria, DC-VA-MD-WV	175	Providence-Warwick, RI-MA	36
Baltimore-Columbia-Towson, MD	148	Tulsa, OK	10
Provo-Orem, UT	131	Charleston-North Charleston, SC	4

Highlighted are those selected as study areas.

Table B5-9. CBSA ranked NO$_X$ emissions for top 20 facility types: chlor-alkali plants and fertilizer plants.

17. Chlor-Alkali Plant		18. Fertilizer Plant	
CBSA	NO$_X$ (tpy)	CBSA	NO$_X$ (tpy)
Lake Charles, LA	6194	Baton Rouge, LA	3793
Wichita, KS	508	Cheyenne, WY	1718
Baton Rouge, LA	170	Gulfport-Biloxi-Pascagoula, MS	322
Cleveland, TN	44	Tampa-St. Petersburg-Clearwater, FL	319
Houston-The Woodlands-Sugar Land, TX	18	Pittsburgh, PA	243
Philadelphia-Camden-Wilmington, PA-NJ-DE-MD	2	Rock Springs, WY	132
		Provo-Orem, UT	106
		Carlsbad-Artesia, NM	74
		Sacramento--Roseville--Arden-Arcade, CA	36
		Cincinnati, OH-KY-IN	19
		Salisbury, MD-DE	16
		Fresno, CA	15
		Houston-The Woodlands-Sugar Land, TX	12
		Riverside-San Bernardino-Ontario, CA	4
		Stockton-Lodi, CA	3
		Chico, CA	1
		Merced, CA	1
		Nashville-Davidson--Murfreesboro--Franklin, TN	1
		Vallejo-Fairfield, CA	1
		Winston-Salem, NC	1

Highlighted are those selected as study areas.

Table B5-10. CBSA ranked NO$_X$ emissions for top 20 facility types: coke battery and plastic, resin, or rubber product plants.

19. Coke Battery		20. Plastic, Resin, or Rubber Products Plant	
CBSA	NO$_X$ (tpy)	CBSA	NO$_X$ (tpy)
Pittsburgh, PA	3502	Baton Rouge, LA	2009
Birmingham-Hoover, AL	1839	Louisville/Jefferson County, KY-IN	778
Chicago-Naperville-Elgin, IL-IN-WI	859	Stockton-Lodi, CA	633
Huntington-Ashland, WV-KY-OH	355	Marshall, TX	465
Buffalo-Cheektowaga-Niagara Falls, NY	153	Beaumont-Port Arthur, TX	459
Erie, PA	106	Houston-The Woodlands-Sugar Land, TX	379
		Springfield, MA	335
		Cincinnati, OH-KY-IN	190
		Lake Charles, LA	167
		Greenville-Anderson-Mauldin, SC	117
		Salt Lake City, UT	115
		Pittsburgh, PA	79
		Philadelphia-Camden-Wilmington, PA-NJ-DE-MD	58
		Buffalo-Cheektowaga-Niagara Falls, NY	54
		Providence-Warwick, RI-MA	46
		Portland-Vancouver-Hillsboro, OR-WA	39
		Dallas-Fort Worth-Arlington, TX	37
		Denver-Aurora-Lakewood, CO	31
		Chicago-Naperville-Elgin, IL-IN-WI	29
		New Haven-Milford, CT	29

Highlighted are those selected as study areas.

5.2 ATTRIBUTES OF AMBIENT NO$_2$ MONITORS USED FOR THE 2010-2015 ANALYSIS

Brief descriptions of the attributes, the sources of data, and any data processing steps follow provided in section 5.2.

1. *Ambient monitor meta-data.* Data sets containing general attributes (e.g., geographic coordinates, local land use, monitor setting, etc.) were obtained from AQS[31] and from querying the AQS database directly.[32] The land-use field indicates the prevalent land use within ¼ mile of the monitoring site.[33] The measurement scale represents the air volumes associated with the monitoring area dimensions.[34] The monitor objective describes the monitor in terms of its attempt to generally characterize health effects, photochemical activity, transport, or welfare effects.[35] Monitors most useful for evaluating public health would be characterized as having a monitor objective of population exposure and/or highest concentration. In addition, county land areas (mi^2) were estimated from the U.S. Census Bureau.[36]

2. *NO$_X$ source emissions data.* For characterizing mobile source emissions associated with each ambient monitor, the same county file described above for evaluating emissions across all CBSAs was used. Presented separately in this analysis are the NO$_X$ emissions from on-road, non-road, and air-craft/marine/locomotive (AML) mobile sources as well as the total NO$_X$ and percent NO$_X$ contributed by all mobile sources combined. For characterizing facility NO$_X$ emissions potentially associated

---

[31] https://www3.epa.gov/airdata/ad_maps.html

[32] https://aqs.epa.gov/api.

[33] Land-use is characterized here as either agricultural (AGRIC), commercial (COMM), desert (DESERT), forest (FOREST), Industrial (INDUS), military reservation (MILIT), mobile (MOBILE), residential (RESID), or unknown (UNK).

[34] Measurement scales for monitors typically are characterized as microscale (in close proximity, up to 100 m from a source), middle (100 m to 0.5 km), neighborhood (500 m to 4 km), or urban (4 to 50 km).

[35] Objectives for monitors are characterized here as population exposure (POP_EXP), highest concentration (HI_CONC), source oriented (SOURCE), general/background (GEN/BKGR), extreme downwind (DOWNWND), maximum ozone concentration (MAX_O3), maximum precursor emissions (MAX_PRE), other (OTH), unknown (UNK), or upwind boackground (UP_BKGR), regional transport (REG TRANS), quality assuramce (QUAL ASSUR).

[36] https://www.census.gov/support/USACdataDownloads.html#LND. For land area, variable 'LND110210D' was used.

with measured concentrations at each monitor, the 'facility-level by pollutant' summary file containing emissions for the entire U.S. was downloaded from the 2011 NEI.[37] For each ambient monitor, all facilities within a 5 km radius were identified, then having in their emissions summed based on facility type. Facility types having aggregated emissions greater than or equal to 10 tpy were retained.

3. *Annual average daily traffic data.* For characterizing ambient monitor distances to roads for all monitors (except for the new near-road monitors), we used shapefiles obtained from the U.S. Department of Transportation (U.S. DOT) Federal Highway Administration (FHA) 2011 Highway Performance Monitoring System (HPMS).[38] Of relevance here were annual average daily traffic (AADT) data for all roadways[39] and their identities, where available. These data were mapped using ArcMap version 10.3.1 and then joined to NO_2 ambient monitors by using the 'Generate near Table' function. This analysis used a 1000-meter search radius to join a monitor to the 100 segments of road identified by the HPMS closest to it, and all data were projected to North American Lambert Conformal Conic, GCS North American 1983. From this output, two metrics were developed. The first was developed using the road closest to the monitor, while the second was developed using the road having the maximum AADT. For both of these metrics, the AADT value, the road distance from the monitor, and the road identity were retained.

Monitor map locations: A plot of all the monitors in each study area is provided using ArcGIS. Individual satellite views of the near-road and area design value monitors were generated using google maps and the respective monitor latitudes (lat) and longitudes (lon).

---

[37] http://aqsdr1.epa.gov/aqsweb/aqstmp/airdata/download_files.html#Meta. Note that for historical monitoring data, the 2011 NEI may not necessarily represent actual emissions when the monitor was in operation during earlier years.

[38] https://www.fhwa.dot.gov/policyinformation/hpms/shapefiles.cfm.

[39] Roads are those characterized as part of the HPMS-defined Federal-Aid System and include 1) interstate, 2) principal arterial – other freeways and expressways, 3) principal arterial – other, minor arterial, 4) major collector, 5) urban minor collector, and 6) other highways that are designated as part of the National Highway System.

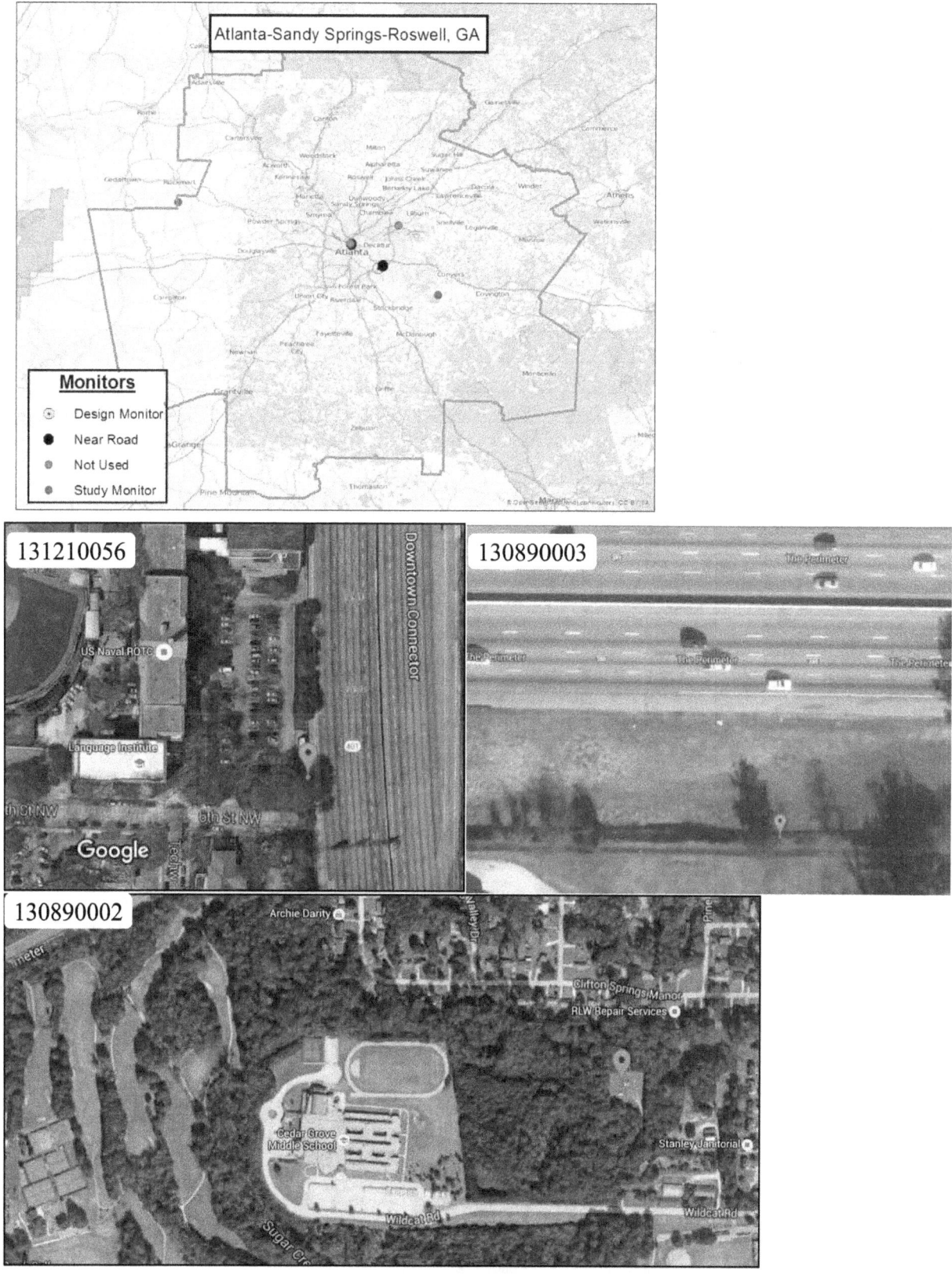

Figure B5-1. Map of all monitors (1990-2015) in the Atlanta study area (top panel) and satellite views of near-road (middle panels) and area design value monitor (bottom panel).

Table B5-11. Attributes of ambient monitors within the Atlanta study area having recent (2010-2015) valid-year NO$_2$ concentrations.

State abbreviation	GA	GA	GA	GA	GA	
County name	DeKalb	DeKalb	Fulton	Paulding	Rockdale	
Site ID	130890002	130890003	131210056	132230003	132470001	
Lat	33.68797	33.69861	33.77832	33.9285	33.59108	
Lon	-84.29048	-84.27261	-84.39142	-85.04534	-84.06529	
year_start	1978	2014	2014	1995	1994	
year_end	2016	2016	2016	2015	2015	
Elevation_m	308	238	286	417	219	
land_use	RESID	COMM	MOBILE	AGRIC	AGRIC	
scale	URBAN	MICRO	MICRO	URBAN	URBAN	
objective1	POP_EXP	HI_CONC	SOURCE	UP_BKGR	MAX_O3	
objective2	MAX_PRE	REG TRANS	POP_EXP	GEN/BKGR	POP_EXP	
objective3	HI_CONC	QUAL ASSUR		POP_EXP	GEN/BKGR	
cnty_landarea_2010	268	268	527	312	130	
AADT information (distance in meters, road/highway/interstate number, AADT value)						
dist_closest	597	30	2	.	.	
road_closest	285	285	85	.	.	
aadt_closest	140820	146000	284920	.	.	
dist_aadtmax	.	642	.	.	.	
road_aadtmax	.	285	.	.	.	
aadt_aadtmax	.	141700	.	.	.	
County Level NOx Emissions (tpy)						
OnRoad_mobile_nox	11230	11230	16404	1780	1529	
NonRoad_mobile_nox	1962	1962	3939	410	329	
AML_mobile_nox	376	376	1560	297	30	
total_nox	14719	14719	23989	2688	2112	
pct_mobile_nox	92.2	92.2	91.3	92.5	89.4	
NOx Emissions by Source Type (summed tons per year (tpy), sources within 5 Km emitting at least 10 tpy)						
Landfill	45	45	.	.	.	
Rail_Yard	.	.	28	.	.	
		Area design value monitor				
		Near-road monitor				

Figure B5-2. Map of all monitors (1990-2015) in the Baltimore study area (top panel) and satellite views of near-road (middle panel) and area design value monitor (bottom panel).

Table B5-12. Attributes of ambient monitors within the Baltimore study area having recent (2010-2015) valid-year NO$_2$ concentrations.

State abbreviation	MD	MD	MD			
County name	Baltimore	Howard	Balt. (City)			
Site ID	240053001	240270006	245100040			
Lat	39.31083	39.14309	39.29773			
Lon	-76.47444	-76.84608	-76.60460			
year_start	1971	2014	1981			
year_end	2016	2016	2016			
Elevation_m	5	10	12			
land_use	RESID	RESID	RESID			
scale	NBHOOD	MICRO	MID			
objective1	POP_EXP	HI_CONC	HI_CONC			
objective2	MAX_PRE	SOURCE	POP_EXP			
objective3	HI_CONC	GEN/BKGR				
cnty_landarea_2010	598	251	81			
AADT information (distance in meters, road/highway/interstate number, AADT value)						
dist_closest	181	16	13			
road_closest	150	95	1470			
aadt_closest	29222	186750	9391			
dist_aadtmax	593	.	655			
road_aadtmax	150	.	83			
aadt_aadtmax	29932	.	102860			
County Level NOx Emissions (tpy)						
OnRoad_mobile_nox	11198	5298	4573			
NonRoad_mobile_nox	3463	991	646			
AML_mobile_nox	1333	216	1385			
total_nox	22136	8474	10421			
pct_mobile_nox	72.3	76.8	63.4			
NOx Emissions by Source Type (summed tons per year (tpy), sources within 5 Km emitting at least 10 tpy)						
Municipal_Waste_Combustor	.	.	1134			
Steam_Heating_Facility	.	.	153			
Sugar_Mill	.	.	126			
Institutional__school__hospital_	.	.	98			
Chemical_Plant	.	.	70			
Rail_Yard	.	.	74			
Electricity_Generation_via_Combu	.	.	40			
not_characterized	36	23	10			
Military_Base	.	.	.			
Wastewater_Treatment_Facility	31	.	.			
Mineral_Processing_Plant	.	.	.			
		Area design value monitor				
		Near-road monitor				

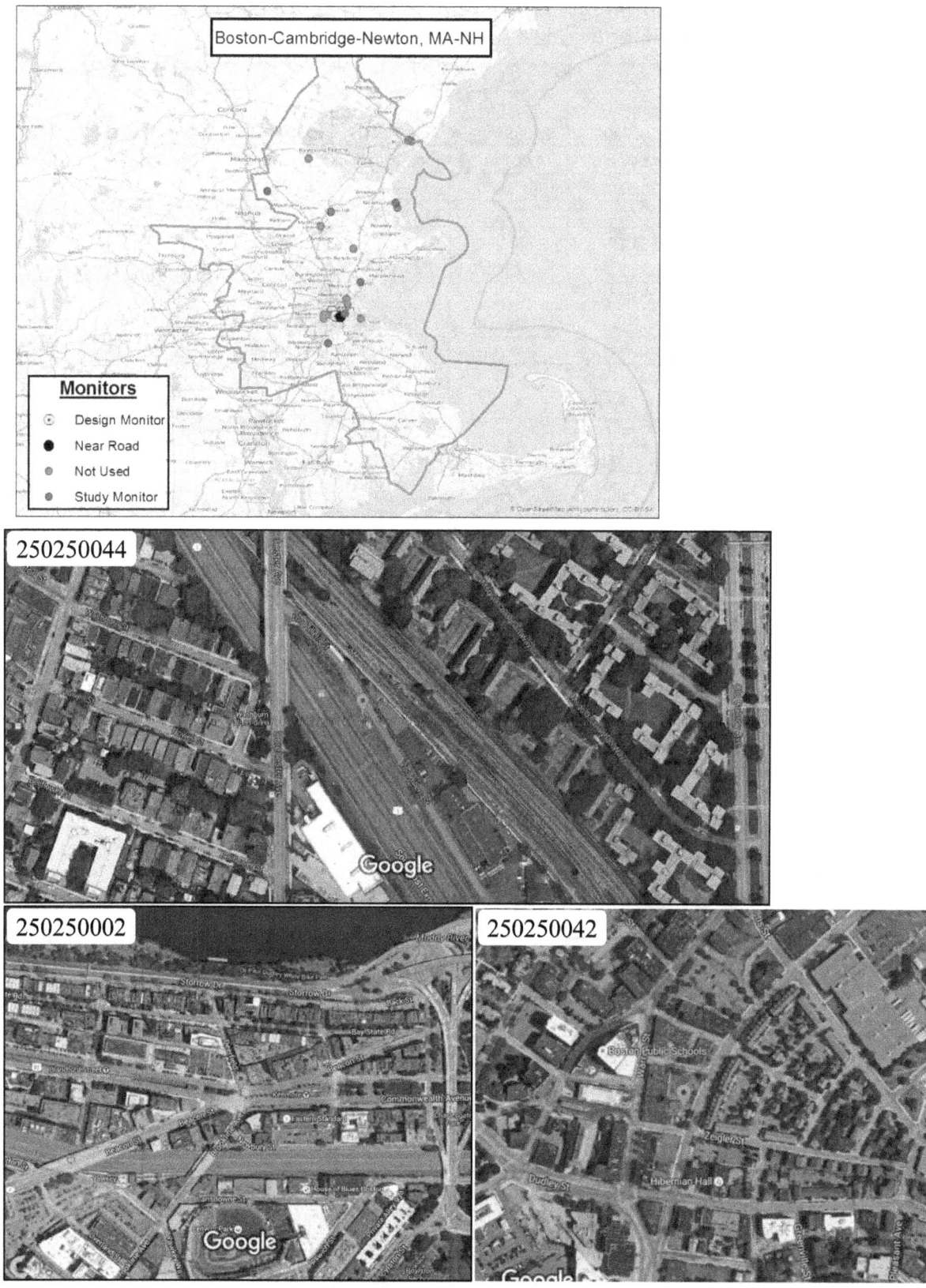

Figure B5-3. Map of all monitors (1990-2015) in the Boston study area (top panel) and satellite views of near-road (middle panel) and area design value monitor (bottom panels).

Table B5-13. Attributes of ambient monitors within the Boston study area having recent (2010-2015) valid-year NO$_2$ concentrations.

State abbreviation	MA	MA	MA	MA	MA	MA
County name	Essex	Essex	Essex	Norfolk	Suffolk	Suffolk
Site ID	250092006	250094005	250095005	250213003	250250002	250250040
Lat	42.47464	42.81441	42.77084	42.21177	42.34887	42.34025
Lon	-70.97082	-70.81778	-71.10229	-71.11397	-71.09716	-71.03835
year_start	1993	2010	2003	2002	1975	1992
year_end	2016	2016	2012	2016	2016	2014
Elevation_m	52	2	0	192	6	0
land_use	COMM	RESID	RESID	FOREST	COMM	INDUS
scale	URBAN	URBAN	NBHOOD	REGION	MICRO	NBHOOD
objective1	POP_EXP	MAX_O3	POP_EXP	GEN/BKGR	HI_CONC	POP_EXP
objective2	MAX_PRE			UP_BKGR	POP_EXP	
objective3	HI_CONC			POP_EXP		
cnty_landarea_2010	493	493	493	396	58	58
AADT information (distance in meters, road/highway/interstate number, AADT value)						
dist_closest	486	.	.	469	7	216
road_closest	363	.	.	138	20	121
aadt_closest	13142	.	.	14059	20942	8618
dist_aadtmax	.	.	.	892	117	663
road_aadtmax	.	.	.	93	90	90
aadt_aadtmax	.	.	.	164568	127957	67804
County Level NOx Emissions (tpy)						
OnRoad_mobile_nox	6568	6568	6568	6685	3201	3201
NonRoad_mobile_nox	2473	2473	2473	2009	2094	2094
AML_mobile_nox	860	860	860	619	5446	5446
total_nox	15750	15750	15750	12261	14015	14015
pct_mobile_nox	62.9	62.9	62.9	76	76.6	76.6
NOx Emissions by Source Type (summed tons per year (tpy), sources within 5 Km emitting at least 10 tpy)						
Airport	.	.	.	.	.	2203
Municipal_Waste_Combustor	705	.	1021	.	.	.
Steam_Heating_Facility	.	.	.	.	182	182
Institutional__school__hospital_	.	.	.	.	209	145
Electricity_Generation_via_Combu	.	.	.	.	253	41
Rail_Yard	.	.	.	.	20	.
Wastewater_Treatment_Facility	.	.	.	.	.	.
Aircraft__Aerospace__or_Related	156	.	.	.	.	.
Fabricated_Metal_Products_Plant	.	.	.	.	42	42
not_characterized	.	.	.	.	27	27
Textile__Yarn__or_Carpet_Plant	.	.	.	.	.	.
Military_Base	.	.	.	.	.	.
Mineral_Processing_Plant	.	.	.	.	.	.
		Area design value monitor				
		Near-road monitor				

Table B5-13, continued. Attributes of ambient monitors within the Boston study area having recent (2010-2015) valid-year NO₂ concentrations.

State abbreviation	MA	MA	NH			
County name	Suffolk	Suffolk	Rockingham			
Site ID	250250042	250250044	330150018			
Lat	42.32950	42.32519	42.86254			
Lon	-71.08260	-71.05606	-71.38017			
year_start	2000	2013	2014			
year_end	2016	2016	2016			
Elevation_m	6	15	123			
land_use	COMM	COMM	RESID			
scale	NBHOOD	MID	REGION			
objective1	POP_EXP	POP_EXP	POP_EXP			
objective2	HI_CONC		GEN/BKGR			
objective3						
cnty_landarea_2010	58	58	695			
AADT information (distance in meters, road/highway/interstate number, AADT value)						
dist_closest	56	10	.			
road_closest	447	93	.			
aadt_closest	9040	198239	.			
dist_aadtmax	876	424	.			
road_aadtmax	125	93	.			
aadt_aadtmax	47853	193777	.			
County Level NOx Emissions (tpy)						
OnRoad_mobile_nox	3201	3201	4691			
NonRoad_mobile_nox	2094	2094	1376			
AML_mobile_nox	5446	5446	533			
total_nox	14015	14015	8767			
pct_mobile_nox	76.6	76.6	75.3			
NOx Emissions by Source Type (summed tons per year (tpy), sources within 5 Km emitting at least 10 tpy)						
Airport	.	.	.			
Municipal_Waste_Combustor	.	.	.			
Steam_Heating_Facility	182	182	.			
Institutional__school__hospital_	222	155	.			
Electricity_Generation_via_Combu	253	161	.			
Rail_Yard	20	.	.			
Wastewater_Treatment_Facility	.	.	.			
Aircraft__Aerospace__or_Related	.	.	.			
Fabricated_Metal_Products_Plant	42	42	.			
not_characterized	27	27	.			
Textile__Yarn__or_Carpet_Plant	.	.	.			
Military_Base	.	.	.			
Mineral_Processing_Plant	.	.	.			
		Area design value monitor				
		Near-road monitor				

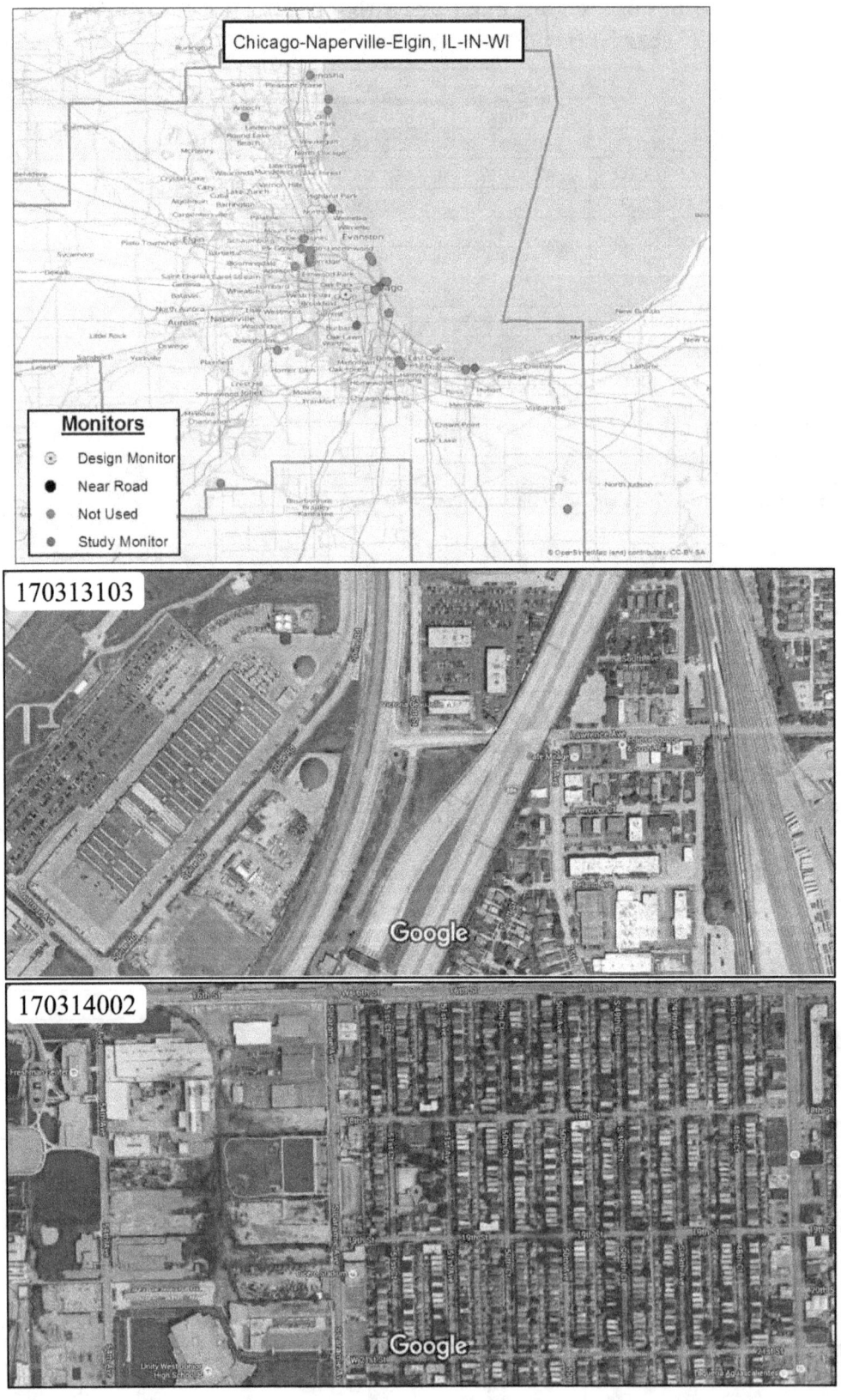

Figure B5-4. Map of all monitors (1990-2015) in the Chicago study area (top panel) and satellite views of near-road (middle panel) and area design value monitor (bottom panel).

Table B5-14. Attributes of ambient monitors within the Chicago study area having recent (2010-2015) valid-year NO$_2$ concentrations.

State abbreviation	IL	IL	IL	IL	IL	IN
County name	Cook	Cook	Cook	Cook	Cook	Lake
Site ID	170310063	170310076	170313103	170314002	170314201	180890022
Lat	41.87768	41.75140	41.96519	41.85524	42.14000	41.60668
Lon	-87.63503	-87.71349	-87.87626	-87.75247	-87.79923	-87.30473
year_start	1988	2001	1997	1982	1997	1995
year_end	2016	2016	2016	2016	2016	2016
Elevation_m	181	186	195	184	198	183
land_use	MOBILE	RESID	MOBILE	RESID	RESID	INDUS
scale	MID	NBHOOD	MID	NBHOOD	URBAN	NBHOOD
objective1	HI_CONC	POP_EXP	HI_CONC	POP_EXP	POP_EXP	POP_EXP
objective2	POP_EXP	HI_CONC	SOURCE	HI_CONC	MAX_O3	HI_CONC
objective3	SOURCE	GEN/BKGR	POP_EXP			SOURCE
cnty_landarea_2010	945	945	945	945	945	499
AADT information (distance in meters, road/highway/interstate number, AADT value)						
dist_closest	45	188	30	629	211	696
road_closest	0	0	12	50	68	90
aadt_closest	9518	22435	43900	32400	34600	34754
dist_aadtmax	842	663	149	766	970	.
road_aadtmax	90	0	294	50	94	.
aadt_aadtmax	253061	45285	190046	34097	135733	.
County Level NOx Emissions (tpy)						
OnRoad_mobile_nox	54900	54900	54900	54900	54900	9294
NonRoad_mobile_nox	19402	19402	19402	19402	19402	2652
AML_mobile_nox	12900	12900	12900	12900	12900	1450
total_nox	113148	113148	113148	113148	113148	38995
pct_mobile_nox	77.1	77.1	77.1	77.1	77.1	34.4
NOx Emissions by Source Type (summed tons per year (tpy), sources within 5 Km emitting at least 10 tpy)						
Steel_Mill	.	.	.	.	.	4336
Airport	.	1150	5261	.	.	.
Electricity_Generation_via_Combu	1118	.	.	1893	.	.
Rail_Yard	185	62	440	229	.	.
Institutional__school__hospital_	232	27	.	31	.	.
Chemical_Plant	.	64	.	115	.	.
Landfill	.	.	.	.	57	.
Industrial_Machinery_or_Equipmen	.	.	.	27	.	.
Steam_Heating_Facility	12	.	.	.	.	.
Wastewater_Treatment_Facility	.	.	.	51	.	.
Food_Products_Processing_Plant	.	36	12	.	.	.
Hot_Mix_Asphalt_Plant	.	.	.	.	.	.
not_characterized	87	.	72	18	.	.
Glass_Plant	.	.	.	.	.	.
Calcined_Pet_Coke_Plant	.	.	.	.	.	.
Automobile_Truck_or_Parts_Plant	.	.	.	.	.	.

		Area design value monitor
		Used as near-road monitor, though not formally designated as such

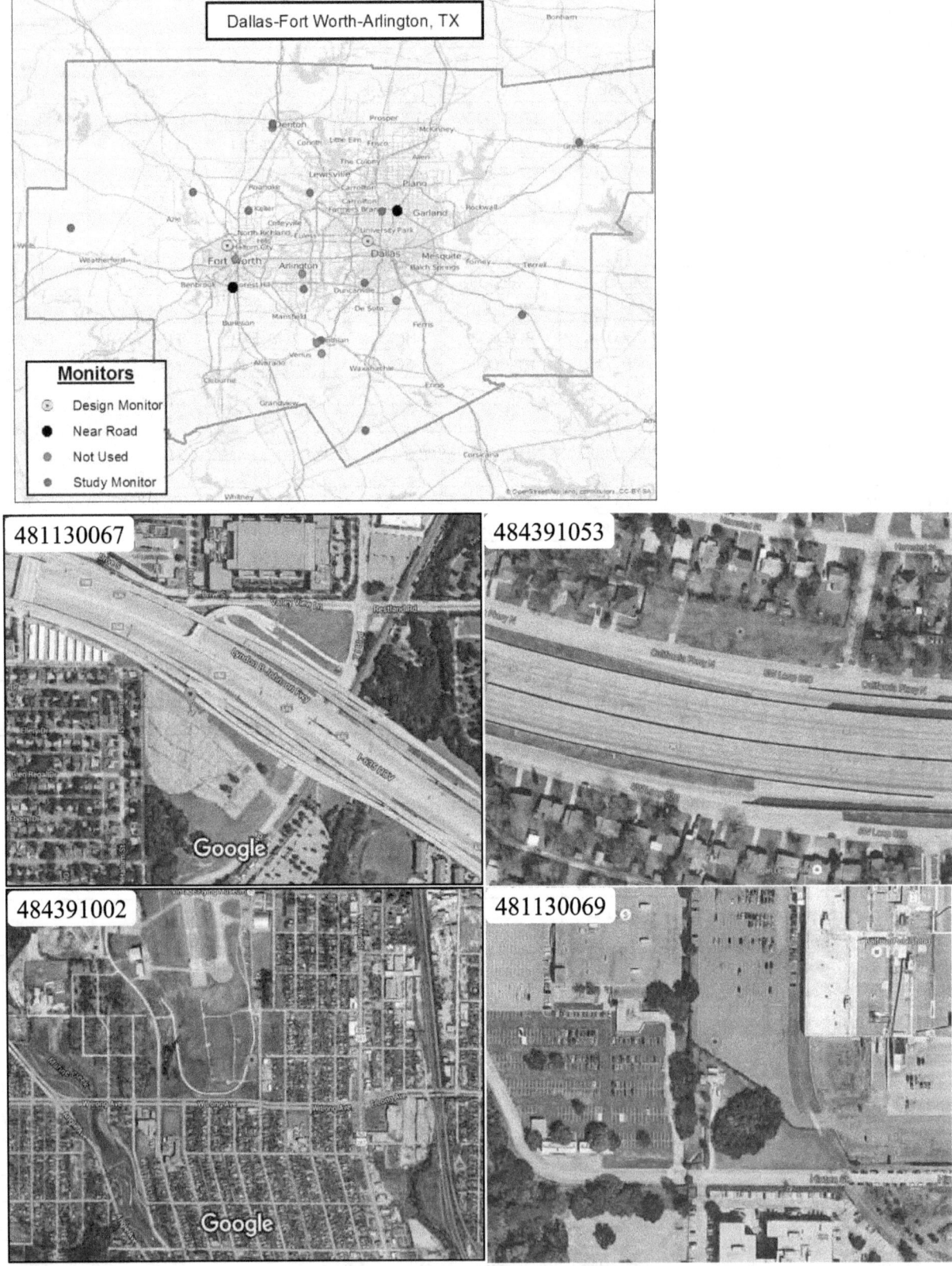

Figure B5-5. Map of all monitors (1990-2015) in the Dallas study area (top panel) and satellite views of near-road (middle panels) and area design value monitor (bottom panels).

Table B5-15. Attributes of ambient monitors within the Dallas study area having recent (2010-2015) valid-year NO₂ concentrations.

State abbreviation	TX	TX	TX	TX	TX	TX
County name	Dallas	Dallas	Dallas	Dallas	Denton	Ellis
Site ID	481130069	481130075	481130087	481131067	481210034	481390016
Lat	32.82006	32.91921	32.67645	32.92118	33.21907	32.48208
Lon	-96.86012	-96.80850	-96.87206	-96.75355	-97.19628	-97.02690
year_start	1986	1998	1994	2014	1998	2003
year_end	2016	2016	2016	2016	2016	2016
Elevation_m	126.5	190.8	206	177	183	195
land_use	COMM	RESID	COMM	COMM	COMM	AGRIC
scale	NBHOOD	NBHOOD	NBHOOD	MICRO	URBAN	NBHOOD
objective1	MAX_PRE	GEN/BKGR	GEN/BKGR	MAX_PRE	POP_EXP	SOURCE
objective2	POP_EXP	POP_EXP	POP_EXP		DOWNWND	GEN/BKGR
objective3	HI_CONC				MAX_O3	REG TRANS
cnty_landarea_2010	871	871	871	871	878	935
AADT information (distance in meters, road/highway/interstate number, AADT value)						
dist_closest	232	435	.	24	.	487
road_closest	0	289	.	635	.	287
aadt_closest	24020	23000	.	235790	.	27000
dist_aadtmax	802	980	.	959	.	.
road_aadtmax	35	635	.	75	.	.
aadt_aadtmax	210680	209000	.	251140	.	.
County Level NOx Emissions (tpy)						
OnRoad_mobile_nox	36623	36623	36623	36623	7555	5173
NonRoad_mobile_nox	8462	8462	8462	8462	1685	854
AML_mobile_nox	1141	1141	1141	1141	1608	272
total_nox	51422	51422	51422	51422	13785	11530
pct_mobile_nox	89.9	89.9	89.9	89.9	78.7	54.6
NOx Emissions by Source Type (summed tons per year (tpy), sources within 5 Km emitting at least 10 tpy)						
Portland_Cement_Manufacturing	.	.	.	.	.	2431
Airport	384	.	.	.	23	.
Rail_Yard	.	.	.	.	.	.
Steel_Mill	.	.	.	.	.	298
Compressor_Station	.	.	.	.	.	.
Electricity_Generation_via_Combu	.	.	.	.	.	.
not_characterized	15	.	.	64	.	.
Bakeries	.	.	.	.	.	.
Breweries_Distilleries_Wineries	.	.	.	.	.	.
Pharmaceutical_Manufacturing	.	.	.	.	.	.
Automobile_Truck_or_Parts_Plant	.	.	.	.	.	.
		Area design value monitor				
		Near-road monitor				

Table B5-15, continued. Attributes of ambient monitors within the Dallas study area having recent (2010-2015) valid-year NO₂ concentrations.

State abbreviation		TX	TX	TX	TX	TX	TX
County name		Ellis	Hunt	Kaufman	Parker	Tarrant	Tarrant
Site ID		481391044	482311006	482570005	483670081	484390075	484391002
Lat		32.17542	33.15309	32.56497	32.86877	32.98789	32.80582
Lon		-96.87019	-96.11557	-96.31769	-97.90593	-97.47718	-97.35657
year_start		2007	2003	2000	2010	2010	1975
year_end		2016	2016	2016	2012	2012	2016
Elevation_m		165	161	128	347	241	203
land_use		AGRIC	RESID	COMM	RESID	RESID	COMM
scale		URBAN	NBHOOD	NBHOOD	NA	NA	NBHOOD
objective1		UP_BKGR	GEN/BKGR	POP_EXP	POP_EXP	MAX_O3	MAX_PRE
objective2			POP_EXP	UP_BKGR	SOURCE	HI_CONC	POP_EXP
objective3			UP_BKGR	GEN/BKGR	GEN/BKGR	POP_EXP	HI_CONC
cnty_landarea_2010		935	840	781	903	864	864
AADT information (distance in meters, road/highway/interstate number, AADT value)							
dist_closest		.	141	253	.	.	450
road_closest		.	69	34	.	.	287
aadt_closest		.	3100	13200	.	.	21000
dist_aadtmax		.	996	563	.	.	.
road_aadtmax		.	34	34	.	.	.
aadt_aadtmax		.	9100	16200	.	.	.
County Level NOx Emissions (tpy)							
OnRoad_mobile_nox		5173	2894	2984	3270	24824	24824
NonRoad_mobile_nox		854	627	765	624	4976	4976
AML_mobile_nox		272	58	194	258	7457	7457
total_nox		11530	4450	5774	5893	45082	45082
pct_mobile_nox		54.6	80.4	68.3	70.5	82.6	82.6
NOx Emissions by Source Type (summed tons per year (tpy), sources within 5 Km emitting at least 10 tpy)							
Portland_Cement_Manufacturing		.	.	.	.	.	.
Airport		.	.	.	.	.	30
Rail_Yard		.	.	.	.	.	510
Steel_Mill		.	.	.	.	.	.
Compressor_Station		.	.	.	.	37	70
Electricity_Generation_via_Combu		.	43	.	.	.	.
not_characterized		.	.	.	.	37	.
Bakeries		.	.	.	.	.	.
Breweries_Distilleries_Wineries		.	.	.	.	.	.
Pharmaceutical_Manufacturing		.	.	.	.	.	.
Automobile_Truck_or_Parts_Plant		.	.	.	.	.	.
		Area design value monitor					
		Near-road monitor					

Table B5-15, continued. Attributes of ambient monitors within the Dallas study area having recent (2010-2015) valid-year NO$_2$ concentrations.

State abbreviation	TX	TX	TX	TX		
County name	Tarrant	Tarrant	Tarrant	Tarrant		
Site ID	484391053	484392003	484393009	484393011		
Lat	32.66472	32.92247	32.98426	32.65636		
Lon	-97.33806	-97.28209	-97.06372	-97.08859		
year_start	2015	2010	2000	2002		
year_end	2016	2012	2016	2016		
Elevation_m	214.9	254	165	183		
land_use	RESID	AGRIC	RESID	COMM		
scale	MICRO	NA	NBHOOD	NBHOOD		
objective1	MAX_PRE	HI_CONC	HI_CONC	HI_CONC		
objective2		POP_EXP	MAX_O3	POP_EXP		
objective3		MAX_O3	POP_EXP			
cnty_landarea_2010	864	864	864	864		
AADT information (distance in meters, road/highway/interstate number, AADT value)						
dist_closest	15	653	494	47		
road_closest	20	0	2499	0		
aadt_closest	184680	9100	44000	36590		
dist_aadtmax	.	832	.	.		
road_aadtmax	.	0	.	.		
aadt_aadtmax	.	24180	.	.		
County Level NOx Emissions (tpy)						
OnRoad_mobile_nox	24824	24824	24824	24824		
NonRoad_mobile_nox	4976	4976	4976	4976		
AML_mobile_nox	7457	7457	7457	7457		
total_nox	45082	45082	45082	45082		
pct_mobile_nox	82.6	82.6	82.6	82.6		
NOx Emissions by Source Type (summed tons per year (tpy), sources within 5 Km emitting at least 10 tpy)						
Portland_Cement_Manufacturing	.	.	.	.		
Airport	.	.	.	11		
Rail_Yard	.	.	.	.		
Steel_Mill	.	.	.	.		
Compressor_Station	89	.	17	.		
Electricity_Generation_via_Combu	.	.	.	.		
not_characterized	.	.	.	.		
Bakeries	24	.	.	.		
Breweries_Distilleries_Wineries	25	.	.	.		
Pharmaceutical_Manufacturing	35	.	.	.		
Automobile_Truck_or_Parts_Plant	.	.	.	.		
		Area design value monitor				
		Near-road monitor				

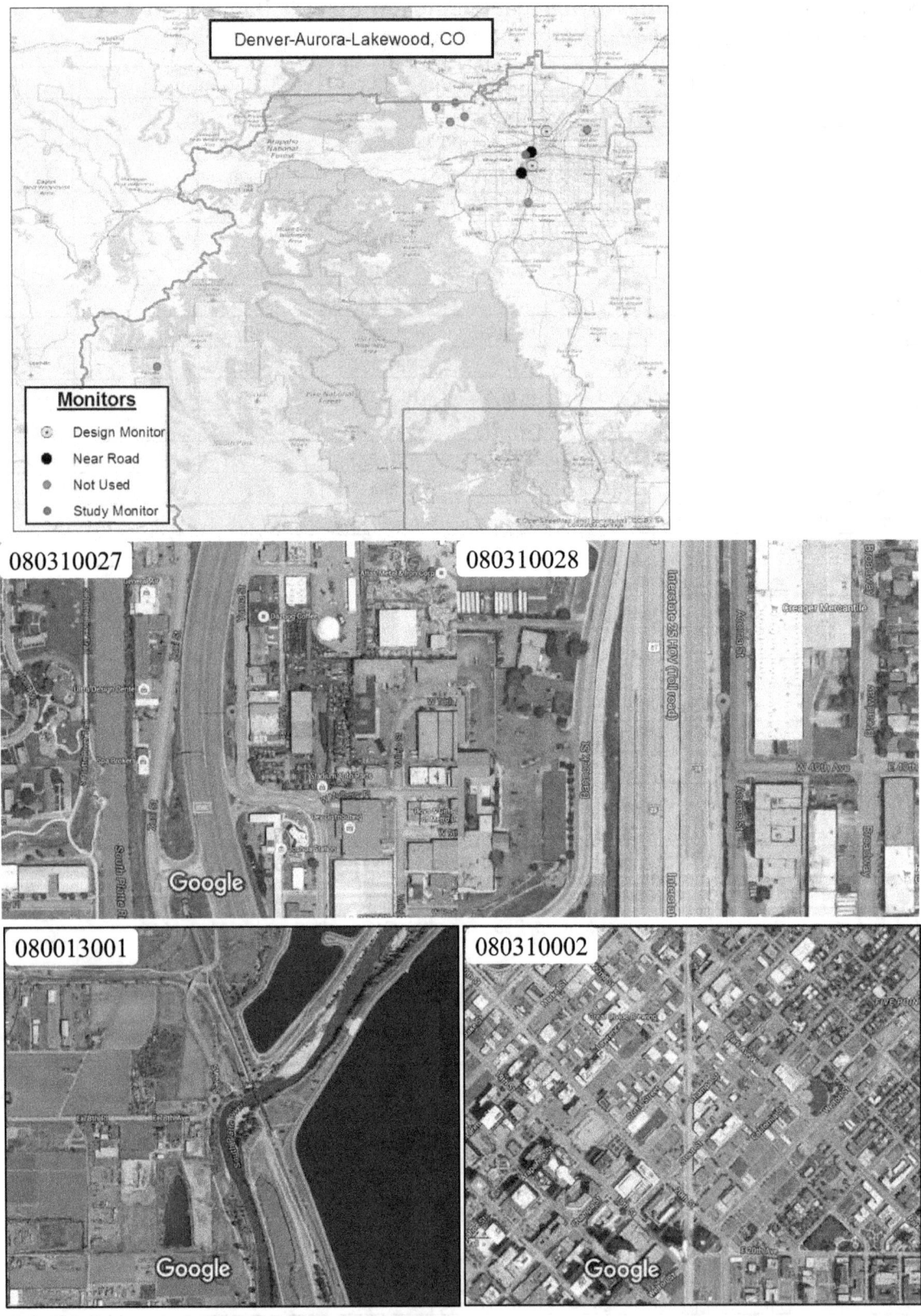

Figure B5-6. Map of all monitors (1990-2015) in the Denver study area (top panel) and satellite views of near-road (middle panels) and area design value monitor (bottom panels).

Table B5-16. Attributes of ambient monitors within the Denver study area having recent (2010-2015) valid-year NO$_2$ concentrations.

State abbreviation	CO	CO	CO	CO	CO	
County name	Adams	Denver	Denver	Denver	Denver	
Site ID	080013001	080310002	080310026	080310027	080310028	
Lat	39.83812	39.75118	39.77949	39.73217	39.78610	
Lon	-104.94984	-104.98763	-105.00518	-105.01530	-104.98860	
year_start	1975	1972	2014	2013	2015	
year_end	2016	2016	2016	2016	2016	
Elevation_m	1554	1593	1602	1583	1587	
land_use	AGRIC	COMM	RESID	COMM	COMM	
scale	URBAN	NBHOOD	NBHOOD	MICRO	MICRO	
objective1	POP_EXP	HI_CONC	POP_EXP	POP_EXP	POP_EXP	
objective2	GEN/BKGR	POP_EXP				
objective3						
cnty_landarea_2010	1168	153	153	153	153	
AADT information (distance in meters, road/highway/interstate number, AADT value)						
dist_closest	996	19	394	9	6	
road_closest	76	0	70	25	25	
aadt_closest	56000	16000	120000	249000	192000	
dist_aadtmax	.	338	399	.	627	
road_aadtmax	.	0	70	.	25	
aadt_aadtmax	.	27000	123000	.	240000	
County Level NOx Emissions (tpy)						
OnRoad_mobile_nox	8763	9618	9618	9618	9618	
NonRoad_mobile_nox	1974	2723	2723	2723	2723	
AML_mobile_nox	838	3556	3556	3556	3556	
total_nox	25245	20042	20042	20042	20042	
pct_mobile_nox	45.9	79.3	79.3	79.3	79.3	
NOx Emissions by Source Type (summed tons per year (tpy), sources within 5 Km emitting at least 10 tpy)						
Electricity_Generation_via_Combu	8996	249	9245	249	9180	
Rail_Yard	.	473	661	188	661	
Petroleum_Refinery	670	.	.	.	670	
Food_Products_Processing_Plant	11	85	96	.	96	
Dry_Cleaner___Perchloroethylene	.	54	54	54	54	
Wastewater_Treatment_Facility	66	.	.	.	66	
Hot_Mix_Asphalt_Plant	50	20	42	20	42	
Institutional__school__hospital_	.	21	11	21	11	
Petroleum_Storage_Facility	14	.	.	.	14	
Bakeries	.	12	.	12	.	
Lumber_Sawmill	.	.	18	.	18	
not_characterized	17	124	124	98	141	
Chemical_Plant	.	.	.	.	.	
Compressor_Station	.	.	.	.	.	
Landfill	.	.	.	.	.	
Brick__Structural_Clay__or_Clay	.	.	.	.	.	
		Area design value monitor				
		Near-road monitor				

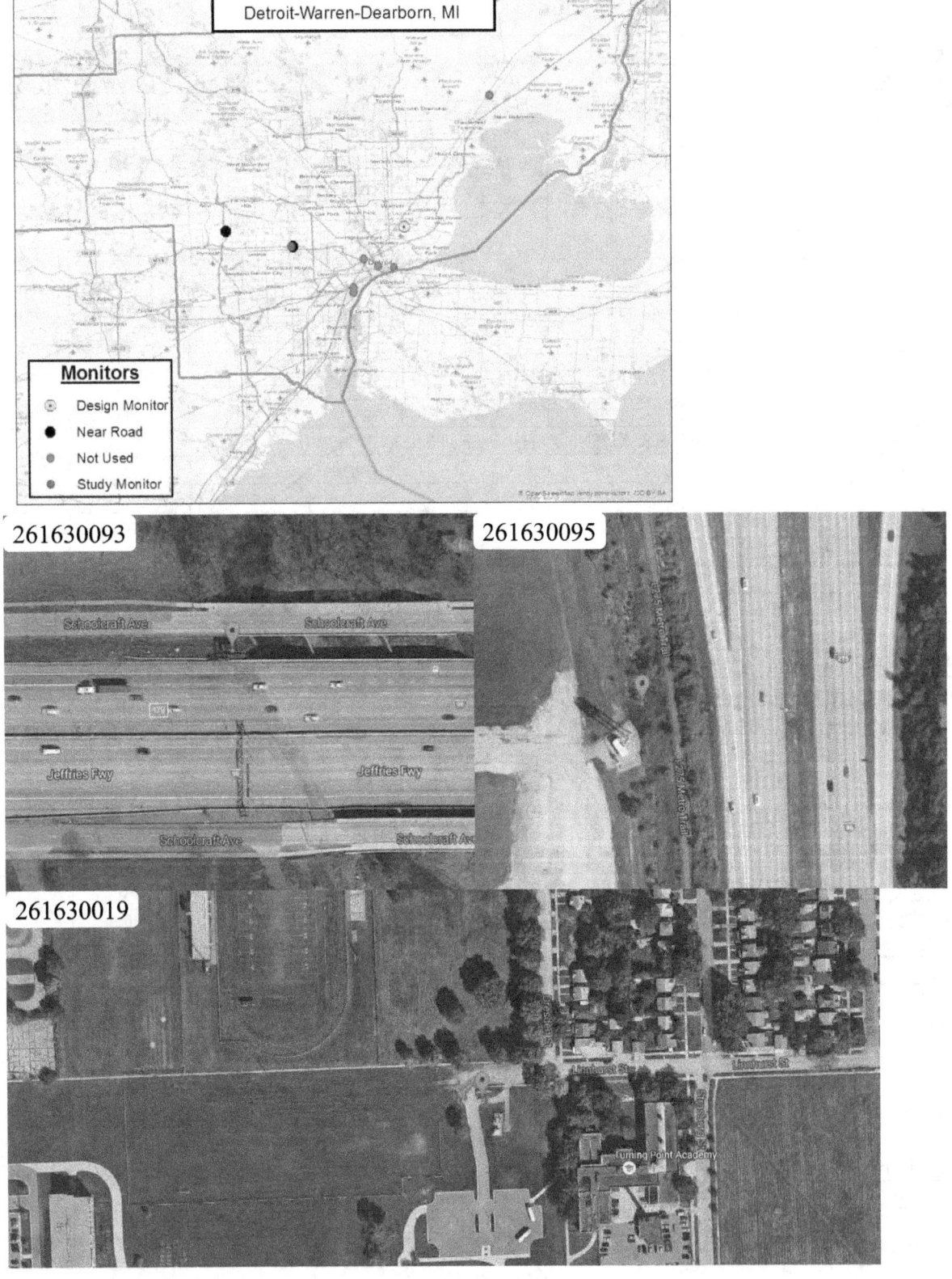

Figure B5-7. Map of all monitors (1990-2015) in the Detroit study area (top panel) and satellite views of near-road (middle panels) and area design value monitor (bottom panel)

Table B5-17. Attributes of ambient monitors within the Detroit study area having recent (2010-2015) valid-year NO₂ concentrations.

State abbreviation	MI	MI	MI	MI		
County name	Wayne	Wayne	Wayne	Wayne		
Site ID	261630019	261630093	261630094	261630095		
Lat	42.43084	42.38600	42.38681	42.42150		
Lon	-83.00014	-83.26619	-83.27051	-83.42504		
year_start	1980	2011	2011	2014		
year_end	2016	2016	2016	2016		
Elevation_m	192	0.5	0.5	0.1		
land_use	RESID	RESID	RESID	COMM		
scale	URBAN	MICRO	MID	MICRO		
objective1	POP_EXP	POP_EXP	DOWNWND	HI_CONC		
objective2	HI_CONC					
objective3	MAX_O3					
cnty_landarea_2010	612	612	612	612		
AADT information (distance in meters, road/highway/interstate number, AADT value)						
dist_closest	337	8	138	49		
road_closest	97	96	96	275		
aadt_closest	11522	140500	131100	172600		
dist_aadtmax	388	767	458	.		
road_aadtmax	0	96	96	.		
aadt_aadtmax	20000	139400	139400	.		
County Level NOx Emissions (tpy)						
OnRoad_mobile_nox	29767	29767	29767	29767		
NonRoad_mobile_nox	7051	7051	7051	7051		
AML_mobile_nox	3496	3496	3496	3496		
total_nox	62423	62423	62423	62423		
pct_mobile_nox	64.6	64.6	64.6	64.6		
NOx Emissions by Source Type (summed tons per year (tpy), sources within 5 Km emitting at least 10 tpy)						
Electricity_Generation_via_Combu	.	.	.	.		
Steel_Mill	.	.	.	.		
Municipal_Waste_Combustor	.	.	.	.		
Mineral_Processing_Plant	.	.	.	.		
Petroleum_Refinery	.	.	.	.		
Wastewater_Treatment_Facility	.	.	.	.		
Steam_Heating_Facility	.	.	.	.		
Automobile_Truck_or_Parts_Plant	95	.	.	.		
Institutional__school__hospital_	.	.	.	.		
Rail_Yard	.	.	.	16		
not_characterized	.	63	63	.		
Landfill	.	.	.	.		
		Area design value monitor				
		Near-road monitor				

B5-29

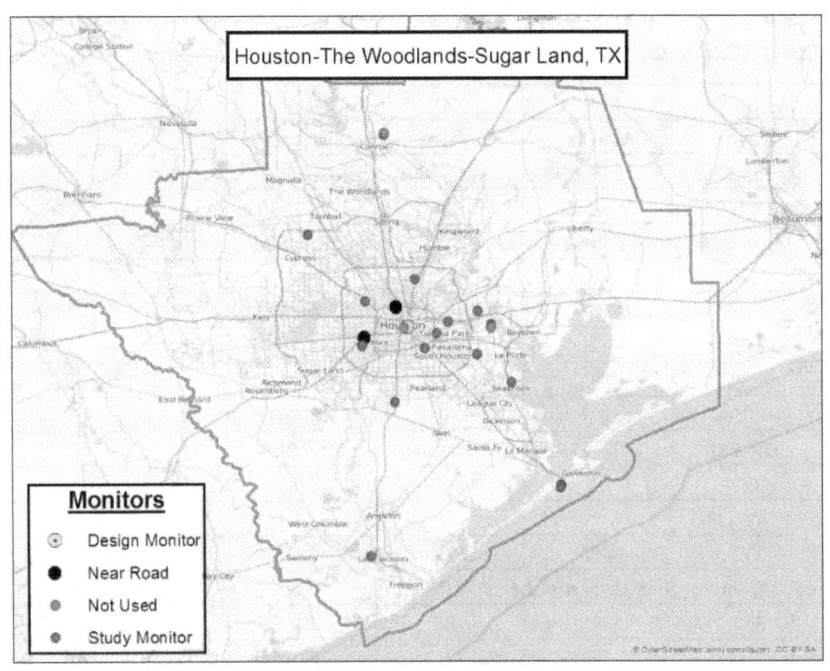

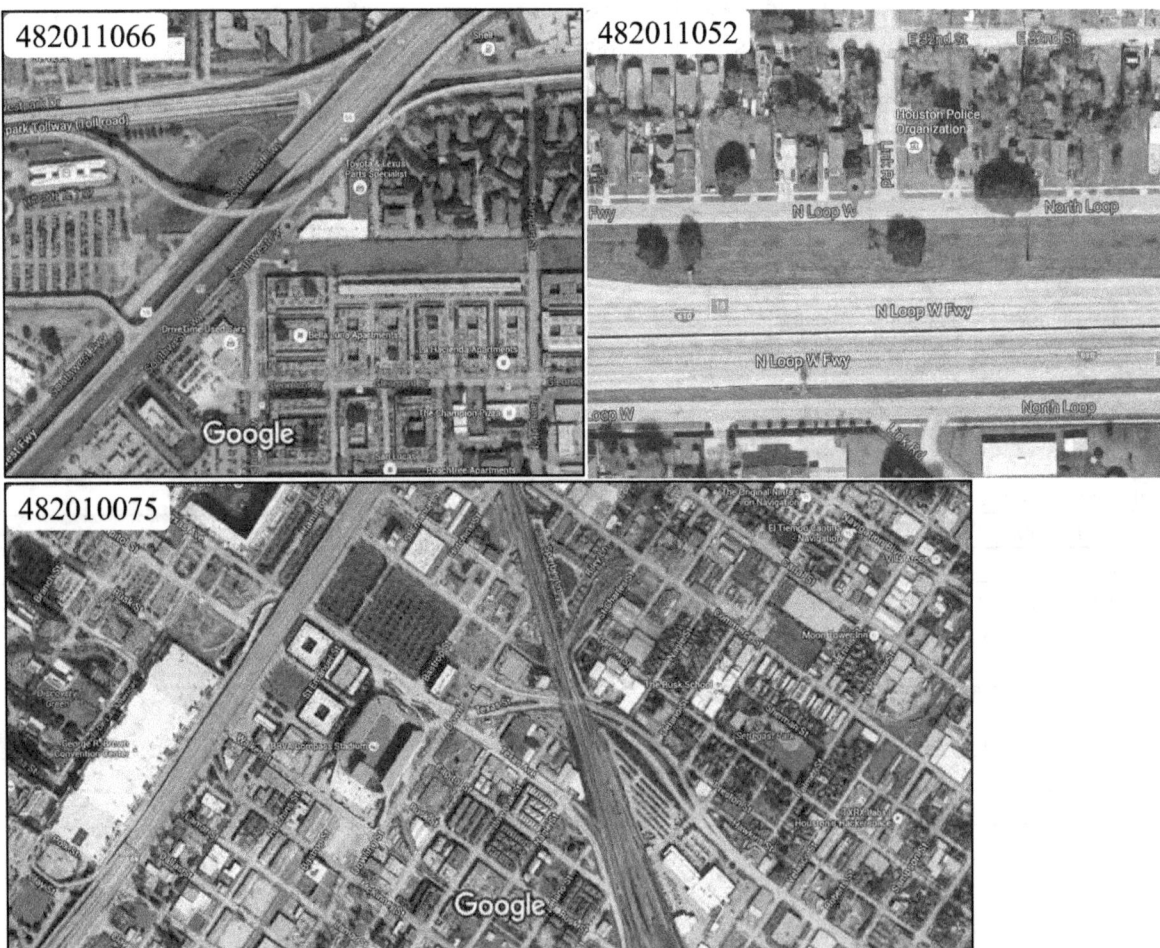

Figure B5-8. Map of all monitors (1990-2015) in the Houston study area (top panel) and satellite views of near-road (middle panels) and area design value monitor (bottom panel).

Table B5-18. Attributes of ambient monitors within the Houston study area having recent (2010-2015) valid-year NO₂ concentrations.

State abbreviation	TX	TX	TX	TX	TX	TX
County name	Brazoria	Brazoria	Galveston	Harris	Harris	Harris
Site ID	480391004	480391016	481671034	482010024	482010026	482010029
Lat	29.52044	29.04376	29.25447	29.90104	29.80271	30.03952
Lon	-95.39251	-95.47295	-94.86129	-95.32614	-95.12550	-95.67395
year_start	2001	2003	2007	1973	2001	1997
year_end	2016	2016	2016	2016	2016	2016
Elevation_m	19	0	5	24.1	11.9	50.6
land_use	RESID	INDUS	COMM	RESID	RESID	RESID
scale	NBHOOD	MID	MID	NBHOOD	MID	URBAN
objective1	POP_EXP	POP_EXP	GEN/BKGR	MAX_O3	POP_EXP	DOWNWND
objective2		SOURCE	UP_BKGR	POP_EXP	HI_CONC	HI_CONC
objective3		HI_CONC	MAX_O3	HI_CONC	MAX_PRE	UP_BKGR
cnty_landarea_2010	1358	1358	378	1703	1703	1703
AADT information (distance in meters, road/highway/interstate number, AADT value)						
dist_closest	541	112	710	608	.	.
road_closest	288	2004	3005	0	.	.
aadt_closest	56070	6900	17400	11020	.	.
dist_aadtmax	.	450	.	.	.	.
road_aadtmax	.	332	.	.	.	.
aadt_aadtmax	.	10800	.	.	.	.
County Level NOx Emissions (tpy)						
OnRoad_mobile_nox	3640	3640	2958	49330	49330	49330
NonRoad_mobile_nox	1176	1176	996	13105	13105	13105
AML_mobile_nox	1960	1960	4215	14455	14455	14455
total_nox	15272	15272	12353	98983	98983	98983
pct_mobile_nox	44.4	44.4	66.1	77.7	77.7	77.7
NOx Emissions by Source Type (summed tons per year (tpy), sources within 5 Km emitting at least 10 tpy)						
Chemical_Plant	.	.	.	.	1092	.
Petroleum_Refinery	.	.	.	.	.	.
Electricity_Generation_via_Combu	.	.	.	.	607	.
Airport	.	.	.	.	.	.
Rail_Yard	.	.	.	.	.	.
Plastic__Resin__or_Rubber_Produc	.	.	.	.	.	.
Food_Products_Processing_Plant	.	.	.	.	.	.
Wastewater_Treatment_Facility	.	.	.	.	.	.
Petroleum_Storage_Facility	.	.	.	.	.	.
Glass_Plant	.	.	.	.	.	.
Chlor_alkali_Plant	.	.	.	.	.	.
Breweries_Distilleries_Wineries	.	.	.	.	.	.
Fertilizer_Plant	.	.	.	.	.	.
Foundries__Iron_and_Steel	.	.	.	.	.	.
Landfill	.	.	.	.	.	.
not_characterized	.	.	.	.	32	.
Foundries__non_ferrous	.	.	.	.	.	.
		Area design value monitor				
		Near-road monitor				

Table B5-18, continued. Attributes of ambient monitors within the Houston study area having recent (2010-2015) valid-year NO₂ concentrations.

State abbreviation	TX	TX	TX	TX	TX	TX
County name	Harris	Harris	Harris	Harris	Harris	Harris
Site ID	482010047	482010055	482010075	482010416	482011015	482011034
Lat	29.83417	29.69573	29.75278	29.68639	29.76165	29.76800
Lon	-95.48917	-95.49922	-95.35028	-95.29472	-95.08139	-95.22058
year_start	1980	1998	2001	2006	2003	1972
year_end	2016	2016	2014	2016	2016	2016
Elevation_m	24	19.5	12	10	5.5	9.1
land_use	RESID	RESID	COMM	RESID	COMM	COMM
scale	URBAN	MID	NBHOOD	NBHOOD	MID	NBHOOD
objective1	POP_EXP	GEN/BKGR	HI_CONC	POP_EXP	SOURCE	POP_EXP
objective2		MAX_PRE	POP_EXP	GEN/BKGR	HI_CONC	GEN/BKGR
objective3		POP_EXP				HI_CONC
cnty_landarea_2010	1703	1703	1703	1703	1703	1703
AADT information (distance in meters, road/highway/interstate number, AADT value)						
dist_closest	78	582	430	400	58	258
road_closest	0	0	59	35	0	10
aadt_closest	18330	23070	190000	19000	1830	161200
dist_aadtmax	269	.	.	962	411	.
road_aadtmax	290	.	.	610	0	.
aadt_aadtmax	204930	.	.	131820	1870	.
County Level NOx Emissions (tpy)						
OnRoad_mobile_nox	49330	49330	49330	49330	49330	49330
NonRoad_mobile_nox	13105	13105	13105	13105	13105	13105
AML_mobile_nox	14455	14455	14455	14455	14455	14455
total_nox	98983	98983	98983	98983	98983	98983
pct_mobile_nox	77.7	77.7	77.7	77.7	77.7	77.7
NOx Emissions by Source Type (summed tons per year (tpy), sources within 5 Km emitting at least 10 tpy)						
Chemical_Plant	.	.	.	681	617	120
Petroleum_Refinery	.	.	.	.	.	.
Electricity_Generation_via_Combu	.	.	.	.	236	.
Airport	.	.	.	763	.	.
Rail_Yard	.	.	269	271	.	.
Plastic__Resin__or_Rubber_Produc	.	.	.	13	153	.
Food_Products_Processing_Plant	.	.	97	.	.	.
Wastewater_Treatment_Facility	.	.	75	.	.	.
Petroleum_Storage_Facility	.	.	.	.	.	37
Glass_Plant	.	.	.	.	.	42
Chlor_alkali_Plant	.	.	.	.	18	.
Breweries_Distilleries_Wineries	.	.	.	.	.	17
Fertilizer_Plant	.	.	.	.	.	12
Foundries__Iron_and_Steel	24	.	.	.	.	.
Landfill	.	.	.	.	28	.
not_characterized	.	.	.	.	71	11
Foundries__non_ferrous	.	.	.	.	.	.
		Area design value monitor				
		Near-road monitor				

Table B5-18, continued. Attributes of ambient monitors within the Houston study area having recent (2010-2015) valid-year NO₂ concentrations.

State abbreviation	TX	TX	TX	TX	TX	TX
County name	Harris	Harris	Harris	Harris	Harris	Montgomery
Site ID	482011035	482011039	482011050	482011052	482011066	483390078
Lat	29.73373	29.67003	29.58305	33.85940	29.72167	30.35030
Lon	-95.25759	-95.12851	-95.01554	-118.20070	-95.49265	-95.42513
year_start	1978	1996	2001	2015	2014	2001
year_end	2016	2016	2016	2016	2016	2016
Elevation_m	12.5	6	0	13.5	13	77
land_use	INDUS	RESID	RESID	RESID	COMM	COMM
scale	NBHOOD	NBHOOD	MID	MICRO	MICRO	URBAN
objective1	MAX_PRE	POP_EXP	POP_EXP	MAX_PRE	MAX_PRE	GEN/BKGR
objective2	POP_EXP	MAX_PRE	SOURCE			POP_EXP
objective3	HI_CONC	SOURCE	HI_CONC			REG TRANS
cnty_landarea_2010	1703	1703	1703	1703	1703	1042
AADT information (distance in meters, road/highway/interstate number, AADT value)						
dist_closest	56	603	.	15	24	.
road_closest	0	0	.	610	59	.
aadt_closest	16170	17980	.	202120	324119	.
dist_aadtmax	768	682	.	344	86	.
road_aadtmax	610	0	.	710	59	.
aadt_aadtmax	125200	31130	.	187000	293930	.
County Level NOx Emissions (tpy)						
OnRoad_mobile_nox	49330	49330	49330	49330	49330	5948
NonRoad_mobile_nox	13105	13105	13105	13105	13105	1208
AML_mobile_nox	14455	14455	14455	14455	14455	448
total_nox	98983	98983	98983	98983	98983	9429
pct_mobile_nox	77.7	77.7	77.7	77.7	77.7	80.6
NOx Emissions by Source Type (summed tons per year (tpy), sources within 5 Km emitting at least 10 tpy)						
Chemical_Plant	681	.	274	24	.	.
Petroleum_Refinery	1328	.	.	.	.	.
Electricity_Generation_via_Combu	141	210	.	14	.	.
Airport	.	.	.	.	.	.
Rail_Yard	21	.	.	.	.	.
Plastic__Resin__or_Rubber_Produc	13	.	18	.	.	.
Food_Products_Processing_Plant	.	.	.	.	.	.
Wastewater_Treatment_Facility	.	.	.	.	.	.
Petroleum_Storage_Facility	37	.	.	.	.	.
Glass_Plant	42	.	.	.	.	.
Chlor_alkali_Plant	.	.	.	.	.	.
Breweries_Distilleries_Wineries	17	.	.	.	.	.
Fertilizer_Plant	.	.	.	.	.	.
Foundries__Iron_and_Steel	.	.	.	13	.	.
Landfill	.	.	.	.	.	.
not_characterized	12	.	17	.	.	.
Foundries__non_ferrous	.	.	.	17	.	.
		Area design value monitor				
		Near-road monitor				

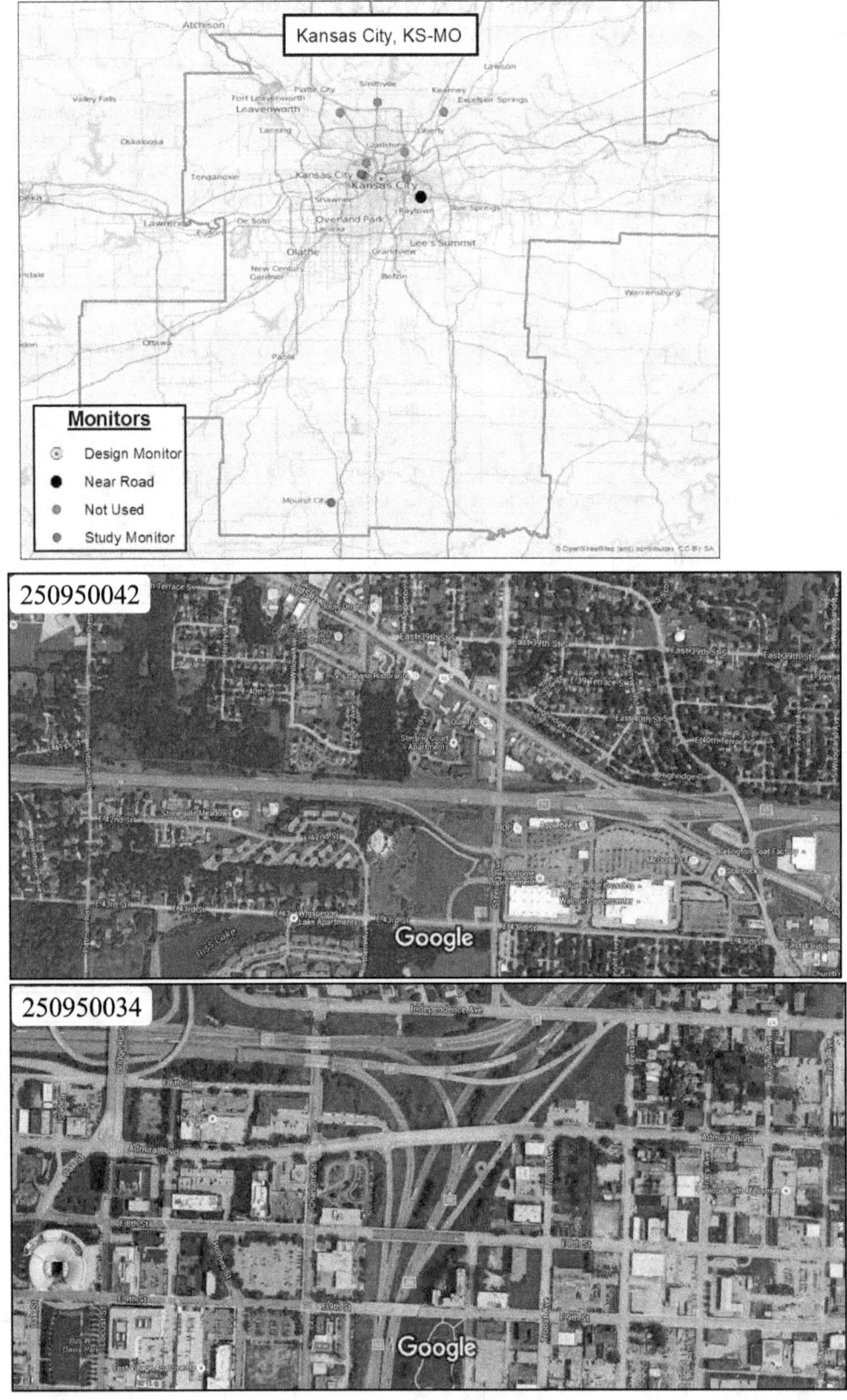

Figure B5-9. Map of all monitors (1990-2015) in the Kansas City study area (top panel) and satellite views of near-road (middle panel) and area design value monitor (bottom panel).

Table B5-19. Attributes of ambient monitors within the Kansas City study area having recent (2010-2015) valid-year NO₂ concentrations.

State abbreviation	KS	KS	MO	MO		
County name	Linn	Wyandotte	Jackson	Jackson		
Site ID	201070002	202090021	290950034	290950042		
Lat	38.13588	39.11722	39.10476	39.04791		
Lon	-94.73199	-94.63561	-94.57080	-94.45051		
year_start	1998	1999	2002	2013		
year_end	2013	2016	2016	2016		
Elevation_m	259	259	296	293		
land_use	AGRIC	RESID	COMM	MOBILE		
scale	REGION	NBHOOD	URBAN	MICRO		
objective1	REG TRANS	POP_EXP	POP_EXP	SOURCE		
objective2	POP_EXP					
objective3						
cnty_landarea_2010	594	152	604	604		
AADT information (distance in meters, road/highway/interstate number, AADT value)						
dist_closest	.	78	80	20		
road_closest	.	2540	29	70		
aadt_closest	.	7460	64982	114495		
dist_aadtmax	.	801	102	656		
road_aadtmax	.	69	70	70		
aadt_aadtmax	.	9910	91327	105483		
County Level NOx Emissions (tpy)						
OnRoad_mobile_nox	392	4241	13680	13680		
NonRoad_mobile_nox	202	449	3213	3213		
AML_mobile_nox	753	2877	3026	3026		
total_nox	10650	16058	28515	28515		
pct_mobile_nox	12.6	47.1	69.9	69.9		
NOx Emissions by Source Type (summed tons per year (tpy), sources within 5 Km emitting at least 10 tpy)						
Electricity_Generation_via_Combu	.	4392	1230	.		
Rail_Yard	.	1112	522	.		
Mineral_Wool_Plant	.	158	.	.		
Automobile_Truck_or_Parts_Plant	.	43	.	.		
Food_Products_Processing_Plant	.	21	21	.		
Chemical_Plant	.	12	.	.		
Wet_Corn_Mill	.	.	32	.		
Municipal_Waste_Combustor	.	11	.	.		
Petroleum_Storage_Facility	.	11	11	.		
not_characterized	.	.	.	.		
Wastewater_Treatment_Facility	.	.	.	.		
Airport	.	.	.	.		
Steam_Heating_Facility	.	.	.	.		
		Area design value monitor				
		Near-road monitor				

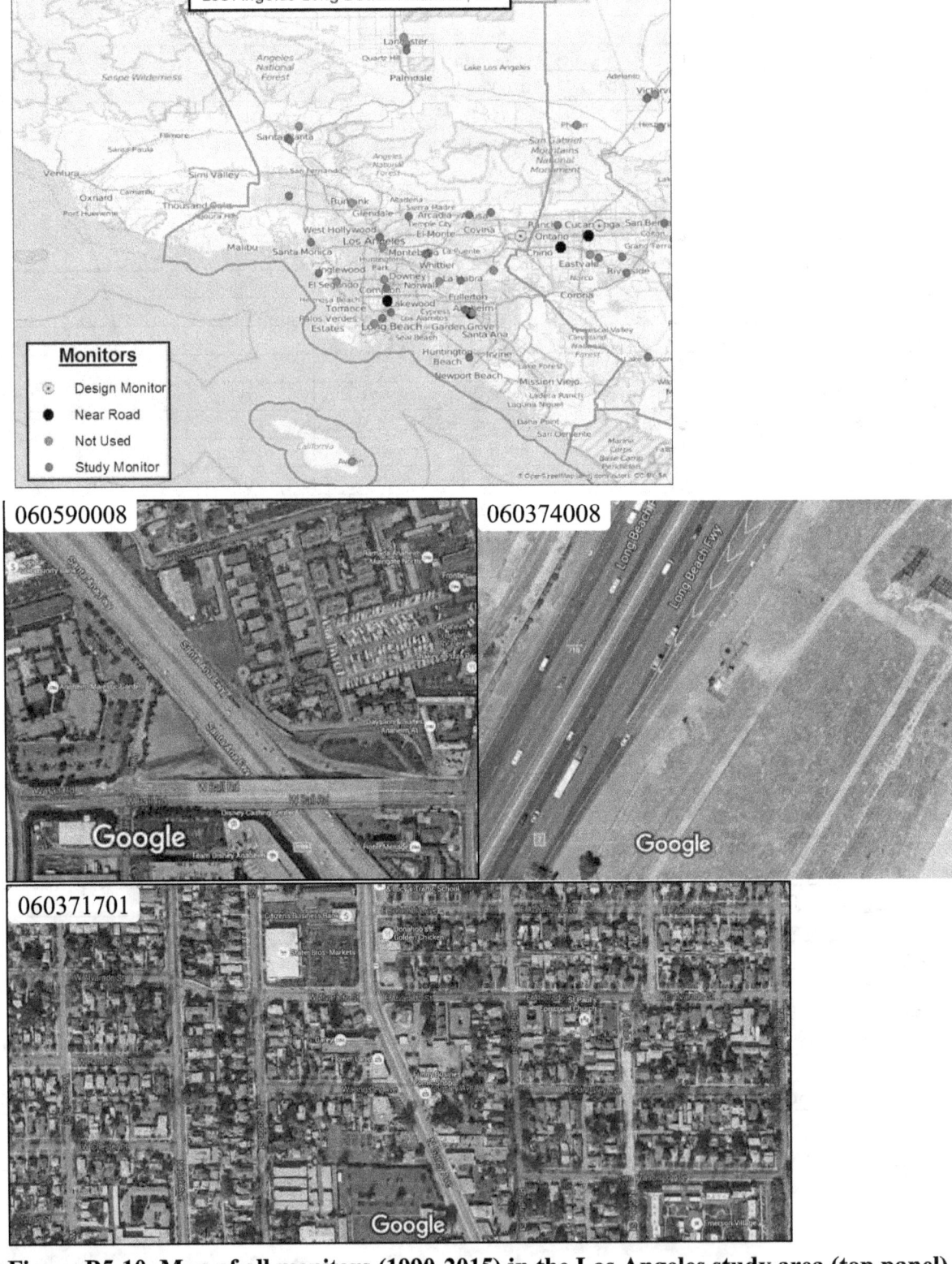

Figure B5-10. Map of all monitors (1990-2015) in the Los Angeles study area (top panel) and satellite views of near-road (middle panels) and area design value monitor (bottom panel).

Table B5-20. Attributes of ambient monitors within the Los Angeles study area having recent (2010-2015) valid-year NO$_2$ concentrations.

State abbreviation	CA	CA	CA	CA	CA	CA
County name	Los Angeles	Los Angeles	Los Angeles	Los Angeles	Los Angeles	Los Angeles
Site ID	060370002	060370016	060370113	060371103	060371201	060371302
Lat	34.13650	34.14435	34.05111	34.06659	34.19925	33.90139
Lon	-117.92391	-117.85036	-118.45636	-118.22688	-118.53276	-118.20500
year_start	1980	1979	1983	1979	1980	2008
year_end	2016	2016	2016	2016	2016	2016
Elevation_m	183	275	91	87	226	27
land_use	RESID	RESID	MOBILE	RESID	COMM	RESID
scale	URBAN	NA	NA	NBHOOD	NA	MID
objective1	UP_BKGR	POP_EXP		POP_EXP	POP_EXP	HI_CONC
objective2	POP_EXP			HI_CONC		POP_EXP
objective3	HI_CONC			GEN/BKGR		
cnty_landarea_2010	4058	4058	4058	4058	4058	4058
AADT information (distance in meters, road/highway/interstate number, AADT value)						
dist_closest	682	.	595	819	.	.
road_closest	210	.	405	110	.	.
aadt_closest	250000	.	295000	135750	.	.
dist_aadtmax	854	.	975	889	.	.
road_aadtmax	210	.	405	5	.	.
aadt_aadtmax	266000	.	313000	227000	.	.
County Level NOx Emissions (tpy)						
OnRoad_mobile_nox	80322	80322	80322	80322	80322	80322
NonRoad_mobile_nox	17796	17796	17796	17796	17796	17796
AML_mobile_nox	17817	17817	17817	17817	17817	17817
total_nox	135857	135857	135857	135857	135857	135857
pct_mobile_nox	85.3	85.3	85.3	85.3	85.3	85.3
NOx Emissions by Source Type (summed tons per year (tpy), sources within 5 Km emitting at least 10 tpy)						
Airport	.	.	.	.	15	.
Petroleum_Refinery	.	.	.	.	.	68
Rail_Yard	.	.	.	200	.	.
Calcined_Pet_Coke_Plant	.	.	.	.	.	.
Wastewater_Treatment_Facility	.	.	.	.	.	.
Electricity_Generation_via_Combu	.	.	.	.	.	14
Oil_or_Gas_Field__On_shore_	.	.	.	.	.	.
Institutional__school__hospital_	.	.	31	.	.	.
Steam_Heating_Facility	.	.	17	.	.	.
Chemical_Plant	.	.	.	.	.	13
Foundries__Iron_and_Steel	.	.	.	.	.	35
Foundries__non_ferrous	.	.	.	.	.	17
Breweries_Distilleries_Wineries	15	.	.	.	.	.
Hot_Mix_Asphalt_Plant	.	.	.	.	.	12
Petroleum_Storage_Facility	.	.	.	.	.	.
Landfill	10	.	.	.	.	.
Food_Products_Processing_Plant	.	.	.	.	.	.
Glass_Plant	.	.	.	.	.	.
Secondary_Lead_Smelting_Plant	.	.	.	.	.	.
not_characterized	.	.	12	15	.	.
Municipal_Waste_Combustor	.	.	.	.	.	.
Aircraft__Aerospace__or_Related	.	.	.	.	.	.
		Area design value monitor				
		Near-road monitor				

Table B5-20, continued. Attributes of ambient monitors within the Los Angeles study area having recent (2010-2015) valid-year NO₂ concentrations.

State abbreviation	CA	CA	CA	CA	CA	CA
County name	Los Angeles	Los Angeles	Los Angeles	Los Angeles	Los Angeles	Los Angeles
Site ID	060371602	060371701	060372005	060374002	060374006	060374008
Lat	34.01194	34.06703	34.13260	33.82376	33.80250	33.85944
Lon	-118.06995	-117.75140	-118.12720	-118.18921	-118.22000	-118.20028
year_start	2005	1980	1981	1980	2009	2015
year_end	2016	2016	2016	2013	2016	2016
Elevation_m	75	270	250	6	10	12
land_use	COMM	COMM	RESID	RESID	INDUS	INDUS
scale	NA	NA	NA	NA	URBAN	MICRO
objective1	MAX_PRE		POP_EXP	POP_EXP	SOURCE	POP_EXP
objective2	POP_EXP			HI_CONC	POP_EXP	SOURCE
objective3						
cnty_landarea_2010	4058	4058	4058	4058	4058	4058
AADT information (distance in meters, road/highway/interstate number, AADT value)						
dist_closest	937	668	.	490	.	9
road_closest	164	10	.	405	.	710
aadt_closest	38000	229000	.	280000	.	192000
dist_aadtmax	969	709	.	960	.	230
road_aadtmax	605	10	.	405	.	710
aadt_aadtmax	255000	239000	.	287000	.	187000
County Level NOx Emissions (tpy)						
OnRoad_mobile_nox	80322	80322	80322	80322	80322	80322
NonRoad_mobile_nox	17796	17796	17796	17796	17796	17796
AML_mobile_nox	17817	17817	17817	17817	17817	17817
total_nox	135857	135857	135857	135857	135857	135857
pct_mobile_nox	85.3	85.3	85.3	85.3	85.3	85.3
NOx Emissions by Source Type (summed tons per year (tpy), sources within 5 Km emitting at least 10 tpy)						
Airport	.	.	.	215	.	.
Petroleum_Refinery	.	.	.	.	1775	.
Rail_Yard	.	.	.	.	22	.
Calcined_Pet_Coke_Plant	.	.	.	.	201	.
Wastewater_Treatment_Facility	.	.	.	.	.	.
Electricity_Generation_via_Combu	.	14	20	.	33	14
Oil_or_Gas_Field__On_shore_	.	.	.	.	.	.
Institutional__school__hospital_	.	.	.	.	.	.
Steam_Heating_Facility	.	.	.	.	.	.
Chemical_Plant	.	.	.	74	74	24
Foundries__Iron_and_Steel	.	.	.	.	.	13
Foundries__non_ferrous	.	.	.	.	.	17
Breweries_Distilleries_Wineries	.	.	.	.	.	.
Hot_Mix_Asphalt_Plant	.	.	.	.	.	.
Petroleum_Storage_Facility	.	.	.	.	11	.
Landfill	.	.	.	.	.	.
Food_Products_Processing_Plant	.	.	.	.	.	.
Glass_Plant	.	.	.	.	.	.
Secondary_Lead_Smelting_Plant	.	.	.	.	.	.
not_characterized	.	.	.	24	65	.
Municipal_Waste_Combustor	.	.	.	.	.	.
Aircraft__Aerospace__or_Related	.	.	.	.	.	.
		Area design value monitor				
		Near-road monitor				

Table B5-20, continued. Attributes of ambient monitors within the Los Angeles study area having recent (2010-2015) valid-year NO$_2$ concentrations.

State abbreviation	CA	CA	CA	CA	CA	CA	CA
County name	Los Angeles	Los Angeles	Los Angeles	Orange	Orange	Orange	Orange
Site ID	060375005	060376012	060379033	060590007	060590008	060591003	060595001
Lat	33.95080	34.38344	34.67139	33.83062	33.81931	33.67464	33.92513
Lon	-118.43043	-118.52840	-118.13146	-117.93845	-117.91876	-117.92568	-117.95264
year_start	2004	2001	2001	2001	2013	1989	1976
year_end	2016	2016	2016	2016	2016	2016	2016
Elevation_m	21	397	725	10	44	0	82
land_use	RESID	COMM	COMM	RESID	MOBILE	RESID	RESID
scale	NBHOOD	NA	MID	URBAN	MICRO	MID	NA
objective1	UP_BKGR	POP_EXP	POP_EXP	POP_EXP	SOURCE	POP_EXP	POP_EXP
objective2							
objective3							
cnty_landarea_2010	4058	4058	4058	791	791	791	791
AADT information (distance in meters, road/highway/interstate number, AADT value)							
dist_closest	.	.	.	432	9	.	819
road_closest	.	.	.	5	5	.	90
aadt_closest	.	.	.	256000	272000	.	46000
dist_aadtmax	.	.	.	735	.	.	999
road_aadtmax	.	.	.	5	.	.	90
aadt_aadtmax	.	.	.	267000	.	.	47000
County Level NOx Emissions (tpy)							
OnRoad_mobile_nox	80322	80322	80322	19742	19742	19742	19742
NonRoad_mobile_nox	17796	17796	17796	6422	6422	6422	6422
AML_mobile_nox	17817	17817	17817	1921	1921	1921	1921
total_nox	135857	135857	135857	31763	31763	31763	31763
pct_mobile_nox	85.3	85.3	85.3	88.4	88.4	88.4	88.4
NOx Emissions by Source Type (summed tons per year (tpy), sources within 5 Km emitting at least 10 tpy)							
Airport	5533	.	.	.	.	.	.
Petroleum_Refinery	.	.	.	.	.	.	.
Rail_Yard	.	.	.	.	.	.	.
Calcined_Pet_Coke_Plant	.	.	.	.	.	.	.
Wastewater_Treatment_Facility	.	.	.	.	.	53	.
Electricity_Generation_via_Combu	63	.	.	.	.	.	.
Oil_or_Gas_Field__On_shore_	21	66	.	.	.	.	.
Institutional__school__hospital_	.	.	.	.	.	.	.
Steam_Heating_Facility	.	.	.	.	.	.	.
Chemical_Plant	.	.	.	.	.	.	.
Foundries__Iron_and_Steel	.	.	.	.	.	.	.
Foundries__non_ferrous	.	.	.	.	.	.	.
Breweries_Distilleries_Wineries	.	.	.	.	.	.	.
Hot_Mix_Asphalt_Plant	.	.	.	.	.	.	.
Petroleum_Storage_Facility	.	.	.	.	.	.	.
Landfill	.	.	.	.	.	.	.
Food_Products_Processing_Plant	.	.	.	.	.	.	.
Glass_Plant	.	.	.	.	.	.	.
Secondary_Lead_Smelting_Plant	.	.	.	.	.	.	.
not_characterized	30	.	.	24	24	.	.
Municipal_Waste_Combustor	.	.	.	.	.	.	.
Aircraft__Aerospace__or_Related	.	.	.	.	.	.	.

		Area design value monitor
		Near-road monitor

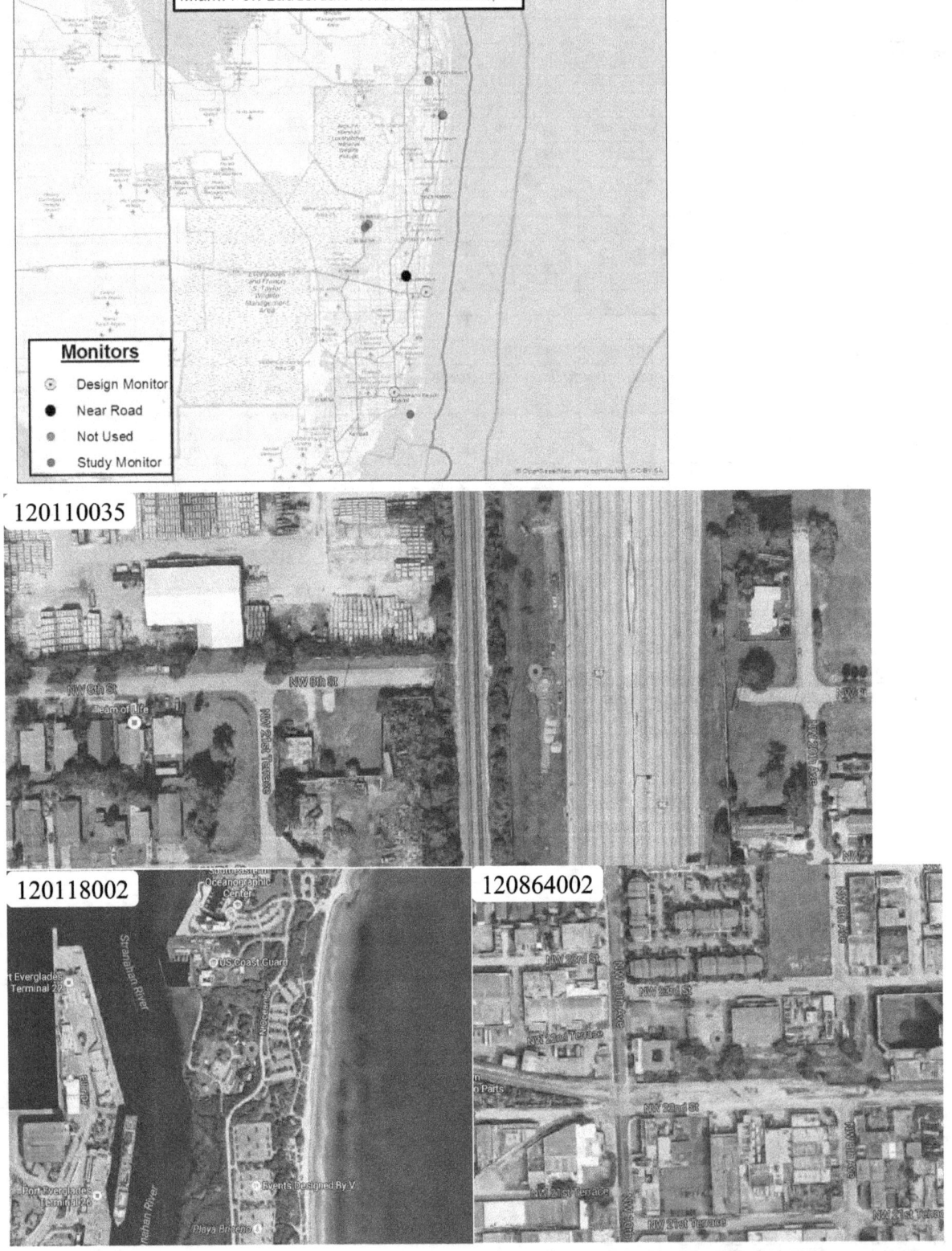

Figure B5-11. Map of all monitors (1990-2015) in the Miami study area (top panel) and satellite views of near-road (middle panels) and area design value monitor (bottom panels).

Table B5-21. Attributes of ambient monitors within the Miami study area having recent (2010-2015) valid-year NO$_2$ concentrations.

State abbreviation	FL	FL	FL	FL	FL	FL
County name	Broward	Broward	Broward	Miami-Dade	Miami-Dade	Palm Beach
Site ID	120110031	120110035	120118002	120860027	120864002	120990020
Lat	26.27236	26.13268	26.08842	25.73338	25.79871	26.59123
Lon	-80.29477	-80.16982	-80.11119	-80.16181	-80.21005	-80.06087
year_start	1997	2015	1990	1984	1977	2008
year_end	2012	2016	2016	2016	2016	2014
Elevation_m	3	3	3	2	5	5
land_use	RESID	RESID	RESID	RESID	COMM	RESID
scale	URBAN	URBAN	URBAN	NBHOOD	NBHOOD	NBHOOD
objective1	HI_CONC	POP_EXP	POP_EXP	POP_EXP	HI_CONC	HI_CONC
objective2	MAX_PRE		HI_CONC	UP_BKGR	POP_EXP	POP_EXP
objective3						
cnty_landarea_2010	1210	1210	1210	1898	1898	1970
AADT information (distance in meters, road/highway/interstate number, AADT value)						
dist_closest	103	30	.	19	475	777
road_closest	0	95	.	0	95	95
aadt_closest	17600	306000	.	30500	225000	204500
dist_aadtmax	145	.	.	52	.	887
road_aadtmax	869	.	.	0	.	95
aadt_aadtmax	56000	.	.	37500	.	224500
County Level NOx Emissions (tpy)						
OnRoad_mobile_nox	22197	22197	22197	27024	27024	16775
NonRoad_mobile_nox	5885	5885	5885	8578	8578	7625
AML_mobile_nox	8072	8072	8072	14830	14830	4059
total_nox	43997	43997	43997	57941	57941	35647
pct_mobile_nox	82.2	82.2	82.2	87	87	79.8
NOx Emissions by Source Type (summed tons per year (tpy), sources within 5 Km emitting at least 10 tpy)						
Airport	.	.	1455	.	.	.
Electricity_Generation_via_Combu	.	.	837	.	.	18
Institutional__school__hospital_	.	.	.	.	10	.
Petroleum_Storage_Facility	.	.	13	.	.	.
Wastewater_Treatment_Facility	.	.	.	48	.	.
		Area design value monitor				
		Near-road monitor				

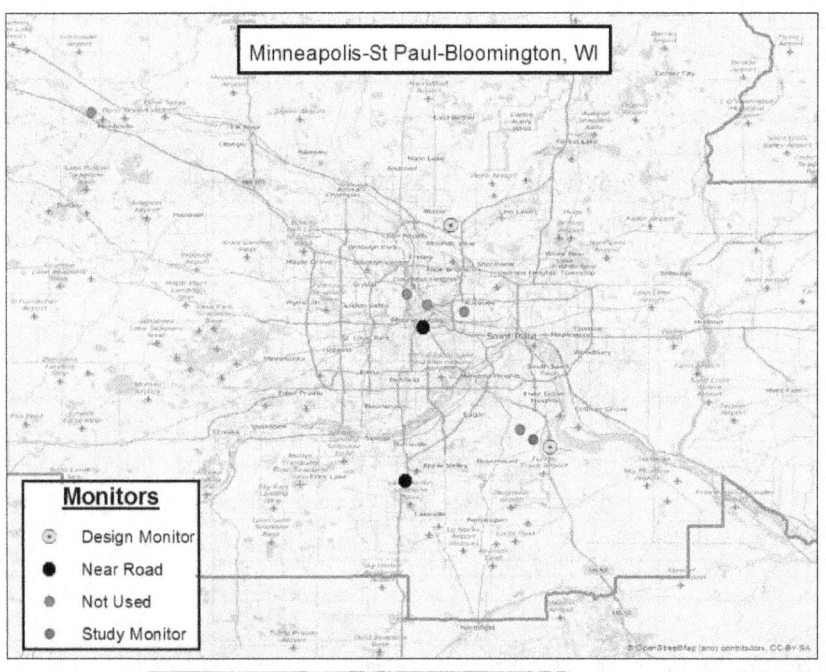

Figure B5-12. Map of all monitors (1990-2015) in the Minneapolis study area (top panel) and satellite views of near-road (middle panels) and area design value monitor (bottom panels).

Table B5-22. Attributes of ambient monitors within the Minneapolis study area having recent (2010-2015) valid-year NO$_2$ concentrations.

State abbreviation	MN	MN	MN	MN	MN	
County name	Anoka	Dakota	Dakota	Dakota	Hennepin	
Site ID	270031002	270370020	270370423	270370480	270530962	
Lat	45.13768	44.76323	44.77553	44.70612	44.96524	
Lon	-93.20761	-93.03255	-93.06299	-93.28580	-93.25476	
year_start	1979	1974	1993	2014	2013	
year_end	2016	2016	2016	2016	2016	
Elevation_m	280	288	272	312	259	
land_use	COMM	INDUS	INDUS	COMM	RESID	
scale	URBAN	NBHOOD	NBHOOD	MID	MID	
objective1	GEN/BKGR	POP_EXP	SOURCE	SOURCE	SOURCE	
objective2	HI_CONC	SOURCE	WELF IMP			
objective3	POP_EXP	WELF IMP	POP_EXP			
cnty_landarea_2010	423	562	562	562	554	
AADT information (distance in meters, road/highway/interstate number, AADT value)						
dist_closest	.	81	.	30	32	
road_closest	.	52	.	35	94	
aadt_closest	.	34741	.	87000	277000	
dist_aadtmax	.	231	.	.	.	
road_aadtmax	.	52	.	.	.	
aadt_aadtmax	.	45818	.	.	.	
County Level NOx Emissions (tpy)						
OnRoad_mobile_nox	6662	7604	7604	7604	21967	
NonRoad_mobile_nox	1507	2061	2061	2061	5495	
AML_mobile_nox	1183	295	295	295	3168	
total_nox	11073	19738	19738	19738	39010	
pct_mobile_nox	84.5	50.5	50.5	50.5	78.5	
NOx Emissions by Source Type (summed tons per year (tpy), sources within 5 Km emitting at least 10 tpy)						
Petroleum_Refinery	.	1296	1296	.	.	
Municipal_Waste_Combustor	.	.	.	.	594	
Electricity_Generation_via_Combu	.	87	87	.	332	
Institutional__school__hospital_	.	.	.	.	174	
Rail_Yard	.	.	.	.	111	
Pulp_and_Paper_Plant	.	.	.	.	339	
Gas_Plant	.	.	63	.	.	
Secondary_Aluminum_Smelting_Refi	.	29	29	.	.	
Steam_Heating_Facility	.	.	.	.	15	
not_characterized	133	49	49	.	.	
Printing_Publishing_Facility	.	.	.	.	.	
Food_Products_Processing_Plant	.	.	.	.	.	
		Area design value monitor				
		Near-road monitor				

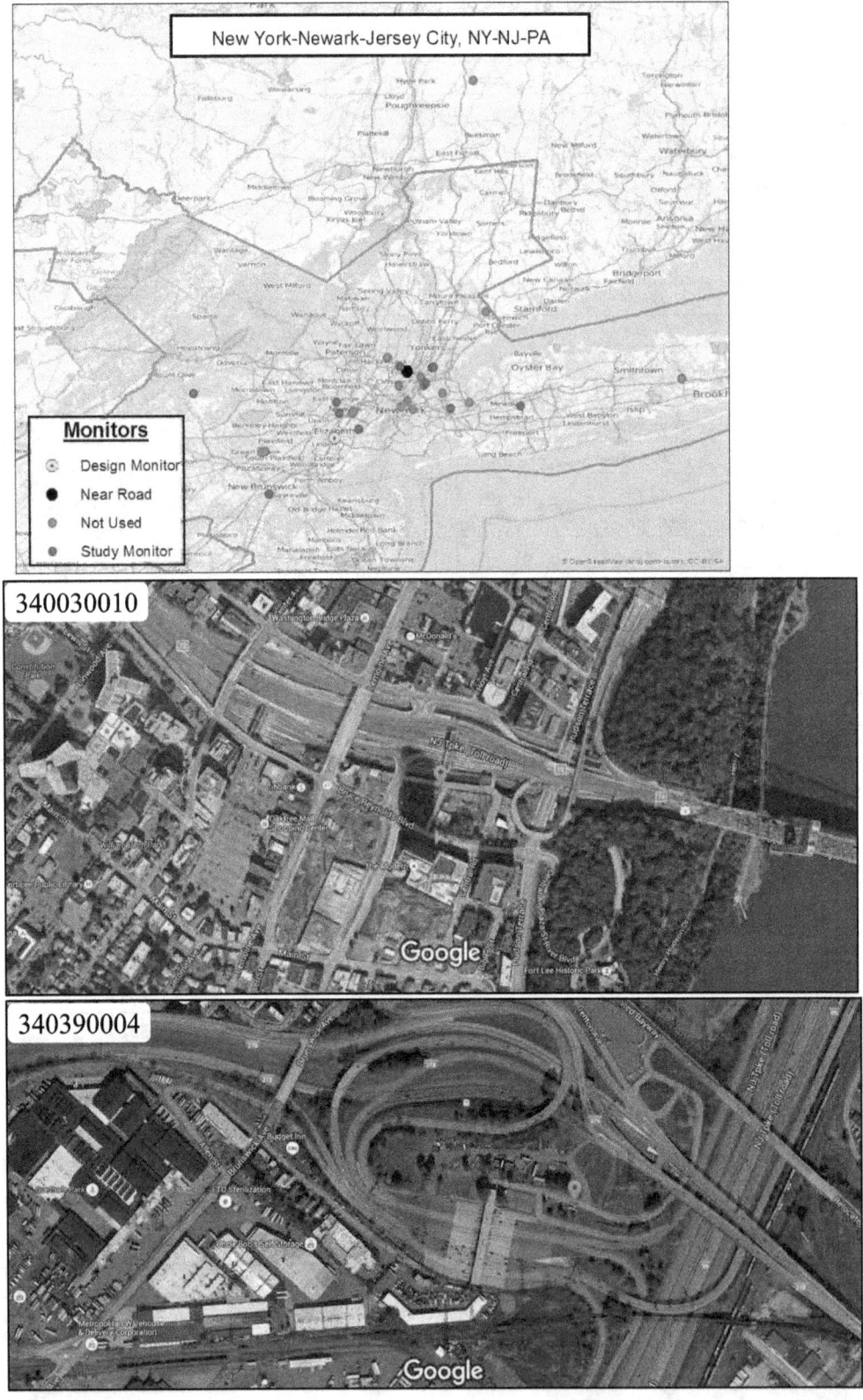

Figure B5-13. Map of all monitors (1990-2015) in the New York/Jersey study area (top panel) and satellite views of near-road (middle panel) and area design value monitor (bottom panel).

Table B5-23. Attributes of ambient monitors within the New York/Jersey study area having recent (2010-2015) valid-year NO₂ concentrations.

State abbreviation	NJ	NJ	NJ	NJ	NJ	NJ
County name	Bergen	Bergen	Essex	Essex	Hudson	Middlesex
Site ID	340030006	340030010	340130003	340131003	340170006	340230011
Lat	40.87044	40.85358	40.72099	40.75750	40.67025	40.46218
Lon	-73.99199	-73.96621	-74.19289	-74.20050	-74.12608	-74.42944
year_start	2007	2014	2011	1980	1983	1994
year_end	2010	2016	2016	2016	2016	2016
Elevation_m	1	87	27	48	3	19
land_use	RESID	COMM	RESID	COMM	COMM	AGRIC
scale	NBHOOD	MICRO	NBHOOD	NBHOOD	URBAN	NBHOOD
objective1	POP_EXP	POP_EXP	POP_EXP	HI_CONC	POP_EXP	POP_EXP
objective2				POP_EXP		UP_BKGR
objective3						GEN/BKGR
cnty_landarea_2010	233	233	126	126	46	309
AADT information (distance in meters, road/highway/interstate number, AADT value)						
dist_closest	332	20	613	70	443	805
road_closest	93	95	27	1203	501	1
aadt_closest	21842	311234	16050	16032	10308	76226
dist_aadtmax	432	.	861	665	.	.
road_aadtmax	95	.	21	444	.	.
aadt_aadtmax	161399	.	97471	201150	.	.
County Level NOx Emissions (tpy)						
OnRoad_mobile_nox	8080	8080	5162	5162	2685	7951
NonRoad_mobile_nox	3313	3313	2070	2070	1652	2922
AML_mobile_nox	1070	1070	3355	3355	3040	805
total_nox	15763	15763	14172	14172	10059	15574
pct_mobile_nox	79.1	79.1	74.7	74.7	73.3	75
NOx Emissions by Source Type (summed tons per year (tpy), sources within 5 Km emitting at least 10 tpy)						
Airport	.	.	2984	.	2984	.
Electricity_Generation_via_Combu	450	.	.	.	53	154
Municipal_Waste_Combustor	.	.	.	.	.	.
Petroleum_Refinery	.	.	.	.	.	.
Institutional__school__hospital_	.	86	74	74	.	.
Wastewater_Treatment_Facility	.	346	.	.	14	.
Breweries_Distilleries_Wineries	.	.	68	.	.	.
Petroleum_Storage_Facility	.	.	.	.	14	.
Pharmaceutical_Manufacturing	.	.	.	.	.	22
Fabricated_Metal_Products_Plant	.	.	.	.	10	.
Compressor_Station	.	.	.	.	.	.
Steam_Heating_Facility	.	.	.	.	.	.
not_characterized	.	36	13	.	59	17

	Area design value monitor
	Near-road monitor

Table B5-23, continued. Attributes of ambient monitors in the New York/Jersey study area having recent (2010-2015) valid-year NO₂ concentrations.

State abbreviation	NJ	NJ	NY	NY	NY	NY
County name	Morris	Union	Bronx	Bronx	Nassau	Queens
Site ID	340273001	340390004	360050110	360050133	360590005	360810124
Lat	40.78763	40.64144	40.81600	40.86790	40.74316	40.73614
Lon	-74.67630	-74.20837	-73.90200	-73.87809	-73.58549	-73.82153
year_start	1982	1979	1999	2006	1980	2001
year_end	2016	2016	2016	2016	2010	2016
Elevation_m	278	5	17	31	27	25
land_use	AGRIC	INDUS	RESID	COMM	COMM	COMM
scale	NA	NBHOOD	URBAN	URBAN	NBHOOD	NBHOOD
objective1	GEN/BKGR	POP_EXP	GEN/BKGR	POP_EXP	POP_EXP	POP_EXP
objective2	POP_EXP	HI_CONC	POP_EXP	GEN/BKGR	GEN/BKGR	GEN/BKGR
objective3			QUAL ASSUR	QUAL ASSUR	HI_CONC	
cnty_landarea_2010	460	103	42	42	285	109
AADT information (distance in meters, road/highway/interstate number, AADT value)						
dist_closest	.	89	81	74	733	280
road_closest	.	278	0	0	0	0
aadt_closest	.	95512	5300	30500	114700	17100
dist_aadtmax	.	240	464	515	854	453
road_aadtmax	.	95	278	0	0	495
aadt_aadtmax	.	242175	99100	112700	122200	166300
County Level NOx Emissions (tpy)						
OnRoad_mobile_nox	5273	4874	4297	4297	13279	11095
NonRoad_mobile_nox	1877	1346	1767	1767	3205	4060
AML_mobile_nox	232	3479	708	708	706	6546
total_nox	9111	13636	9912	9912	23967	29220
pct_mobile_nox	81	71.1	68.3	68.3	71.7	74.3
NOx Emissions by Source Type (summed tons per year (tpy), sources within 5 Km emitting at least 10 tpy)						
Airport	.	.	.	.	.	.
Electricity_Generation_via_Combu	.	525	1048	.	271	.
Municipal_Waste_Combustor	.	.	.	.	981	.
Petroleum_Refinery	.	927	.	.	.	.
Institutional__school__hospital_	.	.	185	249	113	13
Wastewater_Treatment_Facility	.	36	346	.	.	.
Breweries_Distilleries_Wineries	.	.	.	.	.	.
Petroleum_Storage_Facility	.	.	.	.	.	.
Pharmaceutical_Manufacturing	.	.	.	.	.	.
Fabricated_Metal_Products_Plant	.	.	.	.	.	.
Compressor_Station	.	.	11	.	.	.
Steam_Heating_Facility	.	.	.	.	.	.
not_characterized	.	23	266	232	.	207

		Area design value monitor
		Near-road monitor

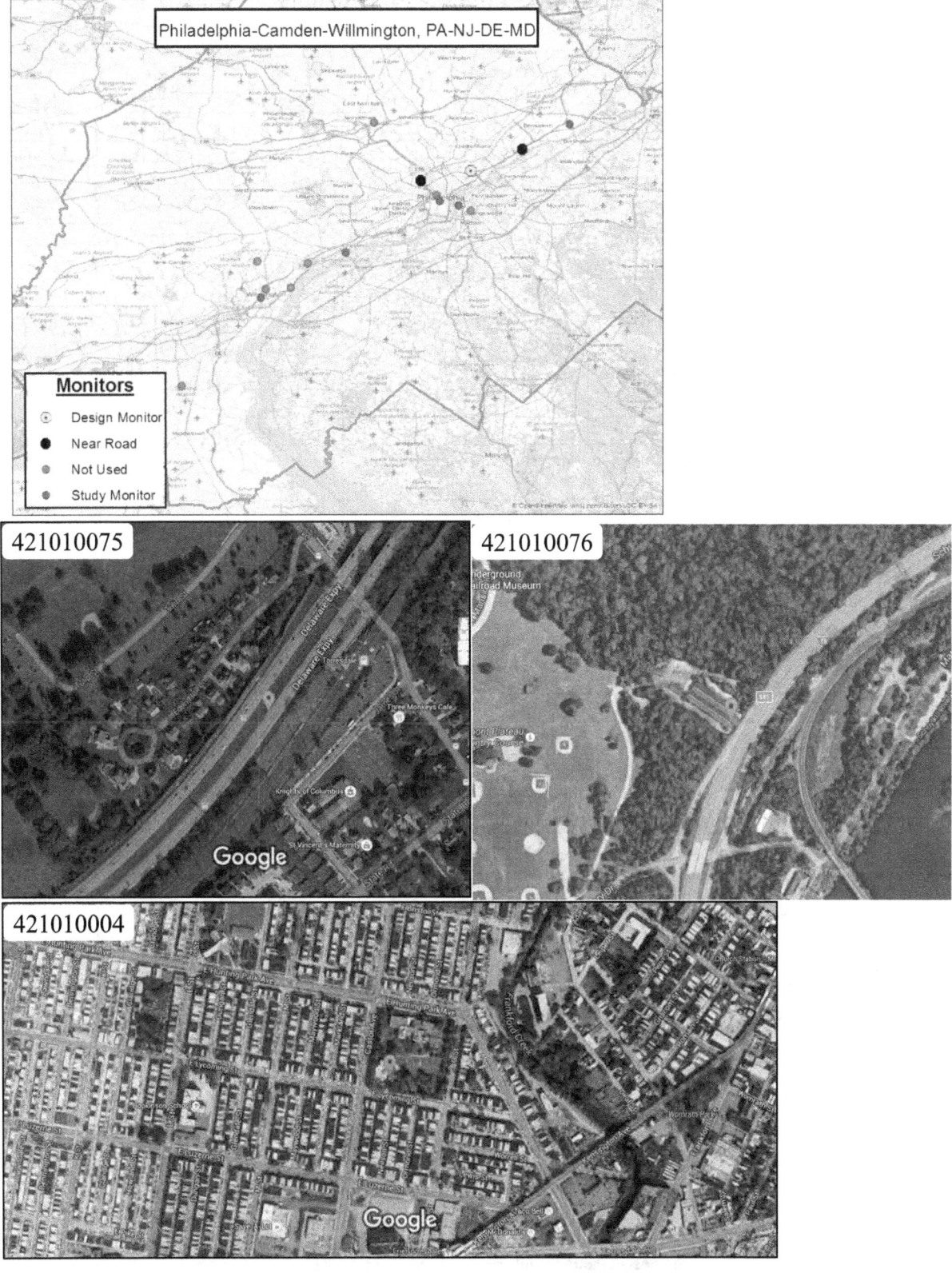

Figure B5-14. Map of all monitors (1990-2015) in the Philadelphia study area (top panel) and satellite views of near-road (middle panels) and area design value monitor (bottom panel).

Table B5-24. Attributes of ambient monitors within the Philadelphia study area having recent (2010-2015) valid-year NO₂ concentrations.

State abbreviation	DE	NJ	PA	PA	PA	PA
County name	New Castle	Camden	Bucks	Delaware	Philadelphia	Philadelphia
Site ID	100032004	340070002	420170012	420450002	421010004	421010047
Lat	39.73944	39.93445	40.10722	39.83556	40.00889	39.94465
Lon	-75.55806	-75.12529	-74.88222	-75.37250	-75.09778	-75.16521
year_start	2000	2012	1973	1973	1976	1981
year_end	2016	2016	2015	2016	2016	2015
Elevation_m	0	4	12	3	22	21
land_use	COMM	INDUS	RESID	INDUS	RESID	RESID
scale	NBHOOD	NBHOOD	NBHOOD	NBHOOD	URBAN	NBHOOD
objective1	POP_EXP	POP_EXP	POP_EXP	POP_EXP	POP_EXP	POP_EXP
objective2	HI_CONC			HI_CONC	HI_CONC	HI_CONC
objective3					QUAL ASSUR	
cnty_landarea_2010	426	221	604	184	134	134
AADT information (distance in meters, road/highway/interstate number, AADT value)						
dist_closest	67	150	389	378	322	2
road_closest	0	1630	413	291	0	291
aadt_closest	18588	9932	31120	14988	11437	20495
dist_aadtmax	612	994	958	813	998	923
road_aadtmax	95	676	95	322	0	611
aadt_aadtmax	102035	59788	76480	38843	18485	24562
County Level NOx Emissions (tpy)						
OnRoad_mobile_nox	6459	5353	7680	5643	10201	10201
NonRoad_mobile_nox	1748	1150	2196	1319	2480	2480
AML_mobile_nox	1805	1088	220	3600	2160	2160
total_nox	13991	9431	12925	17306	21065	21065
pct_mobile_nox	71.6	80.5	78.1	61	70.5	70.5
NOx Emissions by Source Type (summed tons per year (tpy), sources within 5 Km emitting at least 10 tpy)						
Petroleum_Refinery	.	.	.	2146	.	1315
Municipal_Waste_Combustor	20	297	.	1260	.	.
Electricity_Generation_via_Combu	948	26	107	950	.	379
Chemical_Plant	27	14	.	275	183	14
Pulp_and_Paper_Plant	.	.	.	240	92	.
Institutional__school__hospital_	.	.	.	.	30	66
Wastewater_Treatment_Facility	.	14	.	56	.	14
Steam_Heating_Facility	.	24	.	.	.	24
Petroleum_Storage_Facility	.	.	.	.	16	.
Military_Base	.	.	.	.	12	.
Automobile_Truck_or_Parts_Plant	11	.	.	.	.	.
not_characterized	228	13	26	11	.	.
Steel_Mill	.	.	.	.	.	.
Hot_Mix_Asphalt_Plant	.	.	.	.	.	.
Pharmaceutical_Manufacturing	.	.	.	.	.	.
		Area design value monitor				
		Near-road monitor				

Table B5-24, continued. Attributes of ambient monitors within the Philadelphia study area having recent (2010-2015) valid-year NO₂ concentrations, continued.

State abbreviation	PA	PA				
County name	Philadelphia	Philadelphia				
Site ID	421010075	421010076				
Lat	40.05386	39.98883				
Lon	-74.98584	-75.20721				
year_start	2013	2015				
year_end	2016	2016				
Elevation_m	9	4				
land_use	COMM	COMM				
scale	MICRO	NA				
objective1	SOURCE	REG TRANS				
objective2	HI_CONC					
objective3						
cnty_landarea_2010	134	134				
AADT information (distance in meters, road/highway/interstate number, AADT value)						
dist_closest	12	18				
road_closest	95	76				
aadt_closest	124610	154955				
dist_aadtmax	.	.				
road_aadtmax	.	.				
aadt_aadtmax	.	.				
County Level NOx Emissions (tpy)						
OnRoad_mobile_nox	10201	10201				
NonRoad_mobile_nox	2480	2480				
AML_mobile_nox	2160	2160				
total_nox	21065	21065				
pct_mobile_nox	70.5	70.5				
NOx Emissions by Source Type (summed tons per year (tpy), sources within 5 Km emitting at least 10 tpy)						
Petroleum_Refinery	.	.				
Municipal_Waste_Combustor	.	.				
Electricity_Generation_via_Combu	.	.				
Chemical_Plant	.	.				
Pulp_and_Paper_Plant	.	.				
Institutional__school__hospital_	11	76				
Wastewater_Treatment_Facility	.	.				
Steam_Heating_Facility	.	.				
Petroleum_Storage_Facility	.	.				
Military_Base	.	.				
Automobile_Truck_or_Parts_Plant	.	.				
not_characterized	.	.				
Steel_Mill	.	.				
Hot_Mix_Asphalt_Plant	.	.				
Pharmaceutical_Manufacturing	.	.				
		Area design value monitor				
		Near-road monitor				

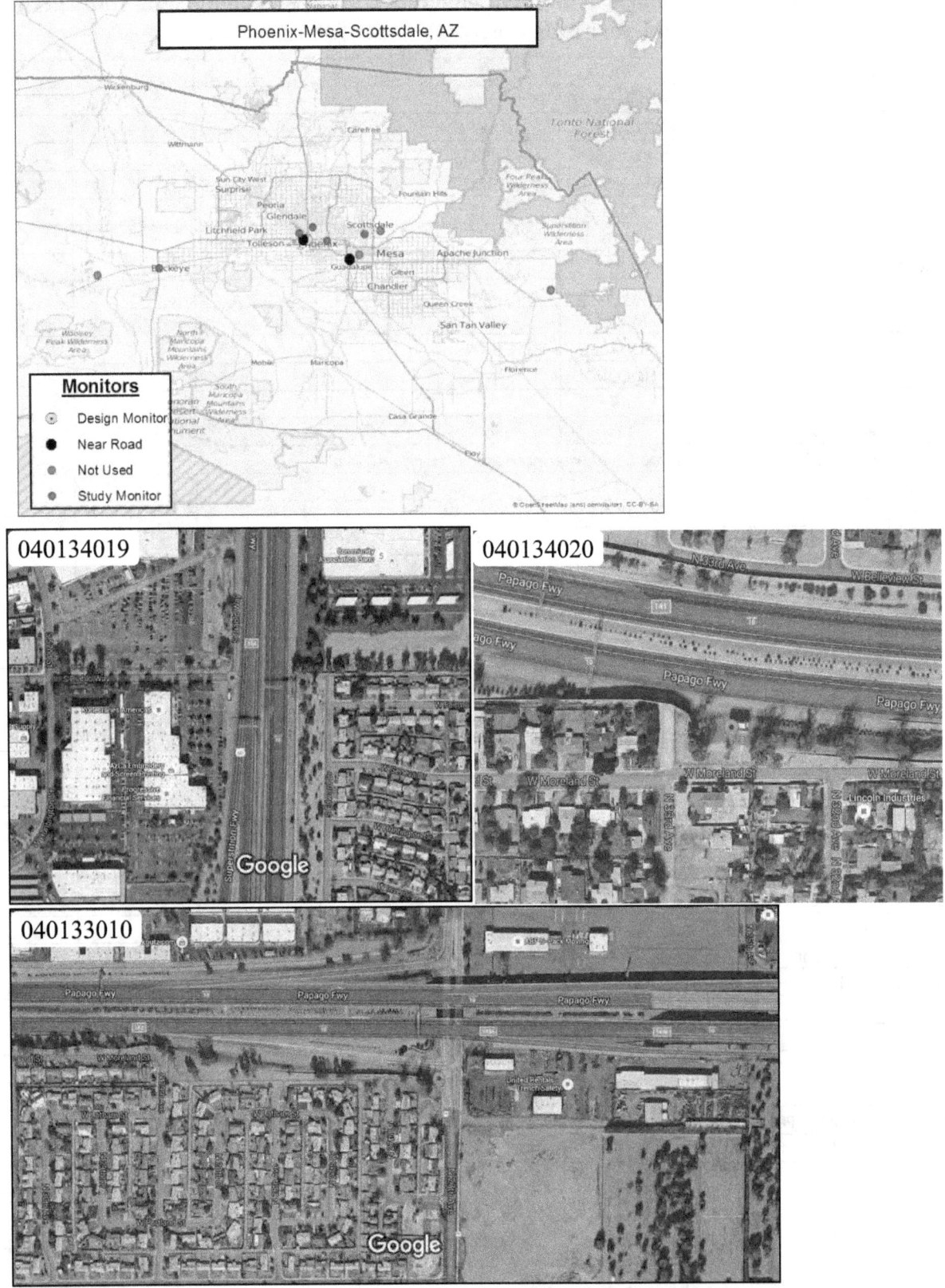

Figure B5-15. Map of all monitors (1990-2015) in the Phoenix study area (top panel) and satellite views of near-road (middle panels) and area design value monitor (bottom panel).

Table B5-25. Attributes of ambient monitors within the Phoenix study area having recent (2010-2015) valid-year NO₂ concentrations.

State abbreviation	AZ	AZ	AZ	AZ	AZ	AZ
County name	Maricopa	Maricopa	Maricopa	Maricopa	Maricopa	Maricopa
Site ID	040130019	040133002	040133003	040133010	040134011	040134019
Lat	33.48385	33.45793	33.47968	33.46093	33.37005	33.39625
Lon	-112.14257	-112.04601	-111.91721	-112.11748	-112.62070	-111.96797
year_start	1990	1967	1975	1993	2004	2014
year_end	2016	2016	2011	2016	2016	2016
Elevation_m	333	339	368	325	258	354
land_use	RESID	RESID	RESID	RESID	AGRIC	COMM
scale	NBHOOD	NBHOOD	URBAN	MID	URBAN	MICRO
objective1	POP_EXP	POP_EXP	POP_EXP	POP_EXP	POP_EXP	SOURCE
objective2	HI_CONC	HI_CONC				
objective3						
cnty_landarea_2010	9200	9200	9200	9200	9200	9200
AADT information (distance in meters, road/highway/interstate number, AADT value)						
dist_closest	356	412	71	87	16	12
road_closest	0	10	0	10	0	10
aadt_closest	32305	248522	20930	294314	4550	320138
dist_aadtmax	806	441	788	.	880	.
road_aadtmax	0	10	0	.	85	.
aadt_aadtmax	33670	255082	41860	.	11562	.
County Level NOx Emissions (tpy)						
OnRoad_mobile_nox	56748	56748	56748	56748	56748	56748
NonRoad_mobile_nox	18998	18998	18998	18998	18998	18998
AMI_mobile_nox	3999	3999	3999	3999	3999	3999
total_nox	88464	88464	88464	88464	88464	88464
pct_mobile_nox	90.1	90.1	90.1	90.1	90.1	90.1
NOx Emissions by Source Type (summed tons per year (tpy), sources within 5 Km emitting at least 10 tpy)						
Airport	.	2657	.	.	.	.
Electricity_Generation_via_Combu	597	.	.	597	.	.
Rail_Yard	58	95	.	118	.	.

		Area design value monitor
		Near-road monitor

Table B5-25, continued. Attributes of ambient monitors within the Phoenix study area having recent (2010-2015) valid-year NO₂ concentrations.

State abbreviation	AZ	AZ				
County name	Maricopa	Maricopa				
Site ID	040134020	040139997				
Lat	33.46155	33.50383				
Lon	-112.12816	-112.09577				
year_start	2015	1993				
year_end	2016	2016				
Elevation_m	326	346				
land_use	RESID	RESID				
scale	MICRO	NBHOOD				
objective1	SOURCE	POP_EXP				
objective2		GEN/BKGR				
objective3						
cnty_landarea_2010	9200	9200				
AADT information (distance in meters, road/highway/interstate number, AADT value)						
dist_closest	20	363				
road_closest	10	0				
aadt_closest	260136	20930				
dist_aadtmax	.	994				
road_aadtmax	.	0				
aadt_aadtmax	.	44896				
County Level NOx Emissions (tpy)						
OnRoad_mobile_nox	56748	56748				
NonRoad_mobile_nox	18998	18998				
AML_mobile_nox	3999	3999				
total_nox	88464	88464				
pct_mobile_nox	90.1	90.1				
NOx Emissions by Source Type (summed tons per year (tpy), sources within 5 Km emitting at least 10 tpy)						
Airport	.	.				
Electricity_Generation_via_Combu	597	.				
Rail_Yard	58	.				
		Area design value monitor				
		Near-road monitor				

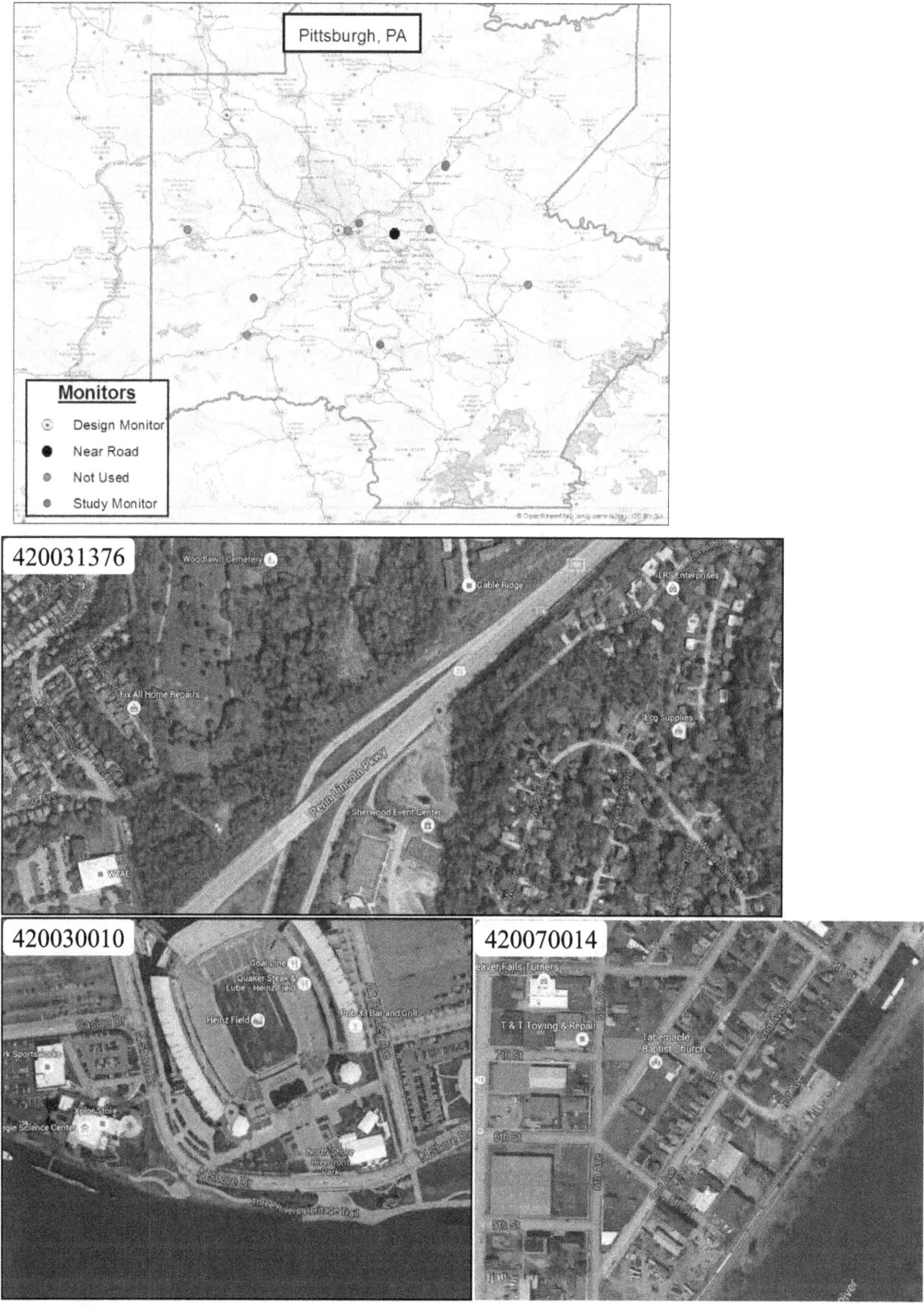

Figure B5-16. Map of all monitors (1990-2015) in the Pittsburgh study area (top panel) and satellite views of near-road (middle panel) and area design value monitor (bottom panels).

Table B5-26. Attributes of ambient monitors within the Pittsburgh study area having recent (2010-2015) valid-year NO$_2$ concentrations.

State abbreviation	PA	PA	PA	PA	PA	PA
County name	Allegheny	Allegheny	Allegheny	Allegheny	Allegheny	Beaver
Site ID	420030008	420030010	420031005	420031008	420031376	420070014
Lat	40.46542	40.44558	40.61395	40.61749	40.43743	40.74780
Lon	-79.96076	-80.01615	-79.72941	-79.72766	-79.86357	-80.31644
year_start	1980	1997	2001	2014	2014	1973
year_end	2014	2013	2014	2016	2016	2016
Elevation_m	256	0	302	302	355	216
land_use	RESID	COMM	RESID	RESID	MOBILE	RESID
scale	NBHOOD	URBAN	NBHOOD	NBHOOD	MICRO	URBAN
objective1	POP_EXP	POP_EXP	POP_EXP	POP_EXP	HI_CONC	POP_EXP
objective2	QUAL ASSUR			MAX_O3	SOURCE	
objective3	GEN/BKGR				POP_EXP	
cnty_landarea_2010	730	730	730	730	730	435
AADT information (distance in meters, road/highway/interstate number, AADT value)						
dist_closest	73	281	.	.	18	250
road_closest	0	0	.	.	376	18
aadt_closest	10385	12007	.	.	87534	9424
dist_aadtmax	622	797	.	.	.	943
road_aadtmax	0	376	.	.	.	18
aadt_aadtmax	27906	40356	.	.	.	9949
County Level NOx Emissions (tpy)						
OnRoad_mobile_nox	13259	13259	13259	13259	13259	2106
NonRoad_mobile_nox	4029	4029	4029	4029	4029	444
AML_mobile_nox	3322	3322	3322	3322	3322	1957
total_nox	35455	35455	35455	35455	35455	21167
pct_mobile_nox	58.1	58.1	58.1	58.1	58.1	21.3
NOx Emissions by Source Type (summed tons per year (tpy), sources within 5 Km emitting at least 10 tpy)						
Steel_Mill	.	.	255	255	.	.
Food_Products_Processing_Plant	212	212	.	.	.	.
Steam_Heating_Facility	166	89	.	.	.	.
Institutional__school__hospital_	36	21	.	.	.	.
Wastewater_Treatment_Facility	.	75	.	.	.	.
Hot_Mix_Asphalt_Plant	18	37	.	.	.	.
Electricity_Generation_via_Combu	33	33	.	.	.	.
Foundries__Iron_and_Steel	13	.	.	.	.	.
Glass_Plant	.	.	.	.	.	.
Chemical_Plant	.	.	.	.	.	.
Coke_Battery	.	.	.	.	.	.
not_characterized	15	15	11	11	12	24
Landfill	.	.	.	.	.	.
		Area design value monitor				
		Near-road monitor				

Table B5-26, continued. Attributes of ambient monitors within the Pittsburgh study area having recent (2010-2015) valid-year NO₂ concentrations.

State abbreviation	PA	PA				
County name	Washington	Washington				
Site ID	421250005	421250200				
Lat	40.14667	40.17056				
Lon	-79.90222	-80.26139				
year_start	1973	1983				
year_end	2016	2008				
Elevation_m	232	334				
land_use	COMM	RESID				
scale	NBHOOD	NBHOOD				
objective1	POP_EXP	POP_EXP				
objective2						
objective3						
cnty_landarea_2010	857	857				
AADT information (distance in meters, road/highway/interstate number, AADT value)						
dist_closest	268	111				
road_closest	88	40				
aadt_closest	7716	12273				
dist_aadtmax	326	799				
road_aadtmax	88	70				
aadt_aadtmax	10524	26252				
County Level NOx Emissions (tpy)						
OnRoad_mobile_nox	3024	3024				
NonRoad_mobile_nox	856	856				
AML_mobile_nox	686	686				
total_nox	10067	10067				
pct_mobile_nox	45.4	45.4				
NOx Emissions by Source Type (summed tons per year (tpy), sources within 5 Km emitting at least 10 tpy)						
Steel_Mill	.	.				
Food_Products_Processing_Plant	.	.				
Steam_Heating_Facility	.	.				
Institutional__school__hospital_	.	.				
Wastewater_Treatment_Facility	.	.				
Hot_Mix_Asphalt_Plant	.	.				
Electricity_Generation_via_Combu	.	.				
Foundries__Iron_and_Steel	.	.				
Glass_Plant	47	.				
Chemical_Plant	12	.				
Coke_Battery	12	.				
not_characterized	.	48				
Landfill	.	18				
		Area design value monitor				
		Near-road monitor				

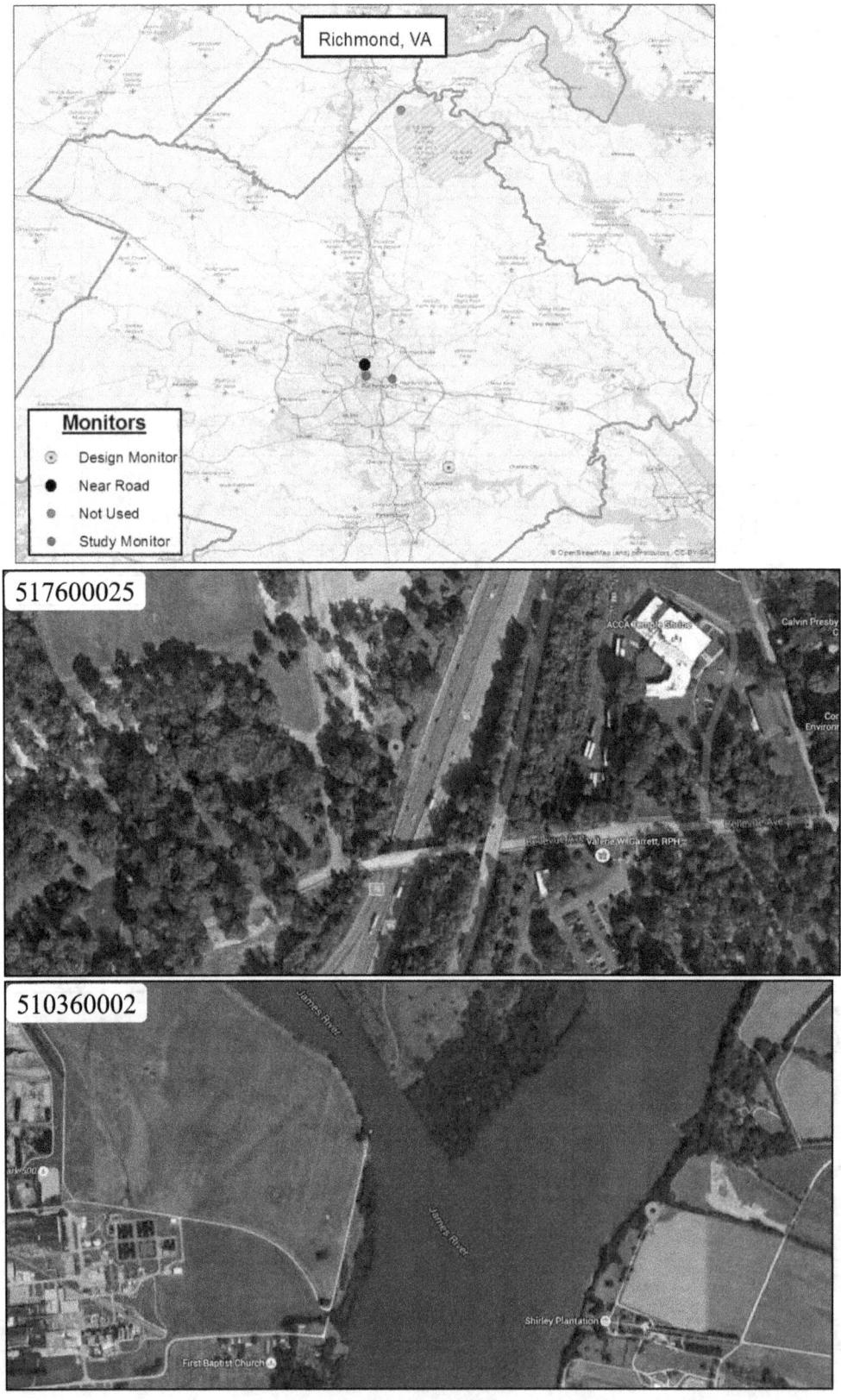

Figure B5-17. Map of all monitors (1990-2015) in the Richmond study area (top panel) and satellite views of near-road (middle panel) and area design value monitor (bottom panel).

Table B5-27. Attributes of ambient monitors within the Richmond study area having recent (2010-2015) valid-year NO$_2$ concentrations.

State abbreviation	VA	VA	VA	VA		
County name	Charles	Henrico	Richmond City	Richmond City		
Site ID	510360002	510870014	517600024	517600025		
Lat	37.34438	37.55652	37.56260	37.59082		
Lon	-77.25925	-77.40027	-77.46500	-77.46933		
year_start	1993	1989	1998	2013		
year_end	2016	2016	2012	2016		
Elevation_m	6	58.5	60	55		
land_use	RESID	RESID	COMM	RESID		
scale	NBHOOD	NBHOOD	NBHOOD	MICRO		
objective1	POP_EXP	POP_EXP	HI_CONC	POP_EXP		
objective2	HI_CONC		POP_EXP	SOURCE		
objective3				HI_CONC		
cnty_landarea_2010	183	234	60	60		
AADT information (distance in meters, road/highway/interstate number, AADT value)						
dist_closest	.	384	271	21		
road_closest	.	64	33	95		
aadt_closest	.	73062	21792	151000		
dist_aadtmax	.	.	893	537		
road_aadtmax	.	.	95	64		
aadt_aadtmax	.	.	134322	136518		
County Level NOx Emissions (tpy)						
OnRoad_mobile_nox	176	4372	2719	2719		
NonRoad_mobile_nox	96	1446	396	396		
AML_mobile_nox	51	695	315	315		
total_nox	404	7312	6928	6928		
pct_mobile_nox	80	89.1	49.5	49.5		
NOx Emissions by Source Type (summed tons per year (tpy), sources within 5 Km emitting at least 10 tpy)						
Pulp_and_Paper_Plant	1243	.	.	.		
Rail_Yard	.	35	115	115		
Institutional__school__hospital_	.	10	10	.		
not_characterized	438	24	24	.		
		Area design value monitor				
		Near-road monitor				

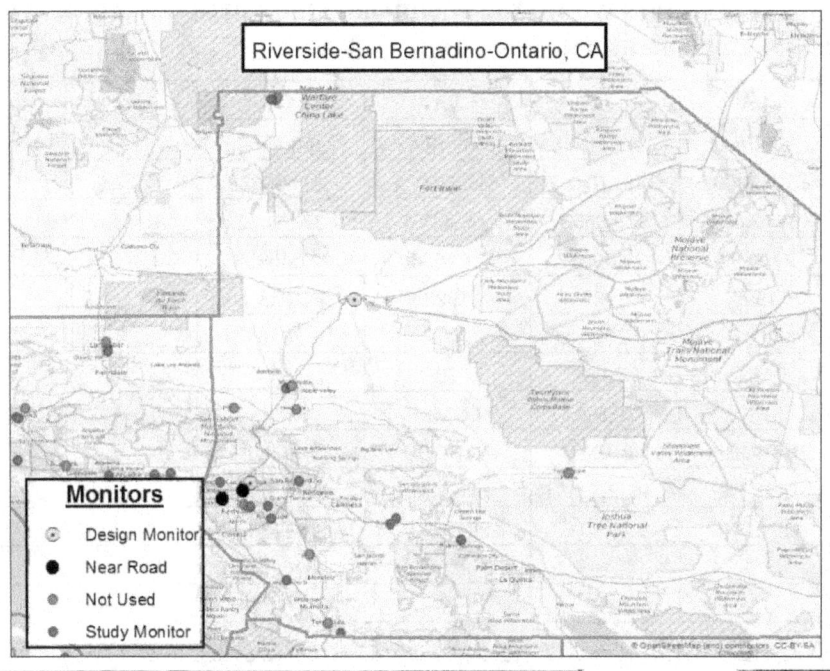

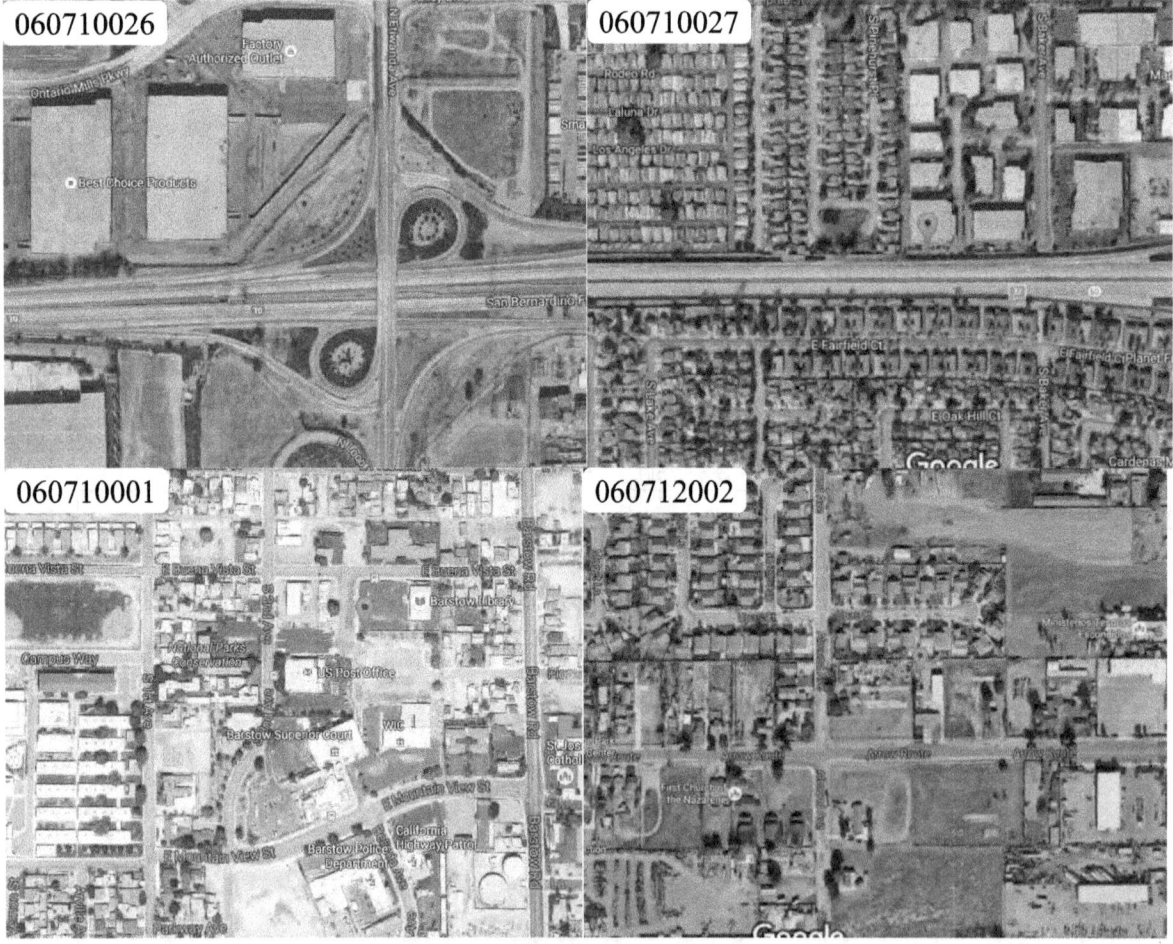

Figure B5-18. Map of all monitors (1990-2015) in the Riverside study area (top panel) and satellite views of two near-road (middle panels) and area design value monitor (bottom panel).

Table B5-28. Attributes of ambient monitors within the Riverside study area having recent (2010-2015) valid-year NO₂ concentrations.

State abbreviation	CA	CA	CA	CA	CA	CA
County name	Riverside	Riverside	Riverside	Riverside	Riverside	Riverside
Site ID	060650009	060650012	060651003	060651016	060655001	060658001
Lat	33.44787	33.92086	33.94603	33.94471	33.85275	33.99958
Lon	-117.08865	-116.85841	-117.40063	-116.83007	-116.54101	-117.41601
year_start	2008	1995	2007	2014	1978	1975
year_end	2016	2016	2014	2016	2016	2016
Elevation_m	366.093	677	249	720	171	250
land_use	RESID	COMM	RESID	COMM	RESID	RESID
scale	NBHOOD	NBHOOD	MICRO	REGION	NA	NA
objective1	GEN/BKGR	POP_EXP	POP_EXP	GEN/BKGR	POP_EXP	HI_CONC
objective2		UP_BKGR	HI_CONC	POP_EXP		POP_EXP
objective3				QUAL		GEN/BKGR
cnty_landarea_2010	7206	7206	7206	7206	7206	7206
AADT information (distance in meters, road/highway/interstate number, AADT value)						
dist_closest	.	387	665	.	.	651
road_closest	.	10	91	.	.	60
aadt_closest	.	108000	168000	.	.	145000
dist_aadtmax	.	925	.	.	.	.
road_aadtmax	.	10	.	.	.	.
aadt_aadtmax	.	118000	.	.	.	.
County Level NOx Emissions (tpy)						
OnRoad_mobile_nox	25868	25868	25868	25868	25868	25868
NonRoad_mobile_nox	5541	5541	5541	5541	5541	5541
AML_mobile_nox	2215	2215	2215	2215	2215	2215
total_nox	37367	37367	37367	37367	37367	37367
pct_mobile_nox	90	90	90	90	90	90
NOx Emissions by Source Type (summed tons per year (tpy), sources within 5 Km emitting at least 10 tpy)						
Mineral_Processing_Plant	.	.	.	.	.	.
Electricity_Generation_via_Combu	.	.	.	.	.	.
Airport	.	.	.	.	91	.
Rail_Yard	.	.	.	.	.	.
Pulp_and_Paper_Plant	.	.	.	.	.	.
Fabricated_Metal_Products_Plant	.	.	.	.	.	.
Food_Products_Processing_Plant	.	.	.	.	.	.
Foundries__non_ferrous	.	.	.	.	.	.
not_characterized	.	.	.	.	11	.
Municipal_Waste_Combustor	.	.	.	.	.	.
Portland_Cement_Manufacturing	.	.	.	.	.	.
Primary_Aluminum_Plant	.	.	.	.	.	.
Secondary_Aluminum_Smelting_Refi	.	.	.	.	.	.
Steel_Mill	.	.	.	.	.	.
Institutional__school__hospital_	.	.	.	.	.	.
		Area design value monitor				
		Near-road monitor				

Table B5-28, continued. Attributes of ambient monitors within the Riverside study area having recent (2010-2015) valid-year NO₂ concentrations.

State abbreviation	CA	CA	CA	CA	CA	CA
County name	Riverside	Riverside	San Bernardino	San Bernardino	San Bernardino	San Bernardino
Site ID	060658005	060659001	060710001	060710026	060710027	060710306
Lat	33.99564	33.67649	34.89501	34.06828	34.03090	34.51001
Lon	-117.49330	-117.33098	-117.02448	-117.52531	-117.61714	-117.33143
year_start	2005	1992	1972	2014	2015	1999
year_end	2016	2016	2016	2016	2016	2016
Elevation_m	250	1440	690	300	258	913
land_use	RESID	RESID	COMM	MOBILE	INDUS	RESID
scale	NA	MID	NA	MICRO	MICRO	NA
objective1	POP_EXP	POP_EXP	REG TRANS	HI_CONC	POP_EXP	REG TRANS
objective2			POP_EXP		SOURCE	POP_EXP
objective3						
cnty_landarea_2010	7206	7206	20057	20057	20057	20057
AADT information (distance in meters, road/highway/interstate number, AADT value)						
dist_closest	.	550	956	50	9	324
road_closest	.	15	15	10	60	18
aadt_closest	.	117000	70000	245300	215000	24750
dist_aadtmax	.	667	.	.	534	788
road_aadtmax	.	15	.	.	60	15
aadt_aadtmax	.	120000	.	.	216000	85000
County Level NOx Emissions (tpy)						
OnRoad_mobile_nox	25868	25868	31067	31067	31067	31067
NonRoad_mobile_nox	5541	5541	5646	5646	5646	5646
AML_mobile_nox	2215	2215	9344	9344	9344	9344
total_nox	37367	37367	68863	68863	68863	68863
pct_mobile_nox	90	90	66.9	66.9	66.9	66.9
NOx Emissions by Source Type (summed tons per year (tpy), sources within 5 Km emitting at least 10 tpy)						
Mineral_Processing_Plant	.	.	.	.	.	.
Electricity_Generation_via_Combu	.	.	.	.	.	.
Airport	.	.	.	.	550	.
Rail_Yard	.	.	223	.	.	.
Pulp_and_Paper_Plant	.	.	.	140	.	.
Fabricated_Metal_Products_Plant	20	.	.	20	.	.
Food_Products_Processing_Plant	.	.	.	.	.	.
Foundries__non_ferrous	.	.	.	26	.	.
not_characterized	.	.	.	243	.	.
Municipal_Waste_Combustor	.	.	.	11	.	.
Portland_Cement_Manufacturing	.	.	.	.	.	.
Primary_Aluminum_Plant	.	.	.	17	.	.
Secondary_Aluminum_Smelting_Refi	.	.	.	20	.	.
Steel_Mill	.	.	.	50	.	.
Institutional__school__hospital_	.	.	.	.	.	.
		Area design value monitor				
		Near-road monitor				

B5-60

Table B5-28, continued. Attributes of ambient monitors within the Riverside study area having recent (2010-2015) valid-year NO₂ concentrations.

State abbreviation	CA	CA	CA	CA		
County name	San Bernardino	San Bernardino	San Bernardino	San Bernardino		
Site ID	060711004	060711234	060712002	060719004		
Lat	34.10374	35.76387	34.10002	34.10688		
Lon	-117.62914	-117.39700	-117.49201	-117.27411		
year_start	1974	1997	1981	1986		
year_end	2016	2016	2016	2016		
Elevation_m	369	545	381	0		
land_use	RESID	DESERT	INDUS	COMM		
scale	NBHOOD	NA	NA	URBAN		
objective1	UP_BKGR		POP_EXP	POP_EXP		
objective2	POP_EXP		HI_CONC	HI_CONC		
objective3	GEN/BKGR					
cnty_landarea_2010	20057	20057	20057	20057		
AADT information (distance in meters, road/highway/interstate number, AADT value)						
dist_closest	.	.	.	.		
road_closest	.	.	.	.		
aadt_closest	.	.	.	.		
dist_aadtmax	.	.	.	.		
road_aadtmax	.	.	.	.		
aadt_aadtmax	.	.	.	.		
County Level NOx Emissions (tpy)						
OnRoad_mobile_nox	31067	31067	31067	31067		
NonRoad_mobile_nox	5646	5646	5646	5646		
AML_mobile_nox	9344	9344	9344	9344		
total_nox	68863	68863	68863	68863		
pct_mobile_nox	66.9	66.9	66.9	66.9		
NOx Emissions by Source Type (summed tons per year (tpy), sources within 5 Km emitting at least 10 tpy)						
Mineral_Processing_Plant	.	1865	.	.		
Electricity_Generation_via_Combu	.	620	.	105		
Airport	.	.	.	.		
Rail_Yard	.	.	.	159		
Pulp_and_Paper_Plant	.	.	.	.		
Fabricated_Metal_Products_Plant	.	.	.	.		
Food_Products_Processing_Plant	10	.	.	.		
Foundries__non_ferrous	.	.	.	.		
not_characterized	.	.	213	.		
Municipal_Waste_Combustor	.	.	.	.		
Portland_Cement_Manufacturing	.	.	.	.		
Primary_Aluminum_Plant	.	.	.	.		
Secondary_Aluminum_Smelting_Refi	.	.	20	.		
Steel_Mill	.	.	50	.		
Institutional__school__hospital_	.	.	.	11		
		Area design value monitor				
		Near-road monitor				

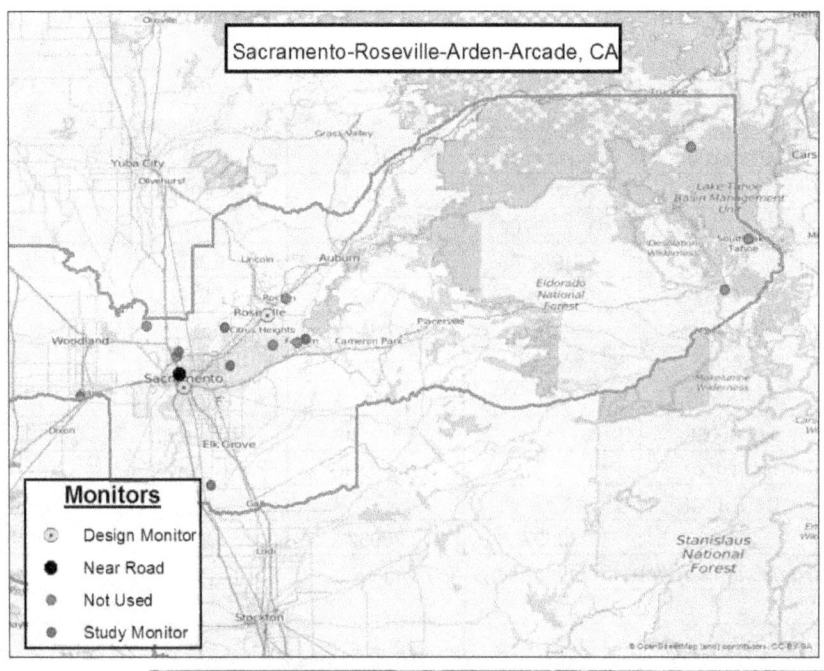

Figure B5-19. Map of all monitors (1990-2015) in the Sacramento study area (top panel) and satellite views of near-road (middle panels) and area design value monitor (bottom panel).

Table B5-29. Attributes of ambient monitors within the Sacramento study area having recent (2010-2015) valid-year NO₂ concentrations.

State abbreviation	CA	CA	CA	CA	CA	CA
County name	Placer	Sacramento	Sacramento	Sacramento	Sacramento	Sacramento
Site ID	060610006	060670002	060670006	060670010	060670011	060670012
Lat	38.74573	38.71209	38.61378	38.55823	38.30259	38.68330
Lon	-121.26631	-121.38109	-121.36801	-121.49298	-121.42084	-121.16446
year_start	1993	1979	1979	1989	1992	1996
year_end	2016	2016	2016	2016	2016	2016
Elevation_m	48	8	8	0	6	98
land_use	MOBILE	RESID	RESID	RESID	AGRIC	RESID
scale	NBHOOD	NA	NBHOOD	NBHOOD	NA	NA
objective1	POP_EXP	POP_EXP	POP_EXP	GEN/BKGR	POP_EXP	POP_EXP
objective2	HI_CONC		HI_CONC	HI_CONC	HI_CONC	HI_CONC
objective3	GEN/BKGR		REG TRANS	POP_EXP	UP_BKGR	MAX_O3
cnty_landarea_2010	1407	965	965	965	965	965
AADT information (distance in meters, road/highway/interstate number, AADT value)						
dist_closest	235	.	.	518	.	.
road_closest	80	.	.	50	.	.
aadt_closest	153000	.	.	246000	.	.
dist_aadtmax	398	.	.	.	.	.
road_aadtmax	80	.	.	.	.	.
aadt_aadtmax	158000	.	.	.	.	.
County Level NOx Emissions (tpy)						
OnRoad_mobile_nox	4760	12361	12361	12361	12361	12361
NonRoad_mobile_nox	1492	3122	3122	3122	3122	3122
AML_mobile_nox	1150	1821	1821	1821	1821	1821
total_nox	9158	19897	19897	19897	19897	19897
pct_mobile_nox	80.8	87	87	87	87	87
NOx Emissions by Source Type (summed tons per year (tpy), sources within 5 Km emitting at least 10 tpy)						
Airport	.	.	.	.	.	.
Rail_Yard	262	.	.	30	.	.
Institutional__school__hospital_	.	.	.	13	.	.
not_characterized	.	.	.	.	.	.
		Area design value monitor				
		Near-road monitor				

B5-63

Table B5-29, continued. Attributes of ambient monitors within the Sacramento study area having recent (2010-2015) valid-year NO₂ concentrations.

State abbreviation	CA	CA	CA			
County name	Sacramento	Sacramento	Yolo			
Site ID	060670014	060670015	061130004			
Lat	38.65078	38.59332	38.53445			
Lon	-121.50677	-121.50380	-121.77340			
year_start	2008	2015	1996			
year_end	2016	2016	2016			
Elevation_m	3	12.8	0			
land_use	COMM	COMM	AGRIC			
scale	NBHOOD	MICRO	NBHOOD			
objective1	POP_EXP	SOURCE	POP_EXP			
objective2						
objective3						
cnty_landarea_2010	965	965	1015			
AADT information (distance in meters, road/highway/interstate number, AADT value)						
dist_closest	.	20	414			
road_closest	.	5	113			
aadt_closest	.	186000	38050			
dist_aadtmax	.	49	.			
road_aadtmax	.	5	.			
aadt_aadtmax	.	190000	.			
County Level NOx Emissions (tpy)						
OnRoad_mobile_nox	12361	12361	2765			
NonRoad_mobile_nox	3122	3122	1323			
AML_mobile_nox	1821	1821	874			
total_nox	19897	19897	6713			
pct_mobile_nox	87	87	73.9			
NOx Emissions by Source Type (summed tons per year (tpy), sources within 5 Km emitting at least 10 tpy)						
Airport	.	.	.			
Rail_Yard	.	.	.			
Institutional__school__hospital_	.	.	20			
not_characterized	.	.	.			
		Area design value monitor				
		Near-road monitor				

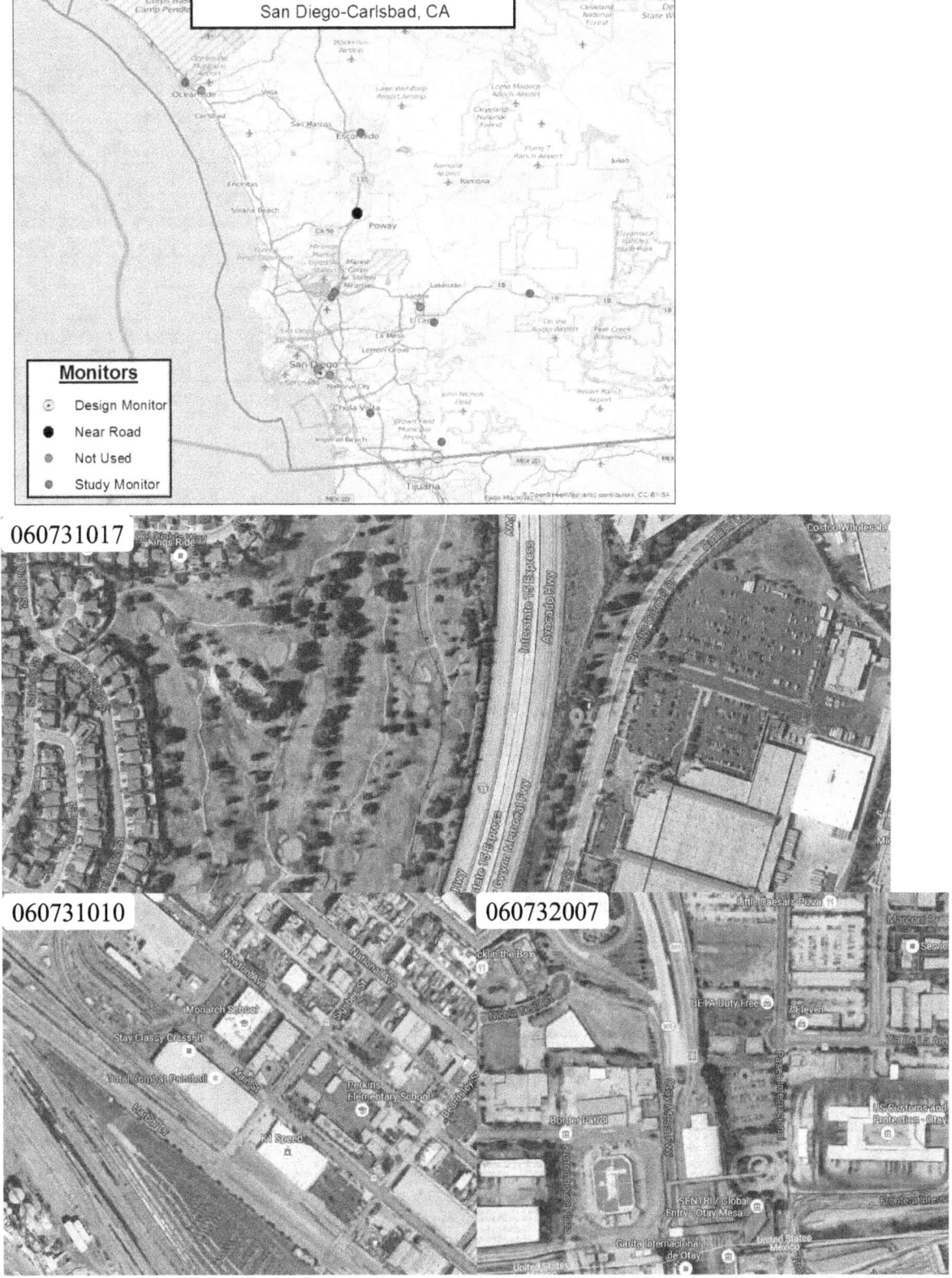

Figure B5-20. Map of all monitors (1990-2015) in the San Diego study area (top panel) and satellite views of near-road (middle panel) and area design value monitor (bottom panels).

Table B5-30. Attributes of ambient monitors within the San Diego study area having recent (2010-2015) valid-year NO$_2$ concentrations.

State abbreviation	CA	CA	CA	CA	CA	CA
County name	San Diego	San Diego	San Diego	San Diego	San Diego	San Diego
Site ID	060730001	060730003	060730006	060731002	060731006	060731008
Lat	32.63123	32.79119	32.83646	33.12771	32.84224	33.21703
Lon	-117.05908	-116.94209	-117.12875	-117.07533	-116.76823	-117.39616
year_start	1973	1971	1975	1973	1981	1997
year_end	2016	2014	2012	2015	2016	2016
Elevation_m	55	143	135	204	603	15
land_use	RESID	COMM	COMM	COMM	RESID	RESID
scale	NBHOOD	NBHOOD	NBHOOD	NBHOOD	URBAN	NBHOOD
objective1	POP_EXP	MAX_PRE	MAX_PRE	POP_EXP	MAX_O3	GEN/BKGR
objective2	GEN/BKGR	POP_EXP	POP_EXP	GEN/BKGR	GEN/BKGR	UP_BKGR
objective3	QUAL	GEN/BKGR	HI_CONC	QUAL	HI_CONC	POP_EXP
cnty_landarea_2010	4207	4207	4207	4207	4207	4207
AADT information (distance in meters, road/highway/interstate number, AADT value)						
dist_closest	982	.	514	187	342	492
road_closest	805	.	52	78	8	5
aadt_closest	190000	.	53000	17700	33500	133000
dist_aadtmax	.	.	933	759	.	.
road_aadtmax	.	.	15	78	.	.
aadt_aadtmax	.	.	174000	32500	.	.
County Level NOx Emissions (tpy)						
OnRoad_mobile_nox	27531	27531	27531	27531	27531	27531
NonRoad_mobile_nox	7428	7428	7428	7428	7428	7428
AML_mobile_nox	3491	3491	3491	3491	3491	3491
total_nox	42700	42700	42700	42700	42700	42700
OnRoad_mobile_nox	27531	27531	27531	27531	27531	27531
NOx Emissions by Source Type (summed tons per year (tpy), sources within 5 Km emitting at least 10 tpy)						
pct_mobile_nox	90	90	90	90	90	90
Electricity_Generation_via_Combu	36	.	57	22	.	.
Military_Base	.	.	22	.	.	19
Pharmaceutical_Manufacturing	.	.	.	.	.	.
Ship_Boat_Manufacturing_or_Repai	.	.	.	.	.	.
not_characterized	.	.	141	15	.	.
		Area design value monitor				
		Near-road monitor				

Table B5-30, continued. Attributes of ambient monitors within the San Diego study area having recent (2010-2015) valid-year NO₂ concentrations.

State abbreviation	CA	CA	CA	CA	CA	
County name	San Diego	San Diego	San Diego	San Diego	San Diego	
Site ID	060731010	060731014	060731016	060731017	060732007	
Lat	32.70149	32.57936	32.84547	32.98544	32.55216	
Lon	-117.14965	-116.92949	-117.12389	-117.08218	-116.93777	
year_start	2005	2014	2011	2015	1990	
year_end	2016	2016	2016	2016	2014	
Elevation_m	3	185	132	218	155	
land_use	COMM	COMM	MILIT	COMM	MOBILE	
scale	NBHOOD	NBHOOD	NBHOOD	NBHOOD	MICRO	
objective1	GEN/BKGR	GEN/BKGR	GEN/BKGR	GEN/BKGR	POP_EXP	
objective2	POP_EXP	POP_EXP	POP_EXP		SOURCE	
objective3			MAX_PRE			
cnty_landarea_2010	4207	4207	4207	4207	4207	
AADT information (distance in meters, road/highway/interstate number, AADT value)						
dist_closest	387	.	404	37	61	
road_closest	5	.	163	15	905	
aadt_closest	161000	.	138000	223000	48500	
dist_aadtmax	964	.	429	801	244	
road_aadtmax	5	.	15	15	905	
aadt_aadtmax	208000	.	174000	225000	52000	
County Level NOx Emissions (tpy)						
OnRoad_mobile_nox	27531	27531	27531	27531	27531	
NonRoad_mobile_nox	7428	7428	7428	7428	7428	
AML_mobile_nox	3491	3491	3491	3491	3491	
total_nox	42700	42700	42700	42700	42700	
OnRoad_mobile_nox	90	90	90	90	90	
NOx Emissions by Source Type (summed tons per year (tpy), sources within 5 Km emitting at least 10 tpy)						
pct_mobile_nox	.	.	57	.	.	
Electricity_Generation_via_Combu	11	.	22	.	.	
Military_Base	38	.	.	.	.	
Pharmaceutical_Manufacturing	13	.	.	.	.	
Ship_Boat_Manufacturing_or_Repai	88	.	141	.	.	
not_characterized	.	.	57	.	.	
		Area design value monitor				
		Near-road monitor				

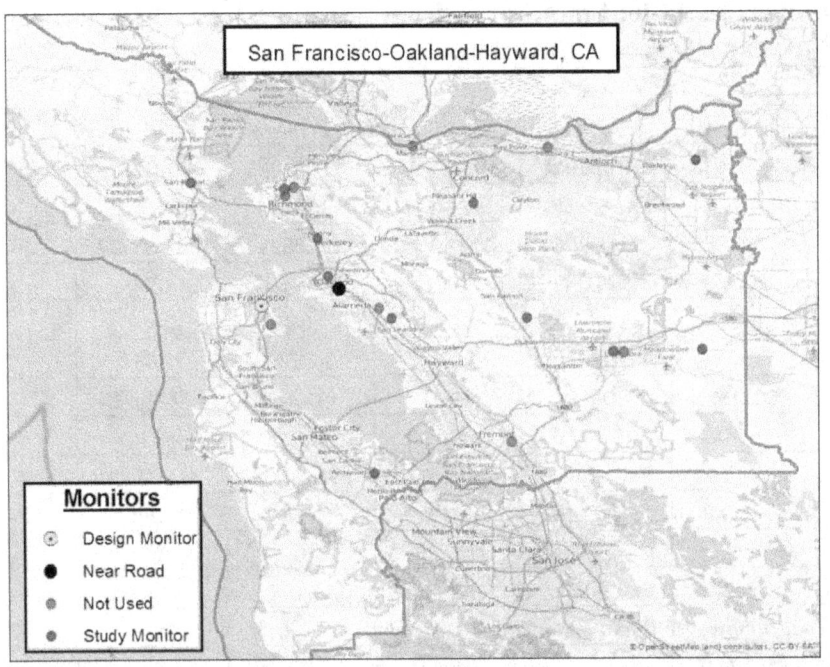

Figure B5-21. Map of all monitors (1990-2010) in the San Francisco study area (top panel) and satellite views of near-road (middle panel) and area design value monitor (bottom panel).

Table B5-31. Attributes of ambient monitors within the San Francisco study area having recent (2010-2015) valid-year NO₂ concentrations.

State abbreviation	CA	CA	CA	CA	CA	CA
County name	Alameda	Alameda	Alameda	Alameda	Alameda	Alameda
Site ID	060010007	060010009	060010011	060010012	060012004	060012005
Lat	37.68753	37.74307	37.81478	37.79362	37.87779	37.68962
Lon	-121.78422	-122.16994	-122.28235	-122.26338	-122.30129	-121.63192
year_start	1999	2007	2009	2014	2007	2011
year_end	2016	2016	2016	2016	2010	2016
Elevation_m	137	11	0	3.9	6	526
land_use	COMM	RESID	RESID	COMM	COMM	AGRIC
scale	NBHOOD	MID	NBHOOD	MICRO	NBHOOD	REGION
objective1	POP_EXP	POP_EXP	POP_EXP	POP_EXP	POP_EXP	REG TRANS
objective2	HI_CONC		SOURCE	SOURCE	SOURCE	DOWNWND
objective3	REG TRANS					
cnty_landarea_2010	739	739	739	739	739	739
AADT information (distance in meters, road/highway/interstate number, AADT value)						
dist_closest	.	36	733	20	467	.
road_closest	.	185	980	880	80	.
aadt_closest	.	21700	113000	216000	256000	.
dist_aadtmax	.	.	.	426	492	.
road_aadtmax	.	.	.	880	80	.
aadt_aadtmax	.	.	.	225000	270000	.
County Level NOx Emissions (tpy)						
OnRoad_mobile_nox	19032	19032	19032	19032	19032	19032
NonRoad_mobile_nox	3158	3158	3158	3158	3158	3158
AML_mobile_nox	3507	3507	3507	3507	3507	3507
total_nox	28381	28381	28381	28381	28381	28381
pct_mobile_nox	90.5	90.5	90.5	90.5	90.5	90.5
NOx Emissions by Source Type (summed tons per year (tpy), sources within 5 Km emitting at least 10 tpy)						
Petroleum_Refinery	.	.	.	.	.	.
Airport	.	941	.	.	.	.
Electricity_Generation_via_Combu	.	.	.	.	48	.
Glass_Plant	.	.	.	327	.	.
Rail_Yard	.	56	150	150	.	.
Wastewater_Treatment_Facility	.	.	53	53	.	.
Foundries__Iron_and_Steel	.	36	.	.	.	.
Pharmaceutical_Manufacturing	.	.	17	.	17	.
not_characterized	.	.	13	.	13	.
Landfill	.	.	.	.	.	.
Hot_Mix_Asphalt_Plant	.	.	.	.	.	.
Chemical_Plant	.	.	.	.	.	.
Institutional__school__hospital_	.	.	.	.	.	.
		Area design value monitor				
		Near-road monitor				

Table B5-31, continued. Attributes of ambient monitors within the San Francisco study area having recent (2010-2015) valid-year NO₂ concentrations.

State abbreviation	CA	CA	CA	CA	CA	CA	CA
County name	Contra Costa	Contra Costa	Contra Costa	Contra Costa	Marin	San Francisco	San Mateo
Site ID	060130002	060131002	060131004	060132007	060410001	060750005	060811001
Lat	37.93601	38.00631	37.96040	37.74365	37.97231	37.76595	37.48293
Lon	-122.02615	-121.64192	-122.35681	-121.93419	-	-122.39904	-
year_start	1980	1981	2002	2011	1967	1985	1967
year_end	2016	2016	2016	2016	2016	2016	2016
Elevation_m	26	-2	20	119	3	5	3
land_use	RESID	AGRIC	COMM	RESID	COMM	INDUS	INDUS
scale	NBHOOD	REGION	MID	URBAN	MID	NBHOOD	NBHOOD
objective1	POP_EXP	GEN/BKGR	SOURCE	POP_EXP	HI_CONC	POP_EXP	POP_EXP
objective2	SOURCE	REG TRANS	POP_EXP	UP_BKGR	POP_EXP	SOURCE	HI_CONC
objective3	GEN/BKGR	HI_CONC		REG TRANS		HI_CONC	QUAL
cnty_landarea_2010	716	716	716	716	520	47	448
AADT information (distance in meters, road/highway/interstate number, AADT value)							
dist_closest	.	.	.	.	115	322	453
road_closest	.	.	.	.	101	280	101
aadt_closest	.	.	.	.	136000	103000	202000
dist_aadtmax	.	.	.	.	506	517	.
road_aadtmax	.	.	.	.	101	101	.
aadt_aadtmax	.	.	.	.	184000	217000	.
County Level NOx Emissions (tpy)							
OnRoad_mobile_nox	9307	9307	9307	9307	2282	4623	5623
NonRoad_mobile_nox	2074	2074	2074	2074	568	1474	1434
AML_mobile_nox	1973	1973	1973	1973	503	5065	3963
total_nox	20714	20714	20714	20714	3967	12406	12208
pct_mobile_nox	64.5	64.5	64.5	64.5	84.5	90	90.3
NOx Emissions by Source Type (summed tons per year (tpy), sources within 5 Km emitting at least 10 tpy)							
Petroleum_Refinery	.	.	835	.	.	.	.
Airport	.	.	.	.	.	.	.
Electricity_Generation_via_Combu	.	.	.	.	.	20	54
Glass_Plant	.	.	.	.	.	.	.
Rail_Yard	.	.	127	.	.	.	.
Wastewater_Treatment_Facility	.	.	.	.	16	74	.
Foundries__Iron_and_Steel	.	.	.	.	.	.	.
Pharmaceutical_Manufacturing	.	.	.	.	.	.	.
not_characterized	.	14	11	.	14	.	.
Landfill	.	.	24	.	.	.	.
Hot_Mix_Asphalt_Plant	.	.	.	.	.	.	.
Chemical_Plant	.	.	.	.	.	.	.
Institutional__school__hospital_	.	.	.	.	.	10	16

		Area design value monitor
		Near-road monitor

B5-70

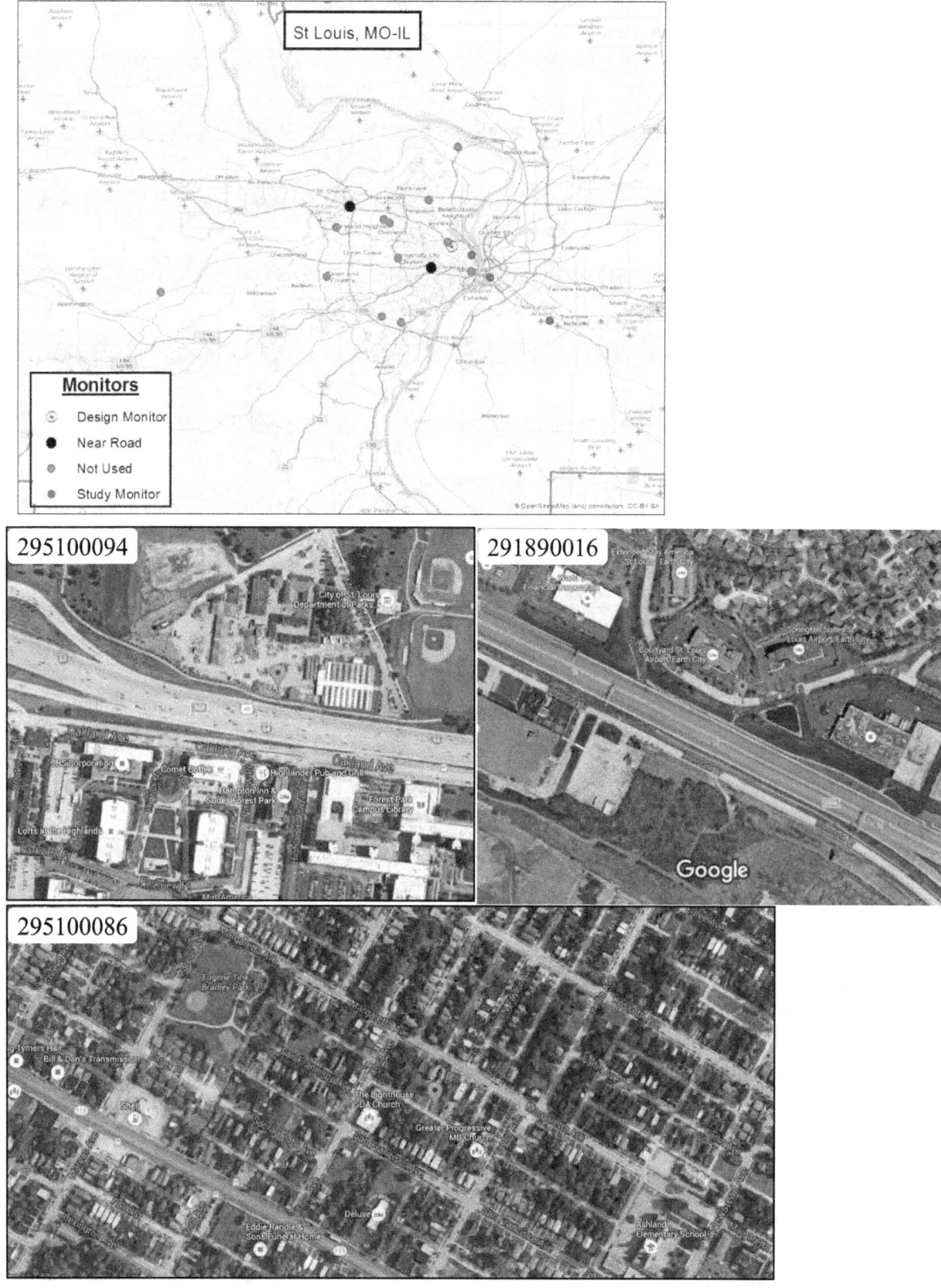

Figure B5-22. Map of all monitors (1990-2015) in the St. Louis study area (top panel) and satellite views of near-road (middle panel) and area design value monitor (bottom panel).

Table B5-32. Attributes of ambient monitors within the St. Louis study area having recent (2010-2015) valid-year NO$_2$ concentrations.

State abbreviation	IL	IL	MO	MO	MO	MO
County name	Saint Clair	Saint Clair	Saint Louis	St. Louis City	St. Louis City	St. Louis City
Site ID	171630010	171630900	291890016	295100085	295100086	295100094
Lat	38.61203	38.52594	38.75264	38.65650	38.67322	38.63106
Lon	-90.16048	-90.03910	-90.44884	-90.19865	-90.23917	-90.28114
year_start	1980	2010	2014	2013	1999	2012
year_end	2016	2015	2016	2016	2016	2016
Elevation_m	125	167	148	137	0	128
land_use	INDUS	COMM	COMM	RESID	RESID	COMM
scale	NBHOOD	NA	MICRO	NBHOOD	NBHOOD	MICRO
objective1	POP_EXP	GEN/BKGR	SOURCE	HI_CONC	POP_EXP	SOURCE
objective2	HI_CONC		POP_EXP	POP_EXP		POP_EXP
objective3	SOURCE			QUAL ASSUR		
cnty_landarea_2010	658	658	508	62	62	62
AADT information (distance in meters, road/highway/interstate number, AADT value)						
dist_closest	45	139	27	221	205	25
road_closest	0	15	70	70	0	64
aadt_closest	2900	25000	161338	98966	12972	159326
dist_aadtmax	773	.	830	571	991	438
road_aadtmax	55	.	270	70	70	64
aadt_aadtmax	57400	.	167602	112323	94733	167347
County Level NOx Emissions (tpy)						
OnRoad_mobile_nox	5026	5026	25063	6296	6296	6296
NonRoad_mobile_nox	1151	1151	4697	663	663	663
AML_mobile_nox	1003	1003	1902	1590	1590	1590
total_nox	8506	8506	39486	10691	10691	10691
pct_mobile_nox	84.4	84.4	80.2	80	80	80
NOx Emissions by Source Type (summed tons per year (tpy), sources within 5 Km emitting at least 10 tpy)						
Breweries_Distilleries_Wineries	467	.	.	.	.	.
Rail_Yard	88	.	.	271	56	309
Electricity_Generation_via_Combu	252	.	.	285	.	.
Wastewater_Treatment_Facility	.	.	89	81	81	.
Institutional__school__hospital_	.	12	.	.	49	70
Ethanol_Biorefineries_Soy_Biodie	49	.	.	60	.	.
Chemical_Plant	27	.	.	.	93	.
Pharmaceutical_Manufacturing	.	.	.	43	43	.
Landfill	58	.	22	.	.	.
Steam_Heating_Facility	20	.	.	.	.	.
Petroleum_Storage_Facility	14	.	.	.	.	.
not_characterized	.	.	.	24	13	12
Aircraft__Aerospace__or_Related	.	.	.	.	.	.
Airport	.	.	.	.	.	.
		Area design value monitor				
		Near-road monitor				

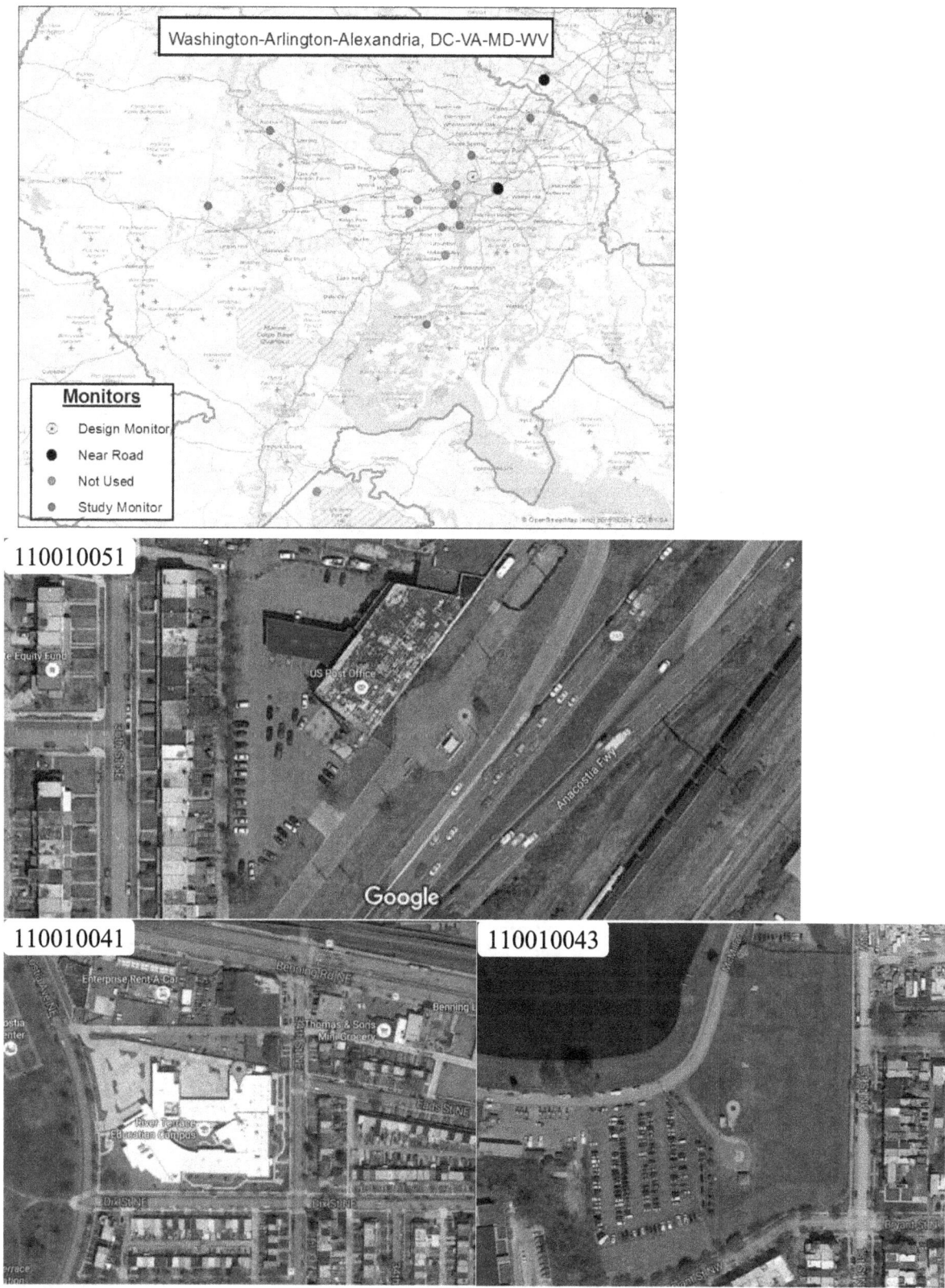

Figure B5-23. Map of all monitors (1990-2015) in the Washington DC study area (top panel) and satellite views of near-road (middle panel) and area design value monitors (bottom panels).

Table B5-33. Attributes of ambient monitors within the Washington DC study area having recent (2010-2015) valid-year NO₂ concentrations.

State abbreviation	DC	DC	DC	DC	DC	MD
County name	DoColumbia	DoColumbia	DoColumbia	DoColumbia	DoColumbia	Prince
Site ID	110010025	110010041	110010043	110010050	110010051	240330030
Lat	38.58323	38.89557	38.92185	38.97009	38.89477	39.05528
Lon	-77.12190	-76.95807	-77.01318	-77.01672	-76.95343	-76.87833
year_start	1980	1993	1993	2012	2015	2005
year_end	2010	2014	2016	2016	2016	2016
Elevation_m	91	8	50	15	25	49
land_use	COMM	RESID	COMM	RESID	COMM	RESID
scale	URBAN	NBHOOD	URBAN	NBHOOD	MICRO	URBAN
objective1	POP_EXP	POP_EXP	POP_EXP	POP_EXP	POP_EXP	POP_EXP
objective2		GEN/BKGR	MAX_PRE	MAX_O3	HI_CONC	GEN/BKGR
objective3		HI_CONC	HI_CONC		SOURCE	UP_BKGR
cnty_landarea_2010	61	61	61	61	61	483
AADT information (distance in meters, road/highway/interstate number, AADT value)						
dist_closest	.	107	337	50	15	699
road_closest	.	0	0	0	295	21
aadt_closest	.	25893	37729	10974	115480	15790
dist_aadtmax	.	377	918	789	.	888
road_aadtmax	.	0	0	0	.	1
aadt_aadtmax	.	112749	38996	29196	.	32451
County Level NOx Emissions (tpy)						
OnRoad_mobile_nox	4739	4739	4739	4739	4739	11955
NonRoad_mobile_nox	2364	2364	2364	2364	2364	2442
AML_mobile_nox	223	223	223	223	223	309
total_nox	9418	9418	9418	9418	9418	21289
pct_mobile_nox	77.8	77.8	77.8	77.8	77.8	69.1
NOx Emissions by Source Type (summed tons per year (tpy), sources within 5 Km emitting at least 10 tpy)						
Airport	.	.	.	.	.	.
Electricity_Generation_via_Combu	.	212	.	.	212	.
Municipal_Waste_Combustor	.	.	.	.	.	.
Steam_Heating_Facility	.	106	301	.	106	.
Military_Base	130	.	.	.	.	.
not_characterized	.	11	.	.	11	20
Institutional__school__hospital_	.	.	27	26	.	.
Rail_Yard	.	16	.	.	16	.
Hot_Mix_Asphalt_Plant	.	.	.	.	.	.

		Area design value monitor
		Near-road monitor

Table B5-33, continued. Attributes of ambient monitors within the Washington DC study area having recent (2010-2015) valid-year NO₂ concentrations.

State abbreviation	VA	VA	VA	VA	VA	
County name	Arlington	Loudoun	Prince	Alexandria	Alexandria	
Site ID	510130020	511071005	511530009	515100009	515100021	
Lat	38.85770	39.02473	38.85287	38.81040	38.80650	
Lon	-77.05922	-77.48925	-77.63462	-77.04435	-77.08640	
year_start	1977	1998	1994	1975	2012	
year_end	2016	2016	2016	2012	2016	
Elevation_m	16	88	117	9	61	
land_use	COMM	RESID	RESID	RESID	COMM	
scale	NA	NBHOOD	URBAN	NA	NBHOOD	
objective1	POP_EXP	POP_EXP	POP_EXP	POP_EXP	POP_EXP	
objective2						
objective3						
cnty_landarea_2010	26	516	336	15	15	
AADT information (distance in meters, road/highway/interstate number, AADT value)						
dist_closest	588	.	.	113	162	
road_closest	1	.	.	400	236	
aadt_closest	51469	.	.	32550	34262	
dist_aadtmax	933	.	.	507	526	
road_aadtmax	395	.	.	90005	95	
aadt_aadtmax	199159	.	.	49465	153908	
County Level NOx Emissions (tpy)						
OnRoad_mobile_nox	1214	2121	3931	682	682	
NonRoad_mobile_nox	1203	2253	1390	127	127	
AML_mobile_nox	1243	1765	297	142	142	
total_nox	4065	6893	6863	2278	2278	
pct_mobile_nox	90	89.1	81.9	41.8	41.8	
NOx Emissions by Source Type (summed tons per year (tpy), sources within 5 Km emitting at least 10 tpy)						
Airport	1311	.	.	.	.	
Electricity_Generation_via_Combu	599	.	.	558	558	
Municipal_Waste_Combustor	.	.	.	.	471	
Steam_Heating_Facility	195	.	.	.	.	
Military_Base	.	.	.	25	.	
not_characterized	.	17	.	15	.	
Institutional__school__hospital_	14	.	.	.	.	
Rail_Yard	.	.	.	.	.	
Hot_Mix_Asphalt_Plant	.	.	.	.	11	
		Area design value monitor				
		Near-road monitor				

5.3 ATTRIBUTES OF HISTORICAL (1990-2009) OR OTHER AMBIENT NO₂ MONITORS NOT USED FOR THE 2010-2015 ANALYSIS

Table B5-34. Attributes of ambient monitors within the Atlanta study area not used for the 2010-2015 analysis.

State abbreviation	GA	GA				
County name	DeKalb	Fulton				
Site ID	130893001	131210048				
Lat	33.84574	33.77933				
Lon	-84.21340	-84.39576				
year_start	1990	1982				
year_end	2006	2009				
Elevation_m	0	290				
land_use	RESID	COMM				
scale	NBHOOD	NBHOOD				
objective1	POP_EXP	HI_CONC				
objective2	GEN/BKGR	POP_EXP				
objective3						
cnty_landarea_2010	268	527				
AADT information (distance in meters, road/highway/interstate number, AADT value)						
dist_closest	.	408				
road_closest	.	0				
aadt_closest	.	22130				
dist_aadtmax	.	430				
road_aadtmax	.	75				
aadt_aadtmax	.	348900				
County Level NOx Emissions (tpy)						
OnRoad_mobile_nox	11230	16404				
NonRoad_mobile_nox	1962	3939				
AML_mobile_nox	376	1560				
total_nox	14719	23989				
pct_mobile_nox	92.2	91.3				
NOx Emissions by Source Type (summed tons per year (tpy), sources within 5 Km emitting at least 10 tpy)						
Landfill	.	.				
Rail_Yard	.	224				
		Monitor not used for 2010-2015 analysis (either historical data or were incomplete)				

Table B5-35. Attributes of ambient monitors within the Baltimore study area not used for the 2010-2015 analysis.

State abbreviation	MD	MD	MD			
County name	Balt. (City)	Balt. (City)	Anne Arundel			
Site ID	245100050	245100051	240030019			
Lat	39.31861	39.28150	39.10111			
Lon	-76.58250	-76.59858	-76.72944			
year_start	1995	1997	1980			
year_end	2001	1999	2003			
Elevation_m	49	5	46			
land_use	RESID	COMM	COMM			
scale	REGION	REGION	URBAN			
objective1	POP_EXP	POP_EXP	UP_BKGR			
objective2	MAX_PRE	HI_CONC	POP_EXP			
objective3	HI_CONC		GEN/BKGR			
cnty_landarea_2010	81	81	415			
AADT information (distance in meters, road/highway/interstate number, AADT value)						
dist_closest	325	305	.			
road_closest	1	1230	.			
aadt_closest	21791	12671	.			
dist_aadtmax	643	387	.			
road_aadtmax	147	1383	.			
aadt_aadtmax	29041	35501	.			
County Level NOx Emissions (tpy)						
OnRoad_mobile_nox	4573	4573	7673			
NonRoad_mobile_nox	646	646	1967			
AML_mobile_nox	1385	1385	3129			
total_nox	10421	10421	20731			
pct_mobile_nox	63.4	63.4	61.6			
NOx Emissions by Source Type (summed tons per year (tpy), sources within 5 Km emitting at least 10 tpy)						
Municipal_Waste_Combustor	.	1134	.			
Steam_Heating_Facility	50	153	.			
Sugar_Mill	.	126	.			
Institutional__school__hospital_	113	81	.			
Chemical_Plant	.	70	.			
Rail_Yard	.	74	.			
Electricity_Generation_via_Combu	25	40	.			
not_characterized	10	10	.			
Military_Base	.	.	52			
Wastewater_Treatment_Facility	.	.	.			
Mineral_Processing_Plant	.	19	.			
		Monitor not used for 2010-2015 analysis (either historical data or were incomplete)				

Table B5-36. Attributes of ambient monitors within the Boston study area not used for the 2010-2015 analysis.

State abbreviation	MA	MA	MA	MA	MA	MA	MA
County name	Essex	Essex	Norfolk	Norfolk	Suffolk	Suffolk	Suffolk
Site ID	250090005	250094004	250210008	250210009	250250021	250250035	250250036
Lat	42.70954	42.79027	42.33343	42.31676	42.37783	42.33343	42.33343
Lon	-71.14589	-70.80835	-71.13283	-71.13283	-71.02714	-71.11616	-71.11616
year_start	1981	1994	1982	1982	1980	1982	1983
year_end	2002	2009	1993	1995	2002	1995	1995
Elevation_m	46	1	0	0	6	0	0
land_use	RESID	RESID	RESID	RESID	RESID	RESID	RESID
scale	NA	URBAN	MID	MICRO	NBHOOD	NA	NA
objective1	HI_CONC	POP_EXP			HI_CONC		
objective2	GEN/BKGR	MAX_O3			POP_EXP		
objective3	POP_EXP	SOURCE					
cnty_landarea_2010	493	493	396	396	58	58	58
AADT information (distance in meters, road/highway/interstate number, AADT value)							
dist_closest	654	.	468	962	19	161	161
road_closest	495	.	9	411	476	129	129
aadt_closest	101413	.	34568	42018	37596	17696	17696
dist_aadtmax	810	.	.	.	27	222	222
road_aadtmax	495	.	.	.	90	9	9
aadt_aadtmax	114788	.	.	.	67804	42578	42578
County Level NOx Emissions (tpy)							
OnRoad_mobile_nox	6568	6568	6685	6685	3201	3201	3201
NonRoad_mobile_nox	2473	2473	2009	2009	2094	2094	2094
AML_mobile_nox	860	860	619	619	5446	5446	5446
total_nox	15750	15750	12261	12261	14015	14015	14015
pct_mobile_nox	62.9	62.9	76	76	76.6	76.6	76.6
NOx Emissions by Source Type (summed tons per year (tpy), sources within 5 Km emitting at least 10 tpy)							
Airport	.	.	.	.	2203	.	.
Municipal_Waste_Combustor	768	.	.	.	.	.	.
Steam_Heating_Facility	.	.	.	.	182	.	.
Institutional__school__hospital_	.	.	258	127	.	258	258
Electricity_Generation_via_Combu	.	.	212	121	353	253	253
Rail_Yard	.	.	20	20	.	20	20
Wastewater_Treatment_Facility	.	.	.	.	.	.	.
Aircraft__Aerospace__or_Related	.	.	.	.	.	.	.
Fabricated_Metal_Products_Plant	.	.	.	.	42	.	.
not_characterized	.	.	.	.	48	.	.
Textile__Yarn__or_Carpet_Plant	26	.	.	.	.	.	.
Military_Base	.	.	.	.	.	.	.
Mineral_Processing_Plant	.	.	.	.	.	.	.
	Monitor not used for 2010-2015 analysis (either historical data or were incomplete)						

Table B5-36, continued. Attributes of ambient monitors within the Boston study area not used for the 2010-2015 analysis.

State abbreviation	MA	MA	MA	NH	NH	NH	NH
County name	Suffolk	Suffolk	Suffolk	Rockingham	Rockingham	Rockingham	Rockingham
Site ID	250250037	250250041	250251003	330150009	330150013	330150014	330150015
Lat	42.61676	42.31737	42.40176	43.07814	43.00009	43.07537	43.08259
Lon	-70.99950	-70.96836	-71.03061	-70.76228	-71.19951	-70.74802	-70.76144
year_start	1977	1999	1984	1977	1997	2003	2001
year_end	1990	2014	1999	2001	2003	2008	2003
Elevation_m	0	10	59	3	0	10.8	3
land_use	COMM	COMM	RESID	COMM	RESID	RESID	COMM
scale	NA	URBAN	URBAN	NA	REGION	NBHOOD	NBHOOD
objective1		POP_EXP	POP_EXP	POP_EXP	HI_CONC	POP_EXP	POP_EXP
objective2		UP_BKGR		HI_CONC	POP_EXP		UP_BKGR
objective3		HI_CONC			UP_BKGR		
cnty_landarea_2010	58	58	58	695	695	695	695
AADT information (distance in meters, road/highway/interstate number, AADT value)							
dist_closest	.	.	212	308	.	.	39
road_closest	.	.	1	0	.	.	0
aadt_closest	.	.	75856	7400	.	.	7400
dist_aadtmax	.	.	.	967	.	.	949
road_aadtmax	.	.	.	95	.	.	95
aadt_aadtmax	.	.	.	69000	.	.	70682
County Level NOx Emissions (tpy)							
OnRoad_mobile_nox	3201	3201	3201	4691	4691	4691	4691
NonRoad_mobile_nox	2094	2094	2094	1376	1376	1376	1376
AML_mobile_nox	5446	5446	5446	533	533	533	533
total_nox	14015	14015	14015	8767	8767	8767	8767
pct_mobile_nox	76.6	76.6	76.6	75.3	75.3	75.3	75.3
NOx Emissions by Source Type (summed tons per year (tpy), sources within 5 Km emitting at least 10 tpy)							
Airport	.	.	2203	.	.	.	.
Municipal_Waste_Combustor	.	.	.	.	.	.	.
Steam_Heating_Facility	.	.	.	.	.	.	.
Institutional__school__hospital_	.	.	.	.	.	.	.
Electricity_Generation_via_Combu	.	.	312	842	.	735	842
Rail_Yard	.	.	.	.	.	.	.
Wastewater_Treatment_Facility	.	14	.	.	.	.	.
Aircraft__Aerospace__or_Related	.	.	.	.	.	.	.
Fabricated_Metal_Products_Plant	.	.	.	.	.	.	.
not_characterized	17	.	22	30	.	.	30
Textile__Yarn__or_Carpet_Plant	.	.	.	.	.	.	.
Military_Base	.	.	.	29	.	29	29
Mineral_Processing_Plant	.	.	.	18	.	18	18

	Monitor not used for 2010-2015 analysis (either historical data or were incomplete)

Table B5-37. Attributes of ambient monitors within the Chicago study area not used for the 2010-2015 analysis.

State abbreviation	IL	IL	IL	IL	IL	IL
County name	Cook	Cook	Cook	Cook	Cook	Cook
Site ID	170310037	170310039	170310064	170310071	170310072	170310075
Lat	41.97948	41.89448	41.79079	41.86670	41.89581	41.96420
Lon	-87.67006	-87.62033	-87.60165	-87.64977	-87.60768	-87.65867
year_start	1981	1978	1992	1992	1995	1997
year_end	1996	1994	2000	1994	2011	2001
Elevation_m	183	180	180	180	181	180
land_use	RESID	COMM	RESID	COMM	COMM	RESID
scale	NA	NBHOOD	NBHOOD	NBHOOD	NBHOOD	NBHOOD
objective1		POP_EXP	POP_EXP	POP_EXP	MAX_PRE	POP_EXP
objective2		HI_CONC			HI_CONC	
objective3					SOURCE	
cnty_landarea_2010	945	945	945	945	945	945
AADT information (distance in meters, road/highway/interstate number, AADT value)						
dist_closest	829	108	377	52	505	144
road_closest	14	0	0	0	0	0
aadt_closest	36980	25225	5263	27906	12414	20978
dist_aadtmax	.	987	498	502	750	913
road_aadtmax	.	41	0	90	41	41
aadt_aadtmax	.	151051	29992	266664	127514	104309
County Level NOx Emissions (tpy)						
OnRoad_mobile_nox	54900	54900	54900	54900	54900	54900
NonRoad_mobile_nox	19402	19402	19402	19402	19402	19402
AML_mobile_nox	12900	12900	12900	12900	12900	12900
total_nox	113148	113148	113148	113148	113148	113148
pct_mobile_nox	77.1	77.1	77.1	77.1	77.1	77.1
NOx Emissions by Source Type (summed tons per year (tpy), sources within 5 Km emitting at least 10 tpy)						
Steel_Mill	.	17	.	.	.	.
Airport	.	.	.	.	.	.
Electricity_Generation_via_Combu	.	.	.	1118	.	.
Rail_Yard	.	.	21	185	.	.
Institutional__school__hospital_	.	232	73	232	97	.
Chemical_Plant	.	.	.	.	.	.
Landfill	.	.	.	.	.	.
Industrial_Machinery_or_Equipmen	.	.	.	.	.	.
Steam_Heating_Facility	.	12	.	12	12	.
Wastewater_Treatment_Facility	.	.	.	.	.	.
Food_Products_Processing_Plant	.	.	.	.	.	.
Hot_Mix_Asphalt_Plant	.	.	.	.	.	.
not_characterized	.	114	.	87	87	.
Glass_Plant	.	.	.	.	.	.
Calcined_Pet_Coke_Plant	.	.	.	.	.	.
Automobile_Truck_or_Parts_Plant	.	.	.	.	.	.
	Monitor not used for 2010-2015 analysis (either historical data or were incmtting incomplete)					

Table B5-37, continued. Attributes of ambient monitors within the Chicago study area not used for the 2010-2015 analysis.

State abbreviation	IL	IL	IL	IL	IL	IL
County name	Cook	Cook	Cook	Cook	Cook	Cook
Site ID	170311002	170311601	170313101	170313102	170313601	170314003
Lat	41.61642	41.66812	41.96528	41.97003	41.99670	42.03892
Lon	-87.55782	-87.99057	-87.87633	-87.87645	-87.87951	-87.89896
year_start	1978	1981	1984	1988	1988	1985
year_end	1991	1991	1997	1990	1990	1990
Elevation_m	179	226	197	197	197	197
land_use	RESID	RESID	MOBILE	COMM	COMM	RESID
scale	NA	NBHOOD	MID	MID	NBHOOD	NA
objective1		POP_EXP	HI_CONC	HI_CONC	GEN/BKGR	POP_EXP
objective2					POP_EXP	
objective3						
cnty_landarea_2010	945	945	945	945	945	945
AADT information (distance in meters, road/highway/interstate number, AADT value)						
dist_closest	103	.	29	24	180	468
road_closest	83	.	12	12	90	12
aadt_closest	19200	.	43900	39100	173242	16005
dist_aadtmax	669	.	152	378	.	827
road_aadtmax	83	.	294	294	.	12
aadt_aadtmax	26600	.	190046	190046	.	21000
County Level NOx Emissions (tpy)						
OnRoad_mobile_nox	54900	54900	54900	54900	54900	54900
NonRoad_mobile_nox	19402	19402	19402	19402	19402	19402
AML_mobile_nox	12900	12900	12900	12900	12900	12900
total_nox	113148	113148	113148	113148	113148	113148
pct_mobile_nox	77.1	77.1	77.1	77.1	77.1	77.1
NOx Emissions by Source Type (summed tons per year (tpy), sources within 5 Km emitting at least 10 tpy)						
Steel_Mill	.	.	.	.	.	.
Airport	.	.	5261	5261	5261	.
Electricity_Generation_via_Combu	.	.	.	.	.	.
Rail_Yard	268	.	440	440	48	.
Institutional__school__hospital_	.	.	.	.	.	19
Chemical_Plant	12	.	.	.	.	.
Landfill	33	.	.	.	.	21
Industrial_Machinery_or_Equipmen	.	.	.	.	.	.
Steam_Heating_Facility	.	.	.	.	.	.
Wastewater_Treatment_Facility	.	.	.	.	.	.
Food_Products_Processing_Plant	.	.	12	12	.	.
Hot_Mix_Asphalt_Plant	.	.	.	.	11	.
not_characterized	.	.	72	83	73	14
Glass_Plant	412	.	.	.	.	.
Calcined_Pet_Coke_Plant	.	145	.	.	.	.
Automobile_Truck_or_Parts_Plant	.	.	.	.	.	.
		Monitor not used for 2010-2015 analysis (either historical data or were incomplete)				

B5-81

Table B5-37, continued. Attributes of ambient monitors within the Chicago study area not used for the 2010-2015 analysis.

State abbreviation	IL	IL	IL	IL	IL	IL
County name	Cook	Cook	Cook	DuPage	Lake	Lake
Site ID	170314004	170314005	170318003	170431003	170971003	170971007
Lat	41.99947	42.00503	41.63142	41.94753	42.44585	42.46757
Lon	-87.87506	-87.90896	-87.56810	-87.92868	-88.10342	-87.81005
year_start	1988	1988	1991	1987	1990	1994
year_end	1991	1990	2002	1990	1991	2002
Elevation_m	197	0	179	201	253	178
land_use	MOBILE	COMM	RESID	COMM	AGRIC	FOREST
scale	MID	NBHOOD	NBHOOD	NBHOOD	URBAN	URBAN
objective1	HI_CONC	HI_CONC	POP_EXP	HI_CONC	HI_CONC	POP_EXP
objective2				POP_EXP		DOWNWND
objective3						HI_CONC
cnty_landarea_2010	945	945	945	328	444	444
AADT information (distance in meters, road/highway/interstate number, AADT value)						
dist_closest	254	403	690	.	644	.
road_closest	90	72	0	.	59	.
aadt_closest	173242	39700	12612	.	13500	.
dist_aadtmax	.	500	806	.	.	.
road_aadtmax	.	90	94	.	.	.
aadt_aadtmax	.	161039	139053	.	.	.
County Level NOx Emissions (tpy)						
OnRoad_mobile_nox	54900	54900	54900	13656	8378	8378
NonRoad_mobile_nox	19402	19402	19402	4770	5355	5355
AML_mobile_nox	12900	12900	12900	1293	878	878
total_nox	113148	113148	113148	24391	20709	20709
pct_mobile_nox	77.1	77.1	77.1	80.8	70.6	70.6
NOx Emissions by Source Type (summed tons per year (tpy), sources within 5 Km emitting at least 10 tpy)						
Steel_Mill	.	.	.	.	.	.
Airport	5261	5261	.	5261	.	.
Electricity_Generation_via_Combu	.	.	.	.	.	.
Rail_Yard	48	.	268	392	.	.
Institutional__school__hospital_	19	.	.	.	.	.
Chemical_Plant	.	.	12	.	.	.
Landfill	.	.	33	.	.	.
Industrial_Machinery_or_Equipmen	.	.	.	.	.	.
Steam_Heating_Facility	.	.	.	.	.	.
Wastewater_Treatment_Facility	.	.	.	.	.	.
Food_Products_Processing_Plant	.	.	.	12	.	.
Hot_Mix_Asphalt_Plant	.	11	.	.	.	.
not_characterized	73	79	.	58	.	.
Glass_Plant	.	.	412	.	.	.
Calcined_Pet_Coke_Plant	.	.	.	.	.	.
Automobile_Truck_or_Parts_Plant	.	.	18	.	.	.
	Monitor not used for 2010-2015 analysis (either historical data or were incomplete)					

Table B5-37, continued. Attributes of ambient monitors within the Chicago study area not used for the 2010-2015 analysis.

State abbreviation	IL	IN	IN	WI	WI	
County name	Will	Jasper	Lake	Kenosha	Kenosha	
Site ID	171971011	180730003	180891016	550590016	550590019	
Lat	41.22154	41.13585	41.60031	42.58585	42.50472	
Lon	-88.19097	-86.98777	-87.33476	-87.87508	-87.80930	
year_start	1995	1975	1989	1979	1990	
year_end	2007	1990	1997	1992	1990	
Elevation_m	181	215	183	220	187	
land_use	AGRIC	AGRIC	RESID	RESID	RESID	
scale	REGION	REGION	NBHOOD	NA	NA	
objective1	GEN/BKGR	HI_CONC	POP_EXP		HI_CONC	
objective2	UP_BKGR	POP_EXP	HI_CONC		POP_EXP	
objective3	POP_EXP				REG TRANS	
cnty_landarea_2010	837	560	499	272	272	
AADT information (distance in meters, road/highway/interstate number, AADT value)						
dist_closest	.	.	187	290	.	
road_closest	.	.	53	158	.	
aadt_closest	.	.	8793	18674	.	
dist_aadtmax	.	.	528	623	.	
road_aadtmax	.	.	90	31	.	
aadt_aadtmax	.	.	35251	46107	.	
County Level NOx Emissions (tpy)						
OnRoad_mobile_nox	9714	2153	9294	2312	2312	
NonRoad_mobile_nox	4244	449	2652	687	687	
AML_mobile_nox	1315	64	1450	284	284	
total_nox	29376	10617	38995	6691	6691	
pct_mobile_nox	52	25.1	34.4	49.1	49.1	
NOx Emissions by Source Type (summed tons per year (tpy), sources within 5 Km emitting at least 10 tpy)						
Steel_Mill	.	.	4336	.	.	
Airport	.	.	.	.	.	
Electricity_Generation_via_Combu	21	.	.	.	.	
Rail_Yard	.	.	.	.	.	
Institutional__school__hospital_	.	.	.	.	.	
Chemical_Plant	.	.	.	.	.	
Landfill	.	.	.	.	.	
Industrial_Machinery_or_Equipmen	.	.	.	.	.	
Steam_Heating_Facility	.	.	.	.	.	
Wastewater_Treatment_Facility	.	.	.	.	.	
Food_Products_Processing_Plant	.	.	.	.	.	
Hot_Mix_Asphalt_Plant	.	.	.	.	.	
not_characterized	.	.	.	.	.	
Glass_Plant	.	.	.	.	.	
Calcined_Pet_Coke_Plant	.	.	.	.	.	
Automobile_Truck_or_Parts_Plant	.	.	.	.	.	
		Monitor not used for 2010-2015 analysis (either historical data or were incomplete)				

Table B5-38. Attributes of ambient monitors within the Dallas study area not used for the 2010-2015 analysis.

State abbreviation	TX	TX	TX	TX	TX	TX	TX
County name	Dallas	Dallas	Denton	Ellis	Ellis	Tarrant	Tarrant
Site ID	481130045	481130055	481210033	481390015	481390017	484390057	484391003
Lat	32.91972	32.61639	33.20623	32.43694	32.47361	32.70694	32.75972
Lon	-96.80806	-96.75694	-97.19557	-97.02500	-97.04250	-97.09361	-97.32806
year_start	1973	1981	1996	1996	2004	1998	1975
year_end	1998	1994	1997	2007	2006	2001	1996
Elevation_m	195	132	0	0	0	0	186
land_use	RESID	UNK	INDUS	AGRIC	RESID	RESID	COMM
scale	URBAN	NA	URBAN	NBHOOD	NA	NBHOOD	NBHOOD
objective1	POP_EXP		GEN/BKGR	UP_BKGR	POP_EXP	POP_EXP	HI_CONC
objective2			HI_CONC	HI_CONC	SOURCE	HI_CONC	POP_EXP
objective3				GEN/BKGR	REG		
cnty_landarea_2010	871	871	878	935	935	864	864
AADT information (distance in meters, road/highway/interstate number, AADT value)							
dist_closest	396	794	.	.	.	118	4
road_closest	289	342	.	.	.	303	347
aadt_closest	23000	13800	.	.	.	39000	35000
dist_aadtmax	970	.	.	.	.	.	838
road_aadtmax	635	.	.	.	.	.	35
aadt_aadtmax	209000	.	.	.	.	.	173390
County Level NOx Emissions (tpy)							
OnRoad_mobile_nox	36623	36623	7555	5173	5173	24824	24824
NonRoad_mobile_nox	8462	8462	1685	854	854	4976	4976
AML_mobile_nox	1141	1141	1608	272	272	7457	7457
total_nox	51422	51422	13785	11530	11530	45082	45082
pct_mobile_nox	89.9	89.9	78.7	54.6	54.6	82.6	82.6
NOx Emissions by Source Type (summed tons per year (tpy), sources within 5 Km emitting at least 10 tpy)							
Portland_Cement_Manufacturing	.	.	.	1087	1087	.	.
Airport	.	.	23	.	.	24	.
Rail_Yard	.	.	.	.	.	151	467
Steel_Mill	.	.	.	298	298	.	.
Compressor_Station	.	.	.	.	26	.	16
Electricity_Generation_via_Combu	.	.	.	164	164	.	.
not_characterized	.	.	.	.	.	.	.
Bakeries	.	.	.	.	.	.	.
Breweries_Distilleries_Wineries	.	.	.	.	.	.	.
Pharmaceutical_Manufacturing	.	.	.	.	.	.	.
Automobile_Truck_or_Parts_Plant	.	.	.	.	.	55	.
	Monitor not used for 2010-2015 analysis (either historical data or were incomplete)						

B5-84

Table B5-39. Attributes of ambient monitors within the Denver study area not used for the 2010-2015 analysis.

State abbreviation	CO	CO	CO	CO	CO	CO	CO
County name	Adams	Arapahoe	Jefferson	Jefferson	Jefferson	Jefferson	Park
Site ID	080017015	080050003	080590006	080590008	080590009	080590010	080930002
Lat	39.84249	39.65721	39.91280	39.87639	39.86193	39.89971	39.24028
Lon	-	-	-	-	-	-	-
year_start	1987	1988	1995	1992	1995	1995	2015
year_end	1994	1996	1999	2001	1999	2001	2016
Elevation_m	1605	1654	1802	1718	1848	1877	3027
land_use	COMM	COMM	INDUS	INDUS	INDUS	AGRIC	FOREST
scale	NBHOOD	NBHOOD	NA	NBHOOD	NBHOOD	NBHOOD	NA
objective1	GEN/BKGR	POP_EXP	GEN/BKGR	GEN/BKGR	GEN/BKGR		GEN/BKGR
objective2	POP_EXP	HI_CONC	POP_EXP				
objective3			HI_CONC				
cnty_landarea_2010	1168	798	764	764	764	764	2194
AADT information (distance in meters, road/highway/interstate number, AADT value)							
dist_closest	.	150	.	.	58	67	330
road_closest	.	85	.	.	72	93	285
aadt_closest	.	70000	.	.	3900	17000	4100
dist_aadtmax	.	341	.	.	.	.	.
road_aadtmax	.	85	.	.	.	.	.
aadt_aadtmax	.	78000	.	.	.	.	.
County Level NOx Emissions (tpy)							
OnRoad_mobile_nox	8763	8397	8825	8825	8825	8825	700
NonRoad_mobile_nox	1974	2016	2226	2226	2226	2226	99
AML_mobile_nox	838	285	259	259	259	259	0
total_nox	25245	13022	14406	14406	14406	14406	1438
pct_mobile_nox	45.9	82.2	78.5	78.5	78.5	78.5	55.6
NOx Emissions by Source Type (summed tons per year (tpy), sources within 5 Km emitting at least 10 tpy)							
Electricity_Generation_via_Combu	.	1999	.	.	19	19	.
Rail_Yard	.	.	.	.	.	.	.
Petroleum_Refinery	.	.	.	.	.	.	.
Food_Products_Processing_Plant	.	.	.	.	.	.	.
Dry_Cleaner___Perchloroethylene	.	.	.	.	.	.	.
Wastewater_Treatment_Facility	.	84	.	.	.	.	.
Hot_Mix_Asphalt_Plant	.	19	.	.	.	.	.
Institutional__school__hospital_	.	16	.	.	.	.	.
Petroleum_Storage_Facility	.	.	.	.	.	.	.
Bakeries	.	.	.	.	.	.	.
Lumber_Sawmill	.	.	.	.	.	.	.
not_characterized	.	.	131	.	.	118	.
Chemical_Plant	12	.	.	.	.	.	.
Compressor_Station	.	.	15	.	.	.	.
Landfill	.	.	.	.	27	.	.
Brick__Structural_Clay__or_Clay	.	16	.	.	.	.	.
	Monitor not used for 2010-2015 analysis (either historical data or were incomplete)						

Table B5-40. Attributes of ambient monitors within the Detroit study area not used for the 2010-2015 analysis.

State abbreviation	MI	MI	MI	MI	MI	MI
County name	Macomb	Wayne	Wayne	Wayne	Wayne	Wayne
Site ID	260990009	261630016	261630029	261630062	261631010	261631011
Lat	42.73139	42.35781	42.33809	42.34087	42.29077	42.28010
Lon	-82.79346	-83.09603	-83.02353	-83.06242	-83.12066	-83.12012
year_start	1993	1974	1988	1993	2015	2015
year_end	1998	2007	1990	1993	2016	2016
Elevation_m	189	191	183	610	0.1	0.1
land_use	COMM	RESID	COMM	RESID	INDUS	INDUS
scale	NA	NBHOOD	MID	NA	MICRO	MICRO
objective1	HI_CONC	POP_EXP	HI_CONC	POP_EXP	MAX_O3	MAX_O3
objective2	POP_EXP	HI_CONC	POP_EXP			
objective3	DOWNWND					
cnty_landarea_2010	479	612	612	612	612	612
AADT information (distance in meters, road/highway/interstate number, AADT value)						
dist_closest	.	111	21	273	147	498
road_closest	.	0	0	0	0	0
aadt_closest	.	4190	19600	11554	8001	6630
dist_aadtmax	.	653	.	518	602	608
road_aadtmax	.	94	.	75	75	0
aadt_aadtmax	.	152200	.	102200	95200	14446
County Level NOx Emissions (tpy)						
OnRoad_mobile_nox	12634	29767	29767	29767	29767	29767
NonRoad_mobile_nox	3670	7051	7051	7051	7051	7051
AML_mobile_nox	108	3496	3496	3496	3496	3496
total_nox	20833	62423	62423	62423	62423	62423
pct_mobile_nox	78.8	64.6	64.6	64.6	64.6	64.6
NOx Emissions by Source Type (summed tons per year (tpy), sources within 5 Km emitting at least 10 tpy)						
Electricity_Generation_via_Combu	.	.	.	.	3865	3865
Steel_Mill	.	.	.	.	2776	2776
Municipal_Waste_Combustor	.	1617	1617	1617	.	.
Mineral_Processing_Plant	.	.	.	.	547	547
Petroleum_Refinery	.	.	.	.	412	412
Wastewater_Treatment_Facility	.	.	.	.	246	246
Steam_Heating_Facility	.	126	126	126	.	.
Automobile_Truck_or_Parts_Plant	.	197	.	197	51	51
Institutional__school__hospital_	.	77	51	77	.	.
Rail_Yard	.	.	.	.	106	106
not_characterized	.	.	12	12	.	.
Landfill	47	.	.	.	.	.
	Monitor not used for 2010-2015 analysis (either historical data or were incomplete)					

Table B5-41. Attributes of ambient monitors within the Houston study area not used for the 2010-2015 analysis.

State abbreviation	TX	TX	TX	TX		
County name	Galveston	Harris	Harris	Montgomery		
Site ID	481670014	482011037	482011041	483390089		
Lat	29.26332	29.75111	29.75167	30.35389		
Lon	-94.85657	-95.36139	-95.08361	-95.42167		
year_start	1996	1978	2002	1999		
year_end	2007	2001	2003	2001		
Elevation_m	0	16	0	0		
land_use	COMM	COMM	COMM	COMM		
scale	MID	NBHOOD	MID	MID		
objective1	UP_BKGR	HI_CONC	GEN/BKGR	GEN/BKGR		
objective2	GEN/BKGR	POP_EXP	HI_CONC	REG TRANS		
objective3	REG TRANS					
cnty_landarea_2010	378	1703	1703	1042		
AADT information (distance in meters, road/highway/interstate number, AADT value)						
dist_closest	.	307	454	.		
road_closest	.	59	0	.		
aadt_closest	.	188200	1830	.		
dist_aadtmax	.	663	.	.		
road_aadtmax	.	45	.	.		
aadt_aadtmax	.	247520	.	.		
County Level NOx Emissions (tpy)						
OnRoad_mobile_nox	2958	49330	49330	5948		
NonRoad_mobile_nox	996	13105	13105	1208		
AML_mobile_nox	4215	14455	14455	448		
total_nox	12353	98983	98983	9429		
pct_mobile_nox	66.1	77.7	77.7	80.6		
NOx Emissions by Source Type (summed tons per year (tpy), sources within 5 Km emitting at least 10 tpy)						
Chemical_Plant	.	.	1552	.		
Petroleum_Refinery	.	.	.	.		
Electricity_Generation_via_Combu	.	.	236	.		
Airport	.	.	.	.		
Rail_Yard	.	269	.	.		
Plastic__Resin__or_Rubber_Produc	.	.	153	.		
Food_Products_Processing_Plant	.	97	.	.		
Wastewater_Treatment_Facility	.	75	.	.		
Petroleum_Storage_Facility	.	.	.	.		
Glass_Plant	.	.	.	.		
Chlor_alkali_Plant	.	.	18	.		
Breweries_Distilleries_Wineries	.	.	.	.		
Fertilizer_Plant	.	.	.	.		
Foundries__Iron_and_Steel	.	.	.	.		
Landfill	.	.	28	.		
not_characterized	.	.	57	.		
Foundries__non_ferrous	.	.	.	.		
	Monitor not used for 2010-2014 analysis (either historical data or were incomplete)					

Table B5-42. Attributes of ambient monitors within the Kansas City study area not used for the 2010-2015 analysis.

State abbreviation	KS	KS	MO	MO	MO	MO	MO
County name	Wyandotte	Wyandotte	Clay	Clay	Clay	Jackson	Platte
Site ID	202090001	202090020	290470005	290470006	290470025	290950038	291650023
Lat	39.11306	39.15139	39.30309	39.33191	39.18392	39.10417	39.30003
Lon	-94.62468	-94.61773	-94.37662	-94.58084	-94.49774	-94.49222	-94.70024
year_start	1977	1992	1981	2002	1977	1994	1977
year_end	1999	1998	2010	2004	2002	1996	2004
Elevation_m	256	228	314	303	283	0	293
land_use	COMM	INDUS	AGRIC	AGRIC	RESID	INDUS	MOBILE
scale	NBHOOD	NA	URBAN	NA	NBHOOD	NA	NA
objective1	POP_EXP	HI_CONC	POP_EXP	POP_EXP	POP_EXP		POP_EXP
objective2	HI_CONC	POP_EXP			HI_CONC		
objective3							
cnty_landarea_2010	152	152	397	397	397	604	420
AADT information (distance in meters, road/highway/interstate number, AADT value)							
dist_closest	193	359	847	290	383	114	.
road_closest	69	69	0	169	435	0	.
aadt_closest	9910	14300	16057	16672	66763	18542	.
dist_aadtmax	717	.	943	290	.	465	.
road_aadtmax	70	.	0	169	.	435	.
aadt_aadtmax	44600	.	20342	18823	.	87490	.
County Level NOx Emissions (tpy)							
OnRoad_mobile_nox	4241	4241	6048	6048	6048	13680	3523
NonRoad_mobile_nox	449	449	769	769	769	3213	552
AML_mobile_nox	2877	2877	797	797	797	3026	1843
total_nox	16058	16058	9065	9065	9065	28515	8838
pct_mobile_nox	47.1	47.1	84	84	84	69.9	67
NOx Emissions by Source Type (summed tons per year (tpy), sources within 5 Km emitting at least 10 tpy)							
Electricity_Generation_via_Combu	4392	3175	.	.	.	1425	.
Rail_Yard	1172	221	.	.	.	308	.
Mineral_Wool_Plant	158	158	.	.	.	.	.
Automobile_Truck_or_Parts_Plant	43	43	.	.	82	.	.
Food_Products_Processing_Plant	21	.	.	.	.	.	.
Chemical_Plant	12	.	.	.	.	48	.
Wet_Corn_Mill	32	32	.	.	.	.	.
Municipal_Waste_Combustor	11	11	.	.	.	.	.
Petroleum_Storage_Facility	11	11	.	.	.	.	.
not_characterized	.	16	.	.	.	.	.
Wastewater_Treatment_Facility	.	.	.	.	.	13	.
Airport	.	.	.	.	.	.	683
Steam_Heating_Facility	.	.	.	.	.	.	18
		Monitor not used for 2010-2015 analysis (either historical data or were incomplete)					

Table B5-43. Attributes of ambient monitors within the Los Angeles study area not used for the 2010-2015 analysis.

State abbreviation	CA	CA	CA	CA	CA	CA	CA
County name	Los Angeles	Los Angeles	Los Angeles	Los Angeles	Los Angeles	Los Angeles	Los Angeles
Site ID	060370019	060370030	060370031	060370206	060371002	060371301	060371601
Lat	33.34225	34.03528	33.78611	33.95835	34.17605	33.92899	34.01407
Lon	-	-	-	-	-	-	-
year_start	1990	2001	2001	1992	1980	1980	1978
year_end	1990	2002	2002	1996	2014	2008	2005
Elevation_m	3	65	0	300	168	27	75
land_use	RESID	RESID	RESID	COMM	COMM	COMM	COMM
scale	NA	NA	NA	MID	NA	NA	NBHOOD
objective1		POP_EXP	POP_EXP		POP_EXP	HI_CONC	MAX_PRE
objective2					GEN/BKGR	POP_EXP	POP_EXP
objective3					UP_BKGR		HI_CONC
cnty_landarea_2010	4058	4058	4058	4058	4058	4058	4058
AADT information (distance in meters, road/highway/interstate number, AADT value)							
dist_closest	.	286	513	.	573	365	378
road_closest	.	5	0	.	5	105	605
aadt_closest	.	229000	29500	.	204000	228000	252000
dist_aadtmax	.	406	.	.	780	410	825
road_aadtmax	.	10	.	.	5	105	605
aadt_aadtmax	.	300000	.	.	215000	233000	256000
County Level NOx Emissions (tpy)							
OnRoad_mobile_nox	80322	80322	80322	80322	80322	80322	80322
NonRoad_mobile_nox	17796	17796	17796	17796	17796	17796	17796
AML_mobile_nox	17817	17817	17817	17817	17817	17817	17817
total_nox	135857	135857	135857	135857	135857	135857	135857
pct_mobile_nox	85.3	85.3	85.3	85.3	85.3	85.3	85.3
NOx Emissions by Source Type (summed tons per year (tpy), sources within 5 Km emitting at least 10 tpy)							
Airport	.	.	.	.	336	.	.
Petroleum_Refinery	.	.	2357	.	.	21	.
Rail_Yard	.	240	22	.	.	.	.
Calcined_Pet_Coke_Plant	.	.	201	.	.	.	.
Wastewater_Treatment_Facility	.	.	91	.	.	.	.
Electricity_Generation_via_Combu	69	19	45	24	36	.	.
Oil_or_Gas_Field__On_shore_	.	.	.	.	.	.	.
Institutional__school__hospital_	.	.	.	.	.	.	.
Steam_Heating_Facility	.	.	.	.	.	.	.
Chemical_Plant	.	.	50	.	.	47	.
Foundries__Iron_and_Steel	.	.	.	.	.	.	.
Foundries__non_ferrous	.	.	.	.	.	15	.
Breweries_Distilleries_Wineries	.	.	.	.	.	.	.
Hot_Mix_Asphalt_Plant	.	.	.	.	.	12	.
Petroleum_Storage_Facility	.	.	11	.	.	.	.
Landfill	.	.	.	51	.	.	108
Food_Products_Processing_Plant	.	21	.	.	.	.	.
Glass_Plant	.	66	.	.	.	.	.
Secondary_Lead_Smelting_Plant	.	31	.	.	.	.	.
not_characterized	.	15	65	.	23	.	.
Municipal_Waste_Combustor	.	.	304	.	.	.	.
Aircraft__Aerospace__or_Related	.	.	.	.	.	.	.
	Monitor not used for 2010-2015 analysis (either historical data or were incomplete)						

Table B5-43, continued. Attributes of ambient monitors within the Los Angeles study area not used for the 2010-2015 analysis.

State abbreviation	CA	CA	CA	CA	CA	CA	CA
County name	Los Angeles	Los Angeles	Los Angeles	Los Angeles	Los Angeles	Los Angeles	Orange
Site ID	060372401	060374101	060375001	060376002	060377001	060379002	060590001
Lat	33.92362	34.42666	33.92288	34.38750	34.71221	34.68999	33.82135
Lon	-	-	-	-	-	-	-
year_start	1980	1989	1979	1989	1980	1990	1971
year_end	1993	1990	2004	2001	1990	2001	2001
Elevation_m	58	383	21	375	709	725	128
land_use	RESID	RESID	COMM	COMM	COMM	COMM	RESID
scale	NA	NBHOOD	NA	MID	NA	MID	URBAN
objective1	HI_CONC	POP_EXP	POP_EXP	POP_EXP		POP_EXP	POP_EXP
objective2			MAX_O3	GEN/BKGR			
objective3							
cnty_landarea_2010	CA	CA	CA	CA	CA	CA	CA
AADT information (distance in meters, road/highway/interstate number, AADT value)							
dist_closest	4058	4058	4058	4058	4058	4058	791
road_closest	.	.	139	.	.	.	463
aadt_closest	.	.	405	.	.	.	5
dist_aadtmax	.	.	239000	.	.	.	267000
road_aadtmax	.	.	779	.	.	.	.
aadt_aadtmax	.	.	405	.	.	.	.
County Level NOx Emissions (tpy)							
OnRoad_mobile_nox	80322	80322	80322	80322	80322	80322	19742
NonRoad_mobile_nox	17796	17796	17796	17796	17796	17796	6422
AML_mobile_nox	17817	17817	17817	17817	17817	17817	1921
total_nox	135857	135857	135857	135857	135857	135857	31763
pct_mobile_nox	85.3	85.3	85.3	85.3	85.3	85.3	88.4
NOx Emissions by Source Type (summed tons per year (tpy), sources within 5 Km emitting at least 10 tpy)							
Airport	.	.	5533	.	.	.	.
Petroleum_Refinery	.	.	649	.	.	.	.
Rail_Yard	.	.	.	.	.	.	.
Calcined_Pet_Coke_Plant	.	.	.	.	.	.	.
Wastewater_Treatment_Facility	.	.	.	.	.	.	.
Electricity_Generation_via_Combu	45	.	.	.	.	.	.
Oil_or_Gas_Field__On_shore_	.	66	.	66	.	.	.
Institutional__school__hospital_	.	.	.	.	.	.	.
Steam_Heating_Facility	.	.	.	.	.	.	.
Chemical_Plant	.	.	.	.	.	.	.
Foundries__Iron_and_Steel	.	.	.	.	.	.	.
Foundries__non_ferrous	.	.	.	.	.	.	.
Breweries_Distilleries_Wineries	.	.	.	.	.	.	.
Hot_Mix_Asphalt_Plant	.	.	.	.	.	.	.
Petroleum_Storage_Facility	.	.	.	.	.	.	.
Landfill	.	.	.	.	.	.	.
Food_Products_Processing_Plant	.	.	.	.	.	.	.
Glass_Plant	.	.	.	.	.	.	.
Secondary_Lead_Smelting_Plant	.	.	.	.	.	.	.
not_characterized	.	.	42	.	.	.	24
Municipal_Waste_Combustor	.	.	.	.	.	.	.
Aircraft__Aerospace__or_Related	.	.	11	.	.	.	.
	Monitor not used for 2010-2015 analysis (either historical data or were incomplete)						

Table B5-44. Attributes of ambient monitors within the Miami study area not used for the 2010-2015 analysis.

State abbreviation	FL	FL	FL			
County name	Broward	Palm Beach	Palm Beach			
Site ID	120110003	120990021	120991004			
Lat	26.28147	26.59381	26.69340			
Lon	-80.28255	-80.05849	-80.09921			
year_start	1990	2015	1979			
year_end	1998	2016	2008			
Elevation_m	3	9	0			
land_use	INDUS	INDUS	RESID			
scale	NBHOOD	URBAN	MID			
objective1	HI_CONC	MAX_O3	HI_CONC			
objective2	POP_EXP	HI_CONC	POP_EXP			
objective3						
cnty_landarea_2010	1210	1970	1970			
AADT information (distance in meters, road/highway/interstate number, AADT value)						
dist_closest	894	.	.			
road_closest	0	.	.			
aadt_closest	17600	.	.			
dist_aadtmax	895	.	.			
road_aadtmax	0	.	.			
aadt_aadtmax	35000	.	.			
County Level NOx Emissions (tpy)						
OnRoad_mobile_nox	22197	16775	16775			
NonRoad_mobile_nox	5885	7625	7625			
AML_mobile_nox	8072	4059	4059			
total_nox	43997	35647	35647			
pct_mobile_nox	82.2	79.8	79.8			
NOx Emissions by Source Type (summed tons per year (tpy), sources within 5 Km emitting at least 10 tpy)						
Airport	.	.	388			
Electricity_Generation_via_Combu	.	18	.			
Institutional__school__hospital_	.	.	.			
Petroleum_Storage_Facility	.	.	.			
Wastewater_Treatment_Facility	.	.	.			
		Monitor not used for 2010-2015 analysis (either historical data or were incomplete)				

Table B5-45. Attributes of ambient monitors within the Minneapolis study area not used for the 2010-2015 analysis.

State abbreviation	MN	MN	MN	MN	MN	
County name	Dakota	Hennepin	Hennepin	Ramsey	Wright	
Site ID	270370428	270530953	270530957	271230864	271710007	
Lat	44.79219	45.00247	45.02108	44.99191	45.32913	
Lon	-93.08549	-93.24772	-93.28217	-93.18328	-93.83608	
year_start	1991	1989	1996	1989	1979	
year_end	1992	1996	2002	2002	1997	
Elevation_m	294	256	0	305	288	
land_use	AGRIC	COMM	INDUS	RESID	AGRIC	
scale	MID	MID	MID	URBAN	NA	
objective1	POP_EXP	POP_EXP	HI_CONC	POP_EXP		
objective2	SOURCE	HI_CONC				
objective3						
cnty_landarea_2010	562	554	554	152	661	
AADT information (distance in meters, road/highway/interstate number, AADT value)						
dist_closest	.	781	5	.	899	
road_closest	.	35	0	.	94	
aadt_closest	.	109340	3441	.	43196	
dist_aadtmax	.	.	107	.	.	
road_aadtmax	.	.	94	.	.	
aadt_aadtmax	.	.	124250	.	.	
County Level NOx Emissions (tpy)						
OnRoad_mobile_nox	7604	21967	21967	9912	3496	
NonRoad_mobile_nox	2061	5495	5495	1700	1041	
AML_mobile_nox	295	3168	3168	1435	698	
total_nox	19738	39010	39010	16428	6143	
pct_mobile_nox	50.5	78.5	78.5	79.4	85.2	
NOx Emissions by Source Type (summed tons per year (tpy), sources within 5 Km emitting at least 10 tpy)						
Petroleum_Refinery	1296	.	.	.	.	
Municipal_Waste_Combustor	.	594	594	.	.	
Electricity_Generation_via_Combu	87	404	72	.	.	
Institutional__school__hospital_	.	145	21	155	.	
Rail_Yard	.	253	358	111	.	
Pulp_and_Paper_Plant	.	.	.	339	.	
Gas_Plant	63	.	.	.	.	
Secondary_Aluminum_Smelting_Refi	.	.	.	.	.	
Steam_Heating_Facility	.	15	.	.	.	
not_characterized	.	11	.	11	.	
Printing_Publishing_Facility	31	.	.	.	.	
Food_Products_Processing_Plant	.	20	.	20	.	
	Monitor not used for 2010-2015 analysis (either historical data or were incomplete)					

Table B5-46. Attributes of ambient monitors within the New York/Jersey study area not used for the 2010-2015 analysis.

State abbreviation	NJ	NJ	NJ	NJ	NJ	NJ
County name	Bergen	Bergen	Essex	Essex	Union	Union
Site ID	340030001	340030005	340130011	340130016	340390008	340395001
Lat	40.80843	40.89858	40.72667	40.72222	40.59803	40.60173
Lon	-73.99236	-74.02990	-74.14374	-74.14694	-74.45381	-74.44107
year_start	1982	2000	1984	2001	1995	1980
year_end	1998	2007	1999	2003	1997	1994
Elevation_m	61	6	3	3	18	19
land_use	RESID	RESID	INDUS	INDUS	RESID	RESID
scale	NA	NBHOOD	NA	NBHOOD	NBHOOD	NBHOOD
objective1		POP_EXP	POP_EXP	POP_EXP	POP_EXP	POP_EXP
objective2			HI_CONC			
objective3						
cnty_landarea_2010	233	233	126	126	103	103
AADT information (distance in meters, road/highway/interstate number, AADT value)						
dist_closest	965	244	218	16	17	343
road_closest	63	4	1	1	28	28
aadt_closest	17535	106918	67959	67959	10995	10995
dist_aadtmax	965	244	920	984	466	.
road_aadtmax	63	4	95	95	28	.
aadt_aadtmax	22502	113560	221613	115451	18357	.
County Level NOx Emissions (tpy)						
OnRoad_mobile_nox	8080	8080	5162	5162	4874	4874
NonRoad_mobile_nox	3313	3313	2070	2070	1346	1346
AML_mobile_nox	1070	1070	3355	3355	3479	3479
total_nox	15763	15763	14172	14172	13636	13636
pct_mobile_nox	79.1	79.1	74.7	74.7	71.1	71.1
NOx Emissions by Source Type (summed tons per year (tpy), sources within 5 Km emitting at least 10 tpy)						
Airport	.	.	2984	2984	.	.
Electricity_Generation_via_Combu	894	.	442	442	.	.
Municipal_Waste_Combustor	.	.	758	758	.	.
Petroleum_Refinery	.	.	.	.	.	.
Institutional__school__hospital_	111	15	74	74	.	.
Wastewater_Treatment_Facility	369	.	14	14	.	.
Breweries_Distilleries_Wineries	.	.	.	68	.	.
Petroleum_Storage_Facility	.	.	.	.	.	.
Pharmaceutical_Manufacturing	.	.	.	.	.	.
Fabricated_Metal_Products_Plant	.	.	10	10	.	.
Compressor_Station	95	.	.	.	.	.
Steam_Heating_Facility	69	.	.	.	.	.
not_characterized	68	.	24	24	.	.
		Monitor not used for 2010-2015 analysis (either historical data or were incomplete)				

B5-93

Table B5-46, continued. Attributes of ambient monitors in the New York/Jersey study area not used for the 2010-2015 analysis.

State abbreviation	NY	NY	NY	NY	NY	NY
County name	Bronx	Bronx	Bronx	Dutchess	Kings	New York
Site ID	360050073	360050080	360050083	360270007	360470011	360610010
Lat	40.81149	40.83606	40.86585	41.78555	40.73277	40.73955
Lon	-73.90958	-73.92009	-73.88083	-73.74136	-73.94722	-73.98569
year_start	1997	1990	1995	1993	1978	1976
year_end	1999	2000	2007	1993	1996	2001
Elevation_m	15	20	24	98	9	38
land_use	RESID	RESID	COMM	AGRIC	INDUS	RESID
scale	NA	NBHOOD	NA	NA	NBHOOD	NBHOOD
objective1	POP_EXP	HI_CONC	GEN/BKGR	POP_EXP	HI_CONC	POP_EXP
objective2	GEN/BKGR	POP_EXP	POP_EXP	GEN/BKGR	GEN/BKGR	HI_CONC
objective3	QUAL ASSUR	GEN/BKGR			POP_EXP	
cnty_landarea_2010	42	42	42	796	71	23
AADT information (distance in meters, road/highway/interstate number, AADT value)						
dist_closest	423	170	82	619	154	57
road_closest	0	0	0	44	0	0
aadt_closest	5300	20300	11500	10400	22000	18200
dist_aadtmax	701	973	700	.	936	946
road_aadtmax	87	95	0	.	495	0
aadt_aadtmax	132100	152100	112700	.	79400	138500
County Level NOx Emissions (tpy)						
OnRoad_mobile_nox	4297	4297	4297	2626	6463	5664
NonRoad_mobile_nox	1767	1767	1767	909	4537	7424
AML_mobile_nox	708	708	708	180	1004	6085
total_nox	9912	9912	9912	5090	19011	33400
pct_mobile_nox	68.3	68.3	68.3	73	63.1	57.4
NOx Emissions by Source Type (summed tons per year (tpy), sources within 5 Km emitting at least 10 tpy)						
Airport	.	.	.	.	.	12
Electricity_Generation_via_Combu	1048	.	.	.	2462	2906
Municipal_Waste_Combustor	.	.	.	.	.	.
Petroleum_Refinery	.	.	.	.	.	.
Institutional__school__hospital_	248	108	249	.	222	201
Wastewater_Treatment_Facility	346	346	.	.	14	14
Breweries_Distilleries_Wineries	.	.	.	.	.	.
Petroleum_Storage_Facility	.	.	.	.	.	.
Pharmaceutical_Manufacturing	.	.	.	.	.	.
Fabricated_Metal_Products_Plant	.	.	.	.	.	.
Compressor_Station	11	11	.	.	.	.
Steam_Heating_Facility	69	69	.	.	381	381
not_characterized	166	166	232	.	170	149
	Monitor not used for 2010-2015 analysis (either historical data or were incomplete)					

Table B5-46, continued. Attributes of ambient monitors in the New York/Jersey study area not used for the 2010-2015 analysis.

State abbreviation	NY	NY	NY	NY	NY	
County name	New York	Queens	Queens	Suffolk	Westchester	
Site ID	360610056	360810097	360810098	361030009	361195003	
Lat	40.75912	40.75527	40.78420	40.82799	41.04500	
Lon	-73.96661	-73.75861	-73.84757	-73.05754	-73.70333	
year_start	1985	1998	1998	1999	1992	
year_end	2008	2001	2006	2010	1993	
Elevation_m	17	0	6	45	15	
land_use	COMM	RESID	RESID	RESID	RESID	
scale	MID	NA	NA	NA	URBAN	
objective1	HI_CONC	GEN/BKGR	GEN/BKGR	POP_EXP	POP_EXP	
objective2	GEN/BKGR		SOURCE	GEN/BKGR		
objective3	POP_EXP					
cnty_landarea_2010	23	109	109	912	431	
AADT information (distance in meters, road/highway/interstate number, AADT value)						
dist_closest	68	189	138	160	.	
road_closest	0	0	0	0	.	
aadt_closest	23300	5600	5000	81700	.	
dist_aadtmax	647	642	550	875	.	
road_aadtmax	0	495	0	495	.	
aadt_aadtmax	135800	150100	15600	166100	.	
County Level NOx Emissions (tpy)						
OnRoad_mobile_nox	5664	11095	11095	16911	8518	
NonRoad_mobile_nox	7424	4060	4060	6367	2766	
AML_mobile_nox	6085	6546	6546	8122	371	
total_nox	33400	29220	29220	39142	16263	
pct_mobile_nox	57.4	74.3	74.3	80.2	71.7	
NOx Emissions by Source Type (summed tons per year (tpy), sources within 5 Km emitting at least 10 tpy)						
Airport	12	.	1694	.	131	
Electricity_Generation_via_Combu	2835	.	274	301	.	
Municipal_Waste_Combustor	.	.	.	.	.	
Petroleum_Refinery	.	.	.	.	.	
Institutional__school__hospital_	300	18	41	.	.	
Wastewater_Treatment_Facility	14	.	108	.	.	
Breweries_Distilleries_Wineries	.	.	.	.	.	
Petroleum_Storage_Facility	.	.	.	.	.	
Pharmaceutical_Manufacturing	.	.	.	.	.	
Fabricated_Metal_Products_Plant	.	.	.	.	.	
Compressor_Station	.	.	11	.	.	
Steam_Heating_Facility	363	.	.	.	.	
not_characterized	202	33	14	.	.	
		Monitor not used for 2010-2015 analysis (either historical data or were incomplete)				

Table B5-47. Attributes of ambient monitors within the Philadelphia study area not used for the 2010-2015 analysis.

State abbreviation	DE	DE	DE	DE	DE	NJ
County name	New Castle	New Castle	New Castle	New Castle	New Castle	Camden
Site ID	100031003	100031007	100031010	100032002	100033001	340070003
Lat	39.76111	39.55130	39.81722	39.75789	39.81233	39.92304
Lon	-75.49194	-75.73200	-75.56389	-75.54603	-75.45520	-75.097617
year_start	1992	1992	2013	1977	1977	1979
year_end	2000	1999	2014	1992	1992	2008
Elevation_m	65	20	0	46	30	7.6
land_use	RESID	AGRIC	AGRIC	COMM	RESID	RESID
scale	NA	NA	NA	NBHOOD	NBHOOD	NBHOOD
objective1	HI_CONC	POP_EXP	POP_EXP	POP_EXP	HI_CONC	POP_EXP
objective2	POP_EXP	GEN/BKGR		HI_CONC		HI_CONC
objective3		UP_BKGR				
cnty_landarea_2010	426	426	426	426	426	221
AADT information (distance in meters, road/highway/interstate number, AADT value)						
dist_closest	253	560	.	374	49	512
road_closest	495	301	.	202	95	605
aadt_closest	73574	21601	.	8387	48906	23085
dist_aadtmax	.	855	.	578	386	838
road_aadtmax	.	301	.	95	95	30
aadt_aadtmax	.	38941	.	100551	118099	57822
County Level NOx Emissions (tpy)						
OnRoad_mobile_nox	6459	6459	6459	6459	6459	5353
NonRoad_mobile_nox	1748	1748	1748	1748	1748	1150
AML_mobile_nox	1805	1805	1805	1805	1805	1088
total_nox	13991	13991	13991	13991	13991	9431
pct_mobile_nox	71.6	71.6	71.6	71.6	71.6	80.5
NOx Emissions by Source Type (summed tons per year (tpy), sources within 5 Km emitting at least 10 tpy)						
Petroleum_Refinery	.	.	.	.	2146	.
Municipal_Waste_Combustor	.	.	.	20	.	297
Electricity_Generation_via_Combu	948	.	.	948	963	26
Chemical_Plant	27	.	.	27	.	14
Pulp_and_Paper_Plant	.	.	.	.	.	.
Institutional__school__hospital_	.	.	.	.	.	.
Wastewater_Treatment_Facility	.	.	.	.	.	14
Steam_Heating_Facility	.	.	.	.	.	.
Petroleum_Storage_Facility	.	.	.	.	.	.
Military_Base	.	.	.	.	.	.
Automobile_Truck_or_Parts_Plant	.	.	.	.	.	.
not_characterized	.	.	180	180	54	13
Steel_Mill	.	.	.	.	166	.
Hot_Mix_Asphalt_Plant	.	.	.	.	.	.
Pharmaceutical_Manufacturing	.	.	.	.	.	.
	Monitor not used for 2010-2015 analysis (either historical data or were incomplete)					

Table B5-47, continued. Attributes of ambient monitors within the Philadelphia study area not used for the 2010-2015 analysis.

State abbreviation	PA	PA				
County name	Montgomery	Philadelphia				
Site ID	420910013	421010029				
Lat	40.11222	39.95733				
Lon	-75.30917	-75.17268				
year_start	1973	1975				
year_end	2008	2005				
Elevation_m	53	25				
land_use	RESID	COMM				
scale	NBHOOD	NBHOOD				
objective1	POP_EXP	POP_EXP				
objective2		HI_CONC				
objective3						
cnty_landarea_2010	483	134				
AADT information (distance in meters, road/highway/interstate number, AADT value)						
dist_closest	569	198				
road_closest	276	0				
aadt_closest	63476	23842				
dist_aadtmax	.	586				
road_aadtmax	.	76				
aadt_aadtmax	.	144789				
County Level NOx Emissions (tpy)						
OnRoad_mobile_nox	9533	10201				
NonRoad_mobile_nox	3076	2480				
AML_mobile_nox	219	2160				
total_nox	16546	21065				
pct_mobile_nox	77.5	70.5				
NOx Emissions by Source Type (summed tons per year (tpy), sources within 5 Km emitting at least 10 tpy)						
Petroleum_Refinery	.	.				
Municipal_Waste_Combustor	735	.				
Electricity_Generation_via_Combu	.	353				
Chemical_Plant	.	.				
Pulp_and_Paper_Plant	.	.				
Institutional__school__hospital_	.	66				
Wastewater_Treatment_Facility	.	.				
Steam_Heating_Facility	.	24				
Petroleum_Storage_Facility	.	.				
Military_Base	.	.				
Automobile_Truck_or_Parts_Plant	.	.				
not_characterized	53	.				
Steel_Mill	85	.				
Hot_Mix_Asphalt_Plant	16	.				
Pharmaceutical_Manufacturing	26	.				
	Monitor not used for 2010-2015 analysis (either historical data or were incomplete)					

Table B5-48. Attributes of ambient monitors within the Phoenix study area not used for the 2010-2015 analysis.

State abbreviation	AZ	AZ	AZ	AZ		
County name	Maricopa	Maricopa	Maricopa	Pinal		
Site ID	040134005	040139993	040139994	040218001		
Lat	33.41240	33.34885	33.48889	33.29347		
Lon	-111.93473	-112.83110	-111.86250	-111.28559		
year_start	2000	1996	1991	2001		
year_end	2003	2004	1999	2006		
Elevation_m	352	265	0	634		
land_use	RESID	INDUS	AGRIC	DESERT		
scale	NA	NA	NA	NA		
objective1	POP_EXP	GEN/BKGR	UP_BKGR	DOWNWND		
objective2				HI_CONC		
objective3				REG TRANS		
cnty_landarea_2010	9200	9200	9200	5366		
AADT information (distance in meters, road/highway/interstate number, AADT value)						
dist_closest	522	.	.	.		
road_closest	0	.	.	.		
aadt_closest	25583	.	.	.		
dist_aadtmax	906	.	.	.		
road_aadtmax	0	.	.	.		
aadt_aadtmax	36152	.	.	.		
County Level NOx Emissions (tpy)						
OnRoad_mobile_nox	56748	56748	56748	9273		
NonRoad_mobile_nox	18998	18998	18998	1575		
AML_mobile_nox	3999	3999	3999	1775		
total_nox	88464	88464	88464	14883		
pct_mobile_nox	90.1	90.1	90.1	84.8		
NOx Emissions by Source Type (summed tons per year (tpy), sources within 5 Km emitting at least 10 tpy)						
Airport	.	.	.	.		
Electricity_Generation_via_Combu	83	343	.	.		
Rail_Yard	.	.	.	.		
	Monitor not used for 2010-2015 analysis (either historical data or were incomplete)					

Table B5-49. Attributes of ambient monitors within the Pittsburgh study area not used for the 2010-2014 analysis.

State abbreviation	PA	PA	PA	PA	PA	
County name	Allegheny	Allegheny	Washington	Washington	Westmoreland	
Site ID	420030003	420030031	421255001	421255200	421290008	
Lat	40.45010	40.44337	40.44528	40.26896	40.30469	
Lon	-79.77096	-79.99029	-80.42083	-80.24400	-79.50567	
year_start	1987	1980	1998	2012	1997	
year_end	1992	2001	2008	2016	2008	
Elevation_m	11	268	335	327	0	
land_use	RESID	COMM	AGRIC	AGRIC	COMM	
scale	NBHOOD	NBHOOD	REGION	NBHOOD	URBAN	
objective1	POP_EXP	POP_EXP	REG TRANS	SOURCE	POP_EXP	
objective2		HI_CONC	GEN/BKGR	POP_EXP	REG TRANS	
objective3			POP_EXP			
cnty_landarea_2010	730	730	857	857	1028	
AADT information (distance in meters, road/highway/interstate number, AADT value)						
dist_closest	563	66	.	.	177	
road_closest	376	579	.	.	30	
aadt_closest	52875	54441	.	.	34536	
dist_aadtmax	.	.	.	.	186	
road_aadtmax	.	.	.	.	30	
aadt_aadtmax	.	.	.	.	44067	
County Level NOx Emissions (tpy)						
OnRoad_mobile_nox	13259	13259	3024	3024	4743	
NonRoad_mobile_nox	4029	4029	856	856	1147	
AML_mobile_nox	3322	3322	686	686	1458	
total_nox	35455	35455	10067	10067	12939	
pct_mobile_nox	58.1	58.1	45.4	45.4	56.8	
NOx Emissions by Source Type (summed tons per year (tpy), sources within 5 Km emitting at least 10 tpy)						
Steel_Mill	.	.	.	.	.	
Food_Products_Processing_Plant	.	212	.	.	.	
Steam_Heating_Facility	.	166	.	.	.	
Institutional__school__hospital_	.	36	.	.	.	
Wastewater_Treatment_Facility	.	.	.	.	.	
Hot_Mix_Asphalt_Plant	.	18	.	.	.	
Electricity_Generation_via_Combu	.	33	.	.	.	
Foundries__Iron_and_Steel	.	13	.	.	.	
Glass_Plant	.	.	.	.	.	
Chemical_Plant	.	.	.	.	.	
Coke_Battery	.	.	.	.	.	
not_characterized	.	15	.	.	.	
Landfill	.	.	.	.	.	
	Monitor not used for 2010-2015 analysis (either historical data or were incomplete)					

Table B5-50. Attributes of ambient monitors within the Richmond study area not used for the 2010-2015 analysis.

State abbreviation	VA	VA				
County name	Caroline	Richmond City				
Site ID	510330001	517600021				
Lat	38.20087	37.56320				
Lon	-77.37742	-77.46721				
year_start	1993	1981				
year_end	2012	1997				
Elevation_m	68	58				
land_use	FOREST	COMM				
scale	URBAN	NA				
objective1	HI_CONC	HI_CONC				
objective2	UP_BKGR					
objective3	GEN/BKGR					
cnty_landarea_2010	528	60				
AADT information (distance in meters, road/highway/interstate number, AADT value)						
dist_closest	.	178				
road_closest	.	161				
aadt_closest	.	22386				
dist_aadtmax	.	342				
road_aadtmax	.	33				
aadt_aadtmax	.	22984				
County Level NOx Emissions (tpy)						
OnRoad_mobile_nox	2472	2719				
NonRoad_mobile_nox	166	396				
AML_mobile_nox	407	315				
total_nox	3486	6928				
pct_mobile_nox	87.3	49.5				
NOx Emissions by Source Type (summed tons per year (tpy), sources within 5 Km emitting at least 10 tpy)						
Pulp_and_Paper_Plant	.	.				
Rail_Yard	.	115				
Institutional__school__hospital_	.	10				
not_characterized	65	24				
		Monitor not used for 2010-2015 analysis (either historical data or were incomplete)				

Table B5-51. Attributes of ambient monitors within the Riverside study area not used for the 2010-2015 analysis.

State abbreviation	CA	CA	CA	CA	CA	CA
County name	Riverside	Riverside	Riverside	San Bernardino	San Bernardino	San Bernardino
Site ID	060650004	060650006	060656001	060710006	060710012	060710014
Lat	34.00700	33.49170	33.78942	35.75995	34.42613	34.51250
Lon	-	-	-	-117.37478	-117.56394	-117.33088
year_start	2007	1991	1988	1979	1987	1991
year_end	2011	1993	1990	1994	1998	1999
Elevation_m	70	341	439	506	4100	876
land_use	RESID	RESID	COMM	INDUS	COMM	RESID
scale	NA	MID	NA	NA	NA	NA
objective1	POP_EXP		POP_EXP		REG TRANS	REG TRANS
objective2						
objective3						
cnty_landarea_2010	7206	7206	7206	20057	20057	20057
AADT information (distance in meters, road/highway/interstate number, AADT value)						
dist_closest	.	64	177	.	.	580
road_closest	.	15	215	.	.	18
aadt_closest	.	152000	82000	.	.	40000
dist_aadtmax	.	.	250	.	.	797
road_aadtmax	.	.	215	.	.	15
aadt_aadtmax	.	.	99000	.	.	85000
County Level NOx Emissions (tpy)						
OnRoad_mobile_nox	25868	25868	25868	31067	31067	31067
NonRoad_mobile_nox	5541	5541	5541	5646	5646	5646
AML_mobile_nox	2215	2215	2215	9344	9344	9344
total_nox	37367	37367	37367	68863	68863	68863
pct_mobile_nox	90	90	90	66.9	66.9	66.9
NOx Emissions by Source Type (summed tons per year (tpy), sources within 5 Km emitting at least 10 tpy)						
Mineral_Processing_Plant	.	.	.	1865	.	.
Electricity_Generation_via_Combu	.	.	.	620	.	.
Airport	.	.	.	.	.	.
Rail_Yard	.	.	.	.	.	.
Pulp_and_Paper_Plant	140	.	.	.	.	.
Fabricated_Metal_Products_Plant	20	.	.	.	.	.
Food_Products_Processing_Plant	.	.	.	.	.	.
Foundries__non_ferrous	.	.	.	.	.	.
not_characterized	.	.	.	.	.	.
Municipal_Waste_Combustor	11	.	.	.	.	.
Portland_Cement_Manufacturing	.	.	.	.	.	.
Primary_Aluminum_Plant	17	.	.	.	.	.
Secondary_Aluminum_Smelting_Refi	.	.	.	.	.	.
Steel_Mill	.	.	.	.	.	.
Institutional__school__hospital_	.	.	.	.	.	.
	Monitor not used for 2010-2015 analysis (either historical data or were incomplete)					

Table B5-51, continued. Attributes of ambient monitors within the Riverside study area not used for the 2010-2015 analysis.

State abbreviation	CA	CA	CA	CA	CA	
County name	San Bernardino	San Bernardino	San Bernardino	San Bernardino	San Bernardino	
Site ID	060710015	060710017	060711101	060714001	060717002	
Lat	35.77495	34.14195	34.14195	34.41807	34.52333	
Lon	-117.36756	-116.05584	-116.05917	-117.28560	-117.30449	
year_start	1993	1993	1992	1979	1981	
year_end	1997	1998	1993	1998	1991	
Elevation_m	498	607	650	1006	895	
land_use	INDUS	MOBILE	AGRIC	RESID	COMM	
scale	NA	NA	NA	NA	NA	
objective1		REG TRANS		REG TRANS		
objective2		POP_EXP		POP_EXP		
objective3				UP_BKGR		
cnty_landarea_2010	20057	20057	20057	20057	20057	
AADT information (distance in meters, road/highway/interstate number, AADT value)						
dist_closest	.	665	669	.	.	
road_closest	.	62	62	.	.	
aadt_closest	.	7100	13000	.	.	
dist_aadtmax	.	708	.	.	.	
road_aadtmax	.	62	.	.	.	
aadt_aadtmax	.	13000	.	.	.	
County Level NOx Emissions (tpy)						
OnRoad_mobile_nox	31067	31067	31067	31067	31067	
NonRoad_mobile_nox	5646	5646	5646	5646	5646	
AML_mobile_nox	9344	9344	9344	9344	9344	
total_nox	68863	68863	68863	68863	68863	
pct_mobile_nox	66.9	66.9	66.9	66.9	66.9	
NOx Emissions by Source Type (summed tons per year (tpy), sources within 5 Km emitting at least 10 tpy)						
Mineral_Processing_Plant	1865	.	.	.	.	
Electricity_Generation_via_Combu	620	.	.	.	.	
Airport	.	.	.	.	.	
Rail_Yard	.	.	.	.	.	
Pulp_and_Paper_Plant	.	.	.	.	.	
Fabricated_Metal_Products_Plant	.	.	.	.	.	
Food_Products_Processing_Plant	.	.	.	.	.	
Foundries__non_ferrous	.	.	.	.	.	
not_characterized	.	.	.	.	.	
Municipal_Waste_Combustor	.	.	.	.	.	
Portland_Cement_Manufacturing	.	.	.	.	33	
Primary_Aluminum_Plant	.	.	.	.	.	
Secondary_Aluminum_Smelting_Refi	.	.	.	.	.	
Steel_Mill	.	.	.	.	.	
Institutional__school__hospital_	.	.	.	.	.	
		Monitor not used for 2010-2015 analysis (either historical data or were incomplete)				

Table B5-52. Attributes of ambient monitors within the Sacramento study area not used for the 2010-2015 analysis.

State abbreviation	CA	CA	CA	CA	CA	CA
County name	El Dorado	El Dorado	El Dorado	Placer	Placer	Sacramento
Site ID	060170009	060170011	060170012	060610007	060613001	060670001
Lat	38.94574	38.94498	38.81161	39.18407	38.78879	38.66713
Lon	-119.96796	-119.97061	-120.03308	-120.12298	-121.21773	-121.25134
year_start	1981	1992	1999	2002	1991	1979
year_end	1992	2004	2004	2004	1996	1993
Elevation_m	1911	1905	2250	1	75	52
land_use	COMM	COMM	FOREST	COMM	RESID	RESID
scale	NA	NA	NA	URBAN	NA	NA
objective1		POP_EXP	HI_CONC	POP_EXP	POP_EXP	POP_EXP
objective2			GEN/BKGR			
objective3						
cnty_landarea_2010	1708	1708	1708	1407	1407	965
AADT information (distance in meters, road/highway/interstate number, AADT value)						
dist_closest	28	86	183	.	321	.
road_closest	50	50	50	.	80	.
aadt_closest	30500	30500	8900	.	89000	.
dist_aadtmax	479	185	.	.	486	.
road_aadtmax	50	50	.	.	80	.
aadt_aadtmax	33000	33000	.	.	113000	.
County Level NOx Emissions (tpy)						
OnRoad_mobile_nox	1886	1886	1886	4760	4760	12361
NonRoad_mobile_nox	721	721	721	1492	1492	3122
AML_mobile_nox	94	94	94	1150	1150	1821
total_nox	3637	3637	3637	9158	9158	19897
pct_mobile_nox	74.3	74.3	74.3	80.8	80.8	87
NOx Emissions by Source Type (summed tons per year (tpy), sources within 5 Km emitting at least 10 tpy)						
Airport	.	.	.	.	.	.
Rail_Yard	.	.	.	.	.	.
Institutional__school__hospital_	.	.	.	.	.	.
not_characterized	27	27	.	.	.	.
		Monitor not used for 2010-2015 analysis (either historical data or were incomplete)				

Table B5-52, continued. Attributes of ambient monitors within the Sacramento study area not used for the 2010-2015 analysis.

State abbreviation	CA	CA	CA			
County name	Sacramento	Sacramento	Sacramento			
Site ID	060670013	060671001	060675002			
Lat	38.63685	38.67490	38.71685			
Lon	-121.51440	-121.18689	-121.59301			
year_start	1998	1979	1989			
year_end	2008	1996	1997			
Elevation_m	5	57	30			
land_use	COMM	COMM	AGRIC			
scale	NA	NA	NBHOOD			
objective1	HI_CONC	HI_CONC	POP_EXP			
objective2	POP_EXP					
objective3						
cnty_landarea_2010	965	965	965			
AADT information (distance in meters, road/highway/interstate number, AADT value)						
dist_closest	704	.	.			
road_closest	5	.	.			
aadt_closest	140000	.	.			
dist_aadtmax	878	.	.			
road_aadtmax	80	.	.			
aadt_aadtmax	144000	.	.			
County Level NOx Emissions (tpy)						
OnRoad_mobile_nox	12361	12361	12361			
NonRoad_mobile_nox	3122	3122	3122			
AML_mobile_nox	1821	1821	1821			
total_nox	19897	19897	19897			
pct_mobile_nox	87	87	87			
NOx Emissions by Source Type (summed tons per year (tpy), sources within 5 Km emitting at least 10 tpy)						
Airport	.	.	586			
Rail_Yard	.	.	.			
Institutional__school__hospital_	.	.	.			
not_characterized	.	.	.			
	Monitor not used for 2010-2015 analysis (either historical data or were incomplete)					

Table B5-53. Attributes of ambient monitors within the San Diego study area not used for the 2010-2015 analysis.

State abbreviation	CA	CA	CA	CA		
County name	San Diego	San Diego	San Diego	San Diego		
Site ID	060730005	060731007	060731009	060731018		
Lat	33.20269	32.70922	32.69866	32.81798		
Lon	-117.36680	-117.15484	-117.13309	-116.96813		
year_start	1979	1989	1999	2014		
year_end	2001	2005	2001	2016		
Elevation_m	37	6	0	119		
land_use	UNK	COMM	RESID	COMM		
scale	NBHOOD	NBHOOD	NA	NBHOOD		
objective1	POP_EXP	HI_CONC	POP_EXP	SOURCE		
objective2		POP_EXP		GEN/BKGR		
objective3				POP_EXP		
cnty_landarea_2010	4207	4207	4207	4207		
AADT information (distance in meters, road/highway/interstate number, AADT value)						
dist_closest	505	575	369	670		
road_closest	5	5	5	67		
aadt_closest	192000	162000	159000	83000		
dist_aadtmax	.	589	686	.		
road_aadtmax	.	5	5	.		
aadt_aadtmax	.	208000	161000	.		
County Level NOx Emissions (tpy)						
OnRoad_mobile_nox	27531	27531	27531	27531		
NonRoad_mobile_nox	7428	7428	7428	7428		
AML_mobile_nox	3491	3491	3491	3491		
total_nox	42700	42700	42700	42700		
pct_mobile_nox	90	90	90	90		
NOx Emissions by Source Type (summed tons per year (tpy), sources within 5 Km emitting at least 10 tpy)						
Electricity_Generation_via_Combu	.	.	.	.		
Military_Base	.	11	11	.		
Pharmaceutical_Manufacturing	.	38	38	.		
Ship_Boat_Manufacturing_or_Repai	.	13	13	.		
not_characterized	.	88	66	.		
	Monitor not used for 2010-2015 analysis (either historical data or were incomplete)					

Table B5-54. Attributes of ambient monitors within the San Francisco study area not used for the 2010-2015 analysis.

State abbreviation	CA	CA	CA	CA	CA	CA
County name	Alameda	Alameda	Alameda	Contra Costa	Contra Costa	Contra Costa
Site ID	060010003	060010010	060011001	060130003	060130010	060131003
Lat	37.68490	37.76023	37.53583	37.94992	38.03122	37.96420
Lon	-121.76590	-122.19358	-121.96182	-122.35719	-122.13288	-122.34030
year_start	1980	2001	1973	1972	2001	1997
year_end	2000	2003	2010	1997	2003	2002
Elevation_m	150	10	17	23	36	15
land_use	COMM	RESID	RESID	COMM	COMM	COMM
scale	NA	NBHOOD	URBAN	NA	NBHOOD	NA
objective1	HI_CONC	POP_EXP	POP_EXP	POP_EXP	POP_EXP	POP_EXP
objective2	POP_EXP			HI_CONC	WELF IMP	
objective3						
cnty_landarea_2010	739	739	739	716	716	716
AADT information (distance in meters, road/highway/interstate number, AADT value)						
dist_closest	.	137	.	.	.	916
road_closest	.	185	.	.	.	80
aadt_closest	.	24000	.	.	.	186000
dist_aadtmax	.	.	.	.	.	.
road_aadtmax	.	.	.	.	.	.
aadt_aadtmax	.	.	.	.	.	.
County Level NOx Emissions (tpy)						
OnRoad_mobile_nox	19032	19032	19032	9307	9307	9307
NonRoad_mobile_nox	3158	3158	3158	2074	2074	2074
AML_mobile_nox	3507	3507	3507	1973	1973	1973
total_nox	28381	28381	28381	20714	20714	20714
pct_mobile_nox	90.5	90.5	90.5	64.5	64.5	64.5
NOx Emissions by Source Type (summed tons per year (tpy), sources within 5 Km emitting at least 10 tpy)						
Petroleum_Refinery	.	941	.	.	.	.
Airport	.	.	.	.	.	.
Electricity_Generation_via_Combu	.	327	.	.	.	.
Glass_Plant	.	.	.	127	65	127
Rail_Yard	.	.	.	.	.	.
Wastewater_Treatment_Facility	.	36	.	.	.	.
Foundries__Iron_and_Steel	.	.	.	.	.	.
Pharmaceutical_Manufacturing	.	.	.	33	.	39
not_characterized	.	.	.	24	.	24
Landfill	.	.	.	.	13	.
Hot_Mix_Asphalt_Plant	.	.	.	.	.	.
Chemical_Plant	.	.	.	.	.	.
Institutional__school__hospital_	.	941	.	.	.	.
	Monitor not used for 2010-2015 analysis (either historical data or were incomplete)					

Table B5-54, continued. Attributes of ambient monitors within the San Francisco study area not used for the 2010-2015 analysis.

State abbreviation	CA	CA				
County name	Contra Costa	San Francisco				
Site ID	060133001	060750006				
Lat	38.02926	37.73380				
Lon	-121.89687	-122.38240				
year_start	1967	2004				
year_end	2008	2005				
Elevation_m	2	82				
land_use	RESID	RESID				
scale	NBHOOD	NBHOOD				
objective1	HI_CONC	POP_EXP				
objective2	POP_EXP					
objective3						
cnty_landarea_2010	716	47				
AADT information (distance in meters, road/highway/interstate number, AADT value)						
dist_closest	.	.				
road_closest	.	.				
aadt_closest	.	.				
dist_aadtmax	.	.				
road_aadtmax	.	.				
aadt_aadtmax	.	.				
County Level NOx Emissions (tpy)						
OnRoad_mobile_nox	9307	4623				
NonRoad_mobile_nox	2074	1474				
AML_mobile_nox	1973	5065				
total_nox	20714	12406				
pct_mobile_nox	64.5	90				
NOx Emissions by Source Type (summed tons per year (tpy), sources within 5 Km emitting at least 10 tpy)						
Petroleum_Refinery	.	.				
Airport	.	.				
Electricity_Generation_via_Combu	350	.				
Glass_Plant	.	.				
Rail_Yard	.	.				
Wastewater_Treatment_Facility	.	74				
Foundries__Iron_and_Steel	.	.				
Pharmaceutical_Manufacturing	.	.				
not_characterized	17	.				
Landfill	.	.				
Hot_Mix_Asphalt_Plant	.	.				
Chemical_Plant	56	.				
Institutional__school__hospital_	.	10				
		Monitor not used for 2010-2015 analysis (either historical data or were incomplete)				

Table B5-55. Attributes of ambient monitors within the St. Louis study area not used for the 2010-2015 analysis.

State abbreviation	MO	MO	MO	MO	MO	MO	MO
County name	Saint Charles	Saint Charles	Saint Louis	Saint Louis	Saint Louis	Saint Louis	Saint Louis
Site ID	291830010	291831002	291890001	291890004	291890006	291890014	291893001
Lat	38.58192	38.87255	38.52172	38.53278	38.61366	38.71090	38.65026
Lon	-90.83531	-90.22649	-90.34371	-90.38243	-90.49594	-90.47590	-90.35046
min_year	1994	1975	1977	1998	1978	2005	1970
max_year	1998	2010	1998	2010	2005	2010	2010
Elevation_m	0	131	183	183	175	193	161
land_use	AGRIC	AGRIC	RESID	RESID	RESID	RESID	COMM
scale	NA	URBAN	NA	NA	NA	NBHOOD	NA
objective1		POP_EXP	POP_EXP	POP_EXP	POP_EXP	POP_EXP	POP_EXP
objective2		MAX_O3		QUAL			
objective3		HI_CONC					
cnty_landarea_2010	560	560	508	508	508	508	508
AADT information (distance in meters, road/highway/interstate number, AADT value)							
dist_closest	.	905	143	131	.	539	75
road_closest	.	67	50	50	.	364	170
aadt_closest	.	26451	30939	19136	.	27533	114639
aadt_aadtmax	.	.	927	187	.	.	.
road_aadtmax	.	.	55	30	.	.	.
dist_aadtmax	.	.	111855	27183	.	.	.
County Level NOx Emissions (tpy)							
OnRoad_mobile_nox	7867	7867	25063	25063	25063	25063	25063
NonRoad_mobile_nox	1714	1714	4697	4697	4697	4697	4697
AML_mobile_nox	580	580	1902	1902	1902	1902	1902
total_nox	18373	18373	39486	39486	39486	39486	39486
pct_mobile_nox	55.3	55.3	80.2	80.2	80.2	80.2	80.2
NOx Emissions by Source Type (summed tons per year (tpy), sources within 5 Km emitting at least 10 tpy)							
Breweries_Distilleries_Wineries	.	.	.	.	.	.	.
Rail_Yard	.	.	.	.	.	.	.
Electricity_Generation_via_Combu	9891	.	.	.	.	.	.
Wastewater_Treatment_Facility	.	.	.	.	.	89	.
Institutional__school__hospital_	.	.	.	.	.	.	20
Ethanol_Biorefineries_Soy_Biodie	.	.	.	.	.	.	.
Chemical_Plant	.	.	.	.	.	.	.
Pharmaceutical_Manufacturing	.	.	.	.	.	.	.
Landfill	.	.	.	.	.	11	.
Steam_Heating_Facility	.	.	.	.	.	.	.
Petroleum_Storage_Facility	.	.	.	.	.	.	.
not_characterized	.	.	.	.	.	.	22
Aircraft__Aerospace__or_Related	.	.	.	.	.	.	.
Airport	.	.	.	.	.	.	.
	Monitor not used for 2010-2015 analysis (either historical data or were incomplete)						

Table B5-55, continued. Attributes of ambient monitors within the St. Louis study area not used for the 2010-2015 analysis.

State abbreviation	MO	MO	MO	MO	MO	MO
County name	Saint Louis	Saint Louis	Saint Louis	Saint Louis	St. Louis City	St. Louis City
Site ID	291895001	291897001	291897002	291897003	295100072	295100080
Lat	38.76616	38.72755	38.72727	38.72096	38.62422	38.65864
Lon	-90.28593	-90.37955	-90.37955	-90.36713	-90.19871	-90.24426
min_year	1978	1976	1990	2001	1981	1982
max_year	2005	1990	2001	2004	2005	1999
Elevation_m	168	168	168	0	154	152
land_use	COMM	RESID	RESID	RESID	COMM	RESID
scale	NA	NA	NA	NBHOOD	NA	NBHOOD
objective1	POP_EXP	POP_EXP	POP_EXP	POP_EXP	POP_EXP	POP_EXP
objective2	QUAL		SOURCE	HI_CONC		HI_CONC
objective3	HI_CONC					
cnty_landarea_2010	508	508	508	508	62	62
AADT information (distance in meters, road/highway/interstate number, AADT value)						
dist_closest	433	77	53	103	70	483
road_closest	0	0	0	0	0	0
aadt_closest	39856	32999	32999	32999	16662	12582
aadt_aadtmax	536	.	.	.	880	.
road_aadtmax	270	.	.	.	55	.
dist_aadtmax	137862	.	.	.	99609	.
County Level NOx Emissions (tpy)						
OnRoad_mobile_nox	25063	25063	25063	25063	6296	6296
NonRoad_mobile_nox	4697	4697	4697	4697	663	663
AML_mobile_nox	1902	1902	1902	1902	1590	1590
total_nox	39486	39486	39486	39486	10691	10691
pct_mobile_nox	80.2	80.2	80.2	80.2	80	80
NOx Emissions by Source Type (summed tons per year (tpy), sources within 5 Km emitting at least 10 tpy)						
Breweries_Distilleries_Wineries	.	.	.	.	467	.
Rail_Yard	.	.	.	.	411	421
Electricity_Generation_via_Combu	.	.	.	.	285	.
Wastewater_Treatment_Facility	.	.	.	.	.	81
Institutional__school__hospital_	.	.	.	.	.	49
Ethanol_Biorefineries_Soy_Biodie	.	.	.	.	49	.
Chemical_Plant	.	.	.	.	27	93
Pharmaceutical_Manufacturing	.	.	.	.	43	43
Landfill	.	.	.	.	58	.
Steam_Heating_Facility	.	.	.	.	20	.
Petroleum_Storage_Facility	.	.	.	.	14	.
not_characterized	.	13	13	13	12	25
Aircraft__Aerospace__or_Related	.	24	24	24	.	.
Airport	.	858	858	858	.	.
	Monitor not used for 2010-2015 analysis (either historical data or were incomplete)					

Table B5-56. Attributes of ambient monitors within the Washington DC study area not used for the 2010-2015 analysis.

State abbreviation	DC	VA	VA	VA	VA	VA
County name	DoColumbia	Fairfax	Fairfax	Fairfax	Fairfax	Fairfax
Site ID	110010017	510590005	510590018	510590031	510591004	510591005
Lat	38.90372	38.89410	38.74232	38.76835	38.86817	38.83738
Lon	-77.05137	-77.46520	-77.07743	-77.18347	-77.14276	-77.16338
min_year	1975	1977	1979	2016	1973	2002
max_year	1996	2009	1997	2016	2001	2009
Elevation_m	20	76	10	73.5	110	116
land_use	COMM	RESID	RESID	COMM	COMM	RESID
scale	NBHOOD	NBHOOD	NA	MICRO	NA	NA
objective1	POP_EXP	POP_EXP	POP_EXP	HI_CONC	POP_EXP	POP_EXP
objective2	HI_CONC			POP_EXP		
objective3						
cnty_landarea_2010	61	391	391	391	391	391
AADT information (distance in meters, road/highway/interstate number, AADT value)						
dist_closest	52	.	580	69	94	57
road_closest	0	.	1	7900	50	244
aadt_closest	29208	.	56482	52319	48102	29542
aadt_aadtmax	395	.	.	411	498	973
road_aadtmax	0	.	.	95	50	244
dist_aadtmax	96163	.	.	231651	59759	35060
County Level NOx Emissions (tpy)						
OnRoad_mobile_nox	4739	8221	8221	8221	8221	8221
NonRoad_mobile_nox	2364	2905	2905	2905	2905	2905
AML_mobile_nox	223	358	358	358	358	358
total_nox	9418	15236	15236	15236	15236	15236
pct_mobile_nox	77.8	75.4	75.4	75.4	75.4	75.4
NOx Emissions by Source Type (summed tons per year (tpy), sources within 5 Km emitting at least 10 tpy)						
Airport	.	.	.	.	.	.
Electricity_Generation_via_Combu	.	.	.	.	.	.
Municipal_Waste_Combustor	.	.	.	.	.	.
Steam_Heating_Facility	301	.	.	.	.	.
Military_Base	.	.	.	.	.	.
not_characterized	.	.	.	.	.	59
Institutional__school__hospital_	27	.	.	.	.	.
Rail_Yard	.	.	.	.	.	.
Hot_Mix_Asphalt_Plant	.	.	.	.	.	11

Monitor not used for 2010-2015 analysis (either historical data or were incomplete)

Table B5-56, continued. Attributes of ambient monitors within the Washington DC study area not used for the 2010-2015 analysis.

State abbreviation	VA	VA				
County name	Fairfax	Fairfax City				
Site ID	510595001	516000005				
Lat	38.93260	38.84483				
Lon	-77.19822	-77.31165				
min_year	1973	1973				
max_year	2009	1992				
Elevation_m	105	125				
land_use	RESID	RESID				
scale	NBHOOD	NA				
objective1	POP_EXP	POP_EXP				
objective2						
objective3						
cnty_landarea_2010	391	6				
AADT information (distance in meters, road/highway/interstate number, AADT value)						
dist_closest	74	304				
road_closest	123	236				
aadt_closest	41718	43148				
aadt_aadtmax	853	.				
road_aadtmax	495	.				
dist_aadtmax	193959	.				
County Level NOx Emissions (tpy)						
OnRoad_mobile_nox	8221	156				
NonRoad_mobile_nox	2905	50				
AML_mobile_nox	358	0				
total_nox	15236	265				
pct_mobile_nox	75.4	77.9				
NOx Emissions by Source Type (summed tons per year (tpy), sources within 5 Km emitting at least 10 tpy)						
Airport	.	.				
Electricity_Generation_via_Combu	.	.				
Municipal_Waste_Combustor	.	.				
Steam_Heating_Facility	.	.				
Military_Base	.	.				
not_characterized	.	.				
Institutional__school__hospital_	.	.				
Rail_Yard	.	.				
Hot_Mix_Asphalt_Plant	.	.				
		Monitor not used for 2010-2015 analysis (either historical data or were incomplete)				

5.4 UPPER (>98TH) PERCENTILE DM1H CONCENTRATION RATIOS: COMPARISON OF AREA DESIGN VALUE MONITOR TO ALL AREA-WIDE MONITORS BY STUDY AREA

The following is a visual representation of the data found in Table B2-8, constructed to facilitate the comparison of the ratios derived from the area design value monitor relative to other area-wide monitors. As a reminder, the near-road monitors typically had only 1 full year of data, the motivation for using the ratios derived from the area design value monitor to adjust the upper (>98th) percentile DM1H concentrations at each near-road monitor.

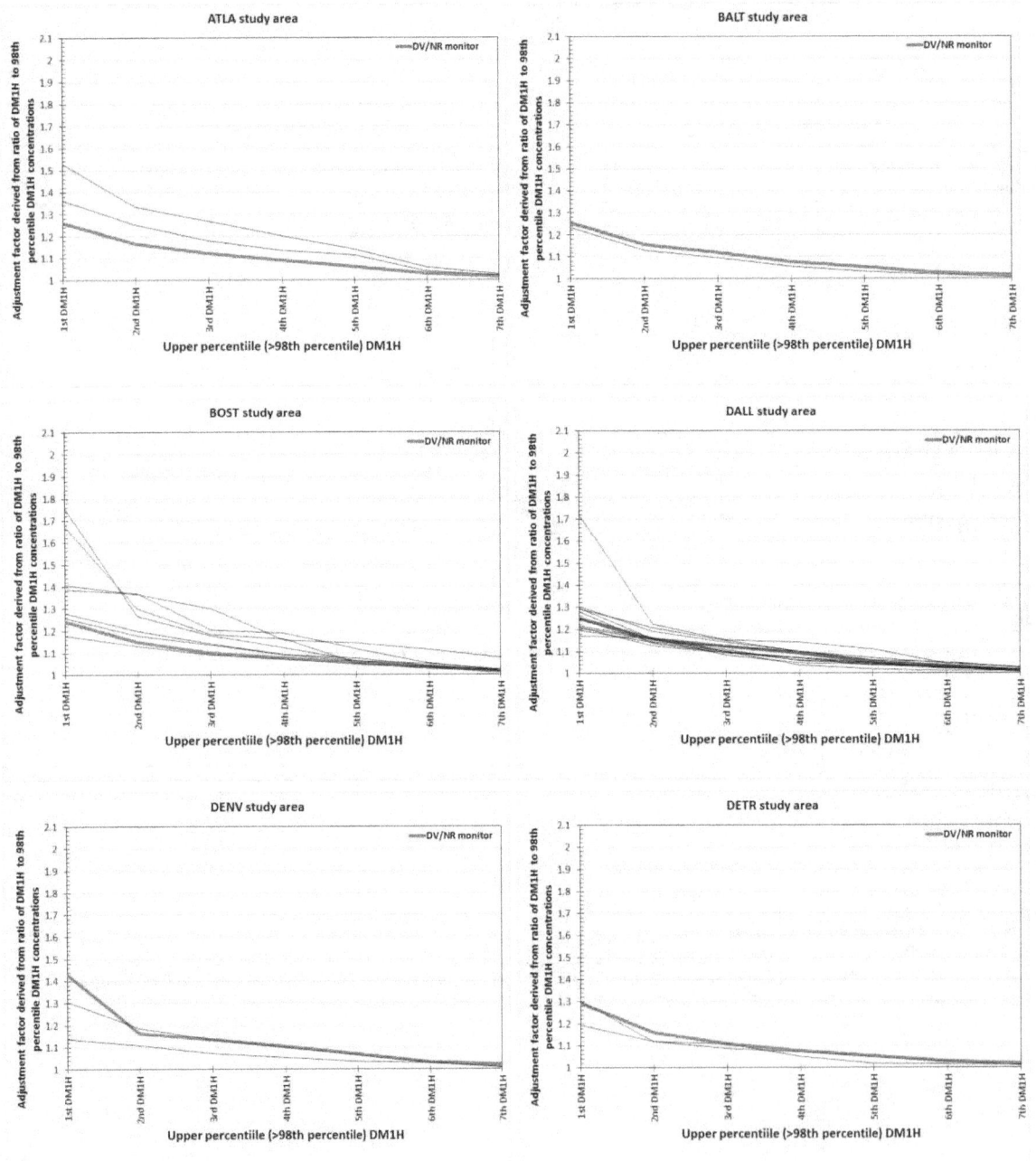

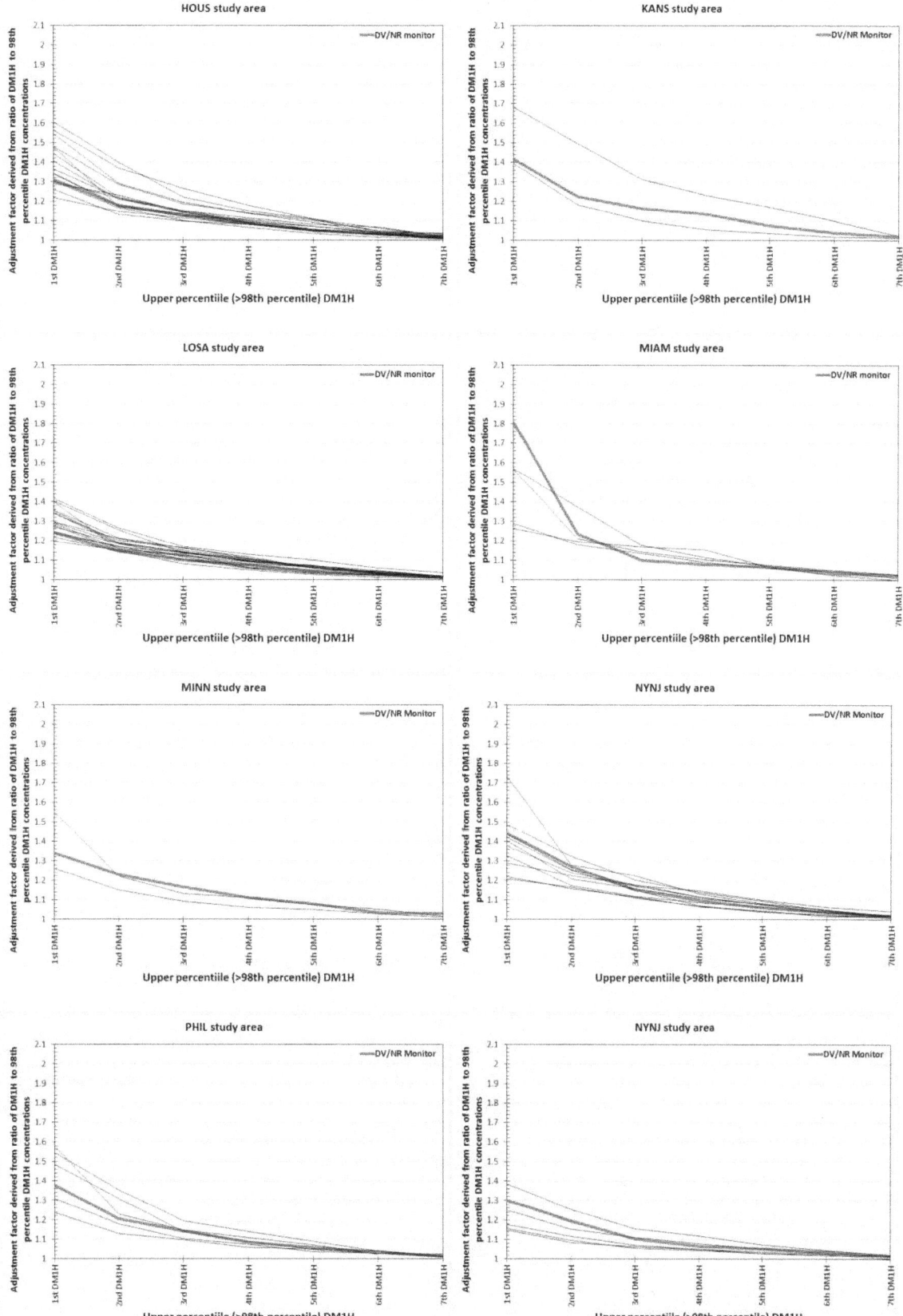

B5-113

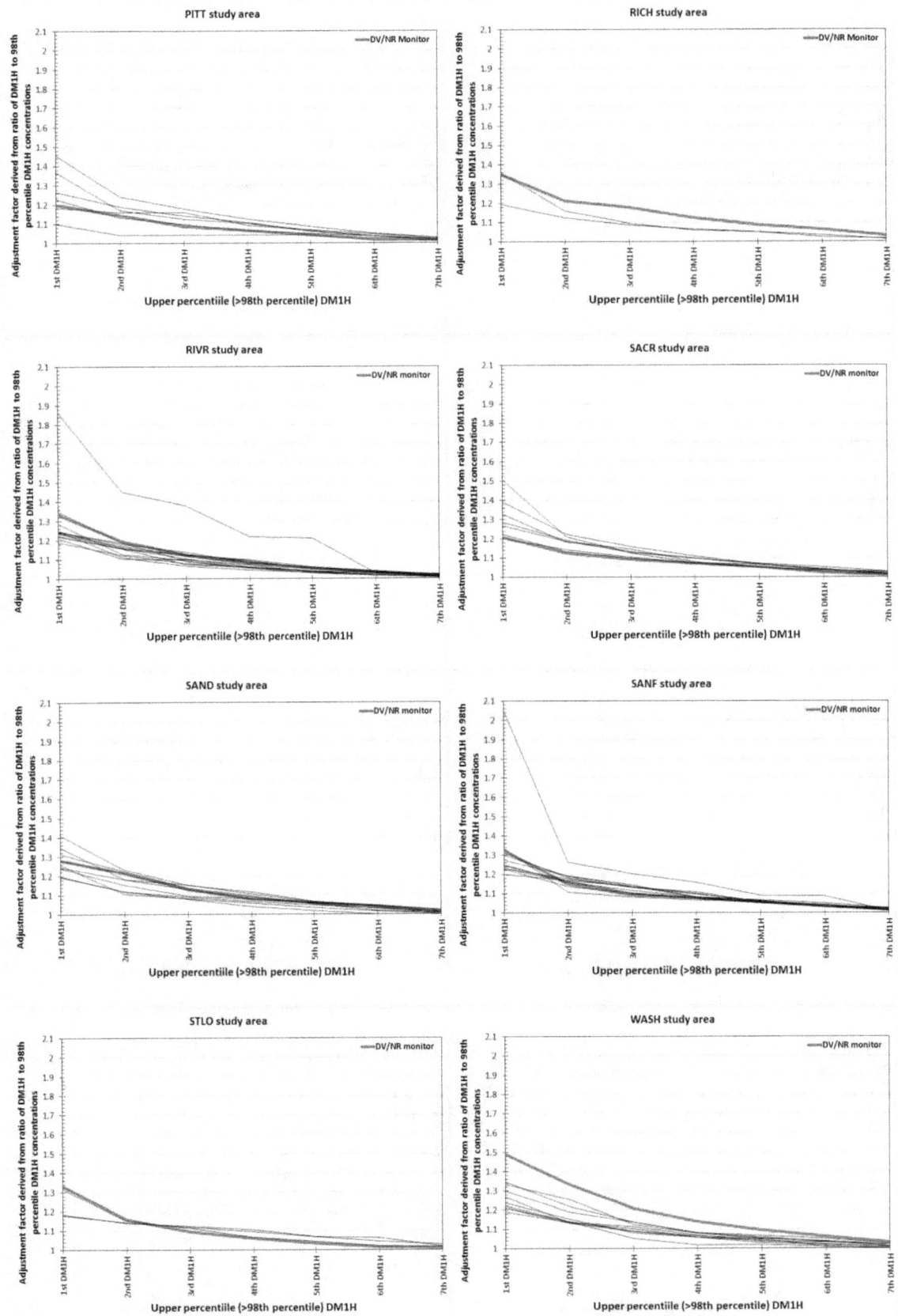

B5-114

5.5 NUMBER OF DAYS PER YEAR NO₂ CONCENTRATIONS WERE AT OR ABOVE 1-HOUR BENCHMARK LEVELS: SITE-YEAR SUMMARY TABLES (2010 – 2015)

Table B5-57. Number of days per year NO$_2$ concentrations were at or above benchmark levels: area-wide and near-road site-year summary table (2010-2015).

CBSA Abbr.	Monitor Type	1-hr Benchmark (ppb)	Number of monitor site-years					Mean number of days per year at or above benchmark level					Maximum number of days per year at or above benchmark level				
			asis	CS1012	CS1113	CS1214	CS1315	asis	CS1012	CS1113	CS1214	CS1315	asis	CS1012	CS1113	CS1214	CS1315
ATLA	area_wide	100	18	9	9	9	9	0	3	5	4	2	0	15	30	16	16
BALT	area_wide	100	11	6	6	6	5	0	4	5	4	4	0	12	12	12	11
BOST	area_wide	100	31	15	15	11	16	0	5	5	6	3	0	15	15	18	12
CHIC	area_wide	100	23	11	0	0	12	0.5	7			4	1	24			12
DALL	area_wide	100	64	34	33	29	30	0	1	2	2	2	0	9	11	16	12
DENV	area_wide	100	11	0	5	6	7	0.5		14	5	4	1		29	10	10
DETR	area_wide	100	10	4	5	6	6	0	6	9	8	4	0	12	23	18	9
HOUS	area_wide	100	85	40	40	44	45	0.5	2	2	2	2	3	12	8	15	14
KANS	area_wide	100	14	9	7	6	5	0	5	5	7	7	0	9	9	11	10
LOSA	area_wide	100	65	30	26	27	35	0.5	5	5	5	4	1	19	22	23	27
MIAM	area_wide	100	20	12	10	8	8	0	5	4	5	6	0	15	12	12	17
MINE	area_wide	100	17	8	9	9	9	0	6	4	4	2	0	22	11	14	10
NYNY	area_wide	100	48	23	22	24	25	0.5	3	3	3	4	1	9	12	12	12
PHIL	area_wide	100	26	13	14	0	13	0.5	3	3		2	1	18	23		13
PHOE	area_wide	100	28	15	14	14	13	0.5	2	4	3	3	1	9	13	12	9
PITT	area_wide	100	25	13	15	0	16	0	2	3		3	0	9	22		14
RICH	area_wide	100	14	9	8	6	5	0	6	5	7	2	0	17	17	15	6
RIVR	area_wide	100	49	24	18	0	25	0.5	1	4	7	1	7	14	25	15	5
SACR	area_wide	100	42	23	21	20	19	0	2	3	4	3	0	10	14	21	17
SAND	area_wide	100	41	22	21	20	19	0.5	1	1	1	0.5	1	9	12	12	4
SANF	area_wide	100	55	29	28	28	26	0.5	1	1	1	2	1	13	23	11	14
STLO	area_wide	100	15	0	8	9	9	0		2	4	1	0		8	14	5
WASH	area_wide	100	33	18	17	15	15	0	4	4	5	6	0	14	20	24	24
ATLA	area_wide	150	18	9	9	9	9	0	0	0	0	0	0	0	0	0	0
BALT	area_wide	150	11	6	6	6	5	0	0	0	0	0	0	0	0	0	0
BOST	area_wide	150	31	15	15	2	16	0	0	0.5	1	0.5	0	1	1	1	1
CHIC	area_wide	150	23	11	0	0	12	0	0			0	0	0			0
DALL	area_wide	150	64	34	33	29	30	0	0	0	0	0	0	0	0	0	0
DENV	area_wide	150	11	0	5	1	7	0	0	0.5	1	0.5	0	0	1	1	1

CBSA Abbr.	Monitor Type	1-hr Benchmark (ppb)	Number of monitor site-years					Mean number of days per year at or above benchmark level					Maximum number of days per year at or above benchmark level				
			asis	CS1012	CS1113	CS1214	CS1315	asis	CS1012	CS1113	CS1214	CS1315	asis	CS1012	CS1113	CS1214	CS1315
DETR	area_wide	150	10	4	5	6	6	0	0	0	0	0	0	0	0	0	0
HOUS	area_wide	150	85	40	40	44	45	0	0.5	0	0	0	0	2	0	0	0
KANS	area_wide	150	14	9	7	6	5	0	0	0	0	0	0	0	0	0	0
LOSA	area_wide	150	65	30	26	27	35	0	0.5	0.5	0.5	0.5	0	1	1	1	1
MIAM	area_wide	150	20	12	10	8	8	0	0.5	0.5	1	0.5	0	1	1	1	1
MINE	area_wide	150	17	8	9	9	9	0	0.5	0.5	0.5	0	0	2	3	2	0
NYNY	area_wide	150	48	23	22	24	25	0.5	0.5	0.5	0.5	0.5	1	1	1	1	1
PHIL	area_wide	150	26	13	14	0	13	0	0.5	0.5		0.5	0	1	1		1
PHOE	area_wide	150	28	15	14	14	13	0	0	0	0	0	0	0	0	0	0
PITT	area_wide	150	25	13	15	0	16	0	0	0		0	0	0	0		0
RICH	area_wide	150	14	9	8	6	5	0	0	0.5	0.5	0	0	0	1	1	0
RIVR	area_wide	150	49	24	18	0	25	0	0.5	0.5		0	0	1	8		0
SACR	area_wide	150	42	23	21	20	19	0	0	0	0	0	0	0	0	0	0
SAND	area_wide	150	41	22	21	20	19	0	0	0	0	0	0	0	0	0	0
SANF	area_wide	150	55	29	28	28	26	0	0	0.5	0	0	0	0	1	0	0
STLO	area_wide	150	15	0	8	9	9	0		0	0	0	0		0	0	0
WASH	area_wide	150	33	18	17	15	15	0	0.5	0.5	0.5	0.5	1	1	1	1	1
ATLA	area_wide	200	18	9	9	9	9	0	0	0	0	0	0	0	0	0	0
BALT	area_wide	200	11	6	6	6	5	0	0	0	0	0	0	0	0	0	0
BOST	area_wide	200	31	15	15	15	16	0	0	0	0	0	0	0	0	0	0
CHIC	area_wide	200	23	11	0	0	12	0	0			0	0	0			0
DALL	area_wide	200	64	34	33	29	30	0	0	0	0	0	0	0	0	0	0
DENV	area_wide	200	11	0	5	6	7	0		0	0	0	0		0	0	0
DETR	area_wide	200	10	4	5	6	6	0	0	0	0	0	0	0	0	0	0
HOUS	area_wide	200	85	40	40	44	45	0	0	0	0	0	0	0	0	0	0
KANS	area_wide	200	14	9	7	6	5	0	0	0	0	0	0	0	0	0	0
LOSA	area_wide	200	65	30	26	27	35	0	0	0	0	0	0	0	0	0	0
MIAM	area_wide	200	20	12	10	8	8	0	0	0	0	0	0	0	0	0	0
MINE	area_wide	200	17	8	9	9	9	0	0	0	0	0	0	0	0	0	0
NYNY	area_wide	200	48	23	22	24	25	0	0	0	0	0	0	0	0	0	0
PHIL	area_wide	200	26	13	14	0	13	0	0	0		0	0	0	0		0
PHOE	area_wide	200	28	15	14	14	13	0	0	0	0	0	0	0	0	0	0
PITT	area_wide	200	25	13	15	0	16	0	0	0		0	0	0	0		0
RICH	area_wide	200	14	9	8	6	5	0	0	0	0	0	0	0	0	0	0
RIVR	area_wide	200	49	24	18	0	25	0	0	0		0	0	0	0		0
SACR	area_wide	200	42	23	21	20	19	0	0	0	0	0	0	0	0	0	0
SAND	area_wide	200	41	22	21	20	19	0	0	0	0	0	0	0	0	0	0
SANF	area_wide	200	55	29	28	28	26	0	0	0	0	0	0	0	0	0	0

CBSA Abbr.	Monitor Type	1-hr Benchmark (ppb)	Number of monitor site-years					Mean number of days per year at or above benchmark level					Maximum number of days per year at or above benchmark level				
			asis	CS1012	CS1113	CS1214	CS1315	asis	CS1012	CS1113	CS1214	CS1315	asis	CS1012	CS1113	CS1214	CS1315
STLO	area_wide	200	15	0	8	9	9	0	0	0	0	0	0	0	0	0	0
WASH	area_wide	200	33	18	17	15	15	0	0	0	0	0	0	0	0	0	0
ATLA	near_road	100	3	0	0	1	3	0			5	6	0			5	8
BALT	near_road	100	2	0	0	1	2	0			5	7	0			5	7
BOST	near_road	100	3	0	1	2	3	0		2	10	6	0		2	18	11
CHIC	near_road	100	5	3	0	0	2	0.5	8			4	1	11			4
DALL	near_road	100	3	0	0	1	3	0			1	2	0			1	5
DENV	near_road	100	4	0	1	2	4	0		4	3	3	0		4	5	5
DETR	near_road	100	6	2	3	3	4	0	11	22	17	7	0	16	31	22	9
HOUS	near_road	100	3	0	0	1	3	0			2	4	0			2	5
KANS	near_road	100	3	0	1	2	3	0		1	3	3	0		1	4	4
LOSA	near_road	100	3	0	0	1	3	0			17	9	0			17	21
MIAM	near_road	100	1	0	0	0	1	0				1	0				1
MINE	near_road	100	4	0	1	2	4	0		7	9	7	0		7	13	11
NYNY	near_road	100	2	0	0	1	2	2			8	8	2			8	8
PHIL	near_road	100	3	0	0	0	3	0				2	0				3
PHOE	near_road	100	3	0	0	1	3	0			3	3	0			3	3
PITT	near_road	100	2	0	0	0	2	0				7	0				11
RICH	near_road	100	3	0	1	2	3	0		1	8	5	0		1	12	9
RIVR	near_road	100	2	0	0	0	2	0				7	0				8
SACR	near_road	100	1	0	0	0	1	0				2	0				2
SAND	near_road	100	1	0	0	0	1	0				4	0				4
SANF	near_road	100	2	0	0	1	2	0			1	2	0			1	2
STLO	near_road	100	4	0	1	2	4	0		5	16	5	0		5	21	10
WASH	near_road	100	1	0	0	0	1	0				4	0				4
ATLA	near_road	150	3	0	0	1	3	0			0	0	0			0	0
BALT	near_road	150	2	0	0	1	2	0			0	0	0			0	0
BOST	near_road	150	3	0	1	0	3	0		0		0	0		0		0
CHIC	near_road	150	5	3	0	0	2	0	0			0	0	0			0
DALL	near_road	150	3	0	0	1	3	0			0	0	0			0	0
DENV	near_road	150	4	0	1	0	4	0		0		0	0		0		0
DETR	near_road	150	6	2	3	3	4	0	1	1	0.5	0	0	1	1	1	0
HOUS	near_road	150	3	0	0	1	3	0			0	0	0			0	0
KANS	near_road	150	3	0	1	2	3	0		0	0	0	0		0	0	0
LOSA	near_road	150	3	0	0	1	3	0			0	0	0			0	0
MIAM	near_road	150	1	0	0	0	1	0				1	0				1
MINE	near_road	150	4	0	1	2	4	0		0	0	0	0		0	0	0
NYNY	near_road	150	2	0	0	1	2	1			3	2	1			3	3

CBSA Abbr.	Monitor Type	1-hr Benchmark (ppb)	Number of monitor site-years					Mean number of days per year at or above benchmark level					Maximum number of days per year at or above benchmark level				
			asis	CS1012	CS1113	CS1214	CS1315	asis	CS1012	CS1113	CS1214	CS1315	asis	CS1012	CS1113	CS1214	CS1315
PHIL	near_road	150	3	0	0	0	3	0				0	0				0
PHOE	near_road	150	3	0	0	1	3	0			0	0	0			0	0
PITT	near_road	150	2	0	0	0	2	0				0	0				0
RICH	near_road	150	3	0	1	2	3	0		0	0	0	0		0	0	0
RIVR	near_road	150	2	0	0	0	2	0				0	0				0
SACR	near_road	150	1	0	0	0	1	0				0	0				0
SAND	near_road	150	1	0	0	0	1	0				0	0				0
SANF	near_road	150	2	0	0	1	2	0			0	0	0			0	0
STLO	near_road	150	4	0	1	2	4	0		0	1	0	0		0	1	0
WASH	near_road	150	1	0	0	0	1	0				0	0				0
ATLA	near_road	200	3	0	0	1	3	0			0	0	0			0	0
BALT	near_road	200	2	0	0	1	2	0			0	0	0			0	0
BOST	near_road	200	3	0	1	2	3	0		0	0	0	0		0	0	0
CHIC	near_road	200	5	3	0	0	2	0	0			0	0	0			0
DALL	near_road	200	3	0	0	1	3	0			0	0	0			0	0
DENV	near_road	200	4	0	1	2	4	0		0	0	0	0		0	0	0
DETR	near_road	200	6	2	3	3	4	0	0	0	0	0	0	0	0	0	0
HOUS	near_road	200	3	0	0	1	3	0			0	0	0			0	0
KANS	near_road	200	3	0	1	2	3	0		0	0	0	0		0	0	0
LOSA	near_road	200	3	0	0	1	3	0			0	0	0			0	0
MIAM	near_road	200	1	0	0	0	1	0				0	0				0
MINE	near_road	200	4	0	1	2	4	0		0	0	0	0		0	0	0
NYNY	near_road	200	2	0	0	1	2	1			0	0	1			0	0
PHIL	near_road	200	3	0	0	0	3	0				0	0				0
PHOE	near_road	200	3	0	0	1	3	0			0	0	0			0	0
PITT	near_road	200	2	0	0	0	2	0				0	0				0
RICH	near_road	200	3	0	1	2	3	0		0	0	0	0		0	0	0
RIVR	near_road	200	2	0	0	0	2	0				0	0				0
SACR	near_road	200	1	0	0	0	1	0				0	0				0
SAND	near_road	200	1	0	0	0	1	0				0	0				0
SANF	near_road	200	2	0	0	1	2	0			0	0	0			0	0
STLO	near_road	200	4	0	1	2	4	0		0	0	0	0		0	0	0
WASH	near_road	200	1	0	0	0	1	0				0	0				0

5.6 NUMBER OF DAYS NO₂ CONCENTRATIONS WERE AT OR ABOVE 1-HOUR BENCHMARK LEVELS: SIMULATED ON-ROAD CONCENTRATIONS (2014-2015 ONLY)

Table B5-58. Number of days NO₂ concentrations were at or above benchmark levels: simulated on-road concentrations for 2014.

CBSA Abbrev.	Near-road monitor site ID	Upwards Adjustment	Year	Number of monitored days in year	1-hour Benchmark (ppb)	Number of days at or above benchmark level	
						asis	CS1315
ATLA	131210056	hi -11%	2014	200	100	0	19
BALT	240270006	hi -19%	2014	275	100	0	24
BOST	250250044	hi -15%	2014	365	100	0	34
CHIC	170313103	hi -19%	2014	360	100	1	21
DALL	481131067	hi -21%	2014	274	100	0	9
DENV	080310027	hi -15%	2014	361	100	1	17
DETR	261630093	hi -15%	2014	364	100	0	24
HOUS	482011066	hi -21%	2014	344	100	0	23
KANS	290950042	hi -19%	2014	358	100	0	18
LOSA	060590008	hi -15%	2014	361	100	0	39
MINE	270530962	hi -21%	2014	365	100	0	51
NYNY	340030010	hi -19%	2014	184	100	4	16
PHIL	421010075	hi -15%	2014	358	100	0	10
PHOE	040134019	hi -15%	2014	322	100	0	22
PITT	420031376	hi -19%	2014	122	100	0	8
RICH	517600025	hi -19%	2014	269	100	0	26
SANF	060010012	hi -19%	2014	334	100	0	13
STLO	295100094	hi -21%	2014	365	100	0	39
ATLA	131210056	low- 7%	2014	200	100	0	17
BALT	240270006	low-12%	2014	275	100	0	20
BOST	250250044	low- 9%	2014	365	100	0	26
CHIC	170313103	low-12%	2014	360	100	1	9
DALL	481131067	low-13%	2014	274	100	U	5
DENV	080310027	low- 9%	2014	361	100	1	12
DETR	261630093	low- 9%	2014	364	100	0	16
HOUS	482011066	low-13%	2014	344	100	0	14
KANS	290950042	low-12%	2014	358	100	0	8
LOSA	060590008	low- 9%	2014	361	100	0	32
MINE	270530962	low-13%	2014	365	100	0	31
NYNY	340030010	low-12%	2014	184	100	4	13
PHIL	421010075	low- 9%	2014	358	100	0	6
PHOE	040134019	low- 9%	2014	322	100	0	10
PITT	420031376	low-12%	2014	122	100	0	6
RICH	517600025	low-12%	2014	269	100	0	17
SANF	060010012	low-12%	2014	334	100	0	8
STLO	295100094	low-13%	2014	365	100	0	30
ATLA	131210056	mid- 9%	2014	200	100	0	19
BALT	240270006	mid-16%	2014	275	100	0	20
BOST	250250044	mid-12%	2014	365	100	0	29
CHIC	170313103	mid-16%	2014	360	100	1	14
DALL	481131067	mid-17%	2014	274	100	0	5
DENV	080310027	mid-12%	2014	361	100	1	16
DETR	261630093	mid-12%	2014	364	100	0	19
HOUS	482011066	mid-17%	2014	344	100	0	18

CBSA Abbrev.	Near-road monitor site ID	Upwards Adjustment	Year	Number of monitored days in year	1-hour Benchmark (ppb)	Number of days at or above benchmark level	
						asis	CS1315
KANS	290950042	mid-16%	2014	358	100	0	11
LOSA	060590008	mid-12%	2014	361	100	0	36
MINE	270530962	mid-17%	2014	365	100	0	35
NYNY	340030010	mid-16%	2014	184	100	4	16
PHIL	421010075	mid-12%	2014	358	100	0	8
PHOE	040134019	mid-12%	2014	322	100	0	13
PITT	420031376	mid-16%	2014	122	100	0	7
RICH	517600025	mid-16%	2014	269	100	0	20
SANF	060010012	mid-16%	2014	334	100	0	11
STLO	295100094	mid-17%	2014	365	100	0	37
ATLA	131210056	hi -11%	2014	200	150	0	0
BALT	240270006	hi -19%	2014	275	150	0	0
BOST	250250044	hi -15%	2014	365	150	0	0
CHIC	170313103	hi -19%	2014	360	150	0	0
DALL	481131067	hi -21%	2014	274	150	0	0
DENV	080310027	hi -15%	2014	361	150	0	1
DETR	261630093	hi -15%	2014	364	150	0	1
HOUS	482011066	hi -21%	2014	344	150	0	0
KANS	290950042	hi -19%	2014	358	150	0	1
LOSA	060590008	hi -15%	2014	361	150	0	1
MINE	270530962	hi -21%	2014	365	150	0	2
NYNY	340030010	hi -19%	2014	184	150	2	4
PHIL	421010075	hi -15%	2014	358	150	0	0
PHOE	040134019	hi -15%	2014	322	150	0	0
PITT	420031376	hi -19%	2014	122	150	0	0
RICH	517600025	hi -19%	2014	269	150	0	1
SANF	060010012	hi -19%	2014	334	150	0	0
STLO	295100094	hi -21%	2014	365	150	0	2
ATLA	131210056	low- 7%	2014	200	150	0	0
BALT	240270006	low-12%	2014	275	150	0	0
BOST	250250044	low- 9%	2014	365	150	0	0
CHIC	170313103	low-12%	2014	360	150	0	0
DALL	481131067	low-13%	2014	274	150	0	0
DENV	080310027	low- 9%	2014	361	150	0	1
DETR	261630093	low- 9%	2014	364	150	0	0
HOUS	482011066	low-13%	2014	344	150	0	0
KANS	290950042	low-12%	2014	358	150	0	0
LOSA	060590008	low- 9%	2014	361	150	0	0
MINE	270530962	low-13%	2014	365	150	0	1
NYNY	340030010	low-12%	2014	184	150	1	4
PHIL	421010075	low- 9%	2014	358	150	0	0
PHOE	040134019	low- 9%	2014	322	150	0	0
PITT	420031376	low-12%	2014	122	150	0	0
RICH	517600025	low-12%	2014	269	150	0	0
SANF	060010012	low-12%	2014	334	150	0	0
STLO	295100094	low-13%	2014	365	150	0	1
ATLA	131210056	mid- 9%	2014	200	150	0	0
BALT	240270006	mid-16%	2014	275	150	0	0
BOST	250250044	mid-12%	2014	365	150	0	0
CHIC	170313103	mid-16%	2014	360	150	0	0
DALL	481131067	mid-17%	2014	274	150	0	0
DENV	080310027	mid-12%	2014	361	150	0	1
DETR	261630093	mid-12%	2014	364	150	0	0
HOUS	482011066	mid-17%	2014	344	150	0	0

CBSA Abbrev.	Near-road monitor site ID	Upwards Adjustment	Year	Number of monitored days in year	1-hour Benchmark (ppb)	Number of days at or above benchmark level	
						asis	CS1315
KANS	290950042	mid-16%	2014	358	150	0	0
LOSA	060590008	mid-12%	2014	361	150	0	0
MINE	270530962	mid-17%	2014	365	150	0	1
NYNY	340030010	mid-16%	2014	184	150	2	4
PHIL	421010075	mid-12%	2014	358	150	0	0
PHOE	040134019	mid-12%	2014	322	150	0	0
PITT	420031376	mid-16%	2014	122	150	0	0
RICH	517600025	mid-16%	2014	269	150	0	1
SANF	060010012	mid-16%	2014	334	150	0	0
STLO	295100094	mid-17%	2014	365	150	0	2
ATLA	131210056	hi -11%	2014	200	200	0	0
BALT	240270006	hi -19%	2014	275	200	0	0
BOST	250250044	hi -15%	2014	365	200	0	0
CHIC	170313103	hi -19%	2014	360	200	0	0
DALL	481131067	hi -21%	2014	274	200	0	0
DENV	080310027	hi -15%	2014	361	200	0	0
DETR	261630093	hi -15%	2014	364	200	0	0
HOUS	482011066	hi -21%	2014	344	200	0	0
KANS	290950042	hi -19%	2014	358	200	0	0
LOSA	060590008	hi -15%	2014	361	200	0	0
MINE	270530962	hi -21%	2014	365	200	0	0
NYNY	340030010	hi -19%	2014	184	200	1	2
PHIL	421010075	hi -15%	2014	358	200	0	0
PHOE	040134019	hi -15%	2014	322	200	0	0
PITT	420031376	hi -19%	2014	122	200	0	0
RICH	517600025	hi -19%	2014	269	200	0	0
SANF	060010012	hi -19%	2014	334	200	0	0
STLO	295100094	hi -21%	2014	365	200	0	0
ATLA	131210056	low- 7%	2014	200	200	0	0
BALT	240270006	low-12%	2014	275	200	0	0
BOST	250250044	low- 9%	2014	365	200	0	0
CHIC	170313103	low-12%	2014	360	200	0	0
DALL	481131067	low-13%	2014	274	200	0	0
DENV	080310027	low- 9%	2014	361	200	0	0
DETR	261630093	low- 9%	2014	364	200	0	0
HOUS	482011066	low-13%	2014	344	200	0	0
KANS	290950042	low-12%	2014	358	200	0	0
LOSA	060590008	low- 9%	2014	361	200	0	0
MINE	270530962	low-13%	2014	365	200	0	0
NYNY	340030010	low-12%	2014	184	200	1	1
PHIL	421010075	low- 9%	2014	358	200	0	0
PHOE	040134019	low- 9%	2014	322	200	0	0
PITT	420031376	low-12%	2014	122	200	0	0
RICH	517600025	low-12%	2014	269	200	0	0
SANF	060010012	low-12%	2014	334	200	0	0
STLO	295100094	low-13%	2014	365	200	0	0
ATLA	131210056	mid-9%	2014	200	200	0	0
BALT	240270006	mid-16%	2014	275	200	0	0
BOST	250250044	mid-12%	2014	365	200	0	0
CHIC	170313103	mid-16%	2014	360	200	0	0
DALL	481131067	mid-17%	2014	274	200	0	0
DENV	080310027	mid-12%	2014	361	200	0	0
DETR	261630093	mid-12%	2014	364	200	0	0
HOUS	482011066	mid-17%	2014	344	200	0	0

CBSA Abbrev.	Near-road monitor site ID	Upwards Adjustment	Year	Number of monitored days in year	1-hour Benchmark (ppb)	Number of days at or above benchmark level	
						asis	CS1315
KANS	290950042	mid-16%	2014	358	200	0	0
LOSA	060590008	mid-12%	2014	361	200	0	0
MINE	270530962	mid-17%	2014	365	200	0	0
NYNY	340030010	mid-16%	2014	184	200	1	2
PHIL	421010075	mid-12%	2014	358	200	0	0
PHOE	040134019	mid-12%	2014	322	200	0	0
PITT	420031376	mid-16%	2014	122	200	0	0
RICH	517600025	mid-16%	2014	269	200	0	0
SANF	060010012	mid-16%	2014	334	200	0	0
STLO	295100094	mid-17%	2014	365	200	0	0

Table B5-59. Number of days NO₂ concentrations were at or above benchmark levels: simulated on-road concentrations for 2015.

CBSA Abbrev.	Near-road monitor site ID	Upwards Adjustment	Year	Number of monitored days in year	1-hour Benchmark (ppb)	Number of days at or above benchmark level	
						asis	CS1315
ATLA	130890003	hi -21%	2015	363	100	0	22
ATLA	131210056	hi -11%	2015	362	100	0	8
BALT	240270006	hi -19%	2015	352	100	0	42
BOST	250250044	hi -15%	2015	362	100	0	28
CHIC	170313103	hi -19%	2015	358	100	0	18
DALL	481131067	hi -21%	2015	365	100	0	22
DALL	484391053	hi -19%	2015	295	100	0	6
DENV	080310027	hi -15%	2015	359	100	0	11
DENV	080310028	hi -15%	2015	92	100	0	16
DETR	261630093	hi -15%	2015	361	100	0	25
DETR	261630095	hi -21%	2015	362	100	0	19
HOUS	482011052	hi -19%	2015	261	100	0	18
HOUS	482011066	hi -21%	2015	365	100	0	14
KANS	290950042	hi -19%	2015	352	100	0	14
LOSA	060374008	hi -15%	2015	255	100	1	17
LOSA	060590008	hi -15%	2015	364	100	0	4
MIAM	120110035	hi -21%	2015	56	100	1	3
MINE	270370480	hi -21%	2015	363	100	0	10
MINE	270530962	hi -21%	2015	362	100	0	30
NYNY	340030010	hi -19%	2015	365	100	1	31
PHIL	421010075	hi -15%	2015	357	100	0	5
PHIL	421010076	hi -19%	2015	163	100	0	1
PHOE	040134019	hi -15%	2015	365	100	0	6
PHOE	040134020	hi -19%	2015	121	100	0	28
PITT	420031376	hi -19%	2015	365	100	0	47
RICH	517600025	hi -19%	2015	353	100	0	43
RIVR	060710026	hi -21%	2015	360	100	2	44
RIVR	060710027	hi -15%	2015	153	100	0	28
SACR	060670015	hi -19%	2015	80	100	0	6
SAND	060731017	hi -21%	2015	269	100	0	17
SANF	060010012	hi -19%	2015	365	100	0	11
STLO	291890016	hi -21%	2015	357	100	0	8
STLO	295100094	hi -21%	2015	364	100	0	26
WASH	110010051	hi -19%	2015	212	100	0	22
ATLA	130890003	low -13%	2015	363	100	0	18
ATLA	131210056	low -7%	2015	362	100	0	6
BALT	240270006	low -12%	2015	352	100	0	21
BOST	250250044	low -9%	2015	362	100	0	23
CHIC	170313103	low -12%	2015	358	100	0	13
DALL	481131067	low -13%	2015	365	100	0	14
DALL	484391053	low -12%	2015	295	100	0	4
DENV	080310027	low -9%	2015	359	100	0	5
DENV	080310028	low -9%	2015	92	100	0	9
DETR	261630093	low -9%	2015	361	100	0	17
DETR	261630095	low -13%	2015	362	100	0	13
HOUS	482011052	low -12%	2015	261	100	0	13
HOUS	482011066	low -13%	2015	365	100	0	8
KANS	290950042	low -12%	2015	352	100	0	6
LOSA	060374008	low -9%	2015	255	100	1	13
LOSA	060590008	low -9%	2015	364	100	0	2

CBSA Abbrev.	Near-road monitor site ID	Upwards Adjustment	Year	Number of monitored days in year	1-hour Benchmark (ppb)	Number of days at or above benchmark level	
						asis	CS1315
MIAM	120110035	low-13%	2015	56	100	1	2
MINE	270370480	low-13%	2015	363	100	0	9
MINE	270530962	low-13%	2015	362	100	0	20
NYNY	340030010	low-12%	2015	365	100	1	18
PHIL	421010075	low- 9%	2015	357	100	0	3
PHIL	421010076	low-12%	2015	163	100	0	1
PHOE	040134019	low- 9%	2015	365	100	0	3
PHOE	040134020	low-12%	2015	121	100	0	17
PITT	420031376	low-12%	2015	365	100	0	29
RICH	517600025	low-12%	2015	353	100	0	33
RIVR	060710026	low-13%	2015	360	100	0	21
RIVR	060710027	low- 9%	2015	153	100	0	17
SACR	060670015	low-12%	2015	80	100	0	5
SAND	060731017	low-13%	2015	269	100	0	12
SANF	060010012	low-12%	2015	365	100	0	5
STLO	291890016	low-13%	2015	357	100	0	4
STLO	295100094	low-13%	2015	364	100	0	13
WASH	110010051	low-12%	2015	212	100	0	12
ATLA	130890003	mid-17%	2015	363	100	0	20
ATLA	131210056	mid- 9%	2015	362	100	0	7
BALT	240270006	mid-16%	2015	352	100	0	32
BOST	250250044	mid-12%	2015	362	100	0	24
CHIC	170313103	mid-16%	2015	358	100	0	15
DALL	481131067	mid-17%	2015	365	100	0	16
DALL	484391053	mid-16%	2015	295	100	0	5
DENV	080310027	mid-12%	2015	359	100	0	7
DENV	080310028	mid-12%	2015	92	100	0	13
DETR	261630093	mid-12%	2015	361	100	0	22
DETR	261630095	mid-17%	2015	362	100	0	15
HOUS	482011052	mid-16%	2015	261	100	0	15
HOUS	482011066	mid-17%	2015	365	100	0	9
KANS	290950042	mid-16%	2015	352	100	0	11
LOSA	060374008	mid-12%	2015	255	100	1	17
LOSA	060590008	mid-12%	2015	364	100	0	3
MIAM	120110035	mid-17%	2015	56	100	1	2
MINE	270370480	mid-17%	2015	363	100	0	9
MINE	270530962	mid-17%	2015	362	100	0	25
NYNY	340030010	mid-16%	2015	365	100	1	26
PHIL	421010075	mid-12%	2015	357	100	0	4
PHIL	421010076	mid-16%	2015	163	100	0	1
PHOE	040134019	mid-12%	2015	365	100	0	4
PHOE	040134020	mid-16%	2015	121	100	0	19
PITT	420031376	mid-16%	2015	365	100	0	36
RICH	517600025	mid-16%	2015	353	100	0	39
RIVR	060710026	mid-17%	2015	360	100	1	31
RIVR	060710027	mid-12%	2015	153	100	0	21
SACR	060670015	mid-16%	2015	80	100	0	6
SAND	060731017	mid-17%	2015	269	100	0	15
SANF	060010012	mid-16%	2015	365	100	0	7
STLO	291890016	mid-17%	2015	357	100	0	5
STLO	295100094	mid-17%	2015	364	100	0	17
WASH	110010051	mid-16%	2015	212	100	0	17
ATLA	130890003	hi -21%	2015	363	150	0	1

CBSA Abbrev.	Near-road monitor site ID	Upwards Adjustment	Year	Number of monitored days in year	1-hour Benchmark (ppb)	Number of days at or above benchmark level	
						asis	CS1315
ATLA	131210056	hi -11%	2015	362	150	0	0
BALT	240270006	hi -19%	2015	352	150	0	0
BOST	250250044	hi -15%	2015	362	150	0	0
CHIC	170313103	hi -19%	2015	358	150	0	0
DALL	481131067	hi -21%	2015	365	150	0	0
DALL	484391053	hi -19%	2015	295	150	0	0
DENV	080310027	hi -15%	2015	359	150	0	0
DENV	080310028	hi -15%	2015	92	150	0	1
DETR	261630093	hi -15%	2015	361	150	0	0
DETR	261630095	hi -21%	2015	362	150	0	1
HOUS	482011052	hi -19%	2015	261	150	0	1
HOUS	482011066	hi -21%	2015	365	150	0	0
KANS	290950042	hi -19%	2015	352	150	0	0
LOSA	060374008	hi -15%	2015	255	150	0	0
LOSA	060590008	hi -15%	2015	364	150	0	0
MIAM	120110035	hi -21%	2015	56	150	0	1
MINE	270370480	hi -21%	2015	363	150	0	0
MINE	270530962	hi -21%	2015	362	150	0	2
NYNY	340030010	hi -19%	2015	365	150	1	2
PHIL	421010075	hi -15%	2015	357	150	0	0
PHIL	421010076	hi -19%	2015	163	150	0	0
PHOE	040134019	hi -15%	2015	365	150	0	0
PHOE	040134020	hi -19%	2015	121	150	0	1
PITT	420031376	hi -19%	2015	365	150	0	1
RICH	517600025	hi -19%	2015	353	150	0	1
RIVR	060710026	hi -21%	2015	360	150	0	1
RIVR	060710027	hi -15%	2015	153	150	0	1
SACR	060670015	hi -19%	2015	80	150	0	0
SAND	060731017	hi -21%	2015	269	150	0	0
SANF	060010012	hi -19%	2015	365	150	0	0
STLO	291890016	hi -21%	2015	357	150	0	0
STLO	295100094	hi -21%	2015	364	150	0	0
WASH	110010051	hi -19%	2015	212	150	0	2
ATLA	130890003	low-13%	2015	363	150	0	0
ATLA	131210056	low- 7%	2015	362	150	0	0
BALT	240270006	low-12%	2015	352	150	0	0
BOST	250250044	low- 9%	2015	362	150	0	0
CHIC	170313103	low-12%	2015	358	150	0	0
DALL	481131067	low-13%	2015	365	150	0	0
DALL	484391053	low-12%	2015	295	150	0	0
DENV	080310027	low- 9%	2015	359	150	0	0
DENV	080310028	low- 9%	2015	92	150	0	1
DETR	261630093	low- 9%	2015	361	150	0	0
DETR	261630095	low-13%	2015	362	150	0	0
HOUS	482011052	low-12%	2015	261	150	0	0
HOUS	482011066	low-13%	2015	365	150	0	0
KANS	290950042	low-12%	2015	352	150	0	0
LOSA	060374008	low- 9%	2015	255	150	0	0
LOSA	060590008	low- 9%	2015	364	150	0	0
MIAM	120110035	low-13%	2015	56	150	0	1
MINE	270370480	low-13%	2015	363	150	0	0
MINE	270530962	low-13%	2015	362	150	0	1
NYNY	340030010	low-12%	2015	365	150	1	1

CBSA Abbrev.	Near-road monitor site ID	Upwards Adjustment	Year	Number of monitored days in year	1-hour Benchmark (ppb)	Number of days at or above benchmark level	
						asis	CS1315
PHIL	421010075	low- 9%	2015	357	150	0	0
PHIL	421010076	low-12%	2015	163	150	0	0
PHOE	040134019	low- 9%	2015	365	150	0	0
PHOE	040134020	low-12%	2015	121	150	0	0
PITT	420031376	low-12%	2015	365	150	0	0
RICH	517600025	low-12%	2015	353	150	0	1
RIVR	060710026	low-13%	2015	360	150	0	1
RIVR	060710027	low- 9%	2015	153	150	0	1
SACR	060670015	low-12%	2015	80	150	0	0
SAND	060731017	low-13%	2015	269	150	0	0
SANF	060010012	low-12%	2015	365	150	0	0
STLO	291890016	low-13%	2015	357	150	0	0
STLO	295100094	low-13%	2015	364	150	0	0
WASH	110010051	low-12%	2015	212	150	0	1
ATLA	130890003	mid-17%	2015	363	150	0	0
ATLA	131210056	mid- 9%	2015	362	150	0	0
BALT	240270006	mid-16%	2015	352	150	0	0
BOST	250250044	mid-12%	2015	362	150	0	0
CHIC	170313103	mid-16%	2015	358	150	0	0
DALL	481131067	mid-17%	2015	365	150	0	0
DALL	484391053	mid-16%	2015	295	150	0	0
DENV	080310027	mid-12%	2015	359	150	0	0
DENV	080310028	mid-12%	2015	92	150	0	1
DETR	261630093	mid-12%	2015	361	150	0	0
DETR	261630095	mid-17%	2015	362	150	0	0
HOUS	482011052	mid-16%	2015	261	150	0	1
HOUS	482011066	mid-17%	2015	365	150	0	0
KANS	290950042	mid-16%	2015	352	150	0	0
LOSA	060374008	mid-12%	2015	255	150	0	0
LOSA	060590008	mid-12%	2015	364	150	0	0
MIAM	120110035	mid-17%	2015	56	150	0	1
MINE	270370480	mid-17%	2015	363	150	0	0
MINE	270530962	mid-17%	2015	362	150	0	1
NYNY	340030010	mid-16%	2015	365	150	1	1
PHIL	421010075	mid-12%	2015	357	150	0	0
PHIL	421010076	mid-16%	2015	163	150	0	0
PHOE	040134019	mid-12%	2015	365	150	0	0
PHOE	040134020	mid-16%	2015	121	150	0	1
PITT	420031376	mid-16%	2015	365	150	0	0
RICH	517600025	mid-16%	2015	353	150	0	1
RIVR	060710026	mid-17%	2015	360	150	0	1
RIVR	060710027	mid-12%	2015	153	150	0	1
SACR	060670015	mid-16%	2015	80	150	0	0
SAND	060731017	mid-17%	2015	269	150	0	0
SANF	060010012	mid-16%	2015	365	150	0	0
STLO	291890016	mid-17%	2015	357	150	0	0
STLO	295100094	mid-17%	2015	364	150	0	0
WASH	110010051	mid-16%	2015	212	150	0	2
ATLA	130890003	hi -21%	2015	363	200	0	0
ATLA	131210056	hi -11%	2015	362	200	0	0
BALT	240270006	hi -19%	2015	352	200	0	0
BOST	250250044	hi -15%	2015	362	200	0	0
CHIC	170313103	hi -19%	2015	358	200	0	0

CBSA Abbrev.	Near-road monitor site ID	Upwards Adjustment	Year	Number of monitored days in year	1-hour Benchmark (ppb)	Number of days at or above benchmark level	
						asis	CS1315
DALL	481131067	hi -21%	2015	365	200	0	0
DALL	484391053	hi -19%	2015	295	200	0	0
DENV	080310027	hi -15%	2015	359	200	0	0
DENV	080310028	hi -15%	2015	92	200	0	0
DETR	261630093	hi -15%	2015	361	200	0	0
DETR	261630095	hi -21%	2015	362	200	0	0
HOUS	482011052	hi -19%	2015	261	200	0	0
HOUS	482011066	hi -21%	2015	365	200	0	0
KANS	290950042	hi -19%	2015	352	200	0	0
LOSA	060374008	hi -15%	2015	255	200	0	0
LOSA	060590008	hi -15%	2015	364	200	0	0
MIAM	120110035	hi -21%	2015	56	200	0	0
MINE	270370480	hi -21%	2015	363	200	0	0
MINE	270530962	hi -21%	2015	362	200	0	0
NYNY	340030010	hi -19%	2015	365	200	0	0
PHIL	421010075	hi -15%	2015	357	200	0	0
PHIL	421010076	hi -19%	2015	163	200	0	0
PHOE	040134019	hi -15%	2015	365	200	0	0
PHOE	040134020	hi -19%	2015	121	200	0	0
PITT	420031376	hi -19%	2015	365	200	0	0
RICH	517600025	hi -19%	2015	353	200	0	0
RIVR	060710026	hi -21%	2015	360	200	0	0
RIVR	060710027	hi -15%	2015	153	200	0	0
SACR	060670015	hi -19%	2015	80	200	0	0
SAND	060731017	hi -21%	2015	269	200	0	0
SANF	060010012	hi -19%	2015	365	200	0	0
STLO	291890016	hi -21%	2015	357	200	0	0
STLO	295100094	hi -21%	2015	364	200	0	0
WASH	110010051	hi -19%	2015	212	200	0	0
ATLA	130890003	low-13%	2015	363	200	0	0
ATLA	131210056	low- 7%	2015	362	200	0	0
BALT	240270006	low-12%	2015	352	200	0	0
BOST	250250044	low- 9%	2015	362	200	0	0
CHIC	170313103	low-12%	2015	358	200	0	0
DALL	481131067	low-13%	2015	365	200	0	0
DALL	484391053	low-12%	2015	295	200	0	0
DENV	080310027	low- 9%	2015	359	200	0	0
DENV	080310028	low- 9%	2015	92	200	0	0
DETR	261630093	low- 9%	2015	361	200	0	0
DETR	261630095	low-13%	2015	362	200	0	0
HOUS	482011052	low-12%	2015	261	200	0	0
HOUS	482011066	low-13%	2015	365	200	0	0
KANS	290950042	low-12%	2015	352	200	0	0
LOSA	060374008	low- 9%	2015	255	200	0	0
LOSA	060590008	low- 9%	2015	364	200	0	0
MIAM	120110035	low-13%	2015	56	200	0	0
MINE	270370480	low-13%	2015	363	200	0	0
MINE	270530962	low-13%	2015	362	200	0	0
NYNY	340030010	low-12%	2015	365	200	0	0
PHIL	421010075	low- 9%	2015	357	200	0	0
PHIL	421010076	low-12%	2015	163	200	0	0
PHOE	040134019	low- 9%	2015	365	200	0	0
PHOE	040134020	low-12%	2015	121	200	0	0

CBSA Abbrev.	Near-road monitor site ID	Upwards Adjustment	Year	Number of monitored days in year	1-hour Benchmark (ppb)	Number of days at or above benchmark level	
						asis	CS1315
PITT	420031376	low-12%	2015	365	200	0	0
RICH	517600025	low-12%	2015	353	200	0	0
RIVR	060710026	low-13%	2015	360	200	0	0
RIVR	060710027	low- 9%	2015	153	200	0	0
SACR	060670015	low-12%	2015	80	200	0	0
SAND	060731017	low-13%	2015	269	200	0	0
SANF	060010012	low-12%	2015	365	200	0	0
STLO	291890016	low-13%	2015	357	200	0	0
STLO	295100094	low-13%	2015	364	200	0	0
WASH	110010051	low-12%	2015	212	200	0	0
ATLA	130890003	mid-17%	2015	363	200	0	0
ATLA	131210056	mid- 9%	2015	362	200	0	0
BALT	240270006	mid-16%	2015	352	200	0	0
BOST	250250044	mid-12%	2015	362	200	0	0
CHIC	170313103	mid-16%	2015	358	200	0	0
DALL	481131067	mid-17%	2015	365	200	0	0
DALL	484391053	mid-16%	2015	295	200	0	0
DENV	080310027	mid-12%	2015	359	200	0	0
DENV	080310028	mid-12%	2015	92	200	0	0
DETR	261630093	mid-12%	2015	361	200	0	0
DETR	261630095	mid-17%	2015	362	200	0	0
HOUS	482011052	mid-16%	2015	261	200	0	0
HOUS	482011066	mid-17%	2015	365	200	0	0
KANS	290950042	mid-16%	2015	352	200	0	0
LOSA	060374008	mid-12%	2015	255	200	0	0
LOSA	060590008	mid-12%	2015	364	200	0	0
MIAM	120110035	mid-17%	2015	56	200	0	0
MINE	270370480	mid-17%	2015	363	200	0	0
MINE	270530962	mid-17%	2015	362	200	0	0
NYNY	340030010	mid-16%	2015	365	200	0	0
PHIL	421010075	mid-12%	2015	357	200	0	0
PHIL	421010076	mid-16%	2015	163	200	0	0
PHOE	040134019	mid-12%	2015	365	200	0	0
PHOE	040134020	mid-16%	2015	121	200	0	0
PITT	420031376	mid-16%	2015	365	200	0	0
RICH	517600025	mid-16%	2015	353	200	0	0
RIVR	060710026	mid-17%	2015	360	200	0	0
RIVR	060710027	mid-12%	2015	153	200	0	0
SACR	060670015	mid-16%	2015	80	200	0	0
SAND	060731017	mid-17%	2015	269	200	0	0
SANF	060010012	mid-16%	2015	365	200	0	0
STLO	291890016	mid-17%	2015	357	200	0	0
STLO	295100094	mid-17%	2015	364	200	0	0
WASH	110010051	mid-16%	2015	212	200	0	0

5.7 NUMBER (AND PERCENT) OF DAYS PER YEAR NO$_2$ CONCENTRATIONS WERE AT OR ABOVE 1-HOUR BENCHMARK LEVELS: CBSA-WIDE SUMMARY TABLES (2010 – 2015)

Table B5-60. Number of monitors used for calculations: area-wide and near-road monitor CBSA-wide summary table (2010-2015).

CBSA Abbrev.	Mean number of monitors per year[a]					Maximum number of monitors per year[a]				
	asis	CS1012	CS1113	CS1214	CS1315	asis	CS1012	CS1113	CS1214	CS1315
ATLA	4	3	3	3	4	5	3	3	4	5
BALT	2	2	2	2	2	3	2	2	3	3
BOST	6	5	5	6	6	7	5	6	6	7
CHIC	5	5			5	6	6			5
DALL	11	11	11	10	11	14	14	14	11	12
DENV	3		2	3	4	5		3	3	5
DETR	3	2	3	3	3	4	3	3	3	4
HOUS	15	13	13	15	16	17	16	16	17	17
KANS	3	3	3	3	3	3	3	3	3	3
LOSA	11	10	9	9	13	17	13	10	12	17
MIAM	4	4	3	3	3	5	5	5	3	3
MINE	4	3	3	4	4	5	3	4	4	5
NYNY	8	8	7	8	9	10	9	8	10	10
PHIL	5	4	5		5	6	5	6		6
PHOE	5	5	5	5	5	6	5	5	6	6
PITT	5	4	5		5	6	5	6		6
RICH	3	3	3	3	3	3	3	3	3	3
RIVR	9	8	6		9	13	11	7		13
SACR	7	8	7	7	7	8	8	8	8	8
SAND	7	7	7	4	2	8	8	7	7	8
SANF	10	10	9	10	9	11	11	11	11	10
STLO	3		3	4	4	5		4	5	5
WASH	6	6	6	5	5	7	7	6	6	6

[a] associated with summary results presented in Table B5-61.

Table B5-61. Mean and maximum number of days per year NO₂ concentrations were at or above 1-hour benchmark levels: area-wide and near-road CBSA-wide summary table (2010-2015).

CBSA Abbrev.	1-hour Benchmark/Metric	Mean number of days per year at or above benchmark level					Maximum number of days per year at or above benchmark level				
		asis	CS1012	CS1113	CS1214	CS1315	asis	CS1012	CS1113	CS1214	CS1315
ATLA	100ppb	0	10	14	12	12	0	15	30	19	23
BALT	100ppb	0	7	9	10	11	0	13	12	14	20
BOST	100ppb	0	17	15	18	14	0	22	23	27	17
CHIC	100ppb	0.5	23			14	1	35			16
DALL	100ppb	0	11	12	14	15	0	15	21	19	18
DENV	100ppb	0.5		22	9	9	1		33	12	12
DETR	100ppb	0	14	31	22	11	0	17	35	26	13
HOUS	100ppb	1	19	14	18	20	3	30	14	24	31
KANS	100ppb	0	13	11	13	11	0	15	15	17	16
LOSA	100ppb	1	27	30	30	30	2	43	41	41	43
MIAM	100ppb	0	14	10	10	12	0	26	13	14	21
MINE	100ppb	0	15	13	14	12	0	23	14	20	16
NYNY	100ppb	2	15	14	18	21	3	21	19	30	30
PHIL	100ppb	0.5	11	13		9	1	22	26		15
PHOE	100ppb	0.5	9	12	12	11	1	9	15	15	14
PITT	100ppb	0	9	14		16	0	12	27		23
RICH	100ppb	0	17	14	19	9	0	30	23	28	11
RIVR	100ppb	2	9	23		8	8	17	37		10
SACR	100ppb	0	13	16	20	14	0	15	19	24	19
SAND	100ppb	0.5	8	9	7	10	1	10	12	12	10
SANF	100ppb	0.5	8	10	10	12	1	13	23	11	19
STLO	100ppb	0		7	17	8	0		8	23	13
WASH	100ppb	0	14	20	20	22	0	24	23	29	30
ATLA	150ppb	0	0	0	0	0	0	0	0	0	0
BALT	150ppb	0	0	0	0	0	0	0	0	0	0
BOST	150ppb	0	1	1	1	0.5	0	1	1	1	1
CHIC	150ppb	0	0			0	0	0			0
DALL	150ppb	0	0	0	0	0	0	0	0	0	0
DENV	150ppb	0	0	1	0.5	0.5	0	0	1	1	1

CBSA Abbrev.	1-hour Benchmark /Metric	Mean number of days per year at or above benchmark level					Maximum number of days per year at or above benchmark level				
		asis	CS1012	CS1113	CS1214	CS1315	asis	CS1012	CS1113	CS1214	CS1315
DETR	150ppb	0	0.5	1	0.5	0	0	1	1	1	0
HOUS	150ppb	0	1	0	0	0	0	2	0	0	0
KANS	150ppb	0	0	0	0	0	0	0	0	0	0
LOSA	150ppb	0	0.5	1	0.5	0.5	0	1	1	1	1
MIAM	150ppb	0	1	1	1	1	0	2	1	2	2
MINE	150ppb	0	1	1	1	0	0	2	3	2	0
NYNY	150ppb	1	1	1	2	2	2	1	2	5	5
PHIL	150ppb	0	1	1		0.5	0	2	2		1
PHOE	150ppb	0	0	0	0	0	0	0	0	0	0
PITT	150ppb	0	0	0		0	0	0	0		0
RICH	150ppb	0	0	0.5	0.5	0	0	0	1	1	0
RIVR	150ppb	0	0.5	3		0	0	1	8		0
SACR	150ppb	0	0	0	0	0	0	0	0	0	0
SAND	150ppb	0	0	0	0	0	0	0	0	0	0
SANF	150ppb	0	0	0.5	0	0	0	0	1	0	0
STLO	150ppb	0	0	0	0.5	0	0		0	1	0
WASH	150ppb	0	0.5	0.5	1	0.5	0	1	1	1	1
ATLA	200ppb	0	0	0	0	0	0	0	0	0	0
BALT	200ppb	0	0	0	0	0	0	0	0	0	0
BOST	200ppb	0	0	0	0	0	0	0	0	0	0
CHIC	200ppb	0	0			0	0	0			0
DALL	200ppb	0	0	0	0	0	0	0	0	0	0
DENV	200ppb	0		0	0	0	0		0	0	0
DETR	200ppb	0	0	0	0	0	0	0	0	0	0
HOUS	200ppb	0	0	0	0	0	0	0	0	0	0
KANS	200ppb	0	0	0	0	0	0	0	0	0	0
LOSA	200ppb	0	0	0	0	0	0	0	0	0	0
MIAM	200ppb	0	0	0	0	0	0	0	0	0	0
MINE	200ppb	0	0	0	0	0	0	0	0	0	0
NYNY	200ppb	0.5	0	0	0	0	1	0	0	0	0
PHIL	200ppb	0	0	0		0	0	0	0		0
PHOE	200ppb	0	0	0	0	0	0	0	0	0	0

CBSA Abbrev.	1-hour Benchmark /Metric	Mean number of days per year at or above benchmark level					Maximum number of days per year at or above benchmark level				
		asis	CS1012	CS1113	CS1214	CS1315	asis	CS1012	CS1113	CS1214	CS1315
PITT	200ppb	0	0	0	0	0	0	0	0	0	0
RICH	200ppb	0	0	0	0	0	0	0	0	0	0
RIVR	200ppb	0	0	0		0	0	0	0		0
SACR	200ppb	0	0	0	0	0	0	0		0	0
SAND	200ppb	0	0	0	0	0	0	0	0	0	0
SANF	200ppb	0	0	0	0	0	0	0	0	0	0
STLO	200ppb	0			0	0	0		0	0	0
WASH	200ppb	0	0	0	0	0	0	0	0	0	0
ATLA	2x100	0	0	0	0	1	0	0	0	0	1
BALT	2x100	0	1	0.5	0.5	1	0	3	1	1	1
BOST	2x100	0	6	6	7	6	0	11	11	13	8
CHIC	2x100	0	6			4	0	9			6
DALL	2x100	0	4	3	5	4	0	5	5	10	5
DENV	2x100	0.5		3	2	4	1		4	5	5
DETR	2x100	0	1	4	9	4	0	2	7	16	6
HOUS	2x100	0	5	5	7	9	0	7	7	9	17
KANS	2x100	0	1	1	2	2	0	2	2	5	5
LOSA	2x100	0	10	8	10	13	0	19	15	20	26
MIAM	2x100	0	4	2	3	4	0	10	3	4	7
MINE	2x100	0	2	2	3	3	0	4	3	7	5
NYNY	2x100	0.5	4	5	6	7	1	7	8	10	10
PHIL	2x100	0.5	2	3		2	1	3	6		3
PHOE	2x100	0	2	3	4	4	0	3	5	5	4
PITT	2x100	0	1	2		2	0	2	4		3
RICH	2x100	0	0	0	0.5	0	0	0	0	1	0
RIVR	2x100	0	0.5	3		3	0	1	6		6
SACR	2x100	0	4	4	6	3	0	7	6	8	6
SAND	2x100	0	0.5	0	0	1	0	1	0	0	1
SANF	2x100	0	0	0	1	2	0	0	0	2	5
STLO	2x100	0		1	3	1	0		3	5	4
WASH	2x100	0	5	3	4	6	0	12	5	6	12

5-132

Table B5-62. Mean and maximum number of data point locations used for calculations: area-wide, near-road, and simulated on-road CBSA-wide summary table (2013-2015).

CBSA Abbrev.	Mean number of data point locations per year		Maximum number of data point locations per year	
	asis	CS1315	asis	CS1315
ATLA	4	5	7	7
BALT	3	3	4	4
BOST	6	7	8	8
CHIC	5	5	6	6
DALL	12	12	14	14
DENV	3	5	7	7
DETR	4	5	6	6
HOUS	15	17	18	18
KANS	3	4	4	4
LOSA	12	14	19	19
MIAM	4	3	5	4
MINE	4	6	7	7
NYNY	9	10	11	11
PHIL	5	6	8	8
PHOE	6	6	8	8
PITT	5	5	6	6
RICH	3	4	4	4
RIVR	9	10	15	15
SACR	7	7	9	9
SAND	7	3	8	8
SANF	10	10	11	11
STLO	4	6	6	6
WASH	6	6	7	7

[a] associated with summary results presented in Table B6-63 and B5-64.

Table B5-63. Mean and maximum number of days per year NO$_2$ concentrations were at or above 1-hour benchmark levels: area-wide, near-road, and simulated on-road CBSA-wide summary table (2013-2015).

CBSA Abbrev.	1-hr Benchmark/Metric	Mean number of days per year at or above benchmark level		Maximum number of days per year at or above benchmark level	
		asis	CS1315	asis	CS1315
ATLA	100ppb	0	20	0	34
BALT	100ppb	0	24	0	35
BOST	100ppb	0	23	0	31
CHIC	100ppb	0.5	19	1	23
DALL	100ppb	0	19	0	22
DENV	100ppb	0.5	15	1	21
DETR	100ppb	0	22	0	30
HOUS	100ppb	1	27	3	31
KANS	100ppb	0	15	0	22
LOSA	100ppb	1	37	2	53
MIAM	100ppb	0.5	12	1	21
MINE	100ppb	0	25	0	37
NYNY	100ppb	2	28	4	38
PHIL	100ppb	0.5	11	1	16
PHOE	100ppb	0.5	19	1	27
PITT	100ppb	0	24	0	43
RICH	100ppb	0	25	0	39
RIVR	100ppb	2	18	8	38
SACR	100ppb	0	16	0	19
SAND	100ppb	0.5	20	1	20
SANF	100ppb	0.5	17	1	20
STLO	100ppb	0	28	0	38
WASH	100ppb	0	26	0	41
ATLA	150ppb	0	0	0	0
BALT	150ppb	0	0	0	0
BOST	150ppb	0	0.5	0	1
CHIC	150ppb	0	0	0	0
DALL	150ppb	0	0	0	0
DENV	150ppb	0	1	0	1
DETR	150ppb	0	0	0	0
HOUS	150ppb	0	0.5	0	1
KANS	150ppb	0	0	0	0
LOSA	150ppb	0	0.5	0	1
MIAM	150ppb	0	1	0	2
MINE	150ppb	0	1	0	1
NYNY	150ppb	1	2	2	6
PHIL	150ppb	0	0.5	0	1
PHOE	150ppb	0	0.5	0	1
PITT	150ppb	0	0	0	0
RICH	150ppb	0	1	0	1
RIVR	150ppb	0	1	0	2
SACR	150ppb	0	0	0	0
SAND	150ppb	0	0	0	0
SANF	150ppb	0	0	0	0
STLO	150ppb	0	1	0	2
WASH	150ppb	0	1	0	2

ATLA	200ppb	0	0	0	0
BALT	200ppb	0	0	0	0
BOST	200ppb	0	0	0	0
CHIC	200ppb	0	0	0	0
DALL	200ppb	0	0	0	0
DENV	200ppb	0	0	0	0
DETR	200ppb	0	0	0	0
HOUS	200ppb	0	0	0	0
KANS	200ppb	0	0	0	0
LOSA	200ppb	0	0	0	0
MIAM	200ppb	0	0	0	0
MINE	200ppb	0	0	0	0
NYNY	200ppb	0.5	1	1	2
PHIL	200ppb	0	0	0	0
PHOE	200ppb	0	0	0	0
PITT	200ppb	0	0	0	0
RICH	200ppb	0	0	0	0
RIVR	200ppb	0	0	0	0
SACR	200ppb	0	0	0	0
SAND	200ppb	0	0	0	0
SANF	200ppb	0	0	0	0
STLO	200ppb	0	0	0	0
WASH	200ppb	0	0	0	0
ATLA	2x100	0	7	0	12
BALT	2x100	0	5	0	9
BOST	2x100	0	11	0	14
CHIC	2x100	0.5	6	1	8
DALL	2x100	0	6	0	9
DENV	2x100	0.5	7	1	10
DETR	2x100	0	10	0	14
HOUS	2x100	0	13	0	17
KANS	2x100	0	5	0	7
LOSA	2x100	0	18	0	32
MIAM	2x100	0	5	0	7
MINE	2x100	0	11	0	14
NYNY	2x100	1	13	2	18
PHIL	2x100	0.5	3	1	4
PHOE	2x100	0	7	0	9
PITT	2x100	0	8	0	17
RICH	2x100	0	6	0	11
RIVR	2x100	0	7	0	19
SACR	2x100	0	4	0	6
SAND	2x100	0	5	0	5
SANF	2x100	0	3	0	7
STLO	2x100	0	8	0	12
WASH	2x100	0	8	0	16

Table B5-64. Percent of days per year NO₂ concentrations were at or above 1-hour benchmark levels: area-wide and near-road CBSA-wide summary table.

CBSA Abbrev.	1-hour Benchmark /Metric	Mean percent of days per year at or above benchmark level					Maximum percent of days per year at or above benchmark level				
		asis	CS1012	CS1113	CS1214	CS1315	asis	CS1012	CS1113	CS1214	CS1315
ATLA	100ppb	0	2.7	3.8	3.4	3.4	0	4.1	8.2	5.2	6.3
BALT	100ppb	0	2	2.4	2.7	3.1	0	3.6	3.3	3.8	5.5
BOST	100ppb	0	4.8	4.1	4.9	3.8	0	6	6.3	7.4	4.7
CHIC	100ppb	0.1	6.2			3.7	0.3	9.6			4.4
DALL	100ppb	0	2.9	3.4	3.7	4.1	0	4.1	5.8	5.2	4.9
DENV	100ppb	0.1		6.1	2.4	2.4	0.3		9	3.3	3.3
DETR	100ppb	0	3.9	8.6	5.9	2.9	0	4.6	9.6	7.1	3.6
HOUS	100ppb	0.1	5.1	3.7	4.9	5.4	0.8	8.2	3.8	6.6	8.5
KANS	100ppb	0	3.7	3	3.6	2.9	0	4.1	4.1	4.6	4.4
LOSA	100ppb	0.1	7.4	8.1	8.3	8.3	0.6	11.8	11.2	11.2	11.8
MIAM	100ppb	0	3.9	2.7	2.8	3.2	0	7.1	3.6	3.8	5.8
MINE	100ppb	0	4	3.7	3.9	3.2	0	6.3	3.8	5.5	4.4
NYNY	100ppb	0.5	4.1	3.7	4.8	5.8	0.8	5.8	5.2	8.2	8.2
PHIL	100ppb	0.1	3.1	3.5		2.4	0.3	6	7.1		4.1
PHOE	100ppb	0.1	2.4	3.4	3.3	3	0.3	2.5	4.1	4.1	3.8
PITT	100ppb	0	2.4	3.7		4.3	0	3.3	7.4		6.3
RICH	100ppb	0	4.8	3.9	5.2	2.5	0	8.2	6.3	7.7	3
RIVR	100ppb	0.5	2.4	6.3		2.3	2.2	4.6	10.1		2.7
SACR	100ppb	0	3.5	4.5	5.5	3.9	0	4.1	5.2	6.6	5.2
SAND	100ppb	0.1	2.2	2.4	1.8	2.7	0.3	2.7	3.3	3.3	2.7
SANF	100ppb	0.1	2.1	2.8	2.7	3.3	0.3	3.6	6.3	3	5.2
STLO	100ppb	0		1.8	4.7	2.3	0		2.2	6.3	3.6
WASH	100ppb	0	3.8	5.4	5.4	6	0	6.6	6.3	8	8.2
ATLA	150ppb	0	0	0	0	0	0	0	0	0	0
BALT	150ppb	0	0	0	0	0	0	0	0	0	0
BOST	150ppb	0	0.1	0.3	0.1	0.1	0	0.3	0.3	0.3	0.3
CHIC	150ppb	0	0			0	0	0			0
DALL	150ppb	0	0	0	0	0	0	0	0	0	0
DENV	150ppb	0	0	0.1	0.1	0.1	0	0	0.3	0.3	0.3

CBSA Abbrev.	1-hour Benchmark /Metric	Mean percent of days per year at or above benchmark level					Maximum percent of days per year at or above benchmark level				
		asis	CS1012	CS1113	CS1214	CS1315	asis	CS1012	CS1113	CS1214	CS1315
DETR	150ppb	0	0.1	0.1	0.1	0	0	0.3	0.3	0.3	0
HOUS	150ppb	0	0.1	0	0	0	0	0.6	0	0	0
KANS	150ppb	0	0	0	0	0	0	0	0	0	0
LOSA	150ppb	0	0.1	0.1	0.1	0.1	0	0.3	0.3	0.3	0.3
MIAM	150ppb	0	0.4	0.3	0.4	0.4	0	0.6	0.3	0.6	0.6
MINE	150ppb	0	0.1	0.3	0.1	0	0	0.6	0.8	0.6	0
NYNY	150ppb	0.1	0.1	0.1	0.5	0.5	0.6	0.3	0.6	1.4	1.4
PHIL	150ppb	0	0.1	0.1		0.1	0	0.6	0.6		0.3
PHOE	150ppb	0	0	0	0	0	0	0	0	0	0
PITT	150ppb	0	0	0		0	0	0	0		0
RICH	150ppb	0	0	0.1	0.1	0	0	0	0.3	0.3	0
RIVR	150ppb	0	0.1	0.7		0	0	0.3	2.2		0
SACR	150ppb	0	0	0	0	0	0	0	0	0	0
SAND	150ppb	0	0	0	0	0	0	0	0	0	0
SANF	150ppb	0	0	0.1	0	0	0	0	0.3	0	0
STLO	150ppb	0		0	0.1	0	0		0	0.3	0
WASH	150ppb	0	0.1	0.1	0.1	0.1	0	0.3	0.3	0.3	0.3
ATLA	200ppb	0	0	0	0	0	0	0	0	0	0
BALT	200ppb	0	0	0	0	0	0	0	0	0	0
BOST	200ppb	0	0	0	0	0	0	0	0	0	0
CHIC	200ppb	0	0			0	0	0			0
DALL	200ppb	0	0		0	0	0	0		0	0
DENV	200ppb	0		0	0	0	0		0	0	0
DETR	200ppb	0	0	0	0	0	0	0	0	0	0
HOUS	200ppb	0	0	0	0	0	0	0	0	0	0
KANS	200ppb	0	0	0	0	0	0	0	0	0	0
LOSA	200ppb	0	0	0	0	0	0	0	0	0	0
MIAM	200ppb	0	0	0	0	0	0	0	0	0	0
MINE	200ppb	0	0	0	0	0	0	0	0	0	0
NYNY	200ppb	0.1	0	0	0	0	0.3	0	0	0	0
PHIL	200ppb	0	0	0	0	0	0	0	0	0	0
PHOE	200ppb	0	0	0	0	0	0	0	0	0	0

CBSA Abbrev.	1-hour Benchmark /Metric	Mean percent of days per year at or above benchmark level					Maximum percent of days per year at or above benchmark level				
		asis	CS1012	CS1113	CS1214	CS1315	asis	CS1012	CS1113	CS1214	CS1315
PITT	200ppb	0	0	0		0	0	0	0	0	0
RICH	200ppb	0	0	0	0	0	0	0	0	0	0
RIVR	200ppb	0	0	0		0	0	0	0		0
SACR	200ppb	0	0	0	0	0	0	0	0	0	0
SAND	200ppb	0	0	0	0	0	0	0	0	0	0
SANF	200ppb	0	0	0	0	0	0	0	0	0	0
STLO	200ppb	0		0	0	0	0		0	0	0
WASH	200ppb	0	0	0	0	0	0	0	0	0	0
ATLA	2X100	0	0	0	0	0.1	0	0	0	0	0.3
BALT	2X100	0	0.3	0.1	0.1	0.1	0	0.8	0.3	0.3	0.3
BOST	2X100	0	1.6	1.7	2	1.7	0	3	3	3.6	2.2
CHIC	2X100	0	1.6			1.1	0	2.5			1.6
DALL	2X100	0	1.1	0.8	1.5	1.2	0	1.4	1.4	2.7	1.4
DENV	2X100	0.1		0.7	0.6	1	0.3		1.1	1.4	1.4
DETR	2X100	0	0.3	1.2	2.4	1	0	0.6	1.9	4.4	1.6
HOUS	2X100	0	1.3	1.5	2	2.6	0	1.9	1.9	2.5	4.7
KANS	2X100	0	0.1	0.1	0.5	0.6	0	0.6	0.6	1.4	1.4
LOSA	2X100	0	2.8	2.1	2.7	3.7	0	5.2	4.1	5.5	7.1
MIAM	2X100	0	1	0.6	0.7	1.2	0	2.7	0.8	1.1	1.9
MINE	2X100	0	0.6	0.5	0.7	0.9	0	1.1	0.8	1.9	1.4
NYNY	2X100	0.1	1.2	1.5	1.7	2	0.3	1.9	2.2	2.7	2.7
PHIL	2X100	0.1	0.6	0.7		0.5	0.3	0.8	1.6		0.8
PHOE	2X100	0	0.5	0.9	1	1	0	0.8	1.4	1.4	1.1
PITT	2X100	0	0.4	0.5		0.6	0	0.6	1.1		0.8
RICH	2X100	0	0	0	0.1	0	0	0	0	0.3	0
RIVR	2X100	0	0.1	0.8		0.8	0	0.3	1.6		1.6
SACR	2X100	0	1.2	1.2	1.7	0.9	0	1.9	1.6	2.2	1.6
SAND	2X100	0	0.1	0	0	0.3	0	0.3	0	0	0.3
SANF	2X100	0	0	0	0.3	0.5	0	0	0	0.6	1.4
STLO	2X100	0		0.3	0.9	0.4	0		0.8	1.4	1.1
WASH	2X100	0	1.4	0.8	1	1.7	0	3.3	1.4	1.6	3.3

5.8 DATA FOR INDIVIDUAL NEAR-ROAD MONITORS 2014 AND 2015: AS IS AND ADJUSTED TO JUST MEET THE EXISTING STANDARD

Table B5-65. Number per year NO₂ concentrations were at or above 1-hour benchmark levels: individual near-road monitor data where monitor was in operation for at least 300 days in year 2014 and/or 2015.

CBSA Abbrev.	Monitor site id	1-hr benchmark	Number of days per year DM1H at or above benchmark level			
			As Is air quality 2014	As Is air quality 2015	Adjusted air quality 2014	Adjusted air quality 2015
ATLA	130890003	100 ppb		0		7
ATLA	131210056	100 ppb		0		3
BALT	240270006	100 ppb		0		7
BOST	250250044	100 ppb	0	0	11	6
CHIC	170313103	100 ppb	1	0	4	4
DALL	481131067	100 ppb		0		5
DENV	080310027	100 ppb	0	0	5	3
DETR	261630093	100 ppb	0	0	9	9
DETR	261630095	100 ppb		0		5
HOUS	482011066	100 ppb	0	0	4	3
KANS	290950042	100 ppb	0	0	4	3
LOSA	060590008	100 ppb	0	0	21	1
MINE	270370480	100 ppb		0		3
MINE	270530962	100 ppb	0	0	11	10
NYNY	340030010	100 ppb		1		8
PHIL	421010075	100 ppb	0	0	3	2
PHOE	040134019	100 ppb	0	0	3	2
PITT	420031376	100 ppb		0		11
RICH	517600025	100 ppb		0		9
RIVR	060710026	100 ppb		0		8
SANF	060010012	100 ppb	0	0	2	1
STLO	291890016	100 ppb		0		1
STLO	295100094	100 ppb	0	0	10	2
ATLA	130890003	150 ppb		0		0
ATLA	131210056	150 ppb		0		0
BALT	240270006	150 ppb		0		0
BOST	250250044	150 ppb	0	0	0	0
CHIC	170313103	150 ppb	0	0	0	0
DALL	481131067	150 ppb		0		0
DENV	080310027	150 ppb	0	0	0	0
DETR	261630093	150 ppb	0	0	0	0
DETR	261630095	150 ppb		0		0
HOUS	482011066	150 ppb	0	0	0	0

CBSA Abbrev.	Monitor site id	1-hr benchmark	Number of days per year DM1H at or above benchmark level			
			As Is air quality 2014	As Is air quality 2015	Adjusted air quality 2014	Adjusted air quality 2015
KANS	290950042	150 ppb	0	0	0	0
LOSA	060590008	150 ppb	0	0	0	0
MINE	270370480	150 ppb		0		0
MINE	270530962	150 ppb	0	0	0	0
NYNY	340030010	150 ppb		1		0
PHIL	421010075	150 ppb	0	0	0	0
PHOE	040134019	150 ppb	0	0	0	0
PITT	420031376	150 ppb		0		0
RICH	517600025	150 ppb		0		0
RIVR	060710026	150 ppb		0		0
SANF	060010012	150 ppb	0	0	0	0
STLO	291890016	150 ppb		0		0
STLO	295100094	150 ppb	0	0	0	0
ATLA	130890003	200 ppb		0		0
ATLA	131210056	200 ppb		0		0
BALT	240270006	200 ppb		0		0
BOST	250250044	200 ppb	0	0	0	0
CHIC	170313103	200 ppb	0	0	0	0
DALL	481131067	200 ppb		0		0
DENV	080310027	200 ppb	0	0	0	0
DETR	261630093	200 ppb	0	0	0	0
DETR	261630095	200 ppb		0		0
HOUS	482011066	200 ppb	0	0	0	0
KANS	290950042	200 ppb	0	0	0	0
LOSA	060590008	200 ppb	0	0	0	0
MINE	270370480	200 ppb		0		0
MINE	270530962	200 ppb	0	0	0	0
NYNY	340030010	200 ppb		0		0
PHIL	421010075	200 ppb	0	0	0	0
PHOE	040134019	200 ppb	0	0	0	0
PITT	420031376	200 ppb		0		0
RICH	517600025	200 ppb		0		0
RIVR	060710026	200 ppb		0		0
SANF	060010012	200 ppb	0	0	0	0
STLO	291890016	200 ppb		0		0
STLO	295100094	200 ppb	0	0	0	0

5.9   COMPARISON OF CURRENT RESULTS WITH 2008 REA

We compared the site-year results generated in this AQC to those generated in the 2008 REA. In general, the estimated number of days per year at or above benchmark levels is similar when evaluating the area-wide and the near-road monitors, considering both the mean and upper percentile values (Table B5-66). The mean number of days per year at or above 100 ppb was about 5, while the maximum number of days per year above that same level ranged from about 10 to 20. There were very few instances where concentrations were at or above the 150 ppb and 200 ppb benchmark levels, regardless of which analytical data set or monitor type was considered.

There are however large differences in the number of days per year at or above benchmark levels when considering the simulated on-road concentrations, whereas the 2008 REA results were consistently higher than those generated in the current AQC. For example, the 2008 REA estimated, on average, 30 to 150 days per year at or above the 100 ppb 1-hour benchmark on-road in the subset of comparable study areas, along with an upper percentile estimate of ranging from about 100 to 300 days per year. In the current AQC, the mean and mean number of days where estimated concentrations were at or above the100 ppb benchmark ranged from about 5 to 20, with a maximum of just around 40 days per year. In addition, the number of days having concentrations at or above the higher benchmark levels (i.e., 150 and 200 ppb) were estimated to occur more frequently when considering the 2008 REA results compared with the current AQC.

This relatively lower range of values estimated in this AQC is to some extent a function of the fewer number of days monitored for the new near-road monitors, with many having what would be considered an incomplete year of monitor data (Table B5-58 and Table B5-59). All data used in the 2008 REA adhered to the standard completeness criteria, while for the current analysis, this restriction was relaxed for the near-road monitors such that we could include the maximum number of days/years of near-road data possible. To a greater extent though, the difference is likely the result of the monitor data used to to simulate the on-road NO_2 concentrations. Here we used the near-road monitors, sited in close proximity to major roads combined with a statistical approach to estimate the on-road concentrations. A similar approach was used in the 2008 REA, though based on the form of the statistical model used, monitors at greater distances from roads (i.e., at least 100 m from a road) were used to estimate the on-road concentrations. The 2008 REA statistical model assumed monitors sited 100 m or greater from a road was a reasonable distance to not have a direct influence from road emissions, though recognized as an important uncertainty in that assessment (2008 REA, section 7.4.6).

In reviewing the current monitor attribute data provided in Table B5-11 through Table B5-33 (and also Table B5-34 through Table B5-56), while it is possible that at a distance of 100 meters or more from a major road these monitors could have limited direct contribution from roadway emissions, there remains the potential for other source emissions to substantially influence concentrations measured at the monitors that were not accounted for before simulating the on-road concentrations. For example, monitor ID 420450002 located in the Philadelphia study area was assumed to be not influenced by roadway emissions based on its siting of 322 m from a major road (Table B5-24) and NO_2 concentrations from this monitor were used to simulate the on-road concentrations. However, in identifying all NO_X emission sources within a 5 km radius of that monitor, it is possible that NO_2 concentrations measured at that monitor could be influenced by emissions from one or more facilities, including NO_X emissions from electricity generation (via combustion), petroleum refineries, and municipal waste combustion. Using this monitor and other monitors[40] that could potentially be influenced by significant facility emissions would tend to overestimate the number of days per year simulated on-road NO_2 concentrations were at or above benchmark levels.

---

[40] For example monitor ID 421010029 also in the Philadelphia study area (Table B5-47) is within 5 km of an electricity generating unit(s) having summed NO_X emissions of 353 tons per year.

Table B5-66. Mean and maximum number of days per year NO₂ concentrations exceed 1-hour benchmark levels: comparison of current air quality characterization with 2008 REA.

Study Area	Source of Results[1]	Monitor Distance from Road or Category[2]	AQ Data Years	Site-Years[3]	Number of days per year at or above benchmark level[4] ≥100 ppb Mean	≥100 ppb p99/max	≥150 ppb Mean	≥150 ppb p99/max	≥200 ppb Mean	≥200 ppb p99/max
Atlanta	Current	Area-wide	2010-12	9	3	15	0	0	0	0
			2011-13	9	5	30	0	0	0	0
			2012-14	9	4	16	0	0	0	0
			2013-15	9	2	16	0	0	0	0
	2008 REA	≥100 m	2001-03	14	2	15	0	1	0	1
		>20m-<100m	2001-03	0	-	-	-	-	-	-
	Current	Near-road	2013-15	3	6	8	0	0	0	0
	2008 REA	≤20m	2001-03	0	-	-	-	-	-	-
	Current	On-road	2014	1	19	-	0	-	0	-
			2015	2	14	20	0	0	0	0
	2008 REA	On-/Near-road	2001-03	1400	46	229	10	104	2	38
Boston	Current	Area-wide	2010-12	15	5	15	0.5	1	0	0
			2011-13	15	5	15	0.5	1	0	0
			2012-14	11	6	18	1	1	0	0
			2013-15	16	3	12	0.5	1	0	0
	2008 REA	≥100 m	2001-03	6	0	1	0	0	0	0
		>20m-<100m	2001-03	14	3	11	0	0	0	0
	Current	Near-road	2013-15	3	6	11	0	0	0	0
	2008 REA	≤20m	2001-03	5	2	6	0	1	0	0
	Current	On-road	2014	1	29	-	0	-	0	-
			2015	1	24	-	0	-	0	-
	2008 REA	On-/Near-road	2001-03	600	26	131	3	25	0	6
Chicago	Current	Area-wide	2010-12	11	7	24	0	0	0	0
			2011-13	0	-	-	-	-	-	-
			2012-14	0	-	-	-	-	-	-
			2013-15	12	4	12	0	0	0	0
	2008 REA	≥100 m	2001-03	9	1	4	0	1	0	0
		>20m-<100m	2001-03	6	7	21	1	3	0	0
	Current	Near-road	2013-15	2	4	4	0	0	0	0
	2008 REA	≤20m	2001-03	4	2	6	0	0	0	0
	Current	On-road	2014	1	14	-	0	-	0	-
			2015	1	15	-	0	-	0	-
	2008 REA	On-/Near-road	2001-03	900	83	257	18	106	4	49
Denver	Current	Area-wide	2010-12	0	-	-	-	-	-	-

Study Area	Source of Results[1]	Monitor Distance from Road or Category[2]	AQ Data Years	Site-Years[3]	Number of days per year at or above benchmark level[4]					
					≥100 ppb		≥150 ppb		≥ 200 ppb	
					Mean	p99/max	Mean	p99/max	Mean	p99/max
			2011-13	5	14	29	0.5	1	0	0
			2012-14	6	5	10	1	1	0	0
			2013-15	7	4	10	0.5	1	0	0
	2008 REA	≥100 m	2001-03	2	2	2	0	0	0	0
		>20m-<100m	2001-03	0	-	-	-	-	-	-
	Current	Near-road	2013-15	4	3	5	0	0	0	0
	2008 REA	≤20m	2001-03	2	7	10	2	3	0	0
	Current	On-road	2014	1	16	-	1	-	0	-
			2015	2	10	13	0.5	1	0	0
	2008 REA	On-/Near-road	2001-03	200	99	269	19	103	4	37
Detroit	Current	Area-wide	2010-12	4	6	12	0	0	0	0
			2011-13	5	9	23	0	0	0	0
			2012-14	6	8	18	0	0	0	0
			2013-15	6	4	9	0	0	0	0
	2008 REA	≥100 m	2001-03	6	3	7	2	7	1	4
		>20m-<100m	2001-03	0	-	-	-	-	-	-
	Current	Near-road	2013-15	4	7	9	0	0	0	0
	2008 REA	≤20m	2001-03	0	-	-	-	-	-	-
	Current	On-road	2014	1	19	-	0	-	0	-
			2015	2	19	22	0	0	0	0
	2008 REA	On-/Near-road	2001-03	600	59	186	13	57	5	28
Los Angeles	Current	Area-wide	2010-12	30	5	19	0.5	1	0	0
			2011-13	26	5	22	0.5	1	0	0
			2012-14	27	5	23	0.5	1	0	0
			2013-15	35	4	27	0.5	1	0	0
	2008 REA	≥100 m	2001-03	51	1	10	0	5	0	0
		>20m-<100m	2001-03	35	1	7	0	1	0	1
	Current	Near-road	2013-15	3	9	21	0	0	0	0
	2008 REA	≤20m	2001-03	9	1	3	0	0	0	0
	Current	On-road	2014	1	36	-	0	-	0	-
			2015	2	10	17	0	0	0	0
	2008 REA	On-/Near-road	2001-03	5100	33	160	6	55	1	20
Miami	Current	Area-wide	2010-12	12	5	15	0.5	1	0	0
			2011-13	10	4	12	0.5	1	0	0
			2012-14	8	5	12	1	1	0	0
			2013-15	8	6	17	0.5	1	0	0

Study Area	Source of Results[1]	Monitor Distance from Road or Category[2]	AQ Data Years	Site-Years[3]	Number of days per year at or above benchmark level[4]					
					≥100 ppb		≥150 ppb		≥ 200 ppb	
					Mean	p99/max	Mean	p99/max	Mean	p99/max
	2008 REA	≥100 m	2001-03	6	4	14	0	2	0	0
		>20m-<100m	2001-03	3	8	15	0	1	0	0
	Current	Near-road	2013-15	1	1	-	1	-	0	-
	2008 REA	≤20m	2001-03	3	1	1	0	0	0	0
	Current	On-road	2015	1	2	-	1	-	0	-
	2008 REA	On-/Near-road	2001-03	600	78	189	19	110	5	48
New York	Current	Area-wide	2010-12	23	3	9	0.5	1	0	0
			2011-13	22	3	12	0.5	1	0	0
			2012-14	24	3	12	0.5	1	0	0
			2013-15	25	4	12	0.5	1	0	0
	2008 REA	≥100 m	2001-03	26	1	5	0	0	0	0
		>20m-<100m	2001-03	13	3	13	0	2	0	0
	Current	Near-road	2013-15	2	8	8	2	3	0	0
	2008 REA	≤20m	2001-03	7	1	2	0	0	0	0
	Current	On-road	2014	1	16	-	4	-	2	-
			2015	1	26	-	1	-	0	-
	2008 REA	On-/Near-road	2001-03	2600	57	226	10	73	3	34
Philadelphia	Current	Area-wide	2010-12	13	3	18	0.5	1	0	0
			2011-13	14	3	23	0.5	1	0	0
			2012-14	0	-	-	-	-	-	-
			2013-15	13	2	13	0.5	1	0	0
	2008 REA	≥100 m	2001-03	14	3	15	0	1	0	1
		>20m-<100m	2001-03	7	5	11	0	0	0	0
	Current	Near-road	2013-15	3	2	3	0	0	0	0
	2008 REA	≤20m	2001-03	0	-	-	-	-	-	-
	Current	On-road	2014	1	8	-	0	-	0	-
			2015	2	3	4	0	0	0	0
	2008 REA	On-/Near-road	2001-03	1400	116	294	27	137	7	68
Phoenix	Current	Area-wide	2010-12	15	2	9	0	0	0	0
			2011-13	14	4	13	0	0	0	0
			2012-14	14	3	12	0	0	0	0
			2013-15	13	3	9	0	0	0	0
	2008 REA	≥100 m	2001-03	5	1	2	0	0	0	0
		>20m-<100m	2001-03	2	0	0	0	0	0	0
	Current	Near-road	2013-15	3	3	3	0	0	0	0
	2008 REA	≤20m	2001-03	3	6	11	0	0	0	0

Study Area	Source of Results[1]	Monitor Distance from Road or Category[2]	AQ Data Years	Site-Years[3]	Number of days per year at or above benchmark level[4]					
					≥100 ppb		≥150 ppb		≥ 200 ppb	
					Mean	p99/max	Mean	p99/max	Mean	p99/max
	Current	On-road	2014	1	13	-	0	-	0	-
			2015	2	12	19	0.5	1	0	0
	2008 REA	On-/Near-road	2001-03	500	153	337	35	206	6	48
St. Louis	Current	Area-wide	2010-12	0	-	-	-	-	-	-
			2011-13	8	2	8	0	0	0	0
			2012-14	9	4	14	0	0	0	0
			2013-15	9	1	5	0	0	0	0
	2008 REA	≥100 m	2001-03	9	3	15	0	1	0	0
		>20m-<100m	2001-03	11	2	11	0	0	0	0
	Current	Near-road	2013-15	4	5	10	0	0	0	0
	2008 REA	≤20m	2001-03	6	3	6	0	1	0	0
	Current	On-road	2014	1	37	-	2	-	0	-
			2015	2	11	17	0	0	0	0
	2008 REA	On-/Near-road	2001-03	900	125	288	28	153	7	51
Washington DC	Current	Area-wide	2010-12	18	4	14	0.5	1	0	0
			2011-13	17	4	20	0.5	1	0	0
			2012-14	15	5	24	0.5	1	0	0
			2013-15	15	6	24	0.5	1	0	0
	2008 REA	≥100 m	2001-03	18	4	11	0	0	0	0
		>20m-<100m	2001-03	10	2	5	0	0	0	0
	Current	Near-road	2013-15	1	4	4	0	0	0	0
	2008 REA	≤20m	2001-03	4	6	9	0	1	0	0
	Current	On-road	2015	1	17	-	2	-	0	-
	2008 REA	On-/Near-road	2001-03	1800	109	310	27	168	7	63

[1] Data compared are from the AQC performed for this current NO$_2$ review and the AQC conducted for the 2008 REA. Shading is added where data can generally be compared for the two sets of results.

[2] For the current AQC, the results are separated into three groups: all area-wide monitors regardless of their distance to major roads, the formally designated near-road monitor, and simulated on-road concentrations (where concentrations from the newly designated near-road monitors served as the basis for the estimation). For the 2008 NO$_2$ REA AQC, three generally similar groups of data are presented: area-wide monitors (though having two categories for describing their distance in meters (m) from a major road, Tables 7-23 (≥100 m) and 7-24 (>20 m to <100 m), near-road monitors (based on monitors sited ≤20 m from a major road, Table 7-25), and simulated on-road concentrations (where concentrations from monitors sited ≥100 m from a major road served as the basis for the estimation, Table 7-28).

[3] In general, the average number of monitors operating per year within the three-year group can estimated by dividing the number of site-years by for the area-wide and near-road data. When using the on-road data, divide by 300.

[4] The mean number of exceedances represents the sum of benchmark exceedances occurring at all monitors in a particular location divided by the total site-years across the three-year monitoring period. The p99 (2008 REA AQC) and maximum (Current AQC) represent an upper estimate of the number of days/year having a benchmark exceedance at a particular monitoring site for a single year within the monitoring period.